Thermally and Optically Stimulated Luminescence

Thermally and Optically Stimulated Luminescence

A Simulation Approach

REUVEN CHEN

Raymond and Beverly Sackler School of Physics and Astronomy, Tel Aviv University, Tel Aviv, Israel

VASILIS PAGONIS

McDaniel College, Westminster, MD, USA

A John Wiley & Sons, Ltd., Publication

Library of Congress Cataloging-in-Publication Data

Chen, R. (Reuven)
  Thermally and optically stimulated luminescence : a simulation approach / Reuven Chen and Vasilis Pagonis.
    p.   cm.
  Includes bibliographical references and index.
  ISBN 978-0-470-74927-2 (hardback)
  1. Thermoluminescence. 2. Thermoluminescence dosimetry. 3. Optically stimulated luminescence dating.
I. Pagonis, Vasilis. II. Title.
  QC479.C47 2011
  612'.014480287–dc22

                                                                2010053722

A catalogue record for this book is available from the British Library.

Print ISBN: 9780470749272
ePDF ISBN: 9781119993773
oBook ISBN: 9781119993766
ePub ISBN: 9781119995760
eMobi: 9781119995777

Set in 10/12pt Times Roman by Thomson Digital, Noida, India
Printed in Great Britain by Antony Rowe Ltd, Chippenham, Wiltshire

Front cover image courtesy of Herb Yeates, http://superfluorescence.com/. Copyright (2007) with permission.

Contents

About the Authors

Reuven Chen

Professor Reuven Chen is a Professor Emeritus at Tel Aviv University. He has been working on thermoluminescence, optically stimulated luminescence and other related topics over the last 48 years. Professor Chen has published approximately 170 scientific papers and two books. He has been a Visiting Professor at several universities in the USA, UK, Canada, Australia, Brazil, France and Hong Kong. At present, he is an Associate Editor of *Radiation Measurements* and referee for several international journals.

Vasilis Pagonis

Professor Vasilis Pagonis is a Professor of Physics at McDaniel College. His research involves working on modeling properties of dosimetric materials and their applications in luminescence dating and radiation dosimetry. Professor Pagonis has published approximately 70 scientific papers, as well as the book *Numerical and Practical Exercises in Thermoluminescence*, published in 2006. He currently holds the Kopp endowed chair in the physical sciences at McDaniel College.

Preface

Thermoluminescence (TL) and optically stimulated luminescence (OSL) are two of the most important techniques used in radiation dosimetry. Hundreds of papers are published every year in the scientific literature on different aspects of TL and OSL. These cover a whole spectrum of subjects, from experimental papers describing various aspects of these phenomena in different materials under different experimental conditions, many times having in mind the potential applications, to publications interested only in the dates reached by these methods and to publications on dosimetry measurements in different environments (e.g., in spaceships). On the other side of the spectrum, one can find work on the physical basis of TL and OSL, in which researchers try to obtain better understanding of the underlying processes. These include the dose dependence of the effects (which may be linear or nonlinear), possible dose-rate dependence, the stability of the effects at ambient temperature (which may include normal and anomalous fading), the dependence of these effects on the relevant defects and impurities, and the nature of the emission spectrum.

The theoretical work on TL and OSL consists, in most cases, of the study of the simultaneous differential rate equations governing the transitions of charge carriers, usually electrons and holes, between the different trapping states associated with impurities and defects in the studied sample, and the conduction and valence bands. These equations are not linear and therefore, in most cases, cannot be solved analytically. In many cases, approximations concerning the trapping parameters and functions are made to reduce the complication, and explicit equations for simplified models such as the first- and second-order kinetics can be written and solved analytically. Here, general solutions are reached from which one can consider how different phenomena may take place, e.g. how the signal fades with time at room temperature under first- or second-order kinetics, and to what extent these predictions agree with specific luminescence experiments for a given material. Obviously, this approach has a strong limitation since one does not know whether the assumptions made hold all along the temperature and time range of a TL/OSL measurement.

This book concentrates on an alternative approach, in which we simulate various experimental situations by numerically solving the relevant coupled differential equations for chosen sets of parameters. Using this approach, several complex situations can be demonstrated such as superlinear and nonmonotonic dose dependencies, dose-rate effects, the occurrence of abnormally high frequency factors and others. Obviously, the shortcoming of this approach is that it does not provide us with general solutions but rather with results associated with specific sets of trapping parameters. However, this kind of demonstration that certain behaviors are commensurate with our understanding of the underlying processes is of great importance. With the present availability of strong computing power and advanced numerical methods, this approach has become very popular during the past 20 years. A second approach that is emphasized throughout this book is demonstrating the possibility of obtaining analytical solutions of the systems of differential equations by using the quasi-equilibrium approximation. Numerous examples are given in which this approach leads to exact analytical solutions which describe accurately the experimental results. This book is designed for practitioners, researchers and graduate students in the field of radiation dosimetry. It is a synthesis of the major developments in modeling and numerical simulations of thermally and optically stimulated processes during the past 50 years.

Chapter 1 is mostly a historical overview of the developments in TL and OSL dosimetry during the past 50 years, followed in Chapter 2 by an overview of the theoretical basis and several quantum aspects of luminescence phenomena, which is based on the energy-band model of solids. Chapter 3 deals with a number of basic experimental measurements relevant to the study of TL and OSL. In Chapter 4 we present the basic kinetic equations governing the TL process, including simple kinetic models based on first- second- general- and mixed-order kinetics. In addition, some aspects of localized versus delocalized electronic transitions during the luminescence process are discussed. The basic methods of evaluating kinetic parameters in TL and OSL experiments are the main topic of Chapter 5, and Chapter 6 addresses a variety of physical phenomena commonly encountered during TL and OSL measurements. The basic theoretical aspects and experimental techniques used in OSL dosimetry are presented in Chapter 7, with a specific emphasis on the relationship between the various models used in obtaining OSL data (LM-OSL, CW-OSL, pseudo LM-OSL, etc.). Chapter 8 addresses a topic of prime importance for radiation dosimetry researchers, namely the dose dependence of TL/OSL signals. Different types of experimentally observed dose behaviors are examined using both an analytical approach and approximate expressions obtained using certain approximations. The topic of TL and OSL simulations for dating applications is presented in Chapter 9, including simulations of recent major developments in TL/OSL dating protocols. Chapter 10 examines the use of several alternative methods for evaluating trapping parameters, based on a variety of advanced numerical methods like Monte-Carlo techniques, genetic algorithms and advanced curve-fitting methods. Chapter 11 contains a more general approach to thermally stimulated phenomena, and several methods of analyzing simultaneous thermal measurements are presented. The processes of thermally stimulated conductivity (TSC), thermally stimulated electron emission (TSEE), optical absorption (OA) and electron spin resonance (ESR) are briefly discussed, in particular in cases where their simultaneous measurements with TL can produce additional information and can be simulated along with the simulation of TL. Chapter 12 deals with applications of luminescence in medical physics and Chapter 13 with the associated phenomenon of radiophotoluminescence. Chapter 14 summarizes theoretical developments and simulation results on the effects of ionization density on TL response, which is a topic of major interest in radiation dosimetry for particles of varying ionization density. Finally Chapter 15 presents various numerical approaches to the exponential integral which appears commonly in TL applications. In addition to the comprehensive list of references, covering all the subjects discussed in the book, we have also included a list of books and review articles published in the literature from 1968 onwards. Finally, in Appendix A, we present some simple examples of computer code that simulate three important models which appear frequently in the book, namely, the one trap-one recombination center (OTOR) model, the interactive multiple trap system (IMTS), and the widely used Bailey model for quartz.

Acknowledgements

We thank our wives Shula Chen and Mary Jo Boylan for their patience, encouragement and sound advice during the years of writing this book. We would like to thank also Dr John L. Lawless for his contributions to some of the analytical aspects discussed in the book. Thanks are also due to Dr Doron Chen for important technical help in preparing the manuscript.

Reuven Chen and Vasilis Pagonis

I would like to thank all my research collaborators, students and friends in the luminescence community for their helpful contributions and stimulating discussions during the development of various luminescence models over the years. Special thanks among these are due to Dr George Kitis of Aristotle University in Greece for his friendship and extensive collaboration over the last 10 years; Dr Ann Wintle for teaching me the importance of seeking perfection in the preparation of manuscripts; to my colleagues Andrew Murray, Mayank Jain and Christina Ankjærgaard in Denmark for their hospitality and many stimulating conversations; and last but not least, special thanks to my co-author, good friend and luminescence mentor Dr Reuven Chen, for teaching me the importance of accuracy and precision during luminescence modeling work.

Vasilis Pagonis

1

Introduction

In this introductory chapter we first provide an overview of the physical mechanism involved in thermoluminescence (TL) and optically stimulated luminescence (OSL) phenomena, followed by a brief historical review of the development of TL and OSL dosimetry. This is followed by a section on the parallel development of luminescence models for TL/OSL phenomena during the past 50 years.

1.1   The Physical Mechanism of TL and OSL Phenomena

The phenomenon of phosphorescence seems to have been discovered first by Vincenzo Casciarolo (see e.g., Arnold [1]), an amateur alchemist in Bologna in 1602 who discovered the "Bologna Phosphorus", the mineral barium sulfide, which was glowing in the dark after exposure to sunlight. An account was later published by Fortunio Liceti in "Litheosphorus, sive de lapide Bononiensi lucem", Utino, 1640. In 1663, Robert Boyle gave the Royal Society one of the first accounts of TL. He described some experiments he had carried out on a diamond, saying "I also brought it to some kind of glimmering light, by taking it into bed with me, and holding it a good while upon a warm part of my naked body" (see e.g. Heckelsberg [2]). The phenomenon of TL had been known since the 17th century, and has been studied intensively since the first half of the 20th century. For example, in 1927, Wick [3] reported on the TL of X-irradiated fluorite and other materials. In 1931, she reported [4] on TL in calcium sulfate doped by manganese and fluorite, following their exposure to radium. She also described the effect of applying pressure on the TL properties of the samples. A preliminary qualitative explanation of the occurrence of TL, based on the band theory of solids was given by Johnson [5] only in 1939. The first quantitative theoretical account based on the model of energy bands in crystals, was given in 1945 in a seminal work by Randall and Wilkins [6]. Basically, TL consists of the excitation of an insulator, usually by ionizing radiation but sometimes by non-ionizing radiation or other means, followed by a "read-out" stage of heating the sample and measuring the light emitted in excess of the

Thermally and Optically Stimulated Luminescence: A Simulation Approach, First Edition. Reuven Chen and Vasilis Pagonis.   © 2011 John Wiley & Sons, Ltd. Published 2011 by John Wiley & Sons, Ltd.

"black-body radiation". In the OSL method, discovered significantly later, the read-out stage consists of releasing the charge carriers, previously excited by irradiation, by illumination with light of an appropriate wavelength; the incident light is capable of releasing trapped charge carriers at the ambient temperature.

The understanding of the phenomenon is associated with the energy-band theory of solids, and has to do with the trapping of charge carriers in the forbidden gap states associated with imperfections in the crystalline material, be it impurities or defects. The trapping states are entities that can capture either electrons or holes during the excitation period and during the read-out stage which, in the TL process is the time when the sample is heated and measurable light is recorded. The energy absorbed during the excitation period causes the production of electrons and holes, which may move around the conduction and valence bands, respectively, and get trapped in electron and hole trapping states. Some of these traps may be rather close to their respective bands, electrons to the conduction band and holes to the valence band, so that within the temperature range of the subsequent heating, they may be thermally released into the band. These entities are usually called "traps".

The trapping states which are farther from their respective bands, in which a recombination of trapped charge carriers and mobile carriers of the opposite sign may take place are usually termed "recombination centers" or just "centers". Thus, during the read-out stage charge carriers, say electrons, may be thermally elevated into the conduction band, where they can move around before recombining with the opposite-sign carriers, say a hole, and emit at least part of the previously absorbed energy in the form of photons. However, some of these recombinations may be radiationless, meaning that the produced energy turns into phonons. It is also possible that recombinations produce photons in a spectral range which is not measurable by the device being used, and for the purpose of our analysis of the results, may be considered as being radiationless.

Note that, although very often one discusses the TL/OSL process as being related to the thermal or optical release of trapped electrons and their subsequent recombination with holes in centers, the inverse situation in which the mobile entity is the positive hole which moves in the valence band and then recombines with a stationary electron in a luminescence center is just as likely to occur. One should also mention the possibility of localized transitions, a situation where the hole and electron trapping states are located in close proximity to each other, and the radiative process takes place by thermal or optical stimulation of one kind of carrier into an excited state which is not in the conduction/valence band, and its subsequent recombination with its opposite-sign companion.

1.2 Historical Development of TL and OSL Dosimetry

The two most important applications of TL and OSL are in the broad fields of radiation dosimetry and geological/archaeological dating. In this section we present a brief outline of the historical development of luminescence techniques in these two broad application areas.

Although the first theoretical work, by Randall and Wilkins and later by Garlick and Gibson was published in the 1940s, the first practical applications of TL were suggested in the 1950s. The applications of TL in radiation dosimetry were initiated in the early 1950s by Daniels [7, 8] who also suggested that natural TL from rocks is related to radioactivity from uranium, thorium and potassium in the material. Later, Kennedy and Knopf [9] discovered natural TL emitted from samples of ancient pottery, which led the way to the work on TL dating of archaeological samples which was developed quickly in the 1960s, first in Oxford by Aitken and his group [10] and later, in dozens of laboratories all over the world. The possible use of optical stimulation instead of thermal stimulation for evaluating the absorbed dose in a sample for dosimetry purposes was first suggested by Antonov-Romanovskiĭ [11] in the mid 1950s and mentioned later by a number of researchers who referred usually to infra-red stimulated luminescence (IRSL). The use of OSL for archaeological and geological dating was suggested in 1985 by Huntley *et al.* [12], and it has been in use in many laboratories since then.

Since the 1950s there has been a continuous extensive search for the "perfect" thermoluminescent dosimetric (TLD) material that will exhibit the ideal linear response over the widest possible range of doses, high sensitivity, excellent reproducibility and stability of the luminescence signal. The historical development, properties and uses of various TLD materials have been summarized in some detail in the book by McKeever *et al.* [13]. The use of TL as a radiation dosimetry technique was first suggested by Farrington Daniels and collaborators at the University of Wisconsin (USA) during the 1950s. Daniels *et al.* [7, 8] first used LiF for radiation dosimetry during atomic bomb testing, and they also studied and considered $CaSO_4:Mn$, sapphire, beryllium oxide and $CaF_2:Mn$ as possible TL dosimeters during the same decade. In the 1960s a variety of new materials were also studied, namely $CaF_2:Dy$, $CaSO_4:Tm$, $CaSO_4:Dy$, CaF_2 and LiF:Mg,Ti. The latter material eventually became one of the most commonly used TLD materials. In the next 20 years various forms of Al_2O_3, CaF_2 and LiF were developed and considered as TLD candidates. Other commonly used and studied TLD materials are $Al_2O_3:C$ and LiF:Mg,Cu,P. The most common applications of TLD materials are in monitoring of personnel radiation exposure, in medical dosimetry, environmental dosimetry, spacecraft, nuclear reactors, mineral prospecting, food irradiation, retrospective dosimetry, and in geological/archaeological dating.

Kortov [14] recently summarized the current status and future trends in the development of materials for TL dosimetry. This author listed the main requirements for practical use of TL dosimeters as: a wide linear dose response, high TL sensitivity per unit of absorbed dose, low signal dependence on the energy of the incident radiation, low signal fading over time, the presence of simple TL curve, luminescence spectrum matching photomultiplier (PM) tube response and appropriate physical characteristics. The author listed the useful dose range and thermal fading properties of the following seven main practical dosimetric materials: LiF:Mg,Ti (TLD-100), LiF:Mg,Cu,P (TLD-100H), ^{6}LiF:Mg,Ti (TLD-600), ^{6}LiF:Mg,Cu,P (TLD-600H), $CaF_2:Dy$ (TLD-200), $CaF_2:Mn$ (TLD-400), and $Al_2O_3:C$ (TLD-500). Kortov [14] also discussed the intrinsic luminescence efficiency η of TL materials; he specifically attributed the high sensitivity of several dosimetric materials to the efficient trapping/detrapping/excitation mechanisms associated with the presence of F-centers.

In a recent comprehensive review of luminescence dosimetry materials Olko [15] summarized the progress of luminescence detectors and dosimetry techniques for personal dosimetry and medical dosimetry. The author discussed traditional personal dosimetry based on OSL, TL and radiophotoluminescence (RPL), and also reviewed more novel luminescence detectors used in clinical dosimetry applications such as radiotherapy, intensity modulated radiotherapy (IMRT) and ion beam radiotherapy. The major advantages of luminescence dosimeters were summarized as: high sensitivity measurement of very low doses, linear dose dependence, good energy response to X-rays, reusability, and sturdiness. However, the review also recognized the problem of decreased response with increasing ionization density of the radiation field. This problem may lead to underestimation of dose after heavy charged particle irradiation. Personal dosimetry is also used widely in the medical sector, with dosimetric films gradually being replaced by TLD, OSL and RPL materials.

The pros and cons of using OSL versus TLD dosimeters have been summarized in McKeever and Moscovitch [16]. Some of the advantages of OSL dosimeters are high efficiency and stable sensitivity, better precision and accuracy, fast read-out, and no thermal annealing steps. However, TL dosimeters have the advantages of high sensitivity, no light sensitivity, simple automated read-out, possibility of neutron dosimetry, and flat photon energy response.

Olko [15] also summarized some newer developments in luminescence detectors: development of a personal neutron dosimeter based on OSL [17], laser-scanned RPL glasses used to measure the dose from fast neutrons by counting tracks of charged recoil particles [18], and fluorescent nuclear track detectors (FNTDs) which allow imaging of individual tracks of heavy charged particles [19, 20]. Oster *et al.* [21] suggested the possibility of using standard LiF:Mg,Ti (TLD-100) and a combined TL/OSL signal to increase the efficiency of detecting high linear energy transfer (LET) particles. Additional novel techniques include the development of a laser-scanned OSL system and TLD systems with a charge-coupled device (CCD) camera [22–24]. Olko [15] identified three active areas for research in new luminescence detectors, namely developing new materials for the medical field, for materials to be used in dosimetry of high LET radiation, and for materials mimicking the radiation response of biological systems. However, this author also identified the absence of luminescence detectors for neutron dosimetry as a major gap in luminescence dosimetry.

The second broad area where TL and OSL dosimetry have found extensive practical applications is in the field of geological and archaeological dating. In a comprehensive review article, Wintle [25] reviewed the historical and technological developments in the field of luminescence dating. During the time period 1957–1979, TL techniques were applied to heated materials, while in the time period 1979–1985 TL dating was extended to older sedimentary samples. The historical developments in the use of TL during this time period include the fine-grain and coarse-grain TL dating techniques, improvements in the calculation and measurement of natural dose rates, applications of TL dating to pottery and fired clay, and authenticity testing of ceramics using predose dating. During these early years, two major problems were identified which hindered successful application of TL dating: the problems of anomalous fading exhibited, e.g., by feldspars; and the phenomenon of supralinearity during dose response measurements. However, there were many attempts to extend the use of TL signals in the study of other materials, such as heated stones, calcite deposits and burnt flint. In many of these areas, TL continues to be a valuable dating tool. Starting in 1979, researchers began exploring the possibility of using TL dating techniques for determining the time of deposition of quartz and feldspar grains. The exploration of new luminescence signals during the period 1979–1985 for the dating of sediment deposition led to the next major phase in luminescence dating, which continues today. During the last 25 years, research in luminescence dating has undergone a dramatic shift, due to the discovery of new luminescence signals which could be zeroed by exposure to sunlight. These new signals led to the development of OSL dating techniques. In 2008, Wintle [25] identified

1999 as the seminal year in which the single aliquot regenerative (SAR) dating procedure was developed; this technique has revolutionized luminescence dating, by providing an accurate and precise tool for routine measurement of equivalent doses. Furthermore, the SAR protocol allows for a completely automated measurement process, resulting in major improvements in the speed of data acquisition and analysis. As a result of these major developments during the past 25 years, OSL has become arguably the most accurate and precise luminescence dating tool in Quaternary geology, as well as a valuable archaeological tool [26].

1.3   Historical Development of Luminescence Models

In this section we present a historical overview of the development of luminescence models, which took place in parallel to the historical development of experimental TL and OSL techniques described in the previous section.

Randall and Wilkins [6] wrote a differential equation governing the TL process and discussed the properties of its solution, by assuming that retrapping is negligible and that the rate of change of trapped carriers is proportional to the concentration of these trapped carriers (first-order kinetics). Garlick and Gibson [27] showed that under different relations between the retrapping and recombination probabilities, the rate of change of the concentration of trapped carriers is proportional to the square of this concentration, i.e. the kinetics is of second order. They wrote the relevant differential equation and studied the properties of its solution. Following a previous suggestion by Hill and Schwed [28], May and Partridge [29] extended this treatment to "general-order" kinetics, namely, cases in which the rate of change of the concentration of trapped carriers is proportional to a non-integer power of their concentration. Although heuristic in nature, the approach has been rather popular in the study of TL. A milestone in the development of luminescence models is the work by Halperin and Braner [30], who introduced a more realistic presentation of a single TL peak. They wrote three simultaneous differential equations governing the traffic of carriers between a trapping state, the conduction band and a recombination center. Since these equations cannot be solved analytically, Halperin and Braner [30], Levy [31] and other authors made some simplifying assumptions, which enabled the solution of the problem in a relatively easy way for some specific circumstances. It is obvious, however, that the only route to follow more complicated cases is by solving numerically the relevant simultaneous differential equations.

During the past 50 years numerous kinetic models have been published which attempt to explain various experimentally observed behaviors in luminescence phenomena. Perhaps the best overview of these models is the paper by McKeever and Chen [32] and the textbook by Chen and McKeever [33]. The approach used in the majority of published TL/OSL papers is to solve numerically the relevant simultaneous differential equations. With modern available software, this is a relatively easy task. One can use reasonable sets of trapping parameters and find how the TL, as well as OSL, signals behave. The obvious disadvantage is that it is usually very hard to draw general conclusions from the simulation. It is possible, however, to demonstrate that certain effects are compatible with specific assumptions concerning the relevant trapping states. For example, nonlinear dose dependencies of TL and OSL have been reported in some materials; even within the one trap-one recombination center (OTOR) model, called by Levy [31] General One Trap (GOT), nonlinear dose dependence can be expected under certain conditions. In addition, different kinds of such nonlinearity can be explained by taking into consideration the occurrence of competitors, the transitions into which are nonradiative. In some extreme cases, this behavior can be shown analytically, but the variety of nonlinear dose dependencies can be demonstrated by simulation through numerical solution of the relevant equations. The simulation should be performed for the excitation stage and for the read-out stage, and properties of the solution can be compared with the experimental results. A comprehensive approach should, however, include both the excitation and read-out stages, with a certain relaxation period in between.

The review article by McKeever and Chen [32] addressed several important questions on the usefulness and need for modeling and numerical simulations of luminescence phenomena. These authors emphasized that one of the most important purposes of modeling is to provide researchers with "a feeling of security"; the use of models can indeed improve our basic understanding of the physical processes being studied. In another familiar example, modeling can provide fundamental answers about the validity of the complex modern protocols used during luminescence dating. In the same review paper, the authors provided a critique of modeling efforts and emphasized the need to test the actual behavior of the proposed models, in order to ascertain what behaviors are possible (or not) within the model. They also pointed out that often, modeling efforts lead to the development of ad hoc models, without regard to how well the model can describe other behaviors observed in the same material. It is our belief that to some extent these two criticisms of modeling efforts have been addressed during the past 20 years, with the development of comprehensive models for a variety of dosimetric materials. As an example of such comprehensive modeling efforts, we mention the recent development of comprehensive models for quartz by several authors [34–36]. Such models have proved to be very useful indeed for explaining a wide variety of experimental behaviors in quartz. As a second example of a comprehensive model, we mention the various models developed to explain the TL and OSL properties of the widely used dosimetric material Al_2O_3:C. Several of these comprehensive models have been shown to be able to describe simultaneously a wide variety of TL/OSL phenomena in this important dosimetric material [37–39].

2

Theoretical Basis of Luminescence Phenomena

Throughout this book, we will be examining how different combinations of rate constants yield different luminescent behaviors. In this chapter, we examine what physics tells us about these constants and their magnitude. We start with a discussion of electron and hole capture rate constants and discuss how the type of trap determines how large the rate constants are expected to be. We then examine thermal equilibrium. Whether or not a TL material ever reaches equilibrium, the theory allows us to relate the magnitude of a rate constant to that of its reverse. Following that, we derive thermal detrapping rate constants from capture rate constants. We will then consider Arrhenius' theory of rate constants and how it relates to both capture and detrapping rate constants. We also consider the role of rate equations in the theory of TL and OSL and their origin in quantum statistics and relaxation theory. We continue by considering emission and absorption of radiation, how emission and absorption rates are related through detailed balance, and discuss estimates for their values. We briefly discuss the theoretical work found in the literature on the association of trapping parameters with certain impurities embedded in given crystals. Finally, we give an account of a number of aspects of possible mechanisms leading to thermal quenching of luminescence.

2.1 Energy Bands and Energy Levels in Crystals

We start with a very brief explanation of the basic properties of a crystal which enable the rich variety of conductivity and luminescence properties. The basic theory of all kinds of luminescence in solids, including TL and OSL has to do with the energy band of solids. The solution of the Schrödinger equation for electrons in a periodic potential yields allowed bands separated by forbidden bands (see e.g. Kittel [40] and Ibach and Lüth [41]). In a pure insulating and semiconducting crystal at absolute zero (0 K), all the bands up to the one called the valence band are full of electrons. The next allowed band, called the conduction

Thermally and Optically Stimulated Luminescence: A Simulation Approach, First Edition. Reuven Chen and Vasilis Pagonis. © 2011 John Wiley & Sons, Ltd. Published 2011 by John Wiley & Sons, Ltd.

band, is empty of electrons and so are the higher allowed bands. The forbidden band between the valence band and the conduction band is called the forbidden band or the gap. Electronic conduction in the crystal can take place only if electrons from the valence band are given enough energy to reach the conduction band. Once an electron is in the conduction band, it can contribute to the electrical conductivity. Moreover, the missing electron in the valence band can be considered as a positive charge carrier, a "hole", and it can move in the crystal, thus contributing to the conduction. At finite temperatures, electrons can be thermally raised from the valence band into the conduction band. However, this may take place at relatively low temperatures in semiconductors which have a relatively narrow band gap, and hardly occurs at all in insulators which are the main subject of the present book, due to their broad band gap. For both perfect semiconductors and insulators with a band gap E_g, optical absorption only takes place for light with photon energies larger than E_g, namely with frequencies above E_g/h where h is the Planck constant.

All real crystals are not ideal in the sense that they always include imperfections, namely defects and impurities. This causes a local change in the otherwise periodical system, and new energy levels are thus produced in the forbidden gap, which makes it possible for electrons and holes to get "trapped". This means that these carriers may possess energies that are forbidden in the ideal crystal. The occurrence of these traps or centers may cause additional optical absorption of light with photon energies significantly lower than the band-to-band energy. Thus, new absorption bands may be observed, which may change the visible color of the crystal. The occurrence of the trapping states in the forbidden gap changes in many cases very drastically the conductivity properties as well as the luminescence features, and practically all the effects discussed in the present book have to do with transitions between such energy levels and the valence and conduction bands. The nature of these trapping states depends on the host material. In the case of impurities, the properties of these point defects also depend on the foreign atoms and ions present in the material and in their location in the host material. As for the defects, the properties of the relevant energy levels depend on the specific defect. The allowed energy levels in the forbidden gap may be discrete or distributed depending on the host lattice and the specific imperfections.

A well known defect type is the Frenkel defects which are interstitial atoms, ions or molecules normally located on the lattice site, which have moved out of their original place. The corresponding vacancies are called Schottky defects. The latter may be the result of a diffusion of the host ions to the surface of the crystal. In some cases high energy radiation may produce a pair of vacancy–interstitial located in rather close proximity to each other, thus forming a defect of a different nature. Another cause for a disturbance in the periodicity of the perfect crystal is the presence of the surface. This can result in trapping levels in the periodic potential, thus yielding usually shallow trapping levels in the surface region.

Low-energy radiation and sometimes even high-energy radiation applied to a sample may not produce new defects, but in most cases play a crucial role in the filling of traps and centers associated with existing impurities and defects. On the other hand, the absorption of photons may photostimulate previously trapped charge carriers into the conduction or valence band, and this leads to a reduction of an expected TL or OSL effect. Practically all the phenomena discussed in this book are related to transitions of this kind. The connection of the effects of excitation and de-excitation to the luminescence phenomena which are the subject matter of the book, are elaborated upon starting with Chapter 4 which describes the most basic way of producing TL and OSL (see in particular Figure 4.1).

2.2   Trapping Parameters Associated with Impurities in Crystals

It is quite obvious that the properties of the traps and recombination centers are directly derived from the nature of the host crystal and the imperfections, impurities and defects, embedded in the crystal. Thus, in principle, knowledge of which imperfections are involved should yield all the relevant parameters. This would mean that in addition to knowledge about the emission spectrum associated with the relevant recombination centers, one might expect to know the activation energies, frequency factors, as well as the recombination and retrapping-probability coefficients associated with the capture cross-sections for retrapping and recombination and the thermal velocity of the carriers. Note that another kind of parameters, namely the total concentrations of traps and centers, is of a different nature since it has to do only with the amounts of the relevant imperfections. One would expect that quantum mechanical theory should yield the values of the trapping parameters. In the literature, there are reports of such quantum mechanical treatments which yield predictions on luminescence properties as well as other related features of the materials with given imperfections. To the best of our knowledge, none of these is directly related to TL and OSL in the sense that one cannot predict, for instance, the dosimetric behavior of a given crystal with given amounts of certain impurities from first principles.

Some related theoretical works are to be mentioned here briefly. Williams [42, 43] describes the theory of luminescence of impurity-activated ionic crystals, using the absolute theory of the absorption and emission of these crystals. The detailed atomic rearrangements following the optical transitions and the equilibrium among accessible atomic configurations of the activator system are determined quantitatively. The author shows that the absorption and emission spectra of KCl:Tl at 298 K can be predicted on theoretical grounds. Norgett *et al.* [44] studied theoretically the electronic structure of the V^- center in MgO, where a hole is trapped at a cation vacancy. Using a model for lattice relaxation calculations with a Coulomb part and a short-range interaction which has overlap and van der Waals components, they can explain the energies of optical transitions of 1.5 and 2.3 eV which involve transitions occurring within an O^- and hole hopping from one oxygen ion to another, respectively. Lagos [45] describes a quantum theory of interstitial impurities and shows that the analytical calculation of a number of effects like phonon-assisted tunneling and optical absorption, which are valid for massive impurities and high concentrations, follow as a direct application. Testa *et al.* [46] calculated the excitation energies of V_k-centers in NaCl by combining an unrestricted Hartree–Fock code with classical potentials to simulate the defect and the distorted lattice around it. In a textbook on solid-state theory, Harrison [47] discusses impurity states in crystals. The tight-binding description is considered, dealing with the situation in which one of the ions in an ionic crystal is replaced by an impurity ion. Impurities, donors and acceptors are also studied in semiconductors. The quantum theory of surface states and impurity states is elaborated upon. In a review paper and, in particular in an extensive book, Stoneham [48, 49] discusses the theory of defects in solids, dealing with the electronic structure of defects in insulators, in particular ionic crystals, and semiconductors. Among other important subjects, he considers different forms of F-centers (F, F', F_t), M-centers, R-centers, V_k-centers, H-centers, etc.

Extensive work by Dorenbos and co-workers [50–53] has been devoted to the evaluation of the trivalent and divalent lanthanide energy level location in different materials used for TL and OSL dosimetric materials such as $CaSO_4:Dy^{3+}$, $SrAl_2O_4:Eu^{2+}$, $YPO_4:Ce^{3+}$ and many others. Dorenbos [50] presented the systematic variation in the energy level positions of divalent lanthanides in wide band gap ionic crystals. He concludes that the width of the charge transfer (CT) band in spectra does not correlate with the width of the valence band. Also, the width of the CT luminescence in Yb^{3+}-doped compounds is about the same as the width of the CT absorption. Finally, there is no significant dependence of the width of the CT band on the type of lanthanide. By comparing the CT energies for different trivalent lanthanides in the same host, constant energy differences were revealed. This may help in predicting the relevant energies. For instance, once the CT bands in Eu^{3+} is known, the energies of other lanthanides can be evaluated. In subsequent papers, Dorenbos and Bos [51] and Bos *et al.* [52] use the same ideas to locate the energy levels in YPO_4 doped by different lanthanide ions and discuss the related TL phenomena. Bos *et al.* [53] state that the trend of predicted trap depths agrees well with experimental results. However, the absolute energy-level positions show a systematic difference of ~ 0.5 eV. They extend the research by studying the excitation spectra of the OSL of a $YPO_4:Ce^{3+},Sm^{3+}$ sample in order to elucidate the discrepancy between the predicted value of 2.5 eV and experimental result of 2.1 eV. From the OSL excitation spectra at different temperatures they deduce the value of the trap depth predicted by the Dorenbos model.

One should note, however, that as far as TL and OSL are concerned, the state of the art at present is such that in most cases one cannot calculate theoretically the important trapping parameters from first principles. Given a certain host crystal and a specific imperfection, be it a defect or impurity, one cannot predict in most cases the relevant values of frequency factors and recombination and retrapping-probability coefficients, and only in few cases the activation energies can be determined using quantum-mechanical considerations. Furthermore, most crystals have different kinds of imperfections, and it is not always clear which one of them is directly (or indirectly) involved in the TL and OSL phenomena. Thus, although it is obvious that the imperfections are always the source of all the luminescence effects discussed in this book, bridging the gap between quantum theory and the theoretical work discussed in this book, which is based mainly on using the relevant sets of rate equations, is still a desirable goal to be dealt with in the future.

2.3   Capture Rate Constants

The lifetime for a free electron is typically of the form $1/A(N - n)$ where A is a capture rate constant and $N - n$ is the concentration of available trapping sites. While both of these quantities can vary by orders of magnitude from one material to another, the range is not unlimited. In practice, trap concentrations N are limited on both the high and low ends. In pure materials, one expects trap concentrations to be small but, in even the more refined single crystals, recombination centers are found with densities of 10^{12} cm^{-3} or more. At the high end, trap concentration is limited by the requirement that there be enough separation that the wavefunctions do not overlap. Deep traps can be closely packed in high band gap materials where values of N as high as 10^{19} cm^{-3} are observed. Next, the observed ranges for capture rate constants will be discussed (see e.g. Rose [54]).

2.3.1   General Considerations

Let us consider an electron (or hole) traveling in a solid at a speed of v. Let us suppose that there is some distance r_c, such that if the electron comes within r_c of a trap, it will be captured. Over a time t, the electron travels a distance vt and it will have been captured if there was a trap anywhere within the volume $vt(\pi r_c^2)$. The probability of capture within time t is then the probability that there was a trap within the volume $vt(\pi r_c^2)$. If traps have a density of N, then the expected number of traps within that volume is $Nvt(\pi r_c^2)$. The capture rate, that is the probability of capture per unit time, is thus $Nv(\pi r_c^2)$. Since we are interested in free electrons with a velocity distribution characterized by a temperature T, we need to average that capture rate over all thermal speeds which yields $N\bar{v}(\pi r_c^2)$ where $\bar{v}$ is the mean thermal speed. Elsewhere in this book, the capture rate for a free electron is written as AN where A is the capture rate constant. It is thus clear that

$$A = \bar{v}(\pi r_c^2). \tag{2.1}$$

Very often, the area πr_c^2 is called a cross-section and denoted by σ. In this case, the rate constant A can be written as

$$A = \bar{v}\sigma. \tag{2.2}$$

The mean thermal speed is given by

$$\bar{v} = \sqrt{\frac{8kT}{\pi m}}, \tag{2.3}$$

where k is Boltzmann's constant, T is temperature, and m is the effective mass of the electron. Effective masses for electrons and holes vary with the crystal structure. Typical values range from $0.05m_e$ to $2m_e$ where $m_e = 9.1 \times 10^{-31}$kg is the mass of an electron in free space. As a rough order of magnitude, 10^7cm s^{-1} is a typical mean thermal speed for either electrons or holes.

To investigate the orders of magnitude involved, let us suppose that the capture radius of a trap is its physical radius. If the trap is the size of an atom, then one might guess that $r_c \sim 2 \times 10^{-8}$cm. It would follow that the cross-section is

$$\sigma = \pi r_c^2 = 3.14 \times (2 \times 10^{-8}\text{cm})^2 \approx 10^{-15}\text{cm}^2. \tag{2.4}$$

Using this σ, a typical capture rate constant would then be

$$A = \bar{v}\sigma \sim 10^7\text{cm s}^{-1} \times 10^{-15}\text{cm}^2 \sim 10^{-8}\text{cm}^3\text{s}^{-1}. \tag{2.5}$$

Experiments show that Equations (2.4) and (2.5) are actually typical magnitudes for capture cross-sections and rate constants, respectively, if the trap is neutral. For example Bemski [55] measured a cross-section of 5×10^{-16} cm^2 for electrons being captured by a neutral Au trap in a silicon crystal at room temperature. Alekseeva *et al.* [56] measured $\sigma = 10^{-15}$ cm^2 for capture of holes by neutral Bi traps in germanium crystals. Lax [57] surveyed experimental data for electron or hole capture by neutral traps and concluded that typical cross-sections range from 10^{-17} to 10^{-15} cm^2 (i.e., A ranging from $\sim 10^{-10}$ to 10^{-8} cm^3 s^{-1}).

When traps have a net charge, very different cross-sections are observed. If the free particle and the trap have charges of the same sign, there is a long-range electrostatic repulsion between them. Consequently, capture is unlikely. Experiments for this case show capture cross-sections of $\sim 10^{-21}$ cm^2 (i.e., $A \sim 10^{-14}$ cm^3 s^{-1}) or smaller [57]. If the free particle and trap have, by contrast, opposite signs, then the free particle will be electrostatically attracted to the trap. In these cases, some very large cross-sections are observed. Values of 10^{-15}–10^{-12} cm^2 are observed.

A trapped electron is at a considerably lower energy state than a free electron. Therefore, as part of the capture process, the electron must lose a significant amount of energy. It is believed that the energy is lost to lattice vibrations (i.e., phonons). Lax [57] has analyzed this process.

In a later work, Mitonneau *et al.* [58] present a method which combines optical absorption and electrical refilling of deep levels, allowing one to measure the majority-carrier capture cross-section for minority carrier traps. They report very large (10^{-15} cm^2) and very small (10^{-21} cm^2) electron capture cross-sections for two levels in GaAs:Cr, and suggest a possible capture mechanism for these cases.

The above considerations apply to capture of an electron or hole by a trap. The reverse process of thermal detrapping is closely related and the rate constants for detrapping can be estimated if the trapping rate constants are known. Before this can be discussed, we need to review thermodynamics, which we shall do in the next section.

2.4   Thermal Equilibrium

In a solid, there are many possible quantum states that an electron could occupy. These states could be in a valence band, in a conduction band, or attached to traps in between. Because there are many such states, we will not deal with them individually but rather consider their distribution $N(E)$ where $N(E)dE$ is the number of such states per unit volume with energies between E and $E + dE$. In thermodynamic equilibrium, the probability that any electron state is occupied by an electron depends on the state's energy E and is given by the formula

$$f(E) = \frac{1}{1 + \exp(E - E_f)/kT}. \tag{2.6}$$

Equation (2.6) is called the Fermi–Dirac distribution or, for short, the Fermi distribution. T is the absolute temperature of the solid, measured in Kelvin. k is the Boltzmann constant and E_f is known as the Fermi energy.

Since we know that $N(E)dE$ is the number of states with energy between E and $E + dE$ and $f(E)$ is the probability that any such state is filled with an electron, it follows that the expected number of electrons occupying a state between E and $E + dE$ is $f(E)N(E)dE$. If we integrate this quantity over all energies, we can find the expected total number of electrons in the solid, N_e

$$N_e = \int N(E) f(E) dE. \tag{2.7}$$

Under normal circumstances, solids are nearly electrically neutral. This means that, for every proton in a nucleus, the solid also has an electron. If one rubs the solid with an electronegative material or an electropositive material, one can generate a charge imbalance in the solid ("static electricity"). Even when such imbalances are present, the difference between the number of electrons and the number of protons is typically much smaller than the population of either alone, and can be ignored for our purposes. This means that the total number of electrons in the solid N_e should match the total number of positive charges in the nuclei.

Consider a solid held at a temperature of absolute zero: $T = 0$. In this case, the Fermi distribution, Equation (2.6) simplifies to

$$f(E) = \begin{cases} 0 & \text{if } E > E_f, \\ 1/2 & \text{if } E = E_f, \\ 1 & \text{if } E < E_f. \end{cases} \tag{2.8}$$

At room temperature, $kT = 1/40\,\text{eV}$ while we deal with materials with band gap energies of 1–12 eV. Consequently, it is useful to consider approximations in which temperature is small. For energies more than a few kT above the Fermi energy, the term $\exp[(E - E_f)/kT]$ is much larger than one. On the other hand, for energies more than a few kT below the Fermi energy, that exponential is much smaller than one. These observations lead to useful approximations for the Fermi–Dirac distribution

$$f(E) = \begin{cases} \exp[-(E_f - E)/kT] & \text{if } kT \ll E - E_f, \\ 1/2 & \text{if } E = E_f, \\ 1 - \exp[(E - E_f)/kT] & \text{if } kT \ll E_f - E. \end{cases} \tag{2.9}$$

This indicates that traps with energies sufficiently below the Fermi level are nearly filled with electrons. We will find it convenient to call these traps "hole-type".

Let us consider a luminescent material with a valence band, a conduction band, and one or more traps as shown in Figure 2.1. If the material is in equilibrium at temperature T, then Fermi–Dirac statistics says that the fraction of trap states N_1 that are filled with electrons is

$$\frac{n_1}{N_1} = f(\mathcal{E}_1) = \frac{1}{1 + \exp(\mathcal{E}_1 - \mathcal{E}_f)/kT} \approx \exp[-(\mathcal{E}_1 - \mathcal{E}_f)/kT], \tag{2.10}$$

where $\mathcal{E}_1$ is the energy of the trap and $\mathcal{E}_f$ is the Fermi level. Finding the population of free electrons, n_c, in the conduction band is slightly more complicated because the free electrons have a range of energies. However, after taking into account the density of states in the conduction band and integrating over energy, one finds

$$\frac{n_c}{N_c} = f(\mathcal{E}_c) = \frac{1}{1 + \exp(\mathcal{E}_c - \mathcal{E}_f)/kT} \approx \exp[-(\mathcal{E}_c - \mathcal{E}_f)/kT], \tag{2.11}$$

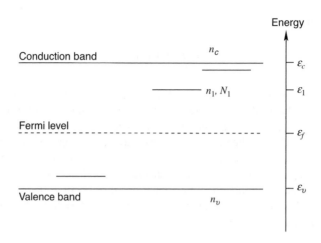

Figure 2.1   *The energy levels of interest for the thermal equilibrium discussion are shown. This material may have one or several traps or centers. For our discussion, though, we only need to refer to one*

where $\mathcal{E}_c$ is the energy of the edge of the conduction band and

$$N_c = \frac{(2\pi m k T)^{3/2}}{h^3},$$ (2.12)

where h is Planck's constant. Now consider the ratio

$$\frac{n_c}{n_1} \approx \frac{N_c}{N_1} \exp[-(\mathcal{E}_c - \mathcal{E}_1)/kT].$$ (2.13)

Note that, in equilibrium, the ratio of the free electron population n_c to the trapped electron population n_1, depends only on material properties and temperature. Also, since the energy difference $\mathcal{E}_c - \mathcal{E}_1$ occurs quite frequently in luminescence studies, it is convenient to define a symbol for it

$$E_1 = \mathcal{E}_c - \mathcal{E}_1.$$ (2.14)

E_1 is the *binding energy* of the electron trap; it is the amount of energy that would be required to raise an electron in the trap up into the conduction band. We can then write

$$\frac{n_c}{n_1} = \frac{N_c}{N_1} \exp(-E_1/kT).$$ (2.15)

2.5   Detailed Balance

Let us consider an electron trap N_1 which can capture electrons from the conduction band with rate $A N_1 n_c$ or the reverse process can occur via thermal excitation; the electron trap can lose electrons back to the conduction band with rate γ. The charge conservation equation can be written as

$$\frac{dn_1}{dt} = AN_1 n_c - \gamma n_1. \tag{2.16}$$

Now, let us consider what happens if this trap is allowed to reach thermal equilibrium at some fixed temperature T. In thermal equilibrium, the concentration of trapped electrons is of course fixed so that $dn_1/dt = 0$. Consequently,

$$AN_1 n_c = \gamma n_1. \tag{2.17}$$

This expresses the balance between capture and loss that occurs in equilibrium. Rearranging,

$$\frac{n_c}{n_1} = \frac{\gamma}{AN_1}. \tag{2.18}$$

Thus, we see that the ratio of free electrons to trapped electrons is determined by the material properties γ, A, and N_1. However, we also know that in equilibrium this same ratio is determined by the Fermi–Dirac distribution as per Equation (2.15). Combining Equation (2.15) with Equation (2.18) yields

$$\frac{\gamma}{AN_1} = \frac{N_c}{N_1} \exp(-E_1/kT). \tag{2.19}$$

or

$$\gamma = AN_c \exp(-E_1/kT). \tag{2.20}$$

One often writes: $\gamma = s \exp(-E_1/kT)$ where s is the frequency factor associated with the trap. In this case,

$$s = AN_c. \tag{2.21}$$

Thus the capture rate constant A is related to the thermal detrapping rate constants, γ, or equivalently to the frequency factor s. The two quantities s and γ are not independent. This relationship applies under the very general condition that the material can be described as having a temperature.

An exception to the requirement that the material has "a temperature" can occur if the capture or detrapping processes have an important step that involves emission or absorption of radiation, and the optical radiation field is not in equilibrium at the same temperature as the material. The materials of interest in this book are often transparent and the optical radiation passing through them is determined by their surroundings which might be at a different temperature. The rate constants associated with radiation processes are discussed in a later section.

2.6   Arrhenius Model

A different way of looking at the capture and de-trapping rate constants was developed by Arrhenius [59]. In his theory a reaction cross-section is given by

$$\sigma = P\sigma_0 \exp(-E_A/kT) \tag{2.22}$$

where σ_0 is the collision cross-section. This is the total cross-section for any collision between the two particles. P is the steric factor which represents the fraction of collisions which result in the reaction of interest to us. The name reflects the concept that the reactants might have to approach each other in a particular orientation for the reaction to proceed. E_A is the activation energy representing the minimum energy needed to overcome a potential barrier.

From the requirements of detailed balance, we find that thermal excitation rates are proportional to the Bolzmann factor $\exp(-E/kT)$. For example, in TL theory the rate constant for thermal de-trapping of an electron or hole is often written as

$$\gamma(T) = s(T)\exp(-E/kT) \tag{2.23}$$

where the pre-exponential $s(T)$ tends to be a weak function of temperature, e.g., $s \sim T^\alpha$ and E is the trap's binding energy. In Arrhenius' theory, E is the activation energy, i.e., the minimum energy required for the electron (or hole) to be freed from the trap and $s = P\sigma_0$. In addition to detrapping, there are cases where trapping (capture) cross-sections are known to have activation energies as shown by Mitonneau *et al.* [58]. For example, they report that the cross-section of chromium in GaAs behaves like

$$\sigma_n = 10^{-17}\exp\left(-0.115/kT\right). \tag{2.24}$$

Different workers described different processes for the escape of trapped electrons (and analogously, of holes). Gibbs [60] gave a rather general account of the conditions leading to the occurrence of the Arrhenius law. He assumed that the system undergoing the thermally activated process consists of weakly interacting subsystems, each of which is subject to a potential of double minimum type. Gibbs showed that the appearance of the Arrhenius exponential is a consequence of the dependence of the barrier climbing rate on the populations of reservoir subsystem levels which, in various aggregates, collectively possess and transfer the requisite energy for the barrier climbing process. Gibbs [60] also discussed the frequency factor and showed that usually, no significant temperature dependence of this factor is expected. This is to be compared with the strong temperature dependence of the exponential Arrhenius term. As pointed out by Curie [61], as soon as a thermal activation is assumed, all the different theories must involve a Boltzmann factor in the probability p. Curie [61] has given a semi-classical interpretation to the problem of activation of trapped electrons.

Consider a trapping model with levels equidistant in energy from each other, the spacing being $h\nu_c$ (i.e. oscillators of characteristic frequency ν_c). By successive absorption of phonons of energy $h\nu_c$, the trap system reaches higher and higher excited levels and the surrounding ions have increasing vibrational energies. When the trap is in a certain state $jh\nu_c$, the electron will be freed. The absorption probability for a transition from the kth to the $(k+1)$th level will be proportional to the number of phonons of frequency $h\nu_c$,

$$A = a\frac{1}{\exp(h\nu_c/kT) - 1}. \tag{2.25}$$

The emission probability [i.e. the probability of returning from $(k+1)$ to k-state] will be

$$R = a\left[1 + \frac{1}{\exp(h\nu_c/kT) - 1}\right]. \tag{2.26}$$

One has a set of j differential equations

$$\frac{dn_0}{dt} = -An_0 + Rn_1, \tag{2.27}$$

$$\frac{dn_1}{dt} = An_0 - (A + R)n_1 + Rn_2, \tag{2.28}$$

$$\cdots\cdots\cdots$$

$$\frac{dn_{j-1}}{dt} = An_{j-2} - (A + R)n_j + Rn_j, \tag{2.29}$$

where n_k is the number of traps in the state k. As explained by Curie [61], the traps are emptied as soon as they reach the jth state, which causes a perturbation of the Boltzmann distribution. The set of equations must be supplemented by

$$\frac{dn_j}{dt} = An_{j-1} - (\alpha + R)n_j, \tag{2.30}$$

where α is the probability of transition of the trapped electron from the jth state into the conduction band.

An equilibrium will be established between the n_k traps in the kth state, in which they decay exponentially with time,

$$n_k = \chi_k \exp(-pt), \tag{2.31}$$

p being the smallest root of the characteristic determinant of the modified set of differential equations. One thus has

$$p = \alpha \frac{\left[\exp(h\nu_c/kT) - 1\right]^2}{\exp(h\nu_c/kT) + \alpha/A - 1} \exp\left[-(j+1)h\nu_c/kT\right]. \tag{2.32}$$

As concluded by Curie [61], this is obviously of the form $p = s\exp(-E/kT)$ where $E = (j+1)h\nu_c$.

A somewhat similar approach has been used by Peacock-López and Suhl [62] who explain the compensation effect by counting the ways in which the heat bath can furnish the energy to surmount a barrier. The reaction rate k_r (s^{-1}) is given by

$$k_r = A\exp(-\beta\epsilon_b), \tag{2.33}$$

where A(s^{-1}) is the pre-exponential frequency factor, $\beta = 1/kT$ and ϵ_b is the activation or barrier energy. The total probability of a reaction is the sum of the probabilities that the reactants absorb one, two, three, etc. excitations with a total energy exceeding the barrier energy. The probability for absorption of exactly n excitations with energy ϵ is the square of the T-matrix element for this process, multiplied by $\rho_n(\varepsilon)$, the density of states of the n excitations with total energy ϵ. It has been shown [62] that the sum over all n behaves like $\exp\left[-(\beta - \beta_c)\varepsilon\right]$, where $\beta_c = 1/kT_c$ is a function of the way the single-excitation density $\rho_1(\varepsilon)$ depends on ε. It is argued that in most cases, $\beta \gg \beta_c$ which brings us back to Equation (2.33).

Böhm *et al.* [63] described a quantum-mechanical treatment of the escape probability from traps in thermally stimulated processes, and the evaluation of the pre-exponential factor. The authors discussed an atomistic model involving multiphonon processes for the calculation of the transition probability of a trapped charge carrier into an excited state. The displacement of lattice atoms from equilibrium position gives rise to electron–lattice interactions. As a result, the temperature dependence was found to be of non-Arrhenius type in some cases. The difference in the probabilities for non radiative transitions, as calculated by the quantum mechanical and by semi-classical methods, was also discussed in this work. The effects of nonradiative transitions on TL was described, and the influence on the position and width of TL peaks was considered. In another paper, Böhm and Scharmann [64] further discussed the quantum-mechanical approach which deals with non radiative transitions due to electron–phonon coupling. This resulted in the possibility of getting an effective value of $s = 10^{15}$ s^{-1} for the frequency factor which is ~2 orders of magnitude higher that the highest value of s expected from the lattice frequency. Some more aspects of the Arrhenius approach were given by Lax [57], Tavgin and Stepanov [65], and by Stepanov [66, 67].

2.7   Rate Equations in the Theory of Luminescence

Throughout this book we utilize the rate equations which govern the different luminescence processes. Although these equations are intuitively well understood, one may wonder about their origin in quantum mechanics and thermodynamics. An important work on the role of rate equations in the theory of luminescent energy transfer was published by Grant [68] in 1971, and included ~120 references to previous work on different aspects of this subject. Grant explains the origin and meaning of the equations that enter the theory of luminescence, and clarifies the relationship between experimental observations and the parameters entering the equations. Grant [68] relates the equations to first principles of quantum statistics and relaxation theory. This has been done in two steps. He first derives equations for microscopic spatial probability distributions, and from these distributions, he derives equations for macroscopically observable probabilities or populations. Although this treatment requires some simplifying assumptions, Grant gives evidence that the rate-equation approximation is indeed appropriate at least for many rare-earth systems in inorganic materials.

Maxia [69–71] explained the rate equations governing the TL process on the basis of non equilibrium thermodynamics. By assuming the phosphor to be held in a cavity full of black-body radiation, and by applying the principle of minimum entropy production, the electron distribution in the conduction band was found to be in a steady quasi-equilibrium with those in trapping states and recombination centers. Maxia [70] also discussed entropy production in diffusion-assisted TL.

Swandic [72] considered statistical fluctuations above and below values predicted by the rate equations, due to the discrete nature of electrons and holes. He presented a stochastic approach to recombination luminescence with retrapping in the steady state. A master equation for a stochastic model with one trapping state and one recombination center was derived. An expansion of this equation by van Kamken's Ω-expansion methods allows one to recover the usual deterministic, macroscopic differential equations and also to obtain the Fokker–Planck equation governing the fluctuations of the various particle numbers. In a later paper [73], the analysis was extended to the general case of several recombination centers and trapping states.

2.8 Radiative Emission and Absorption

Emission of radiation may occur when a trap or center in an excited energy state decays to a lower energy state. Let us call the excited state "2" and the lower state "1". The photon that is emitted has a frequency ω_{21}, measured in radians per second, that obeys the relation

$$\hbar\omega_{21} = E_2 - E_1, \tag{2.34}$$

where E_2 is the energy of the excited level and E_1 is the energy of the lower level, and $\hbar$ is Planck's constant h divided by 2π. According to classical electromagnetic theory, an electron oscillating at frequency ω_{21} radiates at a rate given by

$$\gamma_{cl} = \frac{e^2\omega_{21}^2}{6\pi\epsilon_0 mc^3}, \tag{2.35}$$

where e is the elementary charge, ϵ_0 is the permittivity of free space, c is the speed of light and m is the electron mass. According to quantum mechanics, the photon emission rate is actually

$$A_{21} = \gamma_{cl} \times (3f_{21}), \tag{2.36}$$

where f_{21} is called the oscillator strength and depends on the specifics of the quantum mechanical wave functions for the upper and lower states. For particularly strong transitions, f_{21} is commonly of order one. For weaker transitions, f_{21} is smaller, often much smaller.

Let us consider absorption of light by the two-level energy system. The incident light can be characterized by an intensity, $I(\omega)$, measured in power per unit area per unit radial frequency. If light is absorbed as it passes through a material, its intensity decreases according to

$$\frac{dI(\omega)}{dx} = -n_1\sigma_{12}I(\omega) \tag{2.37}$$

where n_1 is the concentration of state 1 and σ_{12} is the absorption cross-section (in units of cm^2). The cross-section is related to other quantities

$$\sigma_{12} = \frac{\hbar\omega_{12}}{c} B_{12}g(\omega) \tag{2.38}$$

where B_{12} is the Einstein B-coefficient and $g(\omega)$ is the spectral shape of the absorption profile. The strength of the absorption of a photon leading to excitation from state 1 to state 2 varies with the optical frequency, ω. We shall call the shape of this profile $g(\omega)$ where g is normalized,

$$\int_{-\infty}^{\infty} g(\omega)d\omega = 1. \tag{2.39}$$

Absorption and spontaneous emission are reverse processes in the sense discussed above for detailed balance. There is a difference: equilibrium radiation does not obey the Fermi–Dirac statistics as electrons do but rather the black-body law. Einstein determined the relationship that absorption and spontaneous emission rates must obey to be consistent with black-body radiation

$$B_{12} = \frac{g_2}{g_1} \frac{\pi^2 c^3}{\hbar \omega_{12}^3} A_{21}, \tag{2.40}$$

where g_2 and g_1 are the degeneracies of the upper and lower states, respectively.[1] Equation (2.40) is very powerful. If, for example, we know experimentally the absorption length and absorption profile $g(\omega)$, we can, via Equation (2.40) determine the spontaneous emission rate (A_{21}) of the upper level.

2.9 Mechanisms of Thermal Quenching in Dosimetric Materials

The phenomenon of thermal quenching of the stimulated luminescence in quartz has been well known for several decades (see for example, Bøtter-Jensen *et al.* [74], P.44, and references therein). Thermal quenching manifests itself as a reduction of the measured luminescence intensity from quartz as the sample temperature is raised, and has been observed in both TL and OSL experiments on quartz and other dosimetric materials like aluminum oxide, Al_2O_3:C (Wintle [75], Smith *et al.* [76], Spooner [77]; Akselrod *et al.* [78]).

In addition to the well-known effects on the TL and OSL intensities, thermal quenching also affects the apparent luminescence lifetimes in various dosimetric materials. Time-resolved OSL (TR-OSL) measurements have been carried out using samples of both quartz and feldspars, reflecting the importance of these materials for dating and retrospective dosimetry applications [79–85]. During TR-OSL measurements the stimulation is carried out with a brief light pulse and photons are recorded at the luminescence detector, with the decaying signal recorded immediately after. The decay signal can be analyzed using the linear sum of exponential decays, and can therefore be characterized using decay constants. The main advantage of TR-OSL over CW-OSL is that it allows study of such recombination and/or relaxation pathways. Several researchers have studied the temperature dependence of luminescence lifetimes and luminescence intensity from time-resolved luminescence spectra in quartz (see for example, Galloway [86]; Chithambo [87–90] and references therein). Another important dosimetric material, which has been studied using a similar type of experiments in which UV pulses are used for excitation, is Al_2O_3:C. The luminescence lifetimes of both quartz and Al_2O_3:C are affected by thermal quenching phenomena. For example, the luminescence lifetimes for unannealed sedimentary quartz typically are found to be constant at 42 μs for stimulation temperatures between 20 °C and 100 °C, and then to decrease continuously to 8 μs at 200 °C. By comparison, the luminescence lifetimes of Al_2O_3:C at room temperature are 35 ms, and they decrease gradually to zero at temperatures above 200 °C [78].

---

[1] Note that here, as in the literature generally, the symbol g is used for both the spectral line shape, $g(\omega)$, and the unrelated concept of level degeneracies, e.g. g_1 and g_2.

Two main models have been suggested previously for explaining thermal quenching phenomena in a variety of materials, namely the Mott–Seitz and the Schön–Klasens mechanisms (see for example [74], [33] and references therein). These two mechanisms are based on different physical principles. In the case of quartz, the reader is referred to the discussion in Bailey [35] for a detailed comparison of the two mechanisms, in which the author presents detailed experimental arguments which support the common assumption that the prevalent process of thermal quenching in quartz is the Mott–Seitz mechanism, rather than the alternative Schön–Klasens mechanism.

The principles behind the Mott–Seitz mechanism of thermal quenching have been summarized previously, for example in Bøtter-Jensen *et al.* [74], page 44. This mechanism is usually shown schematically using a configurational diagram as in Figure 2.2, and consists of an excited state of the recombination center and the corresponding ground state. In this mechanism, electrons are captured into an excited state of the recombination center, from which they can undergo either one of two competing transitions. The first transition is a direct radiative recombination route resulting in the emission of light and is shown as a vertical arrow in Figure 2.2. The second route is an indirect thermally assisted non-radiative transition into the ground state of the recombination center; the activation energy W for this non-radiative process is also shown in Figure 2.2. The energy given up in the nonradiative recombination is absorbed by the crystal as heat, rather than being emitted as photons. One of the main assumptions of the Mott–Seitz mechanism is that the radiative and nonradiative

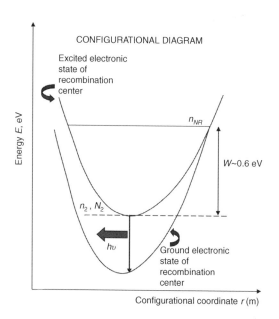

Figure 2.2 *The configurational diagram describing the Mott–Seitz mechanism for thermal quenching. Reprinted from V. Pagonis et al., Modelling the thermal quenching mechanism in quartz based on time-resolved optically stimulated luminescence, J. Lumin. 130, 902. Copyright (2010) with permission from Elsevier*

processes compete within the confines of the recombination center, hence they are referred to as localized transitions.

During TR-OSL experiments a brief stimulating light pulse raises a small number of electrons from the electron trap into the conduction band (CB); some of these electrons are then trapped by the excited state of the recombination center (RC). Electrons trapped in this excited state may then relax to the ground state of the RC through a direct radiative transition (resulting in emission of a photon) or through a non radiative transition (in which the relaxation energy increases the thermally induced vibrations of the lattice, Figure 2.2). In the approach first described by Mott [61, 92], the probability of the nonradiative process A_{NR} (s^{-1}) is assumed to have a temperature dependence described by a Boltzmann factor of the form $\exp(-W/k_B T)$, where W is the activation energy, T is temperature and k_B is Boltzmann's constant, while the radiative probability A_R (s^{-1}) is assumed to be a constant independent of temperature. Thus, the experimentally observed luminescence will be proportional to the luminescence efficiency ratio defined by the relative probabilities of radiative and nonradiative transitions

$$\eta(T) = \frac{A_R}{A_R + A_{NR}\exp(-W/k_B T)} = \frac{1}{1 + \frac{A_{NR}}{A_R}\exp(-W/k_B T)}. \tag{2.41}$$

Experimentally it has been found that the intensity of the CW-OSL or the TL signals from various materials follows a very similar expression to Equation (2.41) with the empirical form

$$I = \frac{I_0}{1 + C\exp(-W/k_B T)} \tag{2.42}$$

where I_0 is the luminescence intensity at low temperatures and C is a dimensionless constant. As the temperature of the quartz sample is increased during stimulation, the experimentally measured luminescence intensity decreases with temperature according to Equation (2.42). By comparison of the empirical Equation (2.42) with the luminescence efficiency Equation (2.41), the dimensionless empirical constant C appearing in Equation (2.42) can be interpreted as the ratio of the non radiative and radiative probabilities. Experimentally it is also found that the luminescence lifetime τ of quartz, Al$_2$O$_3$:C and other materials also shows a similar temperature dependence [78, 90] ,

$$\tau = \frac{\tau_0}{1 + C\exp(-W/k_B T)} \tag{2.43}$$

where τ_0 is the experimentally observed lifetime for the radiative process at low temperatures. As the temperature T of the sample is increased during the optical stimulation, the lifetime of the electrons in the excited state decreases according to Equation 2.43.

In the next two sections we present two models which provide a mathematical description of thermal quenching phenomena in two of the most important dosimetric materials: quartz and Al$_2$O$_3$:C.

2.10   A Kinetic Model for the Mott–Seitz Mechanism in Quartz

Pagonis *et al.* [91] presented a new kinetic model for thermal quenching in quartz based on the Mott–Seitz mechanism. In this model all recombination transitions are localized within the recombination center (in contrast to a delocalized model, where all charge transitions take place to or from the conduction and valence bands, e.g. [35]). Several simulations of typical TR-OSL experiments were carried out using the model, and the results were compared with new measurements of the luminescence lifetimes and luminescence intensity of a sedimentary quartz sample as a function of the stimulation temperature.

Figure 2.3 shows the energy level diagram in the new model developed by Pagonis *et al.* [91]. The model consists of a dosimetric trap shown as level 1, and three levels labeled 2–4 representing energy states within the recombination center. During the transition labeled 1 in Figure 2.3, electrons from a dosimetric trap are raised by optical stimulation into the CB, with some of these electrons being retrapped as shown in transition 2 with a probability A_n. Transition 3 corresponds to an electronic transition from the CB into the excited state located below the conduction band with probability A_{CB}. Transition 5 indicates the direct radiative transition from the excited level into the ground electronic state with probability A_R, and transition 4 indicates the competing thermally assisted route. The probability for this competing thermally assisted process is given by a Boltzmann factor of the form $A_{NR}\exp(-W/k_BT)$, where W represents the activation energy for this process and A_{NR} is a constant representing the nonradiative transition probability. Transition 6 denotes the non radiative process into the ground state, the bold vertical arrow in Figure 2.3 indicating that it is not a discrete energy loss but rather a continuous energy loss resulting in heat. The details of the non radiative process of releasing energy in the model are not

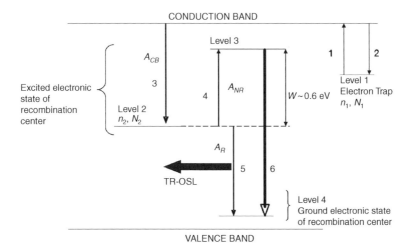

Figure 2.3   *The kinetic model for the Mott–Seitz mechanism of thermal quenching in quartz. Reprinted from V. Pagonis et al., Modelling the thermal quenching mechanism in quartz based on time-resolved optically stimulated luminescence, J. Lumin. 130, 902. Copyright (2010) with permission from Elsevier*

important; what is important in determining the thermal quenching effects is the ratio of the nonradiative and radiative probabilities A_{NR}/A_R, or the constant C and the value of the thermal activation energy W. The competing transitions 4 and 5 in Figure 2.3 are the basis of the description of the thermal quenching process in quartz and are the causes of two simultaneous effects. As the temperature of the sample is increased, electrons are removed from the excited state according to the Boltzmann factor described above. This reduction leads to both a decrease of the intensity of the luminescence signal and to a simultaneous apparent decrease of the luminescence lifetime. Several experiments have reported a range of values of the constants C, W appearing in Equations (2.41–2.43). Typical ranges for the numerical values for these constants in quartz are $W = 0.5$–$0.8\,\text{eV}$ and $C = 10^6 - 6 \times 10^8$ (see for example, Chithambo [87], Table 1 for a tabulation of several results).

The parameters used in the model of Pagonis *et al.* [91] are defined as follows; N_1 is the concentration of electrons in the dosimetric traps (cm^{-3}), n_1 is the corresponding concentration of trapped electrons (cm^{-3}), N_2 and n_2 are the concentrations of electron traps and filled traps correspondingly in the excited level 2 of the recombination center (cm^{-3}), $W = 0.64\,\text{eV}$ is the activation energy for the thermally assisted process (eV), A_n is the conduction band to electron trap transition probability coefficient $(\text{cm}^3\,\text{s}^{-1})$, A_R and A_{NR} are the radiative and non-radiative transition probability coefficients (s^{-1}), and A_{CB} is the transition probability coefficient $(\text{cm}^3\,\text{s}^{-1})$ for the conduction band to excited state transition. The parameter n_c represents the instantaneous concentration of electrons in the CB (cm^{-3}) and P denotes the probability of optical excitation of electrons from the dosimetric trap (s^{-1}). The equations used in the model are as follows

$$\frac{dn_1}{dt} = n_c(N_1 - n_1)A_n - n_1 P, \tag{2.44}$$

$$\frac{dn_c}{dt} = -n_c(N_1 - n_1)A_n + n_1 P - n_c(N_2 - n_2)A_{CB}, \tag{2.45}$$

$$\frac{dn_2}{dt} = n_c(N_2 - n_2)A_{CB} - n_2 A_{NR}\exp(-W/k_B T) - A_R n_2. \tag{2.46}$$

The instantaneous luminescence resulting from the radiative transition is defined as

$$I(t) = A_R n_2. \tag{2.47}$$

It is noted that transitions 4, 5 and 6 in Figure 2.3 are of a localized nature, while transition 3 involves electrons in the CB and hence is delocalized. The difference in the nature of these transitions can also be seen in their mathematical forms in Equations (2.44–2.47). The term $n_c(N_2 - n_2)A_{CB}$ in Equations (2.45) and (2.46) expresses the fact that there are empty electronic states available for electrons in the CB; these states are excited states of the recombination center, in agreement with the general assumptions of the Mott–Seitz mechanism of thermal quenching. The values of the parameters used in the model are as follows: $A_R = 1/42\,\mu\text{s} = 2.38 \times 10^4\,\text{s}^{-1}$, $A_{CB} = 10^{-8}\,\text{cm}^3\,\text{s}^{-1}$, $P = 0.2\,\text{s}^{-1}$, $N_1 = N_2 = 10^{14}\,\text{cm}^{-3}$.

In the absence of any experimental evidence to the contrary, it was assumed that the conduction band empties much faster than the luminescence process in the recombination center, which is assumed to take place with the experimentally observed luminescence lifetime of 42 μs. The authors chose the values of the delocalized transition probability A_{CB} and of the concentration N_2 such that the conduction band empties very quickly when the stimulating light is switched off, on a time scale of 1 μs. Note that the chosen values of A_{CB} and N_2 are about 1 and 3 orders of magnitude higher than those used by Bailey [35], respectively. Without such an increase in either the transition probability and/or the value of N_2, it is not possible to empty the CB sufficiently quickly to explain the experimental TR-OSL results using a model in which the recombination center is thermally quenched. The value of the radiative transition probability is taken from experimentally measured luminescence lifetime of $\tau = 42$ μs at room temperature [84, 90]. The value of the non-radiative probability was treated as an adjustable parameter within the model, and is adjusted to obtain the best possible fits to the experimental data. The initial conditions for the different concentrations are taken as: $n_1(0) = 9 \times 10^{13}$ cm^{-3}, $n_2(0) = 0$, $n_c(0) = 0$.

Pagonis *et al.* [91] simulated a typical TR-OSL experiment in which the optical stimulation is initially on for 200 μs and is subsequently turned off for the same amount of time. The simulation was repeated by varying the stimulation temperature in the range 20–300 °C, with the results shown in Figure 2.4 which shows clearly that as the stimulation temperature increases, the luminescence lifetime for both the rising and falling part of the signal decrease continuously because of thermal quenching. Pagonis *et al.* [91] presented also new experimental data for sedimentary quartz sample WIDG8, obtained using a Risø TL/OSL-20 reader with an integrated pulsing option to control the stimulation light-emitting diodes (LEDs), and with a photon timer attachment to record the TR-OSL data. Blue light stimulation was performed with an LED array emitting at 470 nm, and delivering 50 mW cm^{-2}. In the measurements, a single aliquot of the quartz sample WIDG8 was optically bleached with blue light for 100 s at 260 °C in order to empty all optically active traps, and was subsequently given a beta dose of 5 Gy. It was then preheated for 10s at 260 °C and the

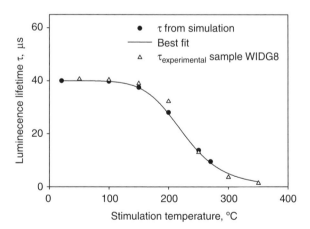

Figure 2.4 *Experimental luminescence lifetimes for sample WIDG8 are compared with the results of the thermal quenching model in Figure 2.3. Reprinted from V. Pagonis et al., Modelling the thermal quenching mechanism in quartz based on time-resolved optically stimulated luminescence, J. Lumin. 130, 902. Copyright (2010) with permission from Elsevier*

TR-OSL was measured by repeatedly turning the optical stimulation on for 50 μs and off for 500 μs . This was followed by an optical bleach for 100s at 260 °C. The whole process was repeated for higher stimulation temperatures, from 20 °C to 320 °C in steps of 20 °C. The luminescence lifetimes obtained by fitting single exponentials to the simulated TR-OSL curves are compared with the experimental luminescence lifetimes obtained for sample WIDG8, as shown in Figure 2.4. Very good agreement can be seen between the simulation and the experimental data.

Pagonis *et al.* also simulated a sequence of 100 000 pulses using the model of Figure 2.3, with the concentrations at the end of each pulse being used as the initial concentrations of the next pulse. The resulting TR-OSL signals in this simulation showed that even after 100 000 pulses, the TR-OSL signal maintains its shape and yields the same luminescence lifetime. Another important result from the model is that it produces reasonable results for very long illumination times of the order of seconds. In such a case one expects that the results of the simulation would resemble what is seen experimentally during a continuous-wave (CW-OSL) experiment. The simulated signal in this case was found to be very similar to an OSL decay curve measured during a CW-OSL experiment. A third important result of the model is that it reproduces the well-known decrease in the TL signal due to thermal quenching. Specifically, when TL glow curves are measured using a variable heating rate, the TL intensity is found to decrease as the heating rate increases, the TL glow peaks become wider and shift towards higher temperatures, in agreement with what is observed during experimental work. Pagonis *et al.* [91] concluded that the Mott–Seitz mechanism for thermal quenching in quartz provides a physical explanation based on localized transitions for several thermal quenching phenomena in quartz. The model can be applied to several types of luminescence experiments involving very different time scales. In the case of TR-OSL experiments the model can describe thermal quenching phenomena within a time scale of microseconds, while in the case of CW-OSL and TL experiments the model provides a satisfactory description of the thermal quenching kinetics within a time scale of tens of seconds.

2.11   The Thermal Quenching Model for Alumina by Nikiforov *et al.*

Nikiforov *et al.* [37] developed a kinetic model which provides a description of the effect of thermal quenching on several luminescence phenomena in Al_2O_3:C. The detailed transitions in the model are shown in Figure 2.5.

The main feature of the model is a description of the thermal quenching mechanism in Al_2O_3:C, which is based on thermal and optical ionization of F-centers. In this model the electron structure of F-centers in aluminum oxide is considered to be similar to the structure of a helium quasi-atom [93]. The ground state is characterized by the $1S$ level, while the excited states are considered to be a singlet ($1P$) and a triplet ($3P$) state. Excitation of an F-center corresponds to the transition $1S \rightarrow 1P$. The F-centers are thought to have the most excited states near the bottom of the CB. Excitation by UV light at 205 nm leads to optical ionization of F-centers and the subsequent capture of electrons from the CB into electron traps. As a result of this excitation the concentration of F^+-centers grows with time. In Figure 2.5 the symbols are as follows: N denotes the main dosimetric trap, M_1 and M_2 stand for deep electron traps, $1P$ and $3P$ are the excited levels of the F-center. Upon optical

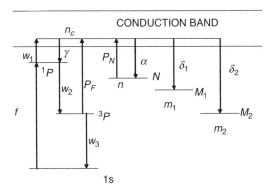

Figure 2.5 *The kinetic model of thermal quenching effects for Al$_2$O$_3$:C. Reprinted from Nikiforov, S.V., Milman, I.I. and Kortov, V.S., Radiat. Meas. 33, 547–551. Copyright (2001) with permission from Elsevier*

excitation in the absorption band, the center is excited to the state 1P via the transition indicated by f in Figure 2.5. Thermal ionization of the excited state $(3P)$ corresponds to the transition indicated as P_F. The probability of thermal ionization of the excited $3P$ state of the F-center is given by the Boltzmann factor $P_F = C \exp(-W/kT)$ where W is the activation energy of thermal quenching and C is a dimensionless constant. Thermal ionization leads to a decrease in the fraction of radiative transitions at the center, and is the direct cause of thermal quenching of luminescence in this material. The optical transition denoted by w_1 results in formation of an F$^+$-center according to $F^* - e = F^+$, while the F$^+$-center can change into an excited F-center upon capture of an electron according to $F^+ + e = F^*$. Free electrons which are formed during the ionization of F-centers, can be captured in dosimetric or deep traps. Luminescence of an F-center (centered at 420 nm) corresponds to the transition w_3 in Figure 2.5. The equations used in the model are as follows

$$\frac{dn}{dt} = -P_N n + \alpha(N - n)n_c, \tag{2.48}$$

$$\frac{dm_1}{dt} = \delta_1(M_1 - m_1)n_c, \tag{2.49}$$

$$\frac{dm_2}{dt} = \delta_2(M_2 - m_2)n_c, \tag{2.50}$$

$$\begin{aligned}
\frac{dn_c}{dt} = \; & P_N n - \alpha(N - n)n_c - \delta_1(M_1 - m_1)n_c - \delta_2(M_2 - m_2)n_c \\
& - \gamma n_{F+} n_c + P_F n_{3P} + w_1 n_{1P},
\end{aligned} \tag{2.51}$$

$$\frac{dn_{F+}}{dt} = -\gamma n_{F+} n_c + P_F n_{3P} + w_1 n_{1P}, \tag{2.52}$$

$$\frac{dn_{3P}}{dt} = -w_3 n_{3P} - P_F n_{3P} + w_2 n_{1P}, \tag{2.53}$$

$$\frac{dn_{1P}}{dt} = f + \gamma n_{F+} n_c - w_1 n_{1P} - w_2 n_{1P}. \tag{2.54}$$

The instantaneous luminescence from the radiative recombination center is defined as

$$I = w_3 n_{3P}. \tag{2.55}$$

Here N, M_1 and M_2 (cm^{-3}), denote the total concentration of dosimetric and deep traps respectively; n, m_1, m_2, n_{1P}, and n_{3P} (cm^{-3}), stand for the concentration of occupied levels N, M_1, M_2, $1P$ and $3P$, respectively; n_{F+} (cm^{-3}) is the concentration of F^+-centers; n_c (cm^{-3}) is the concentration of electrons in the conduction band; δ_1, δ_2, α (cm^3 s^{-1}) are coefficients of carriers capture at the corresponding levels (Figure 2.5); w_1, w_2, w_3 (s^{-1}) denote the transition probability; f (cm^{-3} s^{-1}) is the excitation intensity. The transition probability w_1 is assumed to be independent of temperature since the level $1P$ is located near the CB bottom. The emptying probability for dosimetric traps is described by the expression $s \exp(-E/kT)$ where E is the trap depth of the main dosimetric trap and s is the corresponding frequency factor. In addition to the above equations, the model also ensures the conservation of charge at all times via the equation

$$n + m_1 + m_2 + n_c = n_{F+}. \tag{2.56}$$

The original values of the parameters in the model are as follows: $E = 1.3$ eV, $s = 10^{13}$ s^{-1}, $\alpha = 10^{-14}$ cm^3 s^{-1}, $\delta_1 = 10^{-12}$ cm^3 s^{-1}, $\delta_2 = 10^{-14}$ cm^3 s^{-1}, $\gamma = 10^{-11}$ cm^3 s^{-1}, $N = 10^{13}$ cm^{-3}, $M_1 = 10^{14}$ cm^{-3}, $M_2 = 10^{14}$ cm^{-3}, $w_1 = 1$ s^{-1}, $w_2 = 10$ s^{-1}, $w_3 = 1$ s^{-1}, $f = 10^{10}$ cm^{-3} s^{-1}. These values of the kinetic parameters were used in earlier work by Milman *et al.* [94] to describe specific features of TL in this material (such as the quenching parameters, transition probability, and concentration of traps) and by Nikiforov *et al.* [37] to calculate the temperature dependence of the intensity of stationary photo- and radioluminescence under different occupancy of the deep traps in the model. Variables used in the model were the temperature and the occupancy of deep traps (variable ratios m_1/M_1). The model showed that thermal quenching becomes less efficient when the deep traps are occupied, in agreement with experimental observations. This model has been successful in describing a wide variety of experimental data in Al$_2$O$_3$:C. These experimental results and their agreement with the model will be discussed in detail in later sections of this book.

3

Basic Experimental Measurements

3.1   General Approach to TL and OSL Phenomena

In the present chapter we briefly discuss some experimental considerations concerning the measurements of TL and OSL, and describe some specific experimental results. As pointed out in Chapter 1, the phenomena of thermally stimulated luminescence, also termed thermoluminescence (TL) and optically stimulated luminescence (OSL), are being broadly used in dosimetry as well as in dating of archaeological and geological samples. This application has been first suggested by Daniels *et al.* [8] in the early 1950s. Boyd [95] and Daniels *et al.* [8] studied intensively TL in LiF which later became the main material for dosimetric purposes. These phenomena involve a long-time accumulation and holding of energy absorbed when the sample is exposed to different kinds of radiation, and its subsequent release in the form of emitted light during the heating of the sample (TL) or during its exposure to stimulating light following the original excitation (OSL). Watanabe [96] reported the use of TL in $CaSO_4$:Mn in the measurement of sunlight exposure dose. The use of TL for archaeological dating, and later for geological dating started in the 1960s by Aitken's group in Oxford [10, 97]. The application of OSL for dosimetry has been suggested in the mid 1980s by Huntley *et al.* [98] and has become very quickly a viable competitor to TL [99].

   A technique which may be considered as a hybrid of thermally and optically stimulated luminescence is laser-heated thermoluminescence. The method was first introduced by Bräunlich *et al.* [100] and Gasiot *et al.* [101]. An application for fast neutron dosimetry by heating has been proposed by Mathur *et al.* [102]. Kearfortt and co-workers [103–105] described the laser-heated thermoluminescence when the heating is performed by either continuous or pulsed laser light. Jones *et al.* [106] compared the results of laser and conventional heating in TL dosimetry dose mapping. Gayer and Katzir [107] have described laser fiber-optic controlled heating of samples for TL measurements. Lawless *et al.* [108] also described the use of CO_2 laser heating for getting TL peaks from different materials. Using laser heating has the advantage that in ceramics, there is no damage to the specimen,

Thermally and Optically Stimulated Luminescence: A Simulation Approach, First Edition. Reuven Chen and Vasilis Pagonis.   © 2011 John Wiley & Sons, Ltd. Published 2011 by John Wiley & Sons, Ltd.

and no need to remove small samples from the ceramic material and to place them on the TL oven. Ditcovski *et al.* [109] described the TL stimulated by a non-contact laser fiber-optic system, which brought about programmed linear or nonlinear heating rates.

Other means of excitation of TL or enhancement of TL intensity in samples are known. For example, Ueta *et al.* [110] reported that the intensity of UV-excited TL peaks in KCl increased with deformation. Gabrysh *et al.* [111] investigated the TL induced by high pressure in Al_2O_3 and found a similarity between the γ-induced and pressure-induced glow peaks. Hook and Drickamer [112] found similar results in ZnS:Cu,Cl following the application of high pressure. Also, Joshi *et al.* [113] described deformation-induced TL of NaCl:Ca which exhibits pattern changes under thermal cycling. Kos and Mieke [114] also reported on stress-induced TL, in LiF:Ti. In a similar way, Petralia and Gnani [115] reported on an increase in the TL of X- and γ-excited TLD-100 samples, following plastic deformation. On the other hand, Bradbury and Lilley [116] reported on a decrease of the TL response to radiation of LiF and TLD-100 dosimeter grade LiF following various amounts of plastic deformation. Another effect of plastic deformation has been reported by Petit and Duval [117] who applied pressure on ice, which had previously been UV irradiated at liquid nitrogen temperature (LNT). The resulting TL peak at $\sim$155 K increased significantly with the applied pressure. Reitz and Thomas [118] described the occurrence of TL in CsI crystals following their cooling, and explained it as being due to deformation caused by the cooling. Yet another kind of TL, namely tribothermoluminescence, has been reported by Nyswander and Cohn [119]. They described the effect of TL observed following the grinding of common glass.

Also reported in the literature is self-dose excitation, namely, the excitation induced by radioactive content of the thermoluminescent material. Nambi [120] described this effect in several materials, in particular in the most sensitive TL phosphors such as $CaSO_4$ doped with Dy or Tm, and CaF_2. The self-excitation seemed to be due to minute amounts of radioactive materials such as uranium, thorium and ^{40}K. Self-excitation of TL has also been mentioned by Templer [121, 122] and by Pashchenko *et al.* [123, 124] who, while trying to study the fading characteristics of TL in KCl:Eu^{2+}, found that self-irradiation by ^{40}K needs to be considered.

Araki *et al.* [125] reported on TL in polyethylene due to the application of an electrical field and Levinson *et al.* [126] described a similar effect in semiconducting diamonds. Another version of the field effect of TL has been reported by Miyashita and Henisch [127] who described an enhancement of TL by factors of 100 and more in ZnS:Cu powders.

An interesting effect of TL in photosynthetic materials has been rather broadly described in the literature. The original discovery of TL in chloroplasts by Arnold and Sherwood [128], followed by work on TL in chlorophyll [129] and green plants [130, 131] proved to be a phenomenon common to all photosynthetic bacteria, cyanobacteria, algae and higher plants, which can be observed in isolated membrane particles, intact chloroplasts and unicellular organisms, and whole leaves (see Arnold [1] and Vass [132]). The TL method has become a powerful tool in probing a wide range of PS II redox reactions and their modification by environmental stress effects. A recent review paper on the theory of this kind of TL has been published by Rappaport and Lavergne [133].

OSL, also termed photo-stimulated luminescence (PSL) is the effect in which a solid sample is first irradiated by some ionizing radiation such as α-, β- or γ-rays, X-rays or heavy particles, and then illuminated by stimulating light, thus yielding an emission of measurable light. As in the case of TL, the emitted light is mainly the result of the dose absorbed during the initial irradiation, whereas the stimulating light serves as a trigger to release the absorbed energy. Thus, the role of the optical stimulation in OSL is the same as that of heating in TL (see e.g. [33]). Usually the stimulating light has longer wavelength (lower photon energy) than the emitted light, however, the opposite situation may also take place, namely that lower energy photons are being released when the sample is illuminated by higher energy photons. If this is the case, one has to make sure that the emitted light is related to the initial excitation dose and is not merely a directly excited photoluminescence. Much of the early work on IR stimulation was performed on variously doped sulfides. For example, Keller *et al.* [134] measured the effect in SrS, SrS:Ce,Sm and SrS:Eu,Sm. They reported UV excitation of the samples around 300 nm, IR stimulation between 800 nm and 1400 nm and emission spectrum in the visible light region 450–600 nm, and suggested an energy model to account for their results. Riehl [135] discussed the effect in ZnS and reported that following an excitation at 5 K, stimulation may occur with IR of up to a wavelength of 25μ, which is capable of freeing carriers from shallow traps. This has to be taken into account while performing measurements at the low temperature range, due to IR emitted by the surrounding walls of the measurement compartment; one has therefore to shield the sample from this IR light for measurements performed at low temperatures. Riehl compared the thermal and optical trap depths and mentioned that the ratio is too high to be explained by the ratio of static to high-frequency dielectric constants mentioned in Section 3.4. Baur *et al.* [136] reported on IR stimulated results of the same kind in 420 nm excited ZnO samples. They identified yellow and green emission in both TL and IR-stimulated luminescence, in the temperature range of 5–100 K and showed that whereas the yellow emission is due to donor–acceptor recombination, the green one is related to an emptying of hole traps into the valence band.

The application of OSL for dosimetry was mentioned more than 50 years ago by Antonov-Romanovskiĭ *et al.* [11] and later by Sanborn and Beard [137](see also the Introduction to the book by Cameron *et al.* [138]). These authors concentrated on infra-red stimulated luminescence (IRSL), which made the resolution between the IR stimulation and the visible luminescence easier. Bräunlich *et al.* [139] further discussed the IRSL. For fast neutron detection, the luminescent material SrSe:Eu, Sm can be embedded in transparent plastics and, contrary to the case of TL dosimeters, they can be evaluated without thermal stressing of the plastic. This application in dosimetry has accelerated since the 1980s. Pradhan *et al.* [140], Allen and McKeever [141] and Nanto *et al.* [142, 143] reported on OSL in different dosimetric materials, describing the general OSL properties and their potential use in dosimetry. In these studies, the optical excitation is "continuous wave" (CW), and is either a high power arc source along with a monochromator or filter system to select the excitation wavelength, or a laser operating at the desired wavelength. The luminescence is monitored continuously while the excitation source is on, and narrow-band filters are used in order to discriminate between the excitation light and the emission light, and to prevent scattered excitation light from entering the detector. Since then, several variations of the OSL dosimetry method have been developed, and will be elaborated upon in Chapter 7. Huntley *et al.* [98] suggested the use of the OSL effect for the "optical dating" of sediments. Advanced OSL dating methods have been developed over the years, some are the same as in dosimetry and others specific to dating. These are discussed in Chapter 7 and in Section 9.6.

Pulsed thermoluminescent (PTL) measurements have been described by Manfred *et al.* [144]. Short thermal pulses have been used to excite trapped carriers, leading to radiative recombination. Temperature pulses of 10 and 50 ms were applied to the heater following excitation of Al_2O_3:C by 205 nm light. Curves of intensity versus temperature were thus produced similar to standard TL, except that they shifted to higher temperature. The luminescence of single particles was read multiple times with negligible loss of population, similarly to the technique of pulsed OSL discussed in Section 7.5.

3.2   Excitation Spectra

Information regarding the processes taking place in an irradiated sample can be also extracted from the excitation spectrum. This is mainly related to the excitation of TL or OSL by UV and visible light. "Excitation spectrum" means the dependence of the measured effect on the excitation wavelength. The essence of the measurement, or rather the series of measurements, is to study the relative efficiency of various wavelengths in exciting a certain peak. For example, Nahum and Halperin [145] present the excitation of the 150K TL peak in an insulating (type IIa) diamond, in the range of excitation 200–300 nm. Each point in the graph of the excitation spectrum is related to a separate glow measurement. One usually tries to use the same amount of excitation light in the different wavelengths. A characteristic feature of the given spectrum is the very high and narrow peak at 225 nm which is where the absorption edge in diamond is located. It is quite obvious that when one goes down in wavelength, one gets an increased TL excitation when band to band transition becomes possible. A similar behavior of the 250 K peak in semiconducting diamond has been described by Halperin and Nahum [146] and is shown in Figure 3.1.

The steep decrease in the excitation efficiency at wavelengths lower than 225 nm here (and lower than the absorption edge in the general case) calls for another explanation which

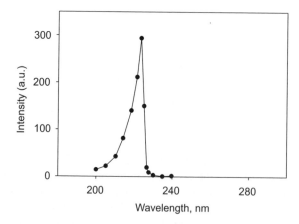

Figure 3.1   *TL excitation spectrum of the 250 K peak in semiconducting diamond near the absorption edge. Reprinted from Halperin, A. and Nahum, J., J. Phys. Chem. Solids 18, 297–306. Copyright (1961) with permission from Elsevier*

will be given below. It is important to note that in order to be able to talk about an "excitation spectrum" without specifying the constant light intensity used at all the wavelengths, one has to be sure that the dose dependence is linear in all the relevant wavelengths. Both superlinearity and sublinearity which may be different at various wavelengths, would limit the significance of the reported information. If such nonlinear effects occur, not identically with different wavelengths, the excitation spectrum may change substantially if it is repeated, say, with the dose of excitation changed by a factor of 10. Another point to be mentioned is that one has to specify in each case whether "the same amount of light" used in the experiment refers to the same number of photons or to a constant energy in various wavelengths.

A superficial examination of the relationship between absorption and TL excitation may suggest that the two effects should increase and decrease always in parallel. This is usually correct for low absorption coefficients. However, when the absorption coefficient μ is large, the effective volume that can be excited is limited to a very narrow layer near the front surface, and a larger value of μ will be associated with lower excitation. For further discussion on this effect and its relationship with the dependence of the excitation on the absorption coefficient, see Section 6.8. This effect may also result in getting minima in the excitation spectrum for wavelengths of maximum absorption, or in other words, one would expect a peak inversion. In this effect, peaks of the absorption spectrum occur at the same wavelengths as the minima of the excitation spectrum. This appears to be due to the penetration depth of the exciting photons. Where the absorption is strong, the photons may not penetrate deeply into the sample, and therefore, the excitation is less efficient. Examples of this behavior for NaCl, KCl and KBr have been reported by Israeli and Kristianpoller [147, 148].

A somewhat similar measurement of the OSL stimulation spectrum, has been reported by Kuhn *et al.* [149]. The optical stimulation of previously excited quartz samples has been found to decrease with wavelength in the range 420–580 nm, with a distinct change in response in the range 500–520 nm.

The OSL excitation or stimulation spectrum for Al_2O_3:C has been studied by several researchers using lasers as well as broad band light sources [150–153]. In a comprehensive study, Whitley and McKeever [151] obtained the OSL stimulation spectrum for a fixed emission wavelength of 420 nm. These authors also presented three-dimensional (3D) plots of the OSL intensity from this material as a function of the stimulation and emission wavelengths. The OSL data showed an intrinsic prominent peak at 205 nm corresponding to electronic transitions from excited F-centers. These authors also reported several features of the emission spectrum which are due to the presence of F^+-centers.

Other materials for which excitation and emission spectra have been studied are synthetic halides, sulfides and oxides. For example, OSL measurements have been reported for Cu-doped NaCl crystals, for RbI doped with a variety of elements, and for the popular dosimetric material CaF_2:Mn. Sulfates and sulfides have been studied to a lesser extent, while several researchers have studied BeO using both CW-OSL and LM-OSL. An excellent review of the spectra measured for these types of materials including several examples of emission and stimulation spectra has been given by Bøtter-Jensen *et al.* [150] (pp.81–95).

The excitation spectra of natural quartz and feldspar samples have been studied extensively for a wide variety of samples and with a variety of experimental techniques, due to the importance of these materials in dating applications. Spooner [77] studied the bleaching of the OSL signal from quartz samples by using the 514.5 nm line in argon-ion lasers and narrow interference filters in the range 400–900 nm. Other types of excitation spectra were obtained by using scanning monochromators at various stimulating wavelengths in the range 400–560 nm [154]. Additional reports have been made of OSL excitation spectra obtained using xenon lamps and interference filters, a variety of laser lines, and also for IR-stimulated signals from a variety of samples ([74]; pp.141–148).

There have been several published reports on direct measurements of excitation spectra in feldspars. As in the similar studies for quartz samples, the OSL stimulation spectra from feldspars have been studied using continuous scanning monochromators, lines from lasers and tunable lasers, and lamps with wide light spectra. Bøtter-Jensen *et al.* [155] used a scanning monochromator and a tungsten halogen lamp with stimulation wavelengths over a wide region of the spectrum 400–1000 nm, and light detection was carried out at 340 nm. In a similar study using a monochromator, a xenon lamp and light detection between 300 nm and 400 nm, Clark and Sanderson [156] reported several excitation bands. These authors also studied the effect of heating on the excitation spectra of their samples. Tunable lasers have also been used in such feldspars studies, and several researchers have reported on a peak near 855 nm. Such studies have suggested the use of IR LEDs (880 nm) for feldspar dating studies. A summary of OSL stimulation data for a variety of feldspar samples has been given by Krbetschek *et al.* [157].

3.3 Emission Spectra

One of the main methods of extracting information from TL curves is to measure the emission spectrum of the measured glow. Many materials contain several types of recombination centers, each of which may be represented by its own characteristic emission wavelength. In addition, materials containing rare earth or transition metal impurities often exhibit several emission wavelengths due to relaxation of ions from several excited states into the ground state. Conventional measurements of TL use detectors which are sensitive over a broad range of wavelengths, and no discrimination between emission colors is made. Broad-band filters are commonly used and therefore potential information concerning the emission process may be lost. Attempts to measure the emission spectrum through spectrometers have been made which led to the plots of intensity versus both temperature and wavelength, the so-called 3D presentation. Different kinds of spectrophotometers have been used, but in most cases, the signal is dispersed over a defined wavelength range and the intensity is measured over each wavelength interval separately and simultaneously. More advanced methods include Fourier transform spectroscopy [158, 159] as well as position-sensitive photodetectors. Here, the light is dispersed over the relevant wavelength range using a grating or prism spectrograph, and the dispersed light is detected using a CCD camera, diode array or position-sensitive photomultiplier. More details can be found in the work by Luff and Townsend [160], Townsend [161] and Piters *et al.* [162]. A review on spectral measurements during TL has been given by Townsend and Kirsh [163]. Some later works on the emission spectrum of TL should be mentioned. Karali *et al.* [164] performed a spectral comparison of Dy, Tm and Dy/Tm in $CaSO_4$ thermoluminescent dosimeters and concluded that the rare earth ions form part of a complex defect which variously provides both the charge trapping and, during heating, the radiative decay. Kuhn *et al.* [149] presented the emission spectrum of quartz from different sources and suggested that the blue/green emission may interfere with the UV emission used for quartz dating. Spooner and Franklin [165] also studied the TL emission spectrum of quartz and found that the change from blue to red emission is accompanied by a significant change in thermal quenching (see, e.g., Section 6.7). Some more examples of measurements of TL emission spectra appear in the literature. Schlesinger and Whippey [166] reported on the energy-level scheme of Gd^{+3} in CaF_2 using TL. Sridaran *et al.* [167] described "monochromatic TL" in X-irradiated KCl:Ag crystals and concluded that tunneling of electrons is responsible for the observed TL glow curves and TL spectral data. Townsend *et al.* [168] described the emission spectrum of LiF:Mg,Ti (TLD-100) during TL, and gave isometric 3D plots of emitted intensity versus wavelength and temperature.

Extensive data on the emission spectra of quartz samples have been reported for TL signals, and to a much smaller extent for OSL signals. The main TL emission bands of quartz in the literature were reviewed in a paper by Krbetschek *et al.* [157]. The emission bands were identified and classified according to their wavelength region as being blue, 420–490 nm, near-UV to violet, 360–420 nm, and the orange-red region of the spectrum, 590–650 nm.

There have also been a smaller number of published OSL emission spectra in quartz. Huntley *et al.* [169] used the 647 nm line from a krypton laser to excite OSL emission spectra and reported a single emission band centered at 365 nm at room temperature. Similar results were obtained by Martini *et al.* [170] who used synthetic quartz crystals with known concentrations of Al and alkali ions.

Huntley *et al.* [171] studied the emission spectra from feldspars using the 514 nm line from an argon laser and the 633 nm line from a He-Ne laser. There is also a wealth of available data on IRSL from feldspars; a detailed presentation of these is beyond the scope of this book, and the reader is referred to the detailed classification and summaries given in Bøtter-Jensen *et al.* [74] (pp. 201–203).

3.4   Bleaching of TL and OSL

The term bleaching in the sense of reduction of the expected TL peak by application of an appropriate illumination following the primary excitation, is taken from the vocabulary of color centers. In that case, a sample is colored by a higher energy excitation, e.g. X-rays, γ-rays, etc., and the additional illumination causes this color to fade away or, in other words, the sample is bleached. The optical bleaching of TL and OSL is used in the application of optical dating of geological samples; the beginning of the period dated is usually assumed to be associated with the exposure of the material to sunshine, which is supposed to have bleached the sample to zero or nearly zero. Thus, the accumulated signal during the following long burial period under a certain radiation field may serve as a measure of the time elapsed.

The basic explanation of the bleaching phenomenon rests upon the understanding that trapped carriers which are candidates for thermal detrapping, can be optically released during the additional illumination at low temperature. The charge carriers thus released may recombine with opposite sign carriers, which may emit light during the application of the bleaching light, thus emitting OSL. However, the decrease of the TL may be accompanied by nonradiative recombinations of the released carriers with opposite sign carriers. Note that bleaching may also be associated with the reduction of the concentration of carriers in centers which, in some cases may be the more important reason for the reduction of TL intensity. No matter whether the illumination is accompanied by the emission of OSL or not, these carriers will be missing during heating and thus, the resulting TL is reduced in size. It is to be noted that the optical and thermal energies needed for the release of trapped carriers are not the same. According to Mott and Gurney [92], the ratio of $E_{optical}$ to $E_{thermal}$ should be equal to k/k_0 where k and k_0 are the dielectric constants of the crystal for static and high frequency dielectric fields, respectively. Thomas and Houston [172] verified this relation for MgO samples in which TL peaks with thermal activation energies of 0.66 and 0.9 eV were bleached by 2.2 and 3.0 eV photons, respectively, where $k/k_0 = 3.3$.

The relative efficiency of various wavelengths in reducing the TL intensities of a given peak can also be measured. The result is known as a "bleaching spectrum". Note that as mentioned above about the excitation spectrum, here too, the definition is unambiguous only in ranges where the bleaching effect is the same, preferably linear, with the bleaching dose for all wavelengths. An example given by Halperin and Chen [173] is given in Figure 3.2. A peak at 150 K in semiconducting diamonds was excited by band-to-band radiation of 225 nm, and could be bleached by longer wavelength UV light as well as by visible and infrared light. Whereas light of wavelength above 400 nm could bleach the peak totally, light of 300–400 nm could bleach it only partially. The somewhat surprising result in this case is that, no matter whether one starts with heavily 225 nm irradiated sample or with an unirradiated sample, a long enough illumination in the 300–400 nm range results in an equilibrium, relatively low level of the 150 K peak intensity which does not change by additional illumination. A similar effect of TL in quartz which is reduced by prolonged illumination, but does not go to zero, was reported by Morris and McKeever [174]. The data obtained

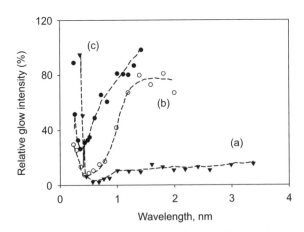

Figure 3.2 *Optical bleaching with various wavelengths of the 150 K glow peak after excitation at 225 nm. The ratio of the doses of the bleaching light was approximately 1:4:40 in curves (a), (b) and (c), respectively. Reprinted from Halperin, A. and Chen, R., Phys. Rev. 148, 839–845. http://prola.aps.org/abstract/PR/v148/i2/p839_1. Copyright (1966) with permission from American Physical Society*

can be described by the summation of a first-order decay and an unbleachable residual. A theoretical account for this effect which includes the optical excitation of electrons from the recombination center into the conduction band during illumination has been given by Chen *et al.* [175], and is discussed in some detail in Section 6.4. A detailed model explaining the bleaching effects in quartz has been given by Morris and McKeever [176] who also performed simulations by solving the coupled rate equations relevant to the model.

Effects of optical bleaching of TL and OSL in different materials have been reported by several authors. Sridaran *et al.* [167] discussed the redistribution of electrons in traps in X-irradiated KCl:Ag by optical bleaching. Subramanian *et al.* [177] reported on bleaching of TL in X-irradiated KCl:Pb samples. Spooner *et al.* [178] described the big difference between the 325 °C and 370 °C peaks in quartz. Whereas the former was bleachable by visible light up to 700 nm, the latter was found to be hard to bleach. Benabdesselam *et al.* [179] studied the bleaching of TL and OSL of chemical vapor deposition (CVD) diamond. Both X-ray excited TL and OSL were found to be bleachable by a broad range of wavelengths from the visible light into the IR. Chruścińska *et al.* [180] investigated optical bleaching of TL in K-feldspars and correlated the differences in the TL bleaching efficiency of different trap groups with significant differences between trap parameters. Secu *et al.* [181] studied the processes involved in the high-temperature TL of $MgF_2:Mn^{2+}$. Based on thermal and optical bleaching, they found that a 420 K peak could be ascribed to a thermally activated recombination between the F-center and a hole center trapped in the neighborhood of the Mn^{2+} impurity. Ambrico *et al.* [182] reported on the TL and OSL of β-irradiated alumina electrode surface, and described the reduction of TL following the measurement of OSL. Dotzler *et al.* [183] described TL and OSL results in X-irradiated and γ-irradiated $RbMgF_3:Eu^{2+}$. Bleaching by an IR LED centered at 940 nm reduced the low temperature peaks below ~130 °C nearly to zero. All peaks disappeared after extended exposure to blue light from a 455 nm LED.

Another possibility, namely, the reduction of the intensity of previously excited luminescence by microwaves, has been suggested by Elle *et al.* [184] who also recorded a similar effect in thermally stimulated electron emission (TSEE) (see Section 11.6). Later, Nagpal *et al.* [185] reported on this effect in $BaSO_4$ and $CaSO_4$ doped by some rare earths and in $MgF_2:Mn$, and suggested the use of this effect for dosimetry of microwave radiation.

4

Thermoluminescence: The Equations Governing a TL Peak

The simplest model for explaining the occurrence of TL includes one trapping state and one kind of recombination center. One of these two entities captures free electrons during the excitation of the sample, and the other captures holes. In most cases, one assumes that one of these entities, the trapping state, is relatively close to the relevant band, so that while the sample is heated, the charge carriers are thermally released into the band. The other entity is assumed to be farther from the relevant band, so that it is thermally stable. An example is shown in Figure 4.1. Here, one assumes that the electron trapping state is close to the conduction band whereas the hole state is relatively far from the valence band. When the sample is heated following excitation, electrons are released from the trapping state, N, into the conduction band and then recombine with captured holes in the recombination center, M. An inverted situation is just as likely to occur. If the hole state is relatively close to the valence band, it can thermally release holes into the valence band during the heating; these holes may recombine with electrons in the relevant level. In this case, the hole capturing state is termed a hole trapping state, and the electron state is termed an electron recombination center. Most authors usually refer to the situation shown in Figure 4.1 as the "normal" situation, but the inverted situation should always be considered. Situations where both electron and hole trapping states can be thermally released simultaneously or nearly simultaneously are also mentioned in the literature and will be discussed below.

4.1 Governing Equations

Figure 4.1 is a schematic energy level diagram showing one trapping state, N and one kind of recombination center, M, and the transitions taking place during the excitation and read-out stages. Here, n (cm^{-3}) and m (cm^{-3}) are the concentrations of electrons in traps and of holes in recombination centers, respectively, and N and M (cm^{-3}) the concentrations of trapping

Thermally and Optically Stimulated Luminescence: A Simulation Approach, First Edition. Reuven Chen and Vasilis Pagonis. © 2011 John Wiley & Sons, Ltd. Published 2011 by John Wiley & Sons, Ltd.

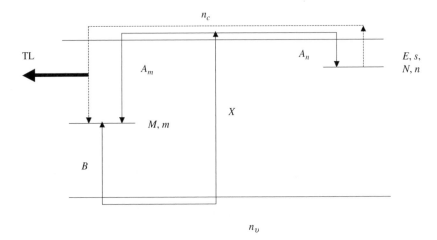

Figure 4.1 *Schematic energy level diagram with one trapping state (N) and one kind of recombination center (M). Transitions occurring during excitation are shown by solid lines, and during heating/optical stimulation, by dashed lines. The thick arrow denotes the emitted TL/OSL*

states and recombination centers. E (eV) and s (s^{-1}) are, respectively, the activation energy and frequency factor associated with the thermal release of electrons from traps. n_c (cm^{-3}) and n_v (cm^{-3}) are the concentrations of free electrons and holes, respectively. A_n (cm^3 s^{-1}) is the retrapping probability coefficient of electrons, A_m (cm^3 s^{-1}) the recombination probability coefficient of electrons, and B (cm^3 s^{-1}) the trapping probability coefficient of holes in centers. X (cm^{-3} s^{-1}) is proportional to the dose-rate of excitation, and actually denotes the rate of production of electron–hole pairs by the excitation irradiation per unit volume per second. The set of simultaneous differential equations governing the process during excitation is

$$\frac{dn}{dt} = A_n(N - n)n_c - s \cdot n \cdot \exp(-E/kT), \tag{4.1}$$

$$\frac{dm}{dt} = B(M - m)n_v - A_m m n_c, \tag{4.2}$$

$$\frac{dn_c}{dt} = X - A_n(N - n)n_c - A_m m n_c, \tag{4.3}$$

$$\frac{dn_v}{dt} = \frac{dn}{dt} + \frac{dn_c}{dt} - \frac{dm}{dt}. \tag{4.4}$$

k (eV K^{-1}) here is the Boltzmann constant. In most cases, when the temperature of the excitation is significantly lower than that of the TL peak, the second term on the right-hand side of Equation (4.1) is negligibly small during the excitation stage. This set of equations cannot be solved analytically. From this point on, one can proceed by making simplifying assumptions to get some general, though approximate behavior, say the dependence of the concentrations of trapped carriers on the dose. An alternative is to solve numerically the set of equations for a chosen set of parameters and for different doses, to obtain a rather accurate dependence on the dose for such specific cases. We will discuss these two complementary routes below.

It is to be noted here that using the terms "trapping probability" or "recombination probability" for A_n, B and A_m (with units of $cm^3 s^{-1}$), although rather common in TL and OSL theory, is not very accurate. In fact, what is meant is that products like $A_m m$ or $A_n (N - n)$, having units of s^{-1}, are probabilities per second. Another way to look at these coefficients is by stating that $A_m = \sigma_m v$ where σ_m is the cross-section for recombination (in cm^2) and $v (cm\, s^{-1})$ is the thermal velocity of the relevant free carrier. We therefore use along this work the terms "probability coefficient" or "recombination coefficient" and "trapping coefficient".

At the end of excitation we end up with finite concentrations of the free electrons, n_c, and free holes, n_v. If we wish to mimic the experimental procedure of TL or OSL, we have to consider a relaxation time between the end of excitation and the beginning of heating or exposure to stimulating light. This is done by setting X to zero and solving Equations (4.1–4.4) for a further period of time, so that at the end, both n_c and n_v are negligibly small. Obviously, the final values of n, m, n_c, and n_v at the end of excitation are used as initial values for the relaxation stage. Finally, for the heating stage, we take the final values of these four functions at the end of the relaxation period as initial values for the heating stage. We utilize a certain heating function, e.g. linear heating $T = T_0 + \beta t$, where β is the constant heating rate. The same set of equations is being solved numerically with $X = 0$ and with the second term on the right-hand side of Equation (4.1) being more and more important as the temperature rises. On the other hand, it is usually assumed that during heating, $n_v \approx 0$. Thus, the first term in Equation (4.2) can be omitted, and with $X = 0$, Equation (4.4) is fulfilled automatically. Following the work by Antonov-Romanovskiĭ [186] and Lushchik [187] (see also the work by Maxia [69] where the mechanism of the TL process is examined on the basis of nonequilibrium thermodynamics), one can see that the equations governing the process during heating will be

$$\frac{dn}{dt} = A_n(N - n)n_c - s \cdot n \cdot \exp(-E/kT), \tag{4.5}$$

$$I(T) = -\frac{dm}{dt} = A_m m n_c, \tag{4.6}$$

$$\frac{dm}{dt} = \frac{dn}{dt} + \frac{dn_c}{dt}. \tag{4.7}$$

The emitted TL light, shown in Figure 4.1 as the thick arrow, is associated with the recombination of free electrons with holes in recombination centers, which is given by Equation (4.6). Note that with the assumption that $n_v = 0$, this is not an additional equation; the equality here is, in fact, Equation (4.2) without the first term on the right-hand side which vanishes when $n_v = 0$. It should also be noted that when OSL rather than TL is studied, exactly the same equations are to be used in the excitation and relaxation periods.

As for the read-out stage, the last term in Equation (4.1) is to be replaced by $-f \cdot n$ where f (s^{-1}) is proportional to the intensity of the stimulating light. Another point to remember is that, in fact, the intensity $I(T)$ is merely proportional to the rate of change of the center concentration. Using an equality sign on the left-hand side of Equation (4.6) means a specific choice of the units of the emitted light intensity.

An important point to be made is, that solving the set of equations for the read-out stage, be it TL or OSL, without the initial solution of the equations during the excitation and just making a guess concerning the initial concentrations (prior to heating or optical stimulation) may not yield reliable results. In the present simple case of one trapping state and one kind of recombination center, one can expect at the end of relaxation that $n_0 = m_0$. However, in more complicated situations where more levels are involved and which will be discussed later, making an arbitrary assumption about n_0 and m_0 as well as for the concentrations of the other relevant trapping states at the end of relaxation may not be compatible with the parameters of these states. This point will also be discussed further. Halperin and Braner [30] have suggested an approximate expression for the dependence of the emitted light on the relevant occupancies of traps and centers. They made the simplifying assumptions, later called the "quasi-equilibrium" or "quasi-steady" assumptions,

$$\left| \frac{dn_c}{dt} \right| \ll \left| \frac{dn}{dt} \right|, \left| \frac{dm}{dt} \right|; \qquad n_c \ll n, \tag{4.8}$$

which state that the rate of change of free carriers is significantly smaller than that of the trapped carriers, and that the concentration of free carriers is significantly smaller than that of trapped carriers. The validity of these assumptions will be discussed below. Making these simplifying assumptions, Halperin and Braner [30] wrote, for the case where carriers travel during read-out through the conduction band,

$$I = -\frac{dm}{dt} = s \cdot n \cdot \exp\left(-E/kT\right) \frac{A_m m}{A_m m + A_n(N - n)}. \tag{4.9}$$

Obviously, this one equation in the two unknown functions, n and m, cannot be solved without making further assumptions. The simplest assumption to make is that recombination is significantly faster than retrapping. This does not necessarily mean that the recombination probability coefficient, A_m, is much larger than the retrapping probability coefficient, A_n, but rather, that $A_m m \gg A_n(N - n)$. With this assumption, the equation gets significantly simpler. However, the derivative is of m, whereas on the right-hand side appear both n and m. In the simplest case, where only one trapping state and one kind of recombination center participate in the process, and using the smallness of n_c from Equation (4.6), we have $n \cong m$, and also $dn/dt \cong dm/dt$, and Equation (4.9) reduces to the well known first-order equation,

$$I = -\frac{dn}{dt} = s \cdot n \cdot \exp\left(-E/kT\right). \tag{4.10}$$

Note that the assumption made that recombination is significantly faster than retrapping does not mean that the retrapping probability is nil. Had this been the case, no trapping into the same trapping states would have taken place during the initial excitation, and no effect could have been expected at all. Also, if there are more electron and/or hole trapping states in the sample which do not take part in the process in the relevant temperature range of occurrence of the TL peak in question, one can write $n = m + c$ where c is the net charge of disconnected electron and hole trapping states, which is constant. As a result, one gets again $dn/dt = dm/dt$, and Equation (4.10) holds true for this somewhat more general case. The solution of Equation (4.10) for a constant heating rate β is the well-known expression

$$I(T) = s \cdot n_0 \cdot \exp\left(-E/kT\right) \exp\left[-(s/\beta) \int_{T_0}^{T} \exp\left(-E/k\theta\right) d\theta\right], \qquad (4.11)$$

where n_0 is the initial trap concentration prior to heating and T_0 the initial temperature; θ is a dummy integration variable representing temperature.

The second-order kinetics suggested by Garlick and Gibson [27] can also be reached from Equation (4.9) by making different simplifying assumptions. To begin with, one has to assume that $m = n$, which is the case when only one trapping state and one kind of recombination center take part in the process. In addition, one can assume that retrapping dominates, namely $A_n(N - n) \gg A_m m$ and that the trap is far from saturation, $N \gg n$. Equation (4.9) reduces to

$$I = -\frac{dn}{dt} = \frac{sA_m}{A_n N} n^2 \exp\left(-E/kT\right). \qquad (4.12)$$

It should be noted here that since both n and m are decreasing functions of time while the sample is heated, if the condition $A_n(N - n) \gg A_m m$ holds at the beginning of the read-out phase, it holds all along the peak and therefore, a peak that starts being of second order, remains so all the way. This is not necessarily the case for the opposite condition $A_n(N - n) \ll A_m m$. As explained by Denks *et al.* [188], one may start with this condition, but as n and m decrease, the condition may be reversed, and the kinetics may change from first to second order.

Another possibility, less likely to occur, is that the retrapping and recombination probability coefficients are exactly equal, $A_m = A_n$. Along with $n = m$, Equation (4.9) reduces to

$$I = -\frac{dn}{dt} = \frac{s}{N} n^2 \exp\left(-E/kT\right). \qquad (4.13)$$

Both of the last two equations can be written as

$$I = -\frac{dn}{dt} = s'n^2 \exp\left(-E/kT\right), \qquad (4.14)$$

where s' is a constant with units of $cm^3 \, s^{-1}$. The solution of Equation (4.14) with a linear heating function is

$$I(T) = \frac{n_0 s'' \exp\left(-E/kT\right)}{\left[1 + (s''/\beta) \int_{T_0}^{T} \exp\left(-E/k\theta\right) d\theta\right]^2} \qquad (4.15)$$

where $s'' = s'n_0$, a quantity with dimensions of s^{-1} and which can be considered to represent a pseudo-frequency factor. Note that whereas the original frequency factor s is constant, the pseudo-frequency factor s'' is not, since it depends on n_0.

When more than one trapping state and one kind of recombination center are involved, obviously more equations of the form of (4.1–4.4) and (4.5–4.7) are to be written and solved simultaneously and sequentially. Kantorovich et al. [189, 190] developed a method of an integral equation derived from sets of simultaneous differential equations governing the traffic of carriers between several trapping states and recombination centers. They also went further by using the integral equation method for continuous distributions of traps and centers [191, 192].

4.1.1 Range of Recombination and Retrapping Probability Coefficients

The physically acceptable values of the mentioned recombination and retrapping probability coefficients, A_m, A_n and B are briefly discussed here. The largest value used (see below) is $A_m = 10^{-5} \text{cm}^3 \text{s}^{-1}$. Values of this size are observed when a free carrier, an electron in the case discussed above, and a carrier in a trap or center have opposite charges. For such Coulomb attractive combinations, experimentally measured probability coefficients are in the range 10^{-5}–$10^{-8} \text{cm}^3 \text{s}^{-1}$ [57, 193]. Smaller probability coefficients, typically 10^{-8}–$10^{-11} \text{cm}^3 \text{s}^{-1}$, are observed in neutral traps [57, 193]. Traps with the same charge as the free particle cause mutual repulsion and consequently have still smaller rate coefficients of $10^{-11} \text{cm}^3 \text{s}^{-1}$, and could be as low as $10^{-15} \text{cm}^3 \text{s}^{-1}$ [57]. An additional test of the reasonableness of the assumed A_n value is provided by a detailed balance which relates the capture rate constant A_n to the pre-exponential factor of the trap, s [193, 194],

$$\frac{s}{A_n} = \frac{\left(2\pi m_e^* kT\right)^{3/2}}{h^3} \frac{2g_0}{g_1},$$

(4.16)

where m_e^* is the effective mass of electrons in the conduction band, g_0 is the degeneracy of an empty trap, g_1 is the degeneracy of a filled trap, k is Boltzmann's constant and h is Planck's constant. A typical value of the right-hand side is 10^{19}cm^{-3}. Thus by Equation (4.16), the value $A_n = 10^{-11} \text{cm}^3 \text{s}^{-1}$ corresponds to a pre-exponential factor s in the neighborhood of 10^8s^{-1} which is well within the range of s that is commonly observed in experiments [194].

4.2 One Trap-One Recombination Center (OTOR) Model

In the specific case when only one trapping state and one kind of recombination center are involved, it is obvious that following excitation, the concentrations of electrons and holes are equal, $n = m$. Without making any further assumptions on the relationship between recombination and retrapping, Equation (4.9) will be written as

$$I = -\frac{dm}{dt} = s \cdot \exp\left(-E/kT\right) \frac{A_m n^2}{A_m n + A_n(N - n)}.$$

(4.17)

This equation was used as early as 1956 by Adirovitch [195] in the study of phosphorescence decay (see also Sections 6.1, 7.7 and 7.9). It has been termed General One Trap (GOT) or OTOR, and sometimes 1T1C model by different authors. Sunta *et al.* [196–198] have solved this equation for the case of TL with a linear-heating function, and compared the results with those received under the assumption of general-order kinetics (see Section 4.3). Levy [199–201] and Hornyak *et al.* [202] studied the properties of the TL governed by Equation (4.17) and its extension to several OTOR peaks. Effects of TL supralinearity seem to result from interactive kinetics equations of several peaks, each of which is of the kind mentioned. Experimental results in CaF_2:Mn are explained using this model [202]. This model is elaborated upon, with regards to OSL and TL in Chapter 7.

4.3   General-order Kinetics

It is quite obvious that the first- and second-order kinetics equations, (4.10) and (4.14), are very special cases of the more general equation (4.9), and even more so if one considers the "real" governing equations for the read-out stage, (4.5–4.7). Obviously, all kinds of intermediate situations are possible. A useful characterization of a single TL peak is its geometrical shape property, measured by the shape factor μ_g [30]. If we denote the maximum temperature by T_m and the low and high half-intensity temperatures by T_1 and T_2, respectively, the high-temperature half width is defined as $\delta = T_2 - T_m$ and the full-width half intensity by $\omega = T_2 - T_1$, then the shape factor is defined by $\mu_g = \delta/\omega$. As shown by Chen [203], the first-order peak is rather asymmetric with $\mu_g \approx 0.42$, and the second-order peak is nearly symmetric with $\mu_g \approx 0.52$. It is quite obvious that since these two cases are rather extreme, different factors can be observed in real TL peaks, intermediate between these two values, and possibly also outside the range of 0.42–0.52. A popular heuristic way to deal with intermediate cases is the "general-order" kinetics, namely by assuming the equation

$$I = -\frac{dn}{dt} = s'n^b \exp\left(-E/kT\right), \tag{4.18}$$

where b may be other than 1 or 2 (see e.g. May and Partridge [29] and Chen [204]). Opanowicz [205] has tried to compare the general OTOR model with the general-order kinetics model, and obtained an expression for the effective kinetic order b which depends on the varying populations of the trap and recombination center. These populations are temperature dependent, and therefore, so is the evaluated b. s' here is a constant with units of $cm^{3(b-1)} s^{-1}$. This reduces to $cm^3 s^{-1}$ for second order, $b = 2$, and to s^{-1} for first order, $b = 1$. In an extension to the above mentioned definition of $s' = s/N$ [see Equations (4.13) and (4.14) above], Rasheedy [206] suggested for the general-order case that $s' = s/N^{b-1}$. The solution of Equation (4.18) for a linear heating function is

$$I = s''n_0 \exp\left(-E/kT\right) \left[1 + (b-1)(s''/\beta) \int_{T_0}^{T} \exp\left(-E/k\theta\right)d\theta\right]^{-b/(b-1)} \tag{4.19}$$

where $s'' = s'n_0^{b-1}$, a quantity with the same units as s (s^{-1}), which is dependent on n_0 when $b \neq 1$. This empirical approach was found to be satisfactory for explaining the different symmetries with μ_g being in the mentioned range between 0.42 for first order and 0.52 for second order. In the more comprehensive situation for one trapping state and one kind of luminescence center, covered by Equations (4.5–4.7), the geometrical shape factor may be outside this range, i.e., effective values of b such that $b < 1$ and $b > 2$ are also to be considered. Actually, values of b between 0.5 and 3.0 are mentioned in the literature. Kanunnikov [207] made an attempt to associate the intermediate cases related to Equation (4.9) where neither recombination nor retrapping are dominating, with the general-order situation. Assuming that, indeed, only one kind of trap and one kind of center are involved, namely $n = m$, Equation (4.9) can be written as

$$I = -\frac{dn}{dt} = s \cdot \exp\left(-E/kT\right)\frac{A_m n^2}{A_m n + A_n(N - n)}.$$ (4.20)

Assuming that the trap is far from saturation, $n \ll N$, Kanunnikov [207] reached an equation for the activation energy E

$$E = \sigma\frac{kT_m^2}{\delta},$$ (4.21)

where δ is the high-temperature half width of the TL peak (see more detailed discussion in Section 5.2), and T_m is the temperature at the maximum. As given in Equations (5.5) and (5.6) below, this equation with $\sigma = 1$ is a good approximation for the first-order case, and with $\sigma = 2$, a reasonable approximation for the second-order case. Kanunnikov [207] showed that the expression

$$\sigma = 1 + \frac{A_n N}{A_n N + A_m n_m},$$ (4.22)

where n_m is the trap occupancy at the temperature of maximum TL, is an acceptable approximation for the use of σ for intermediate cases in Equation (4.21). He therefore suggests to associate this σ with the kinetics order b in Equation (4.18). It is obvious, however, that this approach is heuristic in nature since for a given TL peak, n_m will depend on the initial radiation given to the sample. Thus, the implicit assumption that the value of b in Equations (4.18) and (4.19) and other equations derived thereof is independent of the initial filling of the trap, is somewhat questionable. In particular, although Equation (4.21) reduces to the approximate equations (5.5) and (5.6) for the first- and second-order cases, the association made between b and σ for intermediate kinetics is not fully justified. Takeuchi *et al.* [208] reported on the TL in MgO doped with iron and found that whereas the low-concentration sample showed second-order behavior, the effective order was 1.3 with higher concentration. Xu *et al.* [209], suggested that the general-order approach has a serious deficiency since it deals with a fixed parameter, the kinetics order, to represent an evolving process. Indeed, it is obvious that the general-order approach is merely an approximation to a more complex situation, but it appears to be a pretty good approximation in many instances. An attempt to give the general-order kinetics a physical basis was made by Christodoulides [210]. He suggested the possibility of a continuum of traps, all with the same activation energy but with a distribution of the frequency factors. Starting from Equation (4.18), he found the required distribution of ln s for different values of the order b, which turned out to be a gamma distribution. Rasheedy [211] has suggested a rather complicated way for evaluating the value of b for a single peak, under the assumption that general-order kinetics precisely describes the kinetics.

4.4 Mixed-order Kinetics

In the sets of assumptions leading to second-order kinetics [Equations (4.12–4.14)] a common assertion is that $m = n$, namely, that at any time, the instantaneous concentrations of electrons in traps and holes in centers (or vice versa) are equal. As pointed out by Chen *et al.* [212], this assumption seems actually the least probable to occur since in all real samples, many kinds of defects and impurities are present. In this case, $m = n + n_c + C$ where C is a constant, positive or negative and n_c is the instantaneous concentration of electrons in the conduction band. This constant C represents the number of electrons or holes not taking part in the TL process in the temperature range under consideration, due to their being in deep traps or in low-probability recombination centers. More accurately, C is the difference between the concentrations of these trapped electrons and holes. As is usually assumed, n_c can be neglected as compared with m or n, but taking C to be zero seems much less plausible.

When $C \neq 0$ and in the case of equal probability coefficients for recombination and retrapping, $A_m = A_n$, Equation (4.9) becomes

$$I(t) = -\frac{dn}{dt} = \left(\frac{s}{C+N}\right) n(n+C) \exp\left(-\frac{E}{kT}\right) \qquad (4.23)$$

instead of Equation (4.13), which was derived for $C = 0$. When the situation is that of dominating retrapping, $A_n(N-n) \gg A_m m$, and if the trap is far from saturation i.e. $N \gg n$, Equation (4.9) becomes

$$I(t) = -\frac{dn}{dt} = \left(\frac{sA_m}{NA_n}\right)(n+C) \exp\left(-\frac{E}{kT}\right) \qquad (4.24)$$

instead of Equation (4.12) which is valid when $C = 0$. Both Equations (4.23) and (4.24) can be summed up as

$$I(t) = -\frac{dn}{dt} = s'n(n+C) \exp\left(-\frac{E}{kT}\right), \qquad (4.25)$$

where s' is a constant with units of $\mathrm{cm^3\ s^{-1}}$. It is to be noted that an equation similar to (4.25) for the case of phosphorescence ($T = $ constant) had already been suggested by Mott and Gurney [92]. Equation (4.25) can be considered as a combination of the first- and second-order kinetics. It should be mentioned, however, that the solution of Equation (4.25) is not a combination of the first- and second-order solutions. In this respect, this treatment differs from that of Schlesinger and Menon [213] who assumed, in an empirical way, a weighted combination of first- and second-order solutions.

r to solve Equation (4.25), Chen *et al.* [212] introduced the parameter $\alpha = + C)$, where n_0 is the initial value of n. Equation (4.25) may now be written as

$$I(t) = -\frac{dn}{dt} = s'n\left(n - n_0 + \frac{n_0}{\alpha}\right)\exp\left(-\frac{E}{kT}\right). \tag{4.26}$$

The solution, starting from temperature T_0 with a constant heating rate β is

$$I(T) = \frac{s'C^2\alpha\exp\left[(Cs'/\beta)\int_{T_0}^T\exp(-E/k\theta)d\theta\right]\exp(-E/kT)}{\left\{\exp\left[(Cs'/\beta)\int_{T_0}^T\exp(-E/k\theta)d\theta\right] - \alpha\right\}^2}. \tag{4.27}$$

Similarly to the first-, second- and general-order cases, initially, i.e. at the lower temperatures, this expression behaves like $\exp(-E/kT)$. At higher temperatures, the shape characteristic to the particular value of α is obtained. This solution appears to be intermediate between the first- and second-order expressions in the sense that for $\alpha \cong 0$ ($n_0 \ll C$) it tends to first order, and for $\alpha \cong 1$ ($n_0 \gg C$), to second order. Chen *et al.* [212] have derived from Equation (4.27) numerical values of T_1, T_m and T_2 for given values of E, s', β, α and n_0. Values of μ_g have been readily obtained from these temperatures. The symmetry factor μ_g as a function of α is found to change gradually from the value of 0.42 characteristic of first order when α equals zero, to a value of 0.52 of second order when α equals one, as shown in curve (a) of Figure 4.2.

Only the values of α from zero to one, i.e. for $n_0 \leq m_0$ are shown in Figure 4.2. As for the case with $n_0 > m_0$, i.e. $\alpha > 1$, let us change in Equation (4.25) the variable n into $m = n + C$, and instead of α, use the parameter $\alpha' = \alpha^{-1} = m_0/n_0$. Equation (4.25) readily becomes

$$I(t) = -\frac{dm}{dt} = s'm\left(m - m_0 + \frac{m_0}{\alpha'}\right)\exp\left(-\frac{E}{kT}\right), \tag{4.28}$$

which is strictly the same as Equation (4.26), and hence has the same solution [214]. μ_g as a function of α' here will be the same as μ_g versus α in Equation (4.26). Thus, for the cases where $n_0 > m_0$, μ_g will be obtained by the same curve (a) in Figure (4.2), but using $\alpha^{-1} = m_0/n_0$ instead of α as the variable (and $C' = |C|$).

The main point regarding the curve of μ_g versus α is that, although it has been calculated by choosing certain values for E and s', Chen *et al.* [212] report that changing E and s' modifies the graph only slightly. In other words, μ_g is a relatively strong function of α and a very weak function of E and s'. This and the shape of curve (a) bring to mind the possibility of using experimental values of μ_g to compute the interpolation coefficients in Equations (5.9–5.11) and evaluate the activation energy E by the half intensity formulae as given by Chen [204]. The point is that both in the general-order case and in the mixed-order case, Equations (5.9–5.11) are empirical and the condition for their being useful is that μ_g is a slowly varying function of b in the general-order case and of α in the mixed-order case. Chen *et al.* [212] report the testing of the half-intensity formulae for "synthetic" glow peaks which were computer calculated by Equation (4.27) for given parameters. The values of E thus derived were found to be within 3% of the given values, which represents just as good a fit as with curves computed with the general-order method [204]. As for the value of α itself, it can be deduced from the experimental value of μ_g directly with curve (a) in Figure 4.2. A and C is constant during a glow peak, following a given irradiation. Equation (4.29) can now be written as (see Chen *et al.* [212])

$$I = -\frac{dn}{dt} = Kn^2 \exp\left(-\frac{E}{kT}\right) + KCn \exp\left(-\frac{E}{kT}\right), \qquad (4.32)$$

which is actually the same as Equation (4.25) with K replacing s'. Thus the suggestion made above to use Equations (5.9–5.11) to evaluate E can be used in this case as well.

The expression for the ionic thermo-conductivity σ_i related to the same process can be found by combining Equations (4.30) and (4.31). This yields

$$\sigma_i(T) = \frac{eC(\mu_0/T)\exp(-E/kT)}{\exp\left[(CK/\beta)\int_{T_0}^{T}\exp(-E/k\theta)d\theta\right] - \alpha} \qquad (4.33)$$

with $\alpha = n_0/(n_0 + C)$ or $\alpha = (n_0 + C)/n_0$, whichever is smaller, as explained above; C is the concentration of interstitials trapped by the defects [219]; β is the (constant) heating rate.

Like in the previously described cases, the initial behavior at the lower temperatures is the regular initial rise expression, namely $\sigma(T) \propto \exp(-E/kT)$. Curve (b) in Figure 4.2 shows the variation of the symmetry factor μ_g for the conductivity curve given by Equation (4.33) versus the parameter α. In the same way as in curve (a), the results are rather insensitive to the particular values chosen for E and K. The variation of μ_g with α is seen to be rather moderate, up to $\alpha \simeq 0.9$. However, between $\alpha = 0.9$ and $\alpha = 1$, μ_g changes very quickly and reaches values up to ~ 0.8. This prevents any attempt to use a simple interpolation formula for the evaluation of the parameters from the conductivity curve. The high value of $\mu_g \sim 0.8$ corresponds to conductivity associated with second-order TL. This has been investigated by Saunders [220] and Chen [221] in the similar case of TSC governed by electron or hole transport.

4.5 *Q* and *P* Functions

The discussion so far has concentrated on the use of the quasi-steady approximation which means that the concentrations of free carriers remain quasistationary during the TL process. This assumption helps in decoupling the relevant rate equations, thus enabling the equations mentioned above. Maxia [69] considered the TL emission to be the result of two non-equilibrium phase transitions, the release of electrons into the conduction band and the recombination of free electrons with trapped holes. Maxia applied Gibbs' principle of minimum entropy production to show that the free-carrier concentration indeed remains approximately steady during the TL process. However, as pointed out by Chen and McKeever [33] (see p. 53), the minimum entropy production principle is, in fact, just another way of imposing the limitation of quasi-steady-state on the free carrier concentrations, and the question whether the principle is valid during the TL process remains. Maxia also assumed a linear relationship between the rate of entropy production and the transformation velocities for each of the mentioned reactions, but this assumption is also questionable, and in fact is not valid at the start of the reaction. Shenker and Chen [222] have shown that the simplified expressions for TL reached by using the quasi-equilibrium (QE) approximation are valid solutions only if $|dn_c/dt| < 10^{-3}|dn/dt|$. In the cases examined by them, the ratio of $|dn_c/dt|$ to $|dn/dt|$ varies from 5×10^{-5} to 1.0, with the higher values taking place toward the end of the peak. A study by Kelly *et al.* [223] supports these findings and concludes that the QE approximation is valid for only part of the range of physically plausible parameters. Numerical solutions of the rate equations result in a variety of peak shapes, more than can be predicted from the use of approximate solutions. The uncertainty about the use of the QE approximation results in the pessimistic conclusion that the simplified equations in common use including the kinetic-order assumptions may not be a very good presentation of TL peaks. Kelly *et al.* [223] reached the extremely pessimistic view that application of simplified phenomenological solutions to real systems is almost valueless at quantifying the trapping parameters such as the activation energy and frequency factor. Further discussion on the validity of the quasi-equilibrium assumption was given by Opanowicz and Przybyszewski [224].

Lewandowski and co-workers [225–228] suggested an entirely different approach. Instead of the QE and kinetic-order (KO) assumptions, they defined two physically significant functions, Q and P, respectively. Instead of the QE assumption, Equation 4.8, they present the equality

$$-\frac{dn_c}{dt} = q\frac{dn}{dt}, \tag{4.34}$$

or

$$Q\frac{dm}{dt} = \frac{dn}{dt}, \tag{4.35}$$

where $Q = q + 1$. Q and q are actually functionals, $Q\{T(t)\}$ and $q\{T(t)\}$ and as such, depend on the history of the system. The value of Q is a measure of the degree to which QE is maintained, as a function of T. QE conditions means that $q \approx 0$, or $Q \approx 1$.

Lewandowski *et al.* [225] also defined a KO function, $P(T)$. In Section 4.1, the distinction between first- and second-order kinetics was determined by the ratio $A_n(N - n)/mA_m$. The $P(T)$ function was defined as

$$P(T) = \frac{A_n(N - n)}{mA_m}. \tag{4.36}$$

Note that this ratio is a function of temperature. Slow retrapping is characterized by $P \ll 1$ and fast retrapping by $P \gg 1$. The main importance of defining the Q- and P-functions is that the QE and KO concepts are separated, and that these two functions are allowed to vary with temperature. All the prior analyses assumed QE before the KO approximation was introduced, and required that the approximations be fixed for all temperatures.

Lewandowski and McKeever [225] defined the rate of recombination

$$R_{recom} = -\frac{dm}{dt} = A_n mn_c, \tag{4.37}$$

the rate of thermal excitation

$$R_{ex} = s \cdot n \cdot \exp(-E/kT) \tag{4.38}$$

and the rate of recapture

$$R_{recap} = A_n(N - n)n_c. \tag{4.39}$$

P and Q can now be written in terms of the reaction rates

$$Q = \frac{1}{R_{recom}}(R_{ex} - R_{recap}) \tag{4.40}$$

and

$$P = \frac{R_{ex}}{R_{recom}}. \tag{4.41}$$

The following relationships are valid

$$Q + P = \frac{R_{ex}}{R_{recom}} \tag{4.42}$$

and

$$\frac{Q}{P} = \frac{R_{ex}}{R_{recap}} - 1. \tag{4.43}$$

Lewandowski and McKeever [225] solved numerically the simultaneous rate equations for a set of parameters chosen such that $P \ll 1$, and for different degrees of trap filling ranging from 0.005 to 1.0 and plotted the temperature dependencies $Q(T)$ and $P(T)$. Although $P(T)$ varied significantly with both T and the trap filling, the absolute value of P was always much smaller than unity. Lewandowski and McKeever [225] made some general statements regarding the various assumptions when arriving at simplified solutions:

1. Slow retrapping. This is the case if $Q \approx 1$ and $P \ll 1$. This, in turn, requires that $R_{recom} \gg R_{recap}$ and $R_{ex} \gg R_{recap}$. This relationship may extend over a wide temperature range.
2. Fast retrapping. This takes place if $Q \approx 1$ and $P \gg 1$. This means that we require $R_{ex} \approx R_{recap}$ over all T. Had this been strictly true ($R_{ex} = R_{recap}$), the trap would not empty. Moreover, from Equation (4.40) it is obvious that in order to have $Q \approx 1$, one must have a very small value for R_{recom}, which means that the TL signal will hardly be seen, and also, as shown by Lewandowski *et al.* [226], the accompanying TSC would not have a peak shape but would rather form a step with a long, slowly decaying tail. Thus, when both TL and TSC peaks are observed, it is unlikely that both QE and fast retrapping take place.

A similar situation occurs when $P = 1$ as explained by Lewandowski *et al.* [226]. As explained by Kelly *et al.* [223], comparing the curves reached by the relevant numerical solution of the coupled differential equations with those produced using the Randall–Wilkins (first order) and Garlick–Gibson (second order), shows that whereas the agreement is good in the former, it is rather poor in the latter. Lewandowski *et al.* raise the question of the generality of these conclusions due to the fact that the choice of trapping parameters is an arbitrary selection of a finite number of sets of parameters out of an infinite number of possibilities. They then generalized the discussion by solving the rate equations without any assumption concerning QE and KO. They started with a model of one trap-one recombination center and one deep disconnected trap. To the set of equations (4.5–4.7), a fourth simultaneous equation is to be added,

$$\frac{dh}{dt} = (H - h)A_h n_c, \tag{4.44}$$

where H and h are the concentrations of the disconnected trap and its population, respectively, and A_h the transition probability coefficient into this trap. One gets the solution

$$n = n_0 \exp\left[-\frac{1}{\beta}\int_{T_0}^{T}\left(\frac{Q}{Q+P}\right) s \exp\left(-\frac{E}{k\theta}\right) d\theta\right], \tag{4.45}$$

where β is the constant heating rate. The TL intensity is therefore given by (see p. 58 in Chen and McKeever [33])

$$I(T) = n_0 \left(\frac{s}{Q+P}\right) \exp\left(-\frac{E}{kT}\right) \exp\left[-\frac{1}{\beta}\int_{T_0}^{T}\left(\frac{Q}{Q+P}\right) s \exp\left(-\frac{E}{k\theta}\right) d\theta\right]. \tag{4.46}$$

This is an entirely general equation for TL for the given model. Obviously, it reduces to the Randall–Wilkins equation when $Q \approx 1$ and $P \ll 1$. It is not as obvious how the Garlick–Gibson equation results if $Q \approx 1$ and $P \gg 1$. In spite of its generality, Equation (4.46) is not of practical use since the Q and P functions cannot be determined in most cases. The relatively simple case is $P \ll 1$ which, for most temperatures results in $Q \gg P$ and Equation (4.46) can be written as

$$I(T) = \frac{s n_0}{Q} \exp\left(-\frac{E}{kT}\right) \exp\left[-\frac{s}{\beta}\int_{T_0}^{T}\exp\left(-\frac{E}{k\theta}\right) d\theta\right]. \tag{4.47}$$

If we denote the first-order equation (4.11) by I_{QE}, we get

$$I_{QE}(T) = Q I(T). \tag{4.48}$$

Lewandowski *et al.* [225, 226] made another step forward by using the free carrier recombination lifetime τ_n, defined as the lifetime of a free electron before recombination, given by

$$\tau_n^{-1} = m A_m = (n + h)A_m. \tag{4.49}$$

By further assuming that $h \gg n$, τ_n can be considered as a constant during the thermal emptying of the TL trap. From the definition of Q in Equation (4.35) one may write

$$Q = \frac{\beta \tau_n}{I(T)}\frac{dI(T)}{dT} + 1. \tag{4.50}$$

Obviously, QE is only exact at the TL peak maximum, where $dI(T)/dT = 0$. By combining Equations (4.48) and (4.50), one gets

$$I_{QE}(T) = \beta\tau_n \frac{dI(T)}{dT} + I(T),$$ (4.51)

and since this is a linear differential equation, its solution is

$$I(T) = \frac{1}{\beta\tau_n} \exp\left(-\frac{T}{\beta\tau_n}\right) \int_{T_0}^{T} \exp\left(\frac{\theta}{\beta\tau_n}\right) I_{QE}(\theta)d\theta.$$ (4.52)

Since $I_{QE}(T)$ is the Randall–Wilkins expression, one finally has

$$I(T) = \left(\frac{n_0}{\beta\tau_n}\right) \exp\left(-\frac{T}{\beta\tau_n}\right) \int_{T_0}^{T} \exp\left(\frac{\theta}{\beta\tau_n}\right) s\exp\left(-\frac{E}{k\theta}\right) \times$$ (4.53)

$$\exp\left[-\frac{1}{\beta}\int_{T_0}^{\theta} s\exp\left(-\frac{E}{k\Omega}\right) d\Omega\right] d\theta,$$

where Ω is another dummy variable representing temperature. This complex function represents more accurately the true shape of a slow-retrapping TL peak than does the simple Randall–Wilkins expression. Lewandowski *et al.* [226] tested the validity of Equation (4.53) by comparing the TL curve shapes it produces with those produced using the Randall-Wilkins equation and with those reached by the exact numerical solutions of the rate equations. The authors show some example results. For certain parameters, all three solutions agree, while for others, some slight discrepancies exist between the exact solutions and the Randall-Wilkins expression. In all cases, Equation (4.53) gives the same solution as the numerical analysis. Also, fitting the Randall-Wilkins equation to the numerical data always resulted in an excellent fit, with error of up to 2% in the activation energy E. Thus, the criticism of Kelly *et al.* [223] does not seem justified in the slow-retrapping cases. As for the other situations, namely, $P \approx Q$ and $P \gg Q$, no expressions equivalent to Equation (4.53) have been developed. In these cases, the arguments by Kelly *et al.* may have more validity. Moreover, no generalization of the Q-function is available to the cases of multiple traps and centers, with overlapping TL peaks. In all these cases, the only viable method for determining the expected TL peaks and their behavior is by numerical simulation consisting of solving the relevant sets of simultaneous differential rate equations.

4.6   Localized Transitions

The discussion above has taken into consideration transitions during the heating stage of electrons raised thermally into the conduction band. Beside the mentioned alternative of trapped holes being "raised" thermally into the valence band before recombining with electron recombination centers, another possibility may occur. As suggested by Halperin

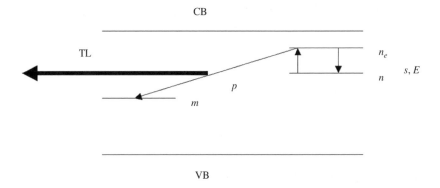

Figure 4.3   *Localized transitions. Schematic energy level diagram of a trap, n, excited state, n_e and recombination center m in the forbidden gap*

and Braner [30], a localized (or geminate) transition may take place. Here, we consider an electron trap and hole center which are in close proximity to each other. The direct transition from the trap into the center is still forbidden, but it is assumed that the electron can be raised thermally from the trap into an excited state within the forbidden gap, from which the radiative transition into the recombination center is allowed. An obvious difference between the two possibilities, the transition through the conduction band discussed before and the present case of localized transition, is that the former may be accompanied by TSC due to the existence of electrons in the conduction band, which is not expected in the latter. A schematic energy level diagram of this situation is shown in Figure 4.3 (see e.g. Chen and Kirsh [229], p. 35).

For the simplest theoretical explanation of TL associated with local transitions, let us consider n (cm^{-3}) which denotes the concentration of trapped electrons, n_e (cm^{-3}) the instantaneous concentration of thermally excited electrons and m, the concentration of holes in centers. Obviously, in this simple situation, $m = n + n_e$. E (eV) is the activation energy and s (s^{-1}) the frequency factor. p (s^{-1}) is the recombination probability coefficient which differs from the coefficient A_m mentioned above for the case of transitions through the conduction band. The main difference between this situation and the previous case of thermal excitation into the conduction band is that once an electron is raised into the excited state, it "sees" only the adjacent recombination center and does not see all the centers of the same kind. It may either recombine with the hole in this close-by center or retrap into its original trap. From the principle of detailed balance (see Mott and Gurney [92]), it is concluded that the probability per second for retrapping is the frequency factor s. The relevant set of equations is

$$I = -\frac{dm}{dt} = p n_e, \tag{4.54}$$

$$-\frac{dn}{dt} = s \cdot n \cdot \exp\left(-E/kT\right) - s n_e, \tag{4.55}$$

$$\frac{dm}{dt} = \frac{dn}{dt} + \frac{dn_e}{dt}. \tag{4.56}$$

As shown by Chen and Kirsh [229], if the quasi-steady conditions hold, namely, $n_e \ll m, n$ and $| dn_e/dt | \ll | dm/dt |, | dn/dt |$, the following expression results

$$I = -\frac{dm}{dt} = \left(\frac{ps}{p+s}\right) m \cdot \exp\left(-E/kT\right). \qquad (4.57)$$

Under these circumstances, the localized transitions are expected to yield a first-order peak with the activation energy E and an effective frequency factor of $\hat{s} = ps/(p+s)$. It can easily be seen that when $s \ll p, \hat{s} \to s$, whereas when $p \ll s, \hat{s} \to p$. For localized transitions one would expect the symmetry factor to be $\mu_g \approx 0.42$. As shown by simulations performed by Bull [230], for $p \ll s$, the quasi-equilibrium condition $n_e \ll m$ does not hold anymore. This results in the symmetry factor being larger, up to 0.48 in the example given by Bull [230], which means that under these circumstances the TL may not look like a first-order peak. Horowitz *et al.* [231] argued that the peak termed 5a in LiF:Mg,Ti arises from localized electron-hole recombination in the trapping center (TC)-luminescence center (LC) pair, believed to be based on Mg^{2+}–Li_{vac} trimers coupled to $Ti(OH)_n$ molecules. More experimental evidence of the local transitions yielding TL is found in the literature. Kirsh *et al.* [232] reported on local transitions and charge transport in TL of calcite. Their kinetic data and the analysis of the emission spectra supports the model by Medlin [233](see also Chapter 11) which ascribes the $80\,°C$ peak to local transitions such as tunneling, and the higher temperature peaks to processes which involve transport of ionic charge carriers. Kirsh and Townsend [234] discuss the TL of X-excited TL in zircon ($ZrSiO_4$). Several peaks are reported, one of which, at $109\,°C$ is believed to be associated with an electron captured by a Dy^{3+} ion and a hole trapped at an adjacent lattice O^{2-}, the recombination of which excites both the SiO_4^{4-} group and the rare earth ion. Zatsepin *et al.* [235] described their results of TL, TSEE, optical absorption and ESR in phosphate glasses by suggesting that localized electronic states participate in the recombination processes.

Another kind of possible localized transitions yielding TL has been discussed by Hagston [236] who considered the role of donor–acceptor pairs in some TL peaks. The concept of TL associated with optically generated donor–acceptor pairs has also been discussed by Ronda *et al.* [237]. In the TL of CdI_2, they ascribe the TL to intrinsic processes, not involving impurities or defects present in the crystal prior to the experiment. Energy transport proceeds via free electrons and holes or by free excitons. These carriers have a finite lifetime due to the formation of localized self-trapped excitons, as a result of strong electron–phonon interactions. At higher excitation densities, the formation of Frenkel defects, namely, interstitial Cd and Cd vacancies due to the decay of self-trapped excitons, becomes important. These optically generated defects play a role as electron or hole traps in the production of TL.

4.7 Semilocalized Transition (SLT) Models of TL

Mandowski [238] developed a complex kinetic model for spatially correlated TL systems, which contains elements from both the localized transitions and simple trap model (STM) of TL. The complexity of this SLT model is due partly to the presence of two distinct activation energies E and E_V shown in Figure 4.4. There are two possible recombination transitions within the model, leading to two TL peaks denoted by L_B and L_C. The L_B peak corresponds to the intra-pair luminescence due to localized transitions, and the L_C peak corresponds to delocalized transitions involving the conduction band. Mandowski [239] found that for certain parameters in the model, the L_C peak can exhibit a nontypical double-peak structure that resembles the TL displacement peaks previously found in spatially correlated systems by Monte-Carlo simulations. This double-peak structure of the L_C peaks was interpreted as follows: the first L_C peak corresponds to an increasing concentration of charges accumulating in the excited local level and undergoing intra-pair recombination. The second L_C peak was interpreted as relating to carriers wandering in the crystal, and undergoing recombination with holes in the recombination center via the conduction band. Mandowski [238] found that the structure of these peaks depends on the retrapping ratio r, which is the ratio of the retrapping and recombination probabilities for an electron in the excited localized state. The results of Mandowski [238] showed that when no retrapping occurs within the T–RC pair (i.e. retrapping ratio $r = 0$), the model produces a single L_C peak with no apparent double-peak structure. When retrapping is present (i.e. values of retrapping ratio ranging from $r = 1$ to $r = 100$), the L_C peaks exhibit a double-peak structure. Mandowski [238] also found that the amount of recombination within the energy level describing the electron–hole pair also influences the shape and relative size of the delocalized peaks. He also studied the effect of the energy gap E_V on the double peak structure of the L_C peak.

The energy diagram for the SLT Model is shown in Figure 4.4, together with the possible electronic transitions. Since several of the transitions shown depend on the instantaneous number of carriers occupying traps and recombination centers, it is not possible to write kinetic equations using global concentrations of charge carriers in each level. In order to overcome this difficulty, Mandowski [238] introduced the following notation to denote the occupation of all states in a single T–RC unit,

$$
\left\{ \begin{array}{c} \overline{n_e} \\ \overline{n} \\ \overline{h} \end{array} \right\}, \tag{4.58}
$$

where $\overline{n_e}$ is the number of electrons in the local excited level, $\overline{n}$ is the number of electrons in the trap level and $\overline{h}$ is the number of holes in the recombination center. By using this notation, the following concentrations of states can be defined

$$
H_0^0 = \left\{ \begin{array}{c} 0 \\ 0 \\ 1 \end{array} \right\}; \quad H_1^0 = \left\{ \begin{array}{c} 0 \\ 1 \\ 1 \end{array} \right\}; \quad H_0^1 = \left\{ \begin{array}{c} 1 \\ 0 \\ 1 \end{array} \right\}; \quad H_1^1 = \left\{ \begin{array}{c} 1 \\ 1 \\ 1 \end{array} \right\} \tag{4.59}
$$

$$
E_0^0 = \left\{ \begin{array}{c} 0 \\ 0 \\ 0 \end{array} \right\}; \quad E_1^0 = \left\{ \begin{array}{c} 0 \\ 1 \\ 0 \end{array} \right\}; \quad E_0^1 = \left\{ \begin{array}{c} 1 \\ 0 \\ 0 \end{array} \right\}; \quad E_1^1 = \left\{ \begin{array}{c} 1 \\ 1 \\ 0 \end{array} \right\}. \tag{4.60}
$$

Mandowski [239] introduced the following simplifying assumptions in the SLT model: (a) The K-transitions shown in Figure 4.4 are such that $K = 0$. (b) The creation of states with two active electrons (H_1^1, E_1^1) is unlikely. As a result of this assumption, the model does

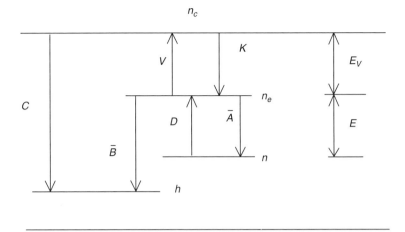

Figure 4.4 *The SLT model proposed by Mandowski [238]. Reprinted from Mandowski, A., Semi-localized transitions model for thermoluminescence, J. Phys. D: Appl. Phys. 38, 17–21. Copyright (2005) with permission from Institute of Physics*

not contain differential equations for H_1^1, E_1^1. (c) Only H_1^0 states are initially filled within the model. By using these assumptions the following set of seven equations were obtained

$$\dot{H}_1^0 = -(S + Cn_e)H_1^0 + \bar{A}H_0^1, \tag{4.61}$$

$$\dot{H}_0^1 = DH_1^0 - (\bar{A} + \bar{B} + V + Cn_c)H_1^0, \tag{4.62}$$

$$\dot{H}_0^0 = VH_0^1 - Cn_cH_0^0, \tag{4.63}$$

$$\dot{E}_1^0 = Cn_cH_1^0 - DE_1^0 + \bar{A}E_0^1, \tag{4.64}$$

$$\dot{E}_0^1 = Cn_cH_0^1 + DE_1^0 - (\bar{A} + V)E_0^1, \tag{4.65}$$

$$\dot{E}_0^0 = \bar{B}H_0^1 + Cn_cH_0^0 + VE_0^1, \tag{4.66}$$

$$\dot{n}_c = -Cn_c(H_1^0 + H_0^1 + H_0^0) + V(H_0^1 + E_0^1). \tag{4.67}$$

The parameters in these equations and the corresponding electronic transitions are shown in Figure 4.4. The activation energy within the T–RC pair is denoted by E, the energy barrier is E_V, and the corresponding frequency factors are ν and ν_V. A linear heating rate is assumed, and the retrapping coefficient r is defined as $r = \bar{A}/\bar{B}$. In addition to these parameters, the global concentration of traps in the crystal is denoted by N (cm^{-3}). The parameters $D(t) = \nu \exp(-E/kT)$ and $V(t) = \nu_V \exp(-E_V/kT)$ describe the thermal excitation probabilities in the SLT model. The variables H_m^n, E_m^n are related to the standard global variables n, n_e, n_c, h used in the localized transition (LT) model and in the STM of TL by the following equations

$$n = H_1^0 + E_1^0, \qquad (4.68)$$

$$n_e = H_0^1 + E_0^1, \qquad (4.69)$$

$$n_c = H_0^0 - E_0^1 - E_1^0, \qquad (4.70)$$

$$h = H_1^0 + H_0^1 + H_0^0. \qquad (4.71)$$

Here n, n_e, n_c and h denote, as usual, the concentrations of carriers in the trap level, excited level, conduction band and recombination center, respectively. The two recombination transitions within the model lead to two possible TL peaks L_B and L_C. The corresponding TL intensities for these two peaks are given by

$$L_C = Cn_c h = Cn_c(H_1^0 + H_0^1 + H_0^0), \qquad (4.72)$$

$$L_B = \bar{B}H_1^0. \qquad (4.73)$$

Figure 4.5 shows an example of L_C peaks calculated by Mandowski [238] for two values of the recombination ratio ($r = 0$ and $r = 1$). The rest of the parameters used in the model are: $E = 0.9$ eV, $\nu = \nu_V = 10^{10}$ s^{-1}, $\beta = 1$K s^{-1}, $C = 10^{-10}$cm^3 s^{-1}, $B = 10^3$ s^{-1}, $E_V = 0.7$ eV. The initial concentration of trapped carriers is assumed to be $H_1^0 = n_0$ and all other initial concentrations are assumed to be zero. Furthermore, it is

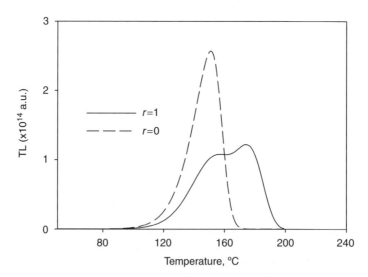

Figure 4.5 *TL glow curves calculated within the SLT model for two values of the retrapping ratio r. Reprinted from Mandowski, A., Semi-localized transitions model for thermoluminescence, J. Phys. D: Appl. Phys. 38, 17–21. Copyright (2005) with permission from Institute of Physics*

assumed that all states are initially full so that $n_0 = N$, where N is the global concentration of traps in the crystal. The results of Figure 4.5 show that for $r = 0$, i.e. when no retrapping occurs within the T–RC pair, the model produces a single L_C peak with no apparent double-peak structure. When values of r between 1 and 100 are used in the model, the L_C peaks have a clear double-peak structure.

Pagonis [240] demonstrated how the two distinct activation energies E and E_V can be extracted from TL glow curves calculated within the SLT model. This author performed a detailed study of the double-peak structure of the L_C peak, and examined whether it is possible to extract the activation energies E and E_V of the SLT model by applying the usual methods of TL analysis to the L_C peaks. By applying a variety of methods of kinetic analysis it was shown that a double-peak structure is also present in the case of L_C peaks corresponding to no retrapping ($r = 0$). The system of seven differential equations (4.61–4.67) was solved by using the differential equation solver in the software package *Mathematica*. Typical CPU time for solving this system of equations on a desktop computer and for producing graphs of the carrier concentrations as a function of temperature was 1–2 min. Pagonis and Kulp [241] have simulated isothermal processes in the SLT model of TL. These authors found that simulated isothermal signals exhibit several unusual time characteristics. Isothermal signals associated with the localized transitions follow first-order kinetics and therefore are described by single decaying exponentials, while isothermal signals associated with delocalized transitions show a non-typical complex structure characterized by several time regions with different decay characteristics. For certain values of the parameters in the SLT model, the isothermal signals exhibited nonmonotonic behavior as a function of time. Pagonis and Kulp [241] concluded that isothermal processes described by the SLT model depend strongly on the presence of semilocalized transitions, in contrast to previous studies using Monte-Carlo simulations, which showed a weak interdependence of these phenomena. These simulations suggested that isothermal experiments offer a sensitive method for detecting the presence of semilocalized transitions in a dosimetric material.

Mandowski *et al.* [242, 243] have reported on the dose dependence and the dose-rate effect on the SLT model. They describe simulations which show the occurrence of the effect, previously found in the STM. These authors also show that the dose-rate effect occurs for both localized and delocalized TL glow peaks.

5

Basic Methods for Evaluating Trapping Parameters

As explained in Section 4.1, even a single TL peak depends on a number of parameters. Certainly, if one deals with a glow curve consisting of several peaks, the number of parameters is significantly larger. When one wishes to simulate a glow curve, the task of getting the curve given the parameters is relatively simple. The reverse problem of getting the trapping parameters from the measured glow curve is significantly more complicated. Luckily the shape of each glow peak is very sensitive to the most important parameter, the activation energy, and therefore, a number of effective methods could be developed for evaluating it. In principle, in order to find a certain trapping parameter, one has to identify a property which is very sensitive to this parameter and very insensitive to all other trapping parameters. The following sections describe such methods for the evaluation of the activation energy, and once it is known then the frequency-factor and the order of kinetics can also be evaluated in some cases. A different method of analysis is described in Section 5.4, where the difference between an experimental curve and a theoretical curve is minimized, thus yielding the best set of parameters. When this method is used for a glow curve with several overlapping peaks, it is termed the deconvolution method (see Section 10.1). The main problem with these methods, in particular when many parameters are involved, is that the minimization procedure does not reach necessarily a global minimum, and therefore, one does not know whether the "correct" set of parameters has been reached. This point is also discussed in Section 5.4.

5.1 The Initial-rise Method

The method considered to be most general, namely, independent of the details of the governing kinetics is the initial-rise method. If we consider equations like (4.10), (4.14) or (4.18), for the first-, second- and general-order kinetics, respectively, at the "initial-rise"

Thermally and Optically Stimulated Luminescence: A Simulation Approach, First Edition. Reuven Chen and Vasilis Pagonis.   © 2011 John Wiley & Sons, Ltd. Published 2011 by John Wiley & Sons, Ltd.

range where the concentration of trapped carriers n has changed only slightly, $n \approx n_0$, we can write the temperature dependence of the emitted TL intensity as

$$I(T) = C \cdot \exp\left(-E/kT\right), \tag{5.1}$$

where the constant C includes n_o, the heating rate β and the order of kinetics b. In fact, even in more complex (and realistic) cases such as that shown in Equation (4.9), if the initial-rise range is defined as that where both n and m change only slightly, the same equation holds where the constant C includes m_o, n_o, N as well as s, A_m and A_n. As a rule of thumb (see e.g. Chen and Haber [244]), it has been suggested that the initial-rise range extends up to the point where the TL intensity reaches $\sim 5\%$ of the maximum. Following Garlick and Gibson [27], we can write Equation (5.1) as

$$\ln(I) = C_1 - E/kT, \tag{5.2}$$

where $C_1 = \ln(C)$. Thus, if $\ln(I)$ is plotted as a function of $1/T$, a straight line is expected in the initial-rise range, with a slope of $-E/k$, from which the activation energy E is readily found. Halperin *et al.* [245] suggested an important improvement of the initial-rise method, which extends the usable range by plotting $\ln(I \cdot n^{-b})$ versus $1/T$ when the kinetics order is known and n is given in Equation (5.3). This method is based on the use of Equation (4.18). Whereas, as pointed out, the plot of $\ln(I)$ versus $1/T$ normally yields a straight line only up to $\sim 5\%$ of the maximum intensity, using the correct value of b in the plot of $\ln(I \cdot n^{-b})$ versus $1/T$ would extend this range substantially, thus enabling the evaluation of the activation energy for low-intensity peaks. In order to be able to use this method, one has to evaluate the value of n for different points along the glow peak. Suppose T_i is a temperature point somewhere along the TL peak and T_f is the temperature at the end of the peak (theoretically, $T_f = \infty$, however any point at the high temperature side of the peak where TL is negligibly small can be used). Assume that t_i and t_f are the corresponding times. From the basic relation $I = -dn/dt$, we immediately get

$$n(T_i) = \int_{t_i}^{t_f} I dt = (1/\beta) \int_{T_i}^{T_f} I dT, \tag{5.3}$$

where β is the constant heating rate. Thus, n is evaluated for different temperatures and therefore, can be plotted as a function of T_i. In cases where the kinetics order b is not known beforehand, the process can be repeated a number of times with different values of b, and the best straight line searched for. An example can be seen in the paper by Halperin *et al.* [245] where such plots are shown for a NaCl peak at 490 K. The best results here are seen for first-order kinetics and the initial-rise range is significantly larger than with the original initial-rise method. Another example was given by Muntoni *et al.* [246]. Ilich [247] suggested a new variant of the initial-rise method. Several numerical examples of calculating the activation energy E from experimental data are given in the book by Pagonis *et al.* [248].

It is quite common to have several overlapping TL peaks. The deconvolution of such a curve, yielding the relevant trapping parameters, will be discussed in Section 10.1. An experimental procedure which does not requires a strong computational power, and yields information on the activation energies of the traps for the peaks in the glow curve, has been discussed in the literature. If, for example, the glow curve consists of a main peak accompanied by satellites at lower temperatures, an effective way of measuring the activation energy of the main peak is by thermal bleaching of the satellites. In this very simple way, the sample is heated up to, say, half of the TL maximum intensity and held at that temperature for a certain period of time. Although this would decrease the overall intensity, it would effectively eliminate the lower-temperature accompanying peaks. This method has been further developed by Nahum and Halperin [249], in such a way that the sample was sequentially heated and cooled several times. For each cycle, $\ln(I)$ versus $1/T$ is plotted and therefore the activation energies of a series of overlapping peaks can be found. The energies thus found can be plotted as a function of the temperature which is, say, the middle of the limited range of each heating. The activation energies as a function of temperature form usually a step-like graph as shown by Gobrecht and Hofmann [250] who termed the method the "fractional glow technique" (FGT). For some additional work along the same lines see the work by Tale [251], Firszt and Oczkowski [252] and Chruścińska [253].

5.2   Peak-shape Methods

The initial-rise method is based on measurements of the TL intensity at the low temperature end of a peak. The method is based on the shape of the curve at this temperature range. Other methods take advantage of the shape of the whole TL peak or of the measured intensities at some specific points along the curve. These are termed "peak-shape" methods. The first method of this family was developed by Grossweiner [254]. Let us denote the temperature at the maximum by T_m, the low and high half-intensity temperatures by T_1 and T_2, respectively, "the low temperature half width" by $\tau = T_m - T_1$, the "high temperature half width" by $\delta = T_2 - T_m$ and the full half width by $\omega = T_2 - T_1$. Grossweiner dealt with first-order kinetics peaks and by using only the first term in the asymptotic series for evaluating the exponential integral, he found

$$E = 1.51 k T_m T_1 / \tau. \tag{5.4}$$

Dussel and Bube [255] have changed the coefficient into 1.41 by using more terms in the relevant asymptotic series (see Chapter 15). Another expression for first-order TL peaks which is based on the high-temperature half width was given by Lushchik [187]

$$E = k T_m^2 / \delta, \tag{5.5}$$

and a similar equation for second-order peaks,

$$E = 2 k T_m^2 / \delta. \tag{5.6}$$

Chen [203] has amended the coefficients in Equations (5.5) and (5.6) as shown below in Section 5.5. Halperin and Braner [30] developed shape methods based on the measurement of the low-temperature half width, τ; these were also amended by Chen [203]. For example, this kind of equation for first-order peaks is

$$E = 1.52 k T_m^2 / \tau - 1.58 (2 k T_m), \tag{5.7}$$

and a similar equation with other coefficients for second-order peaks. In another work, Chen [204] used interpolation on the shape (or symmetry) factor $\mu_g = \delta/\omega$, for evaluating the activation energy of general-order kinetics peaks. He first showed that for first-order peaks $\mu_g \approx 0.42$, and for second-order $\mu_g \approx 0.52$. All the equations of this form were combined into one, namely,

$$E_\alpha = c_\alpha(kT_m^2/\alpha) - b_\alpha(2kT_m), \tag{5.8}$$

where α stands for δ, τ or ω. The relevant coefficients are

$$c_\tau = 1.51 + 3(\mu_g - 0.42); \quad b_\tau = 1.58 + 4.2(\mu_g - 0.42), \tag{5.9}$$

$$c_\delta = 0.976 + 7.3(\mu_g - 0.42); \quad b_\delta = 0, \tag{5.10}$$

$$c_\omega = 2.52 + 10.2(\mu_g - 0.42); \quad b_\omega = 1. \tag{5.11}$$

In cases where the pre-exponential factor s (or s') depends on a power of temperature T^a, where $-2 \leq a \leq 2$, the same equations can be used but the b_α are to be corrected by adding $a/2$ [57, 179, 256–258]. The formulas for first order are reached by inserting $\mu_g = 0.42$, and for second order by using $\mu_g = 0.52$. It should be noted that the reported coefficients are very slightly dependent on the pre-exponential factor, which is not shown in Equations (5.9–5.11), therefore, using the equations as given may result in slight errors. As discussed by Chen [203], theoretically, the δ method gives the best results. However, in practice, δ is rather hard to evaluate accurately. In many cases, smaller "satellite" peaks change the measured values of the three mentioned half-intensity widths. Using thermal cleaning, namely heating the sample up to below the maximum, cooling it back and reheating, may help in getting rid of the low-temperature satellites. This procedure cannot be performed with satellites at the high-temperature side of the peak. In this sense, the τ method has an advantage. The ω method also performs well simply because ω is larger than the other two parameters, and therefore the relative possible error in its evaluation is smaller. Also, T_1 and T_2 are easier to determine accurately than T_m, and ω does not include the latter.

Other peak-shape methods have also been developed. For example, Land [259] has used the two inflection points at the low and high sides of the peak, namely the points where $d^2I/dT^2 = 0$. However, these points are not easy to evaluate experimentally, and therefore the method is not very accurate. Singh *et al.* [260] suggested to utilize temperature points at 2/3 and 4/5 of the maximum intensity, in addition to the half-maximum intensity discussed above.

5.3   Methods of Various Heating Rates

Considering the first-order peak given in Equation (4.11), Randall and Wilkins [6] found the expression for the maximum of a TL glow peak by setting the derivative $dI/dT = 0$. This results in

$$\beta E/(kT_m^2) = s \cdot \exp(-E/kT_m), \tag{5.12}$$

where T_m is the temperature at the maximum. Instead of this conventional way of writing the equation, we can slightly change it into

$$\beta = (sk/E)T_m^2 \exp(-E/kT_m).\tag{5.13}$$

Since the expression on the right-hand side is an increasing function of T_m, increasing the heating rate β increases the right-hand side, and therefore increases T_m. This means that as the heating rate increases, the peak shifts to higher temperature. Booth [261], Bohun [262] and Parfianovitch [263] have independently suggested the use of this shift for evaluating the trapping parameters E and s. Using two different heating rates β_1 and β_2, one gets two maximum temperatures T_{m1} and T_{m2}, respectively. Writing Equation (5.12) twice, for β_1 and β_2, dividing one by the other and rearranging, we obtain

$$E = k\frac{T_{m1}T_{m2}}{T_{m1} - T_{m2}} \ln\left[\frac{\beta_1}{\beta_2}\left(\frac{T_{m2}}{T_{m1}}\right)^2\right].\tag{5.14}$$

The activation energy thus found can be re-inserted into Equation (5.12), and thus the value of the frequency factor s can be found. Hoogenstraaten [264] proposed the use of several heating rates. From Equation (5.12) it is straight forward to see that a plot of $\ln(T_m^2/\beta)$ versus $1/T_m$ should yield a straight line with a slope of $-E/k$, from which E is readily evaluated. Extrapolation of the straight line to $1/T_m \to 0$ gives the value of $\ln(sk/E)$, from which s can be calculated by using the value of E/k found from the slope. It should be mentioned that Chang and Thioulouse [265] have shown that the Hoogenstraaten method is also applicable in cases where the TL results from tunneling pair recombination.

Osada [266] was the first to extend the method to non-linear heating functions. He proved that Equation (5.12) is valid for an exponential heating function

$$T = T_\infty - (T_\infty - T_0)\exp(-\alpha t),\tag{5.15}$$

where α (s^{-1}) is a constant and T_∞ is the final temperature approached asymptotically with time. This heating scheme is what one gets naturally if one lets a cold sample warm up while in contact with an infinite thermal bath at T_∞ (see also Kitis *et al.* [267, 268] for further analytical and experimental work on this subject). The instantaneous heating rate at T_m, which we denote by β_m, should replace β in Equation (5.12), which leads directly to the validity of Equation (5.14) for various exponential heating functions (different values of α and/or T_∞), as well as to the applicability of the Hoogenstraaten method for this case. Chen and Winer [269] showed that Equation (5.12) is correct for any first-order peak measured under any monotonically increasing heating function. To show this, let us rewrite Equation (4.11) as the solution of Equation (4.10), but for a general heating function $T = T(t)$,

$$I(T) = n_0 s \cdot \exp(-E/kT)\exp\left[-s\int_{T_0}^{T}\frac{1}{(dT'/dt)}\exp(-E/kT')dT'\right].\tag{5.16}$$

The assertion that the heating function is increasing ensures that $dT/dt > 0$, thus it is not equal to zero. Setting the derivative of Equation (5.16) to zero yields

$$\beta_m E/(kT_m^2) = s \cdot \exp(-E/kT_m),\tag{5.17}$$

where $\beta_m = (dT/dt)_{T_m}$. Equation (5.17) is a generalization of Equation (5.12) and therefore, the methods of two and several heating rates can be used in the same way with all monotonic heating functions.

It should be noted that these methods of various heating rates described so far, have been developed for first-order peaks only. However, many investigators have used them successfully for more complicated kinetics (see e.g. Singh *et al.* [270]). Chen and Winer [269] showed that the method of various heating rates is applicable for general-order kinetics, which includes the second-order case. However, whereas in the first-order case the relevant equations are proved rigorously, in the general-order case the equations are only a good approximation. Furthermore, Bell and Sizmann [271], dealing with the analogous effect of thermal annealing, considered the kinetic equation

$$-dn/dt = K_0 F(n) \exp(-E/kT), \tag{5.18}$$

where $F(n)$ is a general, smooth function of the trapped carriers concentration n and K_0 a pre-exponential constant. They derived the equation for two heating rates

$$E = k \frac{T_{m1} T_{m2}}{T_{m2} - T_{m1}} \ln \left[\frac{\beta_1}{\beta_2} \left(\frac{T_{m1}}{T_{m2}} \right)^2 \frac{(dF/dn)_{T_{m1}}}{(dF/dn)_{T_{m2}}} \right]. \tag{5.19}$$

If the function $F(n)$ is "normal" in the sense that its first derivative is only slowly varying, leading to $(dF/dn)_1/(dF/dn)_2 \cong 1$, Equation (5.17) holds, and the various heating rates methods can be used.

5.3.1 Temperature Lag and Thermal Gradient

While using the Hoogenstraaten method, it is desirable that the range of heating rates utilized should be as broad as possible. One should remember, however, that there might be a lag between the temperature of the sample and that of the thermometric device (e.g., a thermocouple), which results in errors in the temperature reading. Moreover, in bulk samples with low heating conduction, there may be a thermal gradient within the sample, which may influence the measured glow curve. While changing the heating rate, the size of the error varies. The temperature lag and the thermal gradient have a bearing on all the methods of extracting information and should, therefore, be discussed in some detail. In the presence of temperature lag, a given peak would appear as if a separate curve arises from any isothermal region in the sample. The apparently different peaks, shifted from one another, would superimpose to give a thermally stimulated luminescence curve which usually does not look like a single peak even if the basic process is, say, of first order. Land [259] has given a simulated curve reached under these conditions. As pointed out by Land and by Chvoj *et al.* [272–274], the result is a broadening of the given single peak. These authors have also given a detailed calculation of the expected shape of the curve under these circumstances. For further similar work see Lesz *et al.* [275, 276], Betts *et al.* [277, 278], Kitis *et al.* [279] and Piters and Bos [280]. In many cases, the main aim is to extract the trapping parameters from the measured curve, and this can be done only by eliminating the effect of temperature gradients as much as possible. The obvious precautions which reduce this effect are using a thin sample and relatively slow heating rates. As mentioned above, using thin samples and slow heating rates would also reduce the measured signal. A somewhat better way to cope with this problem is to use, whenever possible, the sample in powder form which is placed, or preferably glued, directly to the oven plate which is heated in a predetermined manner. This is done routinely in TL measurements of quartz and other minerals during dating of archaeological samples. Another method utilized in the reading of TL from archaeological and geological samples is to fill the sample compartment with an inert gas, usually nitrogen or helium, which helps in the homogeneous heating of different parts of the sample. The higher thermal conductivity of helium is an advantage in this regard.

Furetta *et al.* [281] compared the TL results obtained using chips and powder MgB_4O_7, and discussed the thermal gradient effects on the evaluation of the kinetic parameters when using the various heating rates method. Kitis and Tuyn [282, 283] suggested an approximate method to determine the temperature lag and thermal gradients. In addition, some relations are proposed to estimate the effects of temperature lag on the evaluated trapping parameters. For more information on this point, see pages 171–174 in the book by Pagonis *et al.* [248].

5.4 Curve Fitting

Both the initial-rise methods and the peak-shape methods rely on different features of the form of the TL peak. With the progress of computerized data recording methods, curve fitting methods have become more popular. Here, many more points along the measured curve can be used, and the experimental results can be fitted to more complex functions including a larger number of parameters, either in a single peak or in a curve including a number of overlapping peaks. Some problems still remain with complicated curves where even the number of overlapping peaks within the measured glow curve is not known beforehand. We will briefly discuss here the relatively simple situation of curve fitting of a single TL peak, and the much more difficult case of deconvolution of a number of peaks. More details will be given in Section 10.1. Mohan and Chen [284] developed a curve fitting method for first- and second-order peaks, and described its application to synthetic computer-generated peaks. Shenker and Chen [222] showed that the method can be adapted to deal with general-order peaks. Bull *et al.* [285] further showed that even when there are multiple peaks in the glow curve, and the trap levels interact, treatment of the glow curve as the superposition of several individual glow peaks is valid for first-order kinetics, but becomes progressively less valid for higher-order peaks.

The technique of peak fitting starts by establishing or guessing the number of glow peaks present in a glow curve. Let us denote by $f(T)$ the mathematical function of an individual glow peak. In the case of a first-order peak, three parameters are involved, namely, the activation energy, E, the frequency factor, s and the initial concentration of trapped carriers, n_0. If we deal with a first-order peak normalized to, say, maximum intensity of unity, the two parameters E and s are sufficient to fully describe the peak. The situation is similar with a second-order peak where three parameters are needed to define the glow peak. In the cases of general-order and mixed-order peaks, an additional parameter is required, b in the former case and α in the latter. When several glow peaks are involved, the glow curve can be written as

$$I(T) = \sum_{i=1}^{p} \alpha_i f_i(T), \qquad (5.20)$$

where α_i is a scaling factor, $f_i(T)$ is the chosen mathematical function for the description of the individual glow peaks (first, second, general or mixed order), and p is the number of peaks in the glow curve (see e.g. Chen and McKeever [33]). Thus, when p first-order peaks are involved, $3p$ parameters are required whereas if the individual peaks are of general or mixed order, $4p$ parameters should be considered. The process of curve fitting, be it for a single peak or a composite glow curve, basically consists of a first guess of the parameters, evaluating $I(T)$ using Equation (5.20) and comparing it with the experimental curve. The parameters are then changed so that the difference between the experimental and calculated curves is minimized. A rather popular way of doing this is the Marquardt [286] nonlinear least-squares fitting. Let us consider a fitting function $y = f(x, \mathbf{a})$ where $\mathbf{a} = (a_1, a_2, \ldots, a_m)$ is a vector of m parameters. We wish to get the best fit of this function to the n data points $(x_1, y_1, x_2, y_2, \ldots, x_n, y_n)$. We now minimize the quantity

$$S = \sum_{i=1}^{n} \left[\frac{f(x_i, \mathbf{a}) - y_i}{w_i} \right]^2, \qquad (5.21)$$

where w_i is a weighting parameter, often set to unity for all i. S is usually known as the "chi-squared" function. Obviously, one wishes to get a global minimum of this function, thus extracting the best set of parameters. Unfortunately, non-linear functions of this sort usually have many local minima, and practically all the methods of minimization lead to a local minimum which is not necessarily global. Methods like steepest descent, Newton and quasi-Newton are being used for such minimization (see, e.g., Avriel [287]). More recently, methods have been developed for increasing the probability of approaching the global minimum even when the initial guess of the set of parameters is rather far from the final optimum. These include simulated annealing [288], tabu search [289] and genetic algorithms [290]. At the end of the process of minimization of S, one wishes to evaluate the goodness of fit. Due to random variations in the data points, and since the specific kinetics of the peaks and even their number is unknown, and in view of the fact that the minimization process may not lead to the global optimum, one cannot expect to end up with $S = 0$. The final value of S may not be a real measure of the goodness of fit if there is no scaling such that results of two different sets of experimental data can be compared. As mentioned above, when the fitting is for a single peak, a normalization to a maximum intensity of unity may solve the problem, but this is not the case for a multi-peak glow curve. A number of investigators (see, e.g., Horowitz and Yossian [291]) suggested the use of the figure of merit (FOM), defined as

$$\text{FOM} = \sum_{j_{initial}}^{j_{final}} \frac{100 \mid f(x_i, \mathbf{a}_i^*) - y_i \mid}{\Delta}, \qquad (5.22)$$

where $\mathbf{a}^*$ is the final set of parameters and Δ the integral under the curve over the region of interest, from $j_{initial}$ to j_{final}. Since the FOM is normalized by the integral under the curve, the goodness of fit may be compared from one glow curve with another. Fits are considered to be acceptable when the FOM is of a few percent. Several investigators used different versions of glow-curve deconvolution for evaluating trapping parameters and as an aid for more accurate determination of the absorbed dose (see, e.g., Horowitz and Yossian [291]). It should be noted that in order to separate difficulties emanating from experimental inaccuracies and difficulties inherent in the mentioned process of minimizing the function in Equation (5.21), it is advisable to simulate a glow curve consisting of a number of overlapping peaks using a set of chosen parameters, and see if the deconvolution program can retrieve the original values (see, e.g., Kitis *et al.* [268]). A two-stage deconvolution method has been described by Hoogenboom *et al.* [292] who used a nonlinear fitting procedure for the estimation of initial values of the relevant parameters, and a full curve-fitting method for the final evaluation of the trapping parameters. For further discussion on deconvolution see Section 10.1.

An interesting attempt to determine all the trapping parameters by curve fitting in the OTOR case has been made by Sakurai [293]. Since only one concrete example of retrieving the parameters used for the simulation is given, the generality of this method is in some question. The same author [294] expresses doubts about the deconvolution of the TL curve using best fitting methods (see Section 10.1). The reason is that all deconvolution methods assume that the glow curve is simply a superposition of individual glow peaks whereas in reality, the occurrence of trapping of carriers released from one trap into other traps interweaves the different peaks in a much more complicated manner, which may undermine the capability of the method to yield reliable trapping parameters. Sakurai [294] gives a numerical example in which the trapping parameters entering the simulation of the combined glow curve cannot be retrieved to a reasonable accuracy by deconvolution.

5.5   Developing Equations for Evaluating Glow Parameters

Perhaps the oldest and simplest demonstration of the role of simulation in TL theory has to do with developing equations for evaluating the activation energy, summed up as Equation (5.8). Lushchik [187] used the method based on the high-temperature half intensity, δ. He assumed that the area under the curve for $T \geq T_m$ is the same as that of a triangle with the same height and the same δ.

Figure 5.1 shows schematically a TL peak. The triangle assumption can be expressed as $\delta I_m = \beta n_m$, where I_m is the maximum intensity, β is the constant heating rate and n_m is the concentration of carriers at the maximum. This implies that $n_m = \int_{t_m}^{\infty} I dt$. From Equation (4.10) we get for first-order kinetics $I/n = s \cdot \exp(-E/kT)$, which reads at the maximum

$$I_m/n_m = s \cdot \exp\left(-E/kT_m\right). \tag{5.23}$$

Rewriting the maximum condition (5.12), one immediately gets

$$I_m/n_m = \beta E/(kT_m^2). \tag{5.24}$$

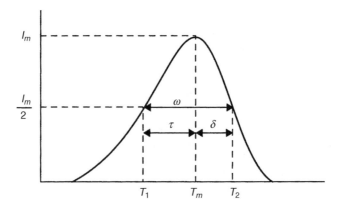

Figure 5.1   *An isolated glow peak showing the width parameters $\omega = T_2 - T_1$; $\delta = T_2 - T_m$; $\tau = T_m - T_1$. Reprinted from Chen, R., Methods for kinetic analysis of thermally stimulated processes, J. Mater. Sci, 11, 1521. Copyright (1976) with permission from Springer*

Using the mentioned triangle assumption yields

$$\beta/\delta = \beta E/(kT_m^2),\tag{5.25}$$

from which the Lushchik formula (5.5) for first-order peak emerges. In a similar way, Lushchik found Equation (5.6) for the activation energy of second-order peaks. Chen [203] has questioned the triangle assumption and tested its validity by simulation of glow peaks in a wide range of activation energies and pre-exponential factors for both first- and second-order kinetics. Instead of the triangle assumption by Lushchik, mentioned above, we can write

$$\delta I_m/\beta n_m = c_\delta,\tag{5.26}$$

and test to what extent c_δ is invariant with variations of E and s. Furthermore, Chen suggested checking by how much the average of c_δ over the range of interest of E and s (or s') differs from unity. For both first and second order, the variations of c_δ with s (or s') and E were confirmed by numerical simulation to be very small, but the values of c_δ were very different for first and second order. The nearly constant value of c_δ for first-order kinetics was found to be 0.976, rather close to the previously assumed value of unity. For second-order peaks, c_δ was determined to be 0.85, again with only minor variations with the activation energies and the pre-exponential factors. For the second-order case, Equation (5.6) should therefore be multiplied by 0.85 so that the amended formula for second order is

$$E = 1.7kT_m^2/\delta.\tag{5.27}$$

Halperin and Braner [30] developed a method based on the measurement of T_m and τ. They also distinguished between first- and second-order cases and in-between instances where carriers are excited into a band (where $s \propto T^2$) and into excited states within the forbidden gap prior to recombination, in which case the pre-exponential factor is independent of the temperature. They utilized Lushchik's assumption along with some other approximations. As shown later by Chen [203], their assumptions are equivalent to the assertion that the *low temperature* half of the peak fulfills the condition that the area under it is equal to that of the triangle with the same height and half width. In the present notation this means that

$$\beta(n_0 - n_m) \approx \tau I_m, \tag{5.28}$$

where n_0 is the initial concentration of trapped carriers and n_m the concentration at the maximum point of intensity I_m. By using the first two terms in the asymptotic series used for the evaluation of the exponential integral (see Chapter 15), Halperin and Braner [30] found that for first-order kinetics and for s independent of T,

$$E = 1.72(kT_m^2/\tau)(1 - 1.58\Delta), \tag{5.29}$$

where $\Delta = 2kT_m/E$. They also developed similar equations for the other three cases, first-order peaks with $s \propto T^2$ and the corresponding two cases for second-order kinetics. This kind of equation has an obvious shortcoming, namely that the quantity Δ appearing in the equation depends on E. One should therefore start with, say, $\Delta = 0$, get a first approximation for E and for Δ, and reinsert it into Equation (5.29). Instead of this iterative route, one can solve Equation (5.29) as a quadratic equation in E. However, using this way, one may face the problem of having two solutions, out of which one has to choose the "correct" one. Another problem is that the triangle approximation [Equation (5.28)] may not be very accurate, and deviations depending on the various E and s values may take place. Chen [203] studied this point, using simulations with a variety of the values of the parameters and gave a better version of Halperin and Braner's formula, namely [see Equation (5.7)]

$$E = 1.52kT_m^2/\tau - 1.58(2kT_m), \tag{5.30}$$

which is noniterative and more accurate than the original version. For the case of $s \propto T^2$, the coefficient 1.58 is replaced as before by 2.58. This amounts to a subtraction of $2kT_m$ from the expression given in Equation (5.30).

As for the equations for evaluating the activation energy for the general-order kinetics, summed up above [Equation (5.8) with the coefficients given in Equations (5.9–5.11)], they have been developed by Chen [204] using simulations and performing interpolation on the shape factor μ_g.

5.6   The Photoionization Cross Section

The most fundamental parameter which experimentalists try to extract from OSL experiments is the photoionization cross-section σ. Bøtter-Jensen *et al.* [74] summarized various optical absorption transitions which are important for dosimetry applications. These include ionization due to band gap transitions, exciton formation, defect ionization, trap ionization and internal intra-center transitions. The case of Al_2O_3:C is mentioned as a typical case of a material in which absorption of a photon with an energy of about 6 eV can cause photoionization of the F-centers into the conduction band.

During optical stimulation by a photon of energy $h\nu$, the absorption coefficient $\alpha(h\nu)$ for an optical transition between a defect and the delocalized bands will depend on the concentration n of the trapped carriers and on the photoionization cross-section σ. This cross-section will depend on the optical ionization threshold E_0 which is a characteristic of the trapped charge carrier, and also on the incident photon energy $h\nu$. In these general terms one can write [74]

$$\alpha(h\nu) = n(E_0)\sigma(h\nu, E_0), \tag{5.31}$$

where the concentration of trapped charge carriers $n(E_0)$ is assumed to depend on the optical ionization threshold E_0.

There are several published expressions for the dependence of the photoionization cross-section $\sigma(h\nu, E_0)$ on the photon energy $h\nu$ and on the ionization threshold E_0 [74]. In the case of dosimetric materials one of the most commonly used expressions is the following expression given by Lucovsky [296, 297], which was derived for deep traps

$$\sigma(h\nu, E_0) \propto \left[\frac{4(h\nu - E_0)E_0}{(h\nu)^2} \right]^{3/2}. \tag{5.32}$$

Several other forms of this equation are given in the book by Bøtter-Jensen *et al.* [74]. There are several methods of estimating the *relative* values of the photoionization cross-section $\sigma(h\nu, E_0)$ from experimental data. Three of these methods were summarized in [74]. The first method is based on keeping the value of the photocurrent (PC) constant during the experiment, while varying the intensity or flux of the stimulating light at various wavelengths. The other two methods are based on maintaining a constant intensity for the stimulating light at various stimulating wavelengths, while monitoring the variation of the PC.

It is possible to obtain estimates of the *absolute* value of the photoionization cross-sections by analyzing the signal obtained during a CW-OSL experiment as suggested by Huntley *et al.* [298]. The OSL intensity is given by [74]

$$I_{OSL} = n\Phi(\lambda)\sigma(\lambda), \tag{5.33}$$

where $\Phi(\lambda)$ represents the illumination flux at wavelength λ, which is the quantity kept constant during CW-OSL experiments. The rate of change $S_{OSL} = dI_{OSL}/dt$ of the CW-OSL intensity is given by

$$\frac{dI_{OSL}}{dt} = S_{OSL} = \frac{dn}{dt}\Phi(\lambda)\sigma(\lambda). \tag{5.34}$$

The OSL intensity in the case of first order kinetics is given by

$$I_{OSL} = \frac{dn}{dt}. \tag{5.35}$$

By combining the last two equations one obtains

$$S_{OSL} = I_{OSL}\Phi(\lambda)\sigma(\lambda) \tag{5.36}$$

or

$$\sigma(\lambda) = \frac{1}{\Phi(\lambda)} \frac{S_{OSL}}{I_{OSL}}. \tag{5.37}$$

This is the desired expression for obtaining absolute values of the photoionization cross-section $\sigma(\lambda)$ from CW-OSL measurements, by using the ratio S_{OSL}/I_{OSL} of the slope S_{OSL} of the CW-OSL decay curve and the CW-OSL intensity I_{OSL}. The flux $\Phi(\lambda)$ is known experimentally.

A rather comprehensive overview of the photoionization cross-sections in quartz has been given by Bailey [35], in connection with his discussion of a comprehensive kinetic model for quartz. The relationship between the optical and thermal properties of quartz is discussed in some depth, including the associated phenomena of thermal assistance and thermal quenching of luminescence.

In some of the earliest reported LM-OSL measurements, Bulur *et al.* [299] studied these signals from quartz samples and performed a deconvolution of the signals using a linear combination of first-order peaks. These authors found a total of four OSL components in their quartz samples. These are the previously known fast, medium and slow components from CW-OSL measurements, and a new slow OSL component which had not been reported previously. Bulur *et al.* also studied the thermal stability and dose responses of the individual components in the LM-OSL signal from the quartz samples. In later work Bulur *et al.* [300] used the LM-OSL technique to study the relationship between the detrapping rate and the intensity of the stimulating light in quartz, Al_2O_3:C, BeO and NaCl samples. Their experiments showed that the usual assumption of a linear relationship between the detrapping rate and the light intensity does not always hold. By studying the initial decay rates of the OSL signals, they found a linear relationship for SiO_2, Al_2O_3:C, BeO. In contrast, a non linear relationship was found in the case of NaCl. The LM-OSL signal for NaCl was interpreted on the basis of this nonlinear relationship, and the authors derived a new mathematical expression which can describe the LM-OSL signal for this material.

Singarayer and Bailey [301] studied the bleaching spectra of OSL signals from quartz samples using light of various wavelengths. These authors commented that earlier work on the subject had consisted of measuring the integrated luminescence versus photon energy, but these previous studies had not attempted to separate the various components which comprise the total integrated OSL signal. Singarayer and Bailey [301] presented the bleaching spectra of the "fast" and "medium" OSL components in quartz samples, and measured the photoionization cross-section for different wavelengths of the incident light. On the basis of their results they suggested a method for optically separating the various OSL components of quartz, based on the wavelength dependence of photoionization cross-sections. They presented graphs of the photoionization cross sections as a function of the stimulating photon energy in the range 2–3.5 eV, and attempted to fit their data to the commonly used Lucovsky equation [296]. They found that the fits were unsatisfactory at low incident photon energies, and obtained better fits using a more complex equation which takes into account the influence of phonons in the optical deexcitation process. They also tabulated values of the cross-sections at different wavelengths for the fast and medium OSL components at room temperature.

Singarayer and Bailey [302] also studied in detail the behavior of the various OSL components found by deconvolution of LM-OSL signals from several different quartz samples. These authors studied a number of sedimentary quartz samples from various locations and found that the LM-OSL peaks appeared in rather similar positions on the time axis. They also reported that most samples exhibited five or six components, and they presented data for the thermal stability, dose response and energy dependence of the photoionization cross section corresponding to each component. The components common to most samples are commonly termed the fast, medium, and slow components $S1$, $S2$ and $S3$. In almost all their samples, the calculated values of the photoionization cross-sections overlapped within error. The averaged values for the fast, medium, and slow components $S1$, $S2$ and $S3$ were of the order of 10^{-17}, 10^{-18}, 10^{-19}, 10^{-20} and 10^{-21} cm^2 correspondingly.

Several experimental studies have attempted to identify the number of OSL components present in LM-OSL measurements, carried out in a variety of quartz samples. These additional studies showed that the number of OSL components varies from sample to sample. While Singarayer and Bailey [302] found that most of their sedimentary quartz samples exhibited five distinct OSL components, Jain *et al.* [303] identified six OSL components from a study of several types of sedimentary quartz samples. Other studies carried out on individual grains showed that the number of OSL components can also vary from grain-to-grain [304, 305]. These authors also demonstrated the presence of an "ultra-fast" OSL component for a Korean sample, appearing before the "fast" component during an LM-OSL experiment. They reported the cross-section for this unusual component as being 10^{-16} cm^2. Jain *et al.* [303] also reported experimental results on the sensitization, thermal stability, recuperation, and IR response from three different samples of sedimentary quartz, as a function of stimulation temperature. These authors tabulated and discussed the relative cross-sections of each of the six OSL components observed in the LM-OSL signal. They labeled their components UltraFast (UF), Fast (F), Medium (M), Slow 1 (S1), Slow 2 (S2), Slow 3 (S3) and Slow 4 (S4).

Kiyak *et al.* [306] studied in detail four sedimentary quartz samples from different sites in Turkey. They applied a computerized deconvolution analysis to the LM-OSL curves, and also found six first-order OSL components. These authors also measured the dose response of each individual component, and presented results on the corresponding photoionization cross-sections. Furthermore, the sensitivity of each component was measured as a function of the activation temperature, in order to investigate the OSL thermal activation curves (OSL-TACs) in these samples. These authors commented that the cross-section values reported in literature vary over four orders of magnitudes, ranging from 10^{-17} cm^2 for the "fast" components up to the 10^{-21} cm^2 for the "slow" components, and tabulated six individual OSL components and termed them C1–C6. They found a one to one correspondence between their six components and the six components listed by Jain *et al.* [303]. Their tabulated photoionization cross-section values were found to be generally in agreement with the values reported by Singarayer and Bailey [302].

In later work on the same four sedimentary samples, Kitis *et al.* [307] studied the thermal stability and OSL signal recuperation properties of these individual OSL components. They reported the presence of a thermally unstable ultra-fast component, which could be essentially removed by heating up to 250–280 °C. Kiyak *et al.* [308] studied thermally activated characteristics of the LM-OSL signals in seven quartz samples of different origin, and compared these OSL results with the corresponding TAC of their TL glow peaks at 110 °C. Results were presented for LM-OSL measurements carried out both at room temperature and at an elevated temperature of 125 °C. As in their previous studies, the authors used a computerized deconvolution procedure to separate the LM-OSL curves into individual components. They reported that the OSL-TAC for all individual LM-OSL components follows the TAC behavior of the respective TL glow peak at 110 °C. Kuhns *et al.* [309] compared CW-OSL and LM-OSL curves from several types of quartz samples, and looked for universality in the OSL properties. Although these authors found several different LM-OSL components in each type of quartz, they did not identify any universal behaviors among the studied samples. They reported on the temperature dependence of the photoionization cross-sections of the individual LM-OSL components.

Choi *et al.* [310] commented that the derivation of trap parameters for each LM-OSL component was somewhat ambiguous in early papers on the subject, and gave a practical guide of how to carry out this type of deconvolution analysis properly. These authors studied several aspects of the analysis of LM-OSL curves for quartz. They discussed the influence of background signals on the fitting procedure, and gave several practical examples of deconvolution of LM-OSL curves using the commercially available software SigmaPlot. They demonstrated the proper way of presenting the detrapping probabilities and the photoionization cross-sections derived from the fitted curves, and suggested ways of identifying each OSL component based on the fitted parameters. Choi *et al.* [310] fitted both experimental and simulated LM-OSL data using first-order expressions containing the number of trapped electrons n_0, the total stimulation time P and the detrapping probability b. The latter quantity is proportional to the photoionization cross-section and the maximum stimulation light intensity (I_0), i.e. $b = \sigma I_0$. Although the value of σ (cm^2) is the fundamental property to be extracted from the analysis of the LM-OSL data, Choi *et al.* [310] suggested that the detrapping probability $b\,(\mathrm{s}^{-1})$ is more convenient to use for curve fitting procedures than the photoionization cross-sections. These authors compared successfully the ratios of the detrapping probabilities for the different OSL components with those published by Jain *et al.* [303]. They also gave explicit equations and numerical examples of how to calculate the photoionization cross-sections.

Chruścińska [311] discussed the optical cross-section of traps. Electron–phonon interactions have been shown to be a possible reason for the dependence of the OSL decay on temperature. The author uses an equation previously suggested by Noras [312] which connects the optical cross section with the optical trap depth, the stimulating photon energy and the Huang–Rhys factor which in the simple model of optical-band shape is a measure of electron–phonon coupling strength. Computer simulations of the OSL process have been carried out for different temperatures using one-trap and two-trap models. CW-OSL as well as LM-OSL have been simulated.

In connection with the LM-OSL deconvolution discussed in this section we also mention the work by Kitis and Pagonis [313]. These authors compared the computerized curve deconvolution techniques used for TL with the corresponding LM-OSL techniques. They derived single LM-OSL peak equations based on variables which can be extracted directly from the experimental OSL curve, for both first- order and general-order LM-OSL peaks. Furthermore, the resolution of the LM-OSL deconvolution technique is also examined in the case of several constituent components, and an experimental procedure is suggested which can be used to separate composite LM-OSL curves into their constituent components.

6

Additional Phenomena Associated with TL

6.1 Phosphorescence Decay

Phosphorescence is the relatively long-time emission of light following excitation, taking place when the sample is held at constant temperature; this as opposed to the short time ($\sim10^{-8}$ s) phenomenon of fluorescence (see e.g., Garlick [314]). The process taking place during phosphorescence decay is the same as in TL, and in fact can be considered as TL at a constant temperature. Thus, Figure 4.1 can explain the relevant transitions. It is a long-time effect usually associated with thermally shallow traps, measured in seconds or minutes, as opposed to fluorescence which is much shorter and is essentially independent of temperature. The validity of terming a trapping state "shallow" obviously depends on the relevant temperature. If the temperature at which the sample is held is, say, liquid nitrogen temperature (LNT), an electron trap 0.1–0.2 eV below the conduction band can be considered to be shallow. This is not a unique definition of a shallow trap, and it depends on the other parameters involved, in particular the frequency factor s. The main point is that the decay time will be between seconds and several minutes. Thus, if the ambient temperature is room temperature, $T \sim300$ K, the depth of the traps relevant to phosphorescence measurements will be substantially higher, for example 0.4–0.5 eV. If significantly higher energies are involved at a given temperature, the number of recombinations per second will be very small, and therefore, the light emission will not be measurable. If on the other hand the traps are too shallow, the decay time following excitation will be very small and the emitted light will be difficult to measure with standard equipment. Also, this fast decaying phosphorescence may be confused with fluorescence which does not involve the intermediate stage of trapping. The excitation process of phosphorescence is exactly the same as in TL and, in fact, so are the thermal stimulation, recombination and retrapping taking place following the excitation, except for the fact that the temperature does not change. Randall and Wilkins in their series of seminal papers on TL [6], presented the concepts of first-order and second-order decay

Thermally and Optically Stimulated Luminescence: A Simulation Approach, First Edition. Reuven Chen and Vasilis Pagonis. © 2011 John Wiley & Sons, Ltd. Published 2011 by John Wiley & Sons, Ltd.

laws of phosphorescence. In the first-order case they assume that excited electrons are still bound and therefore, the rate of change of the concentration n of electrons in traps is given by

$$-\frac{dn}{dt} = \alpha n, \tag{6.1}$$

where α is the decay constant (s^{-1}) of the process, which is constant as long as the temperature does not change. In fact, as explained in Section 4.1 for the case of TL, Equations (4.5–4.7) reduce to the same first-order expression if the quasi-equilibrium condition holds and when recombination dominates. The phosphorescence decay in this case is simply an exponential function

$$I = I_0 \exp(-\alpha t), \tag{6.2}$$

where I_0 is the initial emission intensity at $t = 0$. The value of α can easily be found by a plot of $\ln I$ versus t which should yield a straight line with a slope α. A possible dependence of α on temperature can be easily checked by repeating the measurement at different temperatures. As is obvious from Equation (4.10), $\alpha = s \exp(-E/kT)$. This is the basis of the "isothermal decay" method for evaluating the activation energy E. Once the phosphorescence decay has been recorded at various temperatures, the slopes $\alpha(T)$ can be plotted on a *semilog* scale as a function of $1/T$. This should yield a straight line with a slope of $-E/k$; from this, the activation energy E can be readily evaluated. Unfortunately, the exponential decay is by no means the general case and, more often than not, one does not get a straight line on the plot of $\ln I$ versus time at any temperature. One of the main possible reasons is that retrapping may have an important role and therefore, the kinetics is not of first order (see below). Another possible explanation is that in many cases, more than one trapping state is releasing electrons at a certain temperature and in some cases, one has to deal with a semi-continuum of traps.

Randall and Wilkins [6] also discussed the "bimolecular" reaction which they related to the "hyperbolic decay law". Assuming that the recombination rate is proportional to both the concentrations of the excited electrons and the vacant impurity levels (positive holes), and assuming that these two are equal to each other, one has the equation

$$I(t) = -dn/dt = An^2, \tag{6.3}$$

where A is a constant with units of $cm^3 \, s^{-1}$. Of course, this is the same as Equations (4.12–4.14) which shows immediately that the dependence of A on temperature is proportional to $\exp(-E/kT)$. The solution of Equation (6.3) is

$$I(t) = I_0/(1 + An_0t)^2. \tag{6.4}$$

Chen and Kristianpoller [315] studied the solution of the general-order Equation (4.18) for the isothermal case of phosphorescence. Solving Equation (4.18) with constant temperature, one gets

$$I(t) = s''n_0 \exp(-E/kT)/\left[1 + (b-1)s'' \exp\left(-E/kT\right)t\right]^{b/(b-1)}, \tag{6.5}$$

where b is the empirical order of the kinetics and $s'' = s'n_0^{b-1}$, and where s' is the constant in Equation (4.18). Obviously, Equation (6.5) reduces to Equation (6.4) for $b = 2$.

An isothermal decay of phosphorescence behaving like a $1/t$ function was reported by Randall and Wilkins [6], who described the effect as resulting from a continuum of uniformly distributed trapping states. A similar explanation of the occurrence of a uniform distribution of traps was given by Cook [316] for the decay in calcium tungstate. Another explanation of such a decay was given by Rzepka *et al.* [317] who described the effect in $SrCl_2$ as being due to tunneling recombination between V_k-centers and perturbed F-centers.

6.2   Isothermal Decay of TL Peaks

An effect closely related to phosphorescence decay is the isothermal decay of TL peaks. If a sample is irradiated, a certain TL intensity can be expected if the read-out (heating) process is performed right after excitation. However, if following the same excitation the sample is held at ambient temperature for a certain period of time, some of the trapped carriers may be thermally "lost" and therefore, the TL measured subsequently will be weaker. Rácz [318] has studied the decay of "isothermoluminescence" of molecular solids, and described it by assuming that the recombination mechanism is electron tunneling and that the frequency factor in the rate of tunneling is temperature dependent. The decay could be described in most cases as $I(t) \sim t^{-m}$ with m ranging from 0.7 to 1.5. Delgado *et al.* [319] studied the isothermal decay of peak 4 in LiF:Mg, Cu, P. The decay has been measured at various temperatures in the range of 179–190 °C, presenting always the purely exponential decay of light intensity with time, typical of a first-order process. From the good linear dependence of the logarithm of the decay lifetime ($\ln \tau$) with the inverse of the temperature ($1/T$), also typical of a first-order process, values of the activation energy and the frequency factor have been obtained for this peak. Very high values of these parameters, namely, $E = 2.1 \pm 0.2 \, \text{eV}$ and $s = 10^{21} \, \text{s}^{-1}$ were reported (see discussion in Section 6.3.4). These results are similar to those obtained by the peak-shape methods in the same material [320]. The isothermal decay of the same material and the trapping parameters evaluated have been further studied by Mahajna *et al.* [321]. They discuss the "prompt" and "residual" isothermal decay. The former is simply the isothermal decay of phosphorescence whereas the latter is the decay of TL following different waiting times at ambient temperature before starting the heating process. If the kinetics is of first order, both decays are expected to be exponential, and the lifetime is expected to be $\tau = s^{-1} \exp(E/kT)$ with the same values of E and s in both cases. These authors [321] report similar results while using the two methods, of $E = 1.23 \pm 0.03 \, \text{eV}$ and $s \approx 10^{11} \, \text{s}^{-1}$. This as opposed to values of $E \approx 2 \, \text{eV}$ and $s \approx 10^{20} \, \text{s}^{-1}$ obtained with the peak shape methods. The authors suggest that the trapping structures of peak 4 in LiF:Mg, Cu, P as well as peak 5 in LiF:Mg, Ti(TLD-100) (see Yossian *et al.* [322]) respond differently to isothermal conditions than to linear heating. Additional study along the same lines by Horowitz *et al.* [323] considered the long-term stability in TLD-100 and its correlation with isothermal decay at elevated temperatures. More work on the prompt and residual isothermal decay in LiF:Mg, Cu, P was performed by Kitis *et al.* [324]. Furetta [325] has given a tentative theoretical account of the competition between environmental radiation excitation and thermal fading of the thermoluminescence signal for first- and second-order kinetics situations. Jain *et al.* [326] studied the ITL of quartz at 310 and 320 °C and showed that it can be used for extending the dose range for geological dating up to 1.4 kGy. They also used peak transformed isothermal thermoluminescence

(PT-ITL) curves, similar to the pseudo LM-OSL discussed in Section 7.12. Further work on the ITL of quartz has been given later by Jain *et al.* [327].

6.3 Anomalous Fading and Anomalous Trapping Parameters of TL

This section covers two anomalous TL effects, namely the occurrence of very high activation energies and frequency factors in some materials, and anomalous TL fading. These important phenomena have been explained by Chen *et al.* [328–330] using a model of competition between traps and have been demonstrated using simulation.

6.3.1 Anomalous Fading of TL

An important feature of a TL peak during its application in TL dosimetry and dating is its stability with time. Every peak may undergo thermal fading which normally depends on the trapping parameters of the peak and the temperature at which the sample is held following its excitation, usually room temperature (RT). In the simplest case of first-order kinetics, the relevant parameters are obviously the activation energy E(eV) and the frequency factor s (s^{-1}), and the lifetime for decay at temperature T(K) is $\tau = s^{-1} \exp(E/kT)$ where k is the Boltzmann constant (eV K^{-1}). The trapping parameters E and s of a simple first-order peak can easily be evaluated as explained above, and therefore, the expected decay time τ can be found from this expression. Typically a peak occurring for example at 500 K, can be expected to be rather stable at room temperature ($\sim$300 K).

Quite a number of researchers have found that certain TL peaks behaved in a different manner, and displayed *anomalous fading*. As early as 1950, Bull and Garlick [331] reported that two UV excited peaks in diamond, occurring at 400 and 520K, yield lower light levels if stored at 90 K for 6 h before glowing, than if glowed immediately after excitation. Later, Hoogenstraaten [264] reported the decay of light levels at low temperatures in ZnS samples. He explained the effect as being due to the quantum mechanical tunneling of electrons from traps to recombination centers. Schulman *et al.* [332] reported the effect in CaF_2:Mn samples. Kieffer *et al.* [333] found a similar effect in organic glasses. In all these cases, no temperature dependence of the anomalous fading was reported. As opposed to this, Wintle [75, 334] monitored the anomalous fading in various minerals at different temperatures and discussed its implications regarding the dating of archaeological materials such as feldspars. More work on the anomalous fading of feldspars was carried out by Visocekas and Zink [335, 336]. However, these authors point to a red emission centered at 710 nm which is immune to the effect of anomalous fading and may therefore be considered for use in dating. Jaek *et al.* [337] also discussed the anomalous fading of TL and OSL in alkali feldspars and suggest that the long-time fading (months or years) may be associated with ionic processes. A later work by the same authors [338] describes the TL in feldspars in terms of tunneling and suggests that their procedure of keeping the sample for a month at RT instead of the routine preheating, results in an absence of anomalous fading. More work on the anomalous fading of TL in feldspar was reported by Gong *et al.* [339] and by Li and Li [340].

Bailiff [341] described how PTTL (see Section 6.6) can be used to circumvent anomalous fading in zircon, fluorapatite and plagioclase feldspars. The samples irradiated at RT were subsequently erased by heating to 500 °C, then illuminated by monochromatic light at RT and reheated to 500 °C yielding the phototransferred TL. In the case of zircon, the donor levels associated with the 320 nm illumination were found not to fade and those levels associated with other wavelengths (280 and 240 nm) showed some loss. Somewhat less conclusive results were reported with the other materials examined. Other attempts to circumvent anomalous fading were made by Lamothe and Auclair [342] and Lamothe *et al.* [343] who described a method of extrapolating to zero fading. Another possible correction for the anomalous fading by comparing the IRSL of feldspar and OSL of quartz was reported by Wallinga *et al.* [344]. They found that the anomalous fading of OSL, both blue and IR stimulated, was stronger than that of TL. Some studies of anomalous fading of TL and OSL in sanidine and other feldspars were reported by Wallinga *et al.* [344]. Polymeris *et al.* [345] compared the anomalous fading of TL and OSL in apatite. They found that the anomalous fading of OSL, both blue and IR stimulated, was stronger than that of TL. Some studies of anomalous fading of TL and OSL in sanidine and other feldspars were reported by Spooner [346] and by Visocekas [347]. In the latter work, the anomalous fading of TL was accompanied by red and IR emission of light, attributed to radiative tunneling recombination. This effect of tunneling-related emission of light had previously been found by Visocekas *et al.* [348] who studied the afterglow of $CaSO_4$:Dy and showed that after the initial irradiation, a weak afterglow is observed for a long period of time, with the same emission spectrum as the subsequently measured TL. The peak used for dosimetry at $\sim$250 °C decays down to zero with time at RT and at lower temperatures, practically independently of the temperature. The explanation given was that both the afterglow and the anomalous fading in this material resulted from a quantum mechanical tunneling effect. A similar phenomenon was reported by Visocekas [349] in labradorite. Huntley [12] also commented on the temperature dependence of anomalous fading. Clark and Templer [350] explained the effect as thermally assisted tunneling. Another account on thermally assisted tunneling was given by Vedda *et al.* [351] and de Lima *et al.* [352], who described the effect in calcite.

Further results on anomalous fading of zircon have been reported by Templer [121, 122, 353] who concluded that in zircon and pumice samples above RT, localized transitions are responsible for the anomalous fading whereas at lower temperatures, tunneling predominates. Kitis *et al.* [354] gave a similar explanation for the anomalous fading of TL in natural fluoroapatite. Trautmann *et al.* [355] also reported on localized transitions in K-feldspars, and explained the instability of the luminescence process in terms of localized transitions. Their results indicate that the unstable part of the process has to do with the recombination centers rather than the trapping states. Hasan *et al.* [356] reported on the fading in some meteorites and concluded that meteorites containing feldspars in the low form display anomalous fading. Meisl and Huntley [357] discussed the anomalous fading of feldspars and K-feldspars separated from sediments. These authors report that the activation energy increased from 0.04 to 0.2 eV as the K content increases from 0 to 15%. Tyler and McKeever [358] studied the anomalous fading of TL in oligoclase and examined critically the conclusions of Hasan *et al.* [356]. Their conclusion was that the anomalous decay is more closely described by the local transition model of Templer, rather than by quantum mechanical tunneling. Sears *et al.* [359] reported that in oligoclase, the amount of fading was less for the partially disordered samples than for the fully ordered samples. Tsirliganis *et al.* [360] have reported on anomalous fading of TL and OSL in fluorapatite. They distinguish between blue stimulated luminescence (BSL) and IRSL and between fast and slow components. The anomalous fading of the fast component of BSL and IRSL is almost the same, whereas the anomalous fading of the slow component of IRSL is approximately twice as strong as that of the BSL. Fragoulis and Stoebe [361] discussed the anomalous fading of inclusions in quartz and suggested that the feldspar inclusions may account for a portion of the fading, but other inclusions probably participate in the fading as well. Visocekas [362] studied the tunneling effects in feldspars leading to anomalous fading.

Other explanations have also been given for the anomalous fading of TL in different materials. Moharil *et al.* [363] suggested that the anomalous fading of TL in KBr samples has to do with mobile dislocations. They explain that during their motion, the dislocations "rob" the traps of their electrons. This seems to result in a loss of the filled trap population leading to anomalous fading.

6.3.2   Models of Anomalous Fading

As reported by many researchers, the initial fading rate of anomalous fading is quite rapid, followed by a slower decay at longer times. Visocekas and Geoffroy [364, 365] who studied the fading in calcite concluded that the intensity I of the afterglow observed during anomalous fading follows a hyperbolic law, namely $I \propto t^{-1}$ where t is the time. As explained by Mikhailov [366], a temperature independent hyperbolic dependence on time indicates that the mechanism involved is tunneling. In this case, an electron in a trap of depth E is separated by a distance r from a positive charge, and their recombination would produce the emission of a photon. The rate of tunneling of the electron through the potential barrier separating the two centers is [see e.g., Equation (2.117) in Chen and McKeever [33]]

$$P(r) = P_0 \exp(-\alpha r), \tag{6.6}$$

where P_0 is a frequency factor and α is given by

$$\alpha = \frac{2\sqrt{2m^* E_1}}{\hbar}, \tag{6.7}$$

where E_1 denotes the barrier height at a distance r and m^* is the effective electron mass. This expression is obtained by approximating the electron trap by a square potential well. If the decrease in the number of trapped electrons n obeys first-order kinetics, the luminescence intensity is given by

$$I = -\frac{dn}{dt} = \int_0^\infty n_0^2 \exp[-P(r)t] \, 4\pi r^2 dr, \tag{6.8}$$

where n_0 is the initial concentration of trapped electrons, which is also the total luminescence emitted during the afterglow as can be found by integration (see Delbeq *et al.* [367]). As described by Templer [368], this yields approximately

$$I_{tot} = K/t, \tag{6.9}$$

where K is approximately constant. An extension to this model was given by Huntley [369] who showed that when the tunneling is from an electron trap into a random distribution of recombination centers, a power-law decay of luminescence of the form $I \propto t^{-k}$ can be expected where k varies between 0.95 and 1.5. In particular, results by Cordier *et al.* [370] with an exponent of $k=1.06$ covered nine orders of magnitude and could be explained by this model. A previous work by Debye and Edwards [371] should also be mentioned, which described the long-life phosphorescence by the same decay law, with the value of k varying between 1.00 and 1.13, and the work by Rácz [318] (see Section 6.2) who described isothermoluminescence by the same decay law.

Proceeding with the case of $k = 1$, Equation (6.9) holds if $t \gg t_i$ where t_i is the irradiation time and t is measured from the end of irradiation. If a long irradiation time is taken, then

$$I = \frac{K}{t_i} \int_{-t_i}^{0} \frac{dt'}{t - t'} = \frac{K}{t_i} \ln(1 + t_i/t). \tag{6.10}$$

Since $\ln(1 + x) \approx x$ for small enough x, the difference between Equations (6.9) and (6.10) is negligible for $t > 10t_i$.

As described in Section 6.2, instead of monitoring the phosphorescence emitted during the charge leakage from the trap, one could monitor the residual TL, namely the TL remaining after the charge has escaped. Defining the parameter R as the ratio of the TL at time t to the initial TL at time t_0, one gets

$$R = \frac{TL(t)}{TL(t_0)} = -\frac{\ln(t/t_m)}{\ln(t_m/t_0)}, \tag{6.11}$$

where t_m is the maximum time for which the tunneling mechanism holds. As pointed out by Templer [368], since tunneling is more likely to be from deeper traps, the loss of the TL is more rapid from the high-temperature side of the curve than from the low-temperature side. This is the opposite to what is expected from thermal fading or from fading due to the localized transition model described below, and thus, this feature can be used to distinguish between the two models for fading. Huntley and Lamothe [372] discussed the analogous effect of anomalous fading of OSL and based on the same t^{-1} law gave the time dependence of luminescence intensity as a consequence of optical excitation as

$$I = I_c \left[1 - \kappa \ln(t/t_c) \right], \tag{6.12}$$

where I_c is the intensity at an arbitrary time t_c and κ is a constant, characteristic of the sample, but which depends slightly on the choice of t_c. This logarithmic decay law was first proposed and shown to fit experimental results by Visocekas [373].

Visocekas *et al.* [374] also considered the possibility that the electron has to be thermally excited to an intermediate state with energy E_2 from the bottom of the well (see Figure 2.17 in Chen and McKeever [33]), before tunneling takes place. The tunneling probability will now include a Boltzmann term and the probability of the thermally assisted tunneling will increase with temperature. This seems to be the case in calcite [374] and other materials as mentioned above. If n_2 is the population of the excited state and n_1 is the population of the ground state, one has

$$\frac{n_2}{n_1} = \exp(-E_2/kT),$$ (6.13)

where the total charge density is $n_0 = n_1 + n_2$. The overall tunneling probability is now

$$P(r) = \frac{P_0 \left[\exp(-\alpha_1 r) + \exp(-\alpha_2 r)\right]}{1 + \exp(-E_2/kT)},$$ (6.14)

where $\alpha_1 = 2\sqrt{(2m^* E_1)}/\hbar$ and $\alpha_2 = 2\sqrt{[2m^*(E_1 - E_2)]}/\hbar$, and obviously, a temperature dependence is expected. A review paper on the quantum-mechanical effect of tunneling in afterglow and its association with the anomalous fading was given by Visocekas [375].

As discussed in Section 4.6, one should consider the possibility of localized transitions (see also Chen [295] and p. 35 in Chen and Kirsh [229]). Templer [353, 368] suggested that this effect might be the reason for anomalous fading. The phosphorescence intensity expected from Equation (4.57) is

$$I(t) = n_0 \frac{ps}{p+s} \exp(-E/kT) \exp\left[-\frac{ps}{p+s} t \exp(-E/kT)\right],$$ (6.15)

where all the terms are as defined before. Templer [368] describes the shape of the phosphorescence decay associated with the localized-transition model when the traps are distributed uniformly in energy. For a general distribution of traps, the afterglow intensity is given by

$$I(t) = \int_0^\infty n(E) \frac{p(E)s}{p(E) + s} \exp(-E/kT) \exp\left[-\frac{p(E)s}{p(E) + s} t \exp(-E/kT)\right] dE.$$ (6.16)

For a uniform distribution $n(E) = n_0$, and $p(E) = $ constant, one gets

$$I(t) = \frac{n_0 kT}{t} \left[\exp\left(-\frac{t}{t_{max}}\right) - \exp\left(-\frac{t}{t_{min}}\right)\right],$$ (6.17)

where t_{max} and t_{min} are the trapped charge lifetimes corresponding to the upper and lower limits of the energy distribution, respectively. Except at very short times, the $1/t$ term dominates. If one wishes to consider the irradiation time t_i, one has

$$I(t) = n_0 kT \int_{t_i}^0 \frac{\exp\left[-(t - t')/t_{max}\right] - \exp\left[-(t - t')/t_{min}\right]}{t - t'} dt',$$ (6.18)

where t' is a time dummy variable. Tyler and McKeever [358] reported on the phosphorescence decay from irradiated oligoclase feldspar, after a post-irradiation anneal at 175 °C for 2 min to remove the thermally unstable component. They show that the localized transition model, Equation (6.18), fits the data better than the tunneling model, Equations (6.9) and (6.10). The remaining misfit of the localized-transitions model at long times seems to arise from the assumption of a uniform trap distribution.

If the residual TL is monitored, the fading occurs from the low temperature side of the glow curve for the localized-transition model, as opposed to that expected for tunneling

[368]. In this case the remnant TL parameter R defined by Equation (6.11) is given by

$$R = \frac{Ei(-t/t_{min}) - Ei(-t/t_{max})}{Ei(-t_0/t_{min}) - Ei(-t_0/t_{max})},$$ (6.19)

where $Ei(x)$ is the exponential-integral function (see Chapter 15).

6.3.3   Apparent Anomalous Fading of TL

Chen and Hag-Yahya [329] raised the possibility that, in fact, anomalous fading might be in some instances a normal fading in disguise. They proposed that the visible TL peak may look significantly narrower than otherwise anticipated from the given activation energy E and the frequency factor s of the peak, due to competition of non-radiative centers. A possible reason for a peak to be narrow is competition during heating from nonradiative competitors. In the model shown in Figure 6.1, recombination center M_2 is assumed to be radiative whereas M_1 and M_3 are assumed to be radiationless. This results in the observed peak being significantly narrower than is warranted by the "real" trapping parameters E_2 and s_2. To illustrate this point, we consider a simple equation given by Chen [203], which utilizes the full width of a single first-order TL peak, to evaluate the activation energy,

$$E = 2.29kT_m^2/\omega.$$ (6.20)

Here, $\omega = T_2 - T_1$, where T_1 and T_2 are the half intensity temperatures and T_m is the maximum temperature (in K); k is the Boltzmann constant (eV K^{-1}), and the activation energy E is in eV. It is evident that if by any *artificial* means, the peak looks narrower than otherwise

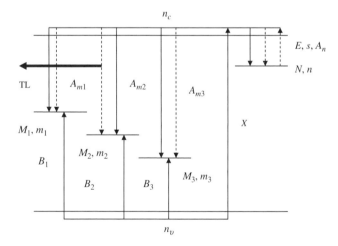

Figure 6.1   *The energy level model consisting of one trapping state and three kinds of recombination centers, used to explain high effective values of the activation energies and frequency factors. The transitions shown as solid lines are those taking place during excitation, and those during read-out are shown as dashed lines. The meaning of the different symbols is given in the text. After Chen and Hag-Yahya [329]. Reprinted from Chen, R., Hag-Yahya, A., A new possible interpretation of the anomalous fading in thermoluminescent materials as normal fading in disguise. Radiat. Meas. 27, 205. Copyright (1997) with permission from Elsevier*

expected, the apparent activation energy evaluated by Equation (6.20) will be larger than the real one. Similar results are expected with other shape methods commonly used for evaluating activation energies [see Equation (5.8)]. Also, the curve fitting methods are utilizing the same features of the TL peak and should therefore yield similar results. Indeed, Chen and Hag-Yahya [328] reported that under these circumstances, similar effective values of E are calculated from the shape methods and the best fit (see next subsection). Obviously, for a certain value of T_m, once we get a narrow peak with a high apparent value of E denoted by E_{app}, the accompanying apparent value of s_{app} will be orders of magnitude higher than the "normal" values. This can be seen from the condition for the maximum intensity in a first-order peak, Equation (5.12), which can be written as

$$s = \beta E / \left(kT_m^2 \right) \exp(E/kT_m),\tag{6.21}$$

and since E appears in the exponent, a large apparent value E_{app} will result in an apparent value s_{app} orders of magnitude higher than the original value of s.

Chen and Hag-Yahya [329] and Chen *et al.* [330] have used these ideas to give an alternative possible explanation for the anomalous fading effect. They developed a kinetic model of a single trapping state and three kinds of recombination centers, shown in Figure 6.1. One trapping state is shown with an activation energy $E(eV)$ and a frequency factor s (s^{-1}). The concentration of the trap is N(cm^{-3}) and its instantaneous occupancy is n(cm^{-3}). Three recombination centers, with concentrations of M_1, M_2, and M_3 (cm^{-3}) and occupancies m_1, m_2 and m_3 (cm^{-3}), are assumed to participate in the process. During the excitation by external irradiation, the rate of pair production is X (cm^{-3} s^{-1}). The time during the excitation is denoted by $t(s)$. Assuming a constant rate of excitation, the dose D of excitation is given by multiplying the dose rate X by the time of excitation t_D, $D = X \cdot t_D$. The concentrations of free electrons and holes are n_c and n_v (cm^{-3}), respectively. The transitions of holes into the centers are governed by the probability coefficients B_1, B_2, and B_3 (cm^3 s^{-1}) and recombinations during excitation are also allowed, governed by the recombination probability coefficients A_{m1}, A_{m2} and A_{m3} (cm^3 s^{-1}). In addition, retrapping of free electrons is allowed, associated with the retrapping probability coefficient A_n (cm^3 s^{-1}). The equations governing the process at the excitation stage are

$$\frac{dm_1}{dt} = -A_{m1} \cdot m_1 \cdot n_c + B_1(M_1 - m_1)n_v,\tag{6.22}$$

$$\frac{dm_2}{dt} = -A_{m2} \cdot m_2 \cdot n_c + B_2(M_2 - m_2)n_v,\tag{6.23}$$

$$\frac{dm_3}{dt} = -A_{m3} \cdot m_1 \cdot n_c + B_3(M_3 - m_3)n_v,\tag{6.24}$$

$$\frac{dn}{dt} = A_n(N - n)n_c,\tag{6.25}$$

$$\frac{dn_v}{dt} = X - B_1(M_1 - m_1)n_v - B_2(M_2 - m_2)n_v - B_3(M_3 - m_3)n_v,\tag{6.26}$$

$$\frac{dn_c}{dt} = \frac{dm_1}{dt} + \frac{dm_2}{dt} + \frac{dm_3}{dt} + \frac{dn_v}{dt} - \frac{dn}{dt}.\tag{6.27}$$

The transitions taking place during the excitation are shown in Figure 6.1 as solid lines. Chen and Hag-Yahya [329] solved these simultaneous differential equations using a fifth-order Runge–Kutta algorithm. Following the excitation, a relaxation period was simulated. At the end of excitation, non-zero concentrations of electrons in the conduction band and holes in the valence band are expected. In order to follow the normal experimental procedure, one should consider a relaxation time until the conduction and valence bands are practically empty. This was done numerically by solving the set of equations (6.22–6.27) while setting the excitation intensity X to zero. The numerical solution of this stage was performed using the same algorithm.

The next stage is that of heating with a linear heating rate. The transitions involved are shown as dashed lines in Figure 6.1. The trapped electrons are now thermally raised into the conduction band and may recombine with holes in the three centers M_1, M_2 and M_3. Also, retrapping of electrons into empty electron trapping states may take place. The equations governing the process are given by

$$-\frac{dm_1}{dt} = A_{m1} \cdot m_1 \cdot n_c, \tag{6.28}$$

$$-\frac{dm_2}{dt} = A_{m2} \cdot m_2 \cdot n_c, \tag{6.29}$$

$$-\frac{dm_3}{dt} = A_{m3} \cdot m_3 \cdot n_c, \tag{6.30}$$

$$\frac{dn}{dt} = -s \cdot n \cdot \exp(-E/kT) + A_n(N - n)n_c, \tag{6.31}$$

$$\frac{dn_c}{dt} = \frac{dm_1}{dt} + \frac{dm_2}{dt} + \frac{dm_3}{dt} - \frac{dn}{dt}. \tag{6.32}$$

For this stage also the numerical solution was performed using the same fifth-order Runge–Kutta algorithm.

Only the transition into one of the recombination centers M_2 is assumed to be radiative whereas the other two are assumed to be radiationless. Thus, the intensity of emitted measurable TL is proportional to the rate of change of m_2 and assuming the proportionality factor to be unity (which sets certain units for the emitted light), one can write the emitted light $L(T)$ as

$$L(T) = -\frac{dm_2}{dt}. \tag{6.33}$$

Chen and Hag-Yahya have chosen a plausible set of parameters and solved numerically the above mentioned sets of equations for the sequence of irradiation, relaxation and heating. The parameters chosen for the trapping state were $E = 0.7\,\text{eV}$ and $s = 10^6\,\text{s}^{-1}$. The decay time at room temperature for these traps is $\tau = s^{-1} \exp(E/kT) = 5.75 \times 10^5\text{s}$, which is just under 1 week. The other parameters chosen were $A_n = 1 \times 10^{-16}\,\text{cm}^3\,\text{s}^{-1}$, $A_{m1} = 1.2 \times 10^{-12}\,\text{cm}^3\,\text{s}^{-1}$, $A_{m2} = 2.8 \times 10^{-13}\,\text{cm}^3\,\text{s}^{-1}$, $A_{m3} = 7.5 \times 10^{-14}\,\text{cm}^3\,\text{s}^{-1}$, $B_1 = 1 \times 10^{-15}\,\text{cm}^3\,\text{s}^{-1}$, $B_2 = 1.1 \times 10^{-15}\,\text{cm}^3\,\text{s}^{-1}$, $B_3 = 1.2 \times 10^{-15}\,\text{cm}^3\,\text{s}^{-1}$, $N = 1.2 \times 10^{15}\,\text{cm}^{-3}$ and $M_1 = M_2 = M_3 = 4 \times 10^{14}\,\text{cm}^{-3}$. The radiation intensity in the appropriate units is $X = 5 \times 10^{14}\,\text{cm}^{-3}\,\text{s}^{-1}$, the irradiation time is 200 s and the constant

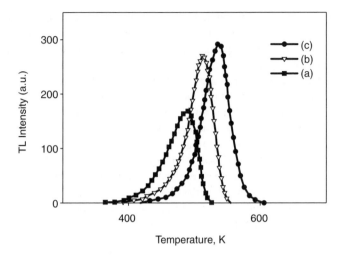

Figure 6.2 *The rates of transitions into the three competing centers as a function of temperature. (a) $-dm_1/dt$; (b) $-dm_2/dt$; (c) $-dm_3/dt$, as simulated by the numerical solution of the sets of simultaneous differential equations. The values of the chosen parameters are given in the text. Reprinted from Chen, R. and Hag-Yahya, A., A new possible interpretation of the anomalous fading in thermoluminescent materials as normal fading in disguise. Radiat. Meas. 27, 205. Copyright (1997) with permission from Elsevier*

heating rate was chosen to be $5\,\mathrm{K\,s^{-1}}$. Figure 6.2 depicts the curves of $-dm_1/dt$, $-dm_2/dt$, and $-dm_3/dt$ [curves (a), (b) and (c), respectively]. As suggested above, only the middle curve is assumed to be measurable.

Figure 6.3 shows only curve (b) of Figure 6.2; this looks exactly like a simple TL peak. The maximum temperature here is 515 K (242 °C). The symmetry factor is defined as $\mu_g = \delta/\omega$ where $\delta = T_2 - T_m$, $\omega = T_2 - T_1$ and where T_m is the temperature at the maximum and T_1, T_2 are the low and high temperatures at half intensity, respectively. The value found in this example was $\mu_g = 0.424$, which is typical of first-order kinetics. Using Equation (6.20) for evaluating E from a first-order peak yields $E_{app} = 1.3\,\mathrm{eV}$. Using the equation for the maximum, Equation (6.21) of a first-order peak, results in the apparent frequency factor $s_{app} = 1.6 \times 10^{12}\,\mathrm{s^{-1}}$. With these values, the apparent lifetime at room temperature is found to be $\tau_{app} = s_{app}^{-1} \times \exp(E_{app}/kT) = 4.3 \times 10^9$ s, or $\sim$137 years. In this example the ratio between the apparent and real lifetimes is therefore $\sim$10^4.

Chen *et al.* [329, 330] conclude that one of the possible explanations for the anomalous fading represents the competition during heating of the radiative center with non radiative centers. Such competition makes the measured TL peak significantly narrower than otherwise expected. Thus, peak shape methods yield an apparent activation energy much larger than the "real" one, and an apparent frequency factor which may be several orders of magnitude larger than the original one. As a result, the apparent life time may be very significantly larger than the real one. Therefore, when one expects the former and observes the latter, the measured fading is considered to be anomalous. As mentioned above, one can consider this effect as a normal fading in disguise.

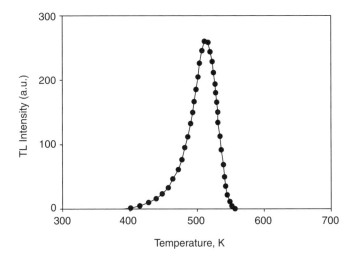

Figure 6.3 *Curve (b) from Figure 6.2 is assumed to be the only radiative, measurable TL emission. The chosen set of parameters is the same as in Figure 6.2. Reprinted from Chen, R. and Hag-Yahya, A., A new possible interpretation of the anomalous fading in thermoluminescent materials as normal fading in disguise. Radiat. Meas. 27, 205. Copyright (1997) with permission from Elsevier*

6.3.4　Interpretation of High Activation Energies and Frequency Factors in TL

As pointed out above, the values reported in the literature for the frequency factor s were in the range of $10^8 - 10^{13}$ s^{-1} (see also Haynes and Hornbeck [376] and Lax [108]). According to Mott and Gurney [92], s should be of the order of magnitude of the Debye frequency, which has to do with the number of times per second that a trapped electron interacts with the phonons. Lower values of the frequency factor were also reported in the literature. In a number of published TL peaks, values of s have been reported that are much higher than the expected "physical" values of up to $\sim 10^{13}$ s^{-1}. The best known case is that of peak 5 at $\sim$480 K in LiF:Mg, Ti (TLD-100). The experimental values of E and s appearing in the literature vary a lot, depending on the method used for evaluation of the parameters. Sometimes, different authors reported significantly different results even while using the same method. Zimmerman *et al.* [377], Blak and Watanabe [378], Johnson [379] and Yossian *et al.* [322] used isothermal decay methods and found values of $E \cong 1.25$ eV and $s \cong 10^{10}-2 \times 10^{12}$ s^{-1}. However, Taylor and Lilley [380] reported an activation energy of 2.06 eV and a frequency factor of 2×10^{20} s^{-1} for the same peak. Gorbics *et al.* [381] used the method of various heating rates and reported $E = 2.4$ eV and $s = 1.7 \times 10^{24}$ s^{-1}, whereas Pohlit [382] reported an exceedingly high values of $E = 3.62$ eV and $s = 10^{42}$ s^{-1}. Fairchild *et al.* [383] suggested that the kinetics of this peak and other peaks with unusually large s values might actually be complex and that the apparent first-order behavior be a special case, or approximation, of a more complex kinetics situation. Townsend *et al.* [384] suggested a theory of TL in terms of the more complex model of defect aggregates and alternative paths for the electrons which are either radiative or non-radiative, which could possibly account for the high E and s values of peak 5 in LiF:Mg, Ti. McKeever [385] suggested that peak 5 of LiF:Mg, Ti is related to trimer dissociation, where the trimers may have originated during heating. McKeever proposed that the activation energy of peak 5 is the sum of the trimer binding energy and a term related to the release of charge from the dissociated dipoles. Finally, an international effort to compare results of the glow curve in LiF by numerical deconvolution of the glow curve [386] showed practically unanimously the same very high values of E and s for peak 5.

Tatake et al. [387] also reported anomalously high frequency factors of up to 10^{22} s^{-1}. They studied TL in photosynthetic materials frozen in liquid nitrogen, illuminated and then warmed in the dark (see Section 3.1) and found higher-than-expected values of the activation energies and exceedingly high frequency factors. DeVault et al. [388] explained the effect in terms of the enthalpy change ΔH of the equilibrium which tends to add to the activation energy. Similarly, the entropy changes ΔS of the equilibrium tend to add to the entropy of activation, giving the large apparent frequency factors.

Chen and Hag-Yahya [328] demonstrated, by numerical simulation, the possible occurrence of very high E and s values as a result of competing centers. The model used is practically the same as described in Subsection 6.3.3. Curves very similar to those in Figuers 6.2 and 6.3 were simulated using the following parameters: $E = 1.2$ eV, $s = 2.5 \times 10^{11}$ s^{-1}, $A_n = 10^{-16}$ cm^3 s^{-1}, $N = 1.2 \times 10^{15}$ cm^{-3}, $A_{m1} = 1.2 \times 10^{-12}$ cm^3 s^{-1}, $A_{m2} = 2.8 \times 10^{-13}$ cm^3 s^{-1}, $A_{m3} = 7.5 \times 10^{-14}$ cm^3 s^{-1}, $B_1 = 10^{-15}$ cm^3 s^{-1}, $B_2 = 1.1 \times 10^{-15}$ cm^3 s^{-1}, $B_3 = 1.2 \times 10^{-15}$ cm^3 s^{-1}, $M_1 = M_2 = M_3 = 4 \times 10^{-14}$ cm^{-3}, $X = 10^{15}$ cm^{-3} s^{-1}, $t_D = 35$ s and $\beta = 6$ K s^{-1}. Curves very similar to those in Figures 6.2 and 6.3 were presented by Chen and Hag-Yahya [328]. The peak assumed to be the radiative one looks like a normal first-order peak with $\mu_g = 0.42$. Using Equation (6.20) for evaluating the activation energy, an apparent value of 2.24 eV was found. Substituting this into Equation (6.21) yields $s = 9.3 \times 10^{21}$ s^{-1}. Obviously these apparent values are very significantly larger than the original ones, and are in the same range as the above mentioned ones in LiF:Mg, Ti [380]. It should be noted that the adjacent peaks associated with M_1 and M_3 may not be radiationless as mentioned above. It is sufficient that there is an effective way of separating the peaks, e.g. by deconvolution, that the apparent high values of the parameters will be found; this may be the case in LiF:Mg, Ti (TLD-100), where a number of peaks are known to occur above and below peak 5 at 480 K.

6.3.5   Anomalous Inability to Excite TL

As can be expected from the theory of TL, a glow peak can usually be excited at any temperature below that of the peak. If the irradiation temperature is very close to that of the peak, then one can expect a rather fast decay. In this case, the intensity of the peak strongly depends on the time elapsed from the end of the irradiation. Otherwise, excitation can take place at any temperature below that of the peak. There are, however, two experimental examples in the literature where a TL peak cannot be excited in a certain temperature range below the peak, although it is found that the peak can be excited by the same irradiation at yet lower temperatures. This effect was reported by Sayer and Souder [389] in CaWO$_4$ and by Winer et al. [390] in semiconducting diamonds.

The glow curve of semiconducting diamonds (type IIb) UV-excited at LNT consists of two peaks, at $\sim$150 K and $\sim$250 K. An example is shown in curve (a) of Figure 6.4 which is the glow curve following irradiation with 360 nm at 77 K, taken from Winer et al. [390].

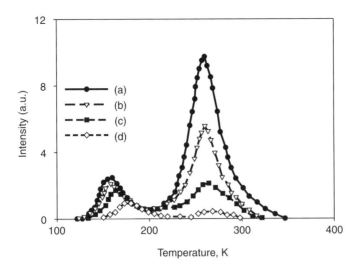

Figure 6.4 *Glow curves of semiconducting diamonds excited by 360 nm at LNT (a), 130 K (b), 140 K (c) and 150 K (d). Reprinted from Winer, S.A.A., Kristianpoller, N. and Chen, R., Luminescence of Crystals, Molecules and Solutions, F. Williams, Editor, Plenum, 473–477. Copyright (1973) with kind permission of Spinger Science and Business Media*

Curves (b), (c) and (d) are obtained after the same dose of irradiation at 130 K, 140 K and 150 K, respectively. As could be expected, the higher the temperature of excitation, the lower the intensity of the low temperature peak; note that the maximum shifts to higher temperature. The surprising point is that the peak at ∼250 K decreases as well, and it can hardly be excited above 150 K.

The maximum intensity as a function of excitation temperature is shown in curve (a) of Figure 6.5. For comparison, similar measurements were performed when the excitation was of band-to-band transitions using 225 nm UV light. These results are shown in curve (b) of Figure 6.5 and the behavior here is "normal" in the sense that the intensity decreases only when the temperature of excitation approaches that of the peak. As explained in Section 8.1, the 225 nm UV light raises electrons directly from the valence band into the conduction band, whereas with the longer wavelength UV light, a multistage transition of electrons from the valence band into the conduction band takes place. The strong superlinear dose dependence in the latter case is associated with this multistage transition whereas the band-to-band transition by the 225 nm photons yields a linear dose dependence. In the present context, it appears that one of the intermediate stages also associated with the 150 K peak is unstable at this temperature, and therefore excitation of the 250 K peak at 150 K and above is not possible.

The results by Sayer and Souder [389] on $CaWO_4$ doped with Cu or Na are surprisingly similar to those in semiconducting diamonds. Two main peaks at ∼160 K and ∼260 K are reported for excitations by UV light at 80 K. However, the excitation here is by 253.7 nm light which enables band-to-band excitation. The authors show that in both Cu- and Na-doped samples, the 260 K peak could not be excited above ∼180 K. Similar results were reported for X-ray excitation which, of course, could raise electrons from the valence band

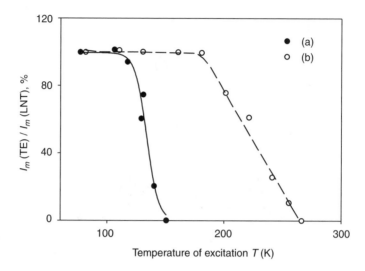

Figure 6.5   *The dependence of the maximum intensity I_m of the 250 K blue peak in IIb diamond on the temperature of excitation for (a) 360 nm; (b) 225 nm radiation. Reprinted from Winer, S.A.A., Kristianpoller, N. and Chen, R., Luminescence of Crystals, Molecules and Solutions, F. Williams, Editor, Plenum, 473–477. Copyright (1973) with kind permission of Spinger Science and Business Media*

into the conduction band. The explanation given by these authors for the anomalous behavior is associated with a band-to-band electron–hole recombination which usually is considered to be of low probability (see e.g. Pilkuhn [391]). The two peaks are assumed to be related to an electron and a hole trap, and at temperatures >160 K, the instability of the former reduces the concentration of the valence band holes via the mentioned band-to-band recombination. This, in turn, prevents the supply of holes to the deeper hole trap. Obviously, the analogy with semiconducting diamonds is not complete because in the latter, the effect occurs only at the low energy UV range (say, ∼360 nm), rather far from the band gap energy (5.5 eV, 225 nm).

6.4   Competition Between Excitation and Bleaching of TL

Exposure to sunlight has been considered to be the cause for zeroing TL in sea sediments (see Wintle and Huntley [392]). Huntley [393] pointed out that some traps in minerals are emptied very quickly by sunlight, whereas others require longer times and some traps may not be completely emptied by sunlight at all. Singhvi *et al.* [394] also reported that exposure to sunlight causes the bleaching of sediments to a residual value of TL rather than to zero, even when the exposure is very long. Spooner *et al.* [178] have studied the bleaching of TL in quartz and found different behaviors for different wavelengths. Whereas 320 nm light was very efficient in bleaching the TL peaks, 370 nm light was less effective and 500 nm light was even less effective. Light of 500 nm wavelength brought about a decrease of the TL intensity to a residual value only after a very long exposure. Similar results

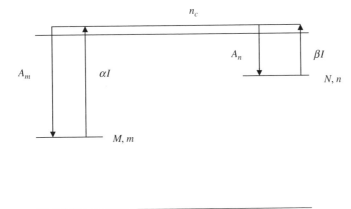

Figure 6.6 *Diagrammatic representation of the simple kinetic model for the excitation and bleaching of a TL phosphor under continuous light irradiation. Reprinted from Chen, R., Hornyak, W.F. and Mathur, V.K., Competition between excitation and bleaching of thermoluminescence, J. Phys. D: Appl. Phys. 23, 724–728. Copyright (1990) with permission from Institute of Physics*

have been reported by Shlukov and Shakhovetz [395] on a different kind of quartz. These authors explained their results in terms of equilibrium between bleaching and excitation produced by the same UV light. In order to show that excitation is a factor in reaching this equilibrium, these researchers de-activated their samples thermally and then illuminated them by the same UV light source. This resulted in the excitation of the same TL peaks. Thus, Shlukov and Shakhovetz conclude that the equilibrium value is reached as a result of competing UV excitation and bleaching processes, no matter whether one starts from a highly excited sample or with an entirely unexcited one. A much older work by Halperin and Chen [173] on TL in semiconducting diamonds reported on a similar effect. In this work, a TL peak at ~150 K was reported to be excited by a band-to-band transition using 225 nm (5.5 eV) UV light. Longer-wavelength light was found to bleach this peak. When light of wavelength >400 nm was used, complete bleaching would result. However, when light in the wavelength range of 400 nm > λ > 300 nm was used, bleaching could only be achieved down to a residual value even after a long exposure. When an unexcited sample was illuminated with light in this same wavelength range, the same build-up equilibrium value was reached after a long exposure. Here the effect was also explained to be a result of competition between excitation and bleaching caused by light of a given wavelength.

   Chen *et al.* [175] gave a simple theoretical account of this kind of competition. The simplest possible model for a TL peak, namely OTOR (see also Section 7.2), suffices to explain the effect of competition between excitation and bleaching. As shown in Figure 6.6, the phosphor material is assumed to possess a single type of electron trap and a single type of recombination center. When high-energy radiation is being used, the excitation process is usually explained as the excitation of an electron from the valence band to the conduction band in a band-to-band transition. The electrons in the conduction band may be trapped at the electron trap sites, whereas the remaining holes in the valence band may be trapped at the recombination centers. A subsequent heating would thermally raise the electrons from the trap into the conduction band, from where they can recombine with trapped holes, yielding the TL.

In the situation where lower-energy light is used, band-to-band transitions are not possible. Photons may still be able to elevate electrons from the neutral recombination centers into the conduction band, leaving a trapped hole behind. This transition is expected to be much less probable than the excitation of electrons from the valence band to the conduction band by high energy radiation, but it is of great importance in the present context. The electrons thus raised into the conduction band can either be trapped in an electron trap or recombine with a hole at a recombination center. In a similar way, this light may release a trapped electron from an electron trap. Once in the conduction band, this electron can either be retrapped or recombine with a hole at the recombination center. The relative effectiveness of raising electrons from occupied electron traps and recombination centers empty of holes should depend on the specific properties of the electron trap and recombination center involved, as well as on the wavelength of the light used. Thus, it is conceivable that certain wavelengths will favor the excitation whereas others will preferentially produce de-excitation. The equilibrium values expected after a long exposure will depend on these factors. It is to be noted that, although reference so far has been to a situation in which a narrow wavelength range is used for bleaching or excitation, the model should be able to cover also the more general situation of a broad range of wavelengths, such as for a sunlight exposure. In this case, the whole spectrum itself would be expected to have an effective probability for raising electrons from electron traps and from recombination centers.

A schematic diagram of the energy levels involved and the relevant designations for the various quantities is shown in Figure 6.6. Here n, n_c and m, all in units of cm^{-3}, are the occupation numbers of electrons at traps, electrons in the conduction band and holes at centers, respectively. Their initial values will be denoted by n_0, n_{c0} and m_0 (cm^{-3}), which may be all zero in an unexcited sample, or non-zero values in an excited sample. N and M (cm^{-3}) are, respectively, the total concentrations of electron-trap and recombination-center sites. A_m and A_n are the recombination and retrapping-probability coefficients (cm^3 s^{-1}). I is the intensity of light illuminating the sample in arbitrary units. α and β are the efficiencies of raising electrons from the recombination centers and electron traps, respectively. Their dimensions are such that αI and βI have dimensions of s^{-1}. The kinetic equations governing the process are

$$\frac{dm}{dt} = \alpha I(M - m) - A_m m n_c, \tag{6.34}$$

$$\frac{dn}{dt} = -\beta I n + A_n(N - n)n_c, \tag{6.35}$$

$$\frac{dn_c}{dt} = \frac{dm}{dt} - \frac{dn}{dt}. \tag{6.36}$$

6.4.1   Analytical Considerations

Before describing the numerical procedure and results, some simple analytical conclusions should be noted. In the framework of the OTOR model, including only a single trapping state and a single kind of recombination center,

$$m = n + n_c. \tag{6.37}$$

Furthermore, as long as the light intensity used is not extremely large, n_c is substantially smaller than n and therefore, $m \cong n$. This turns out to be the case for all the results of the simulations described below. Clearly equilibrium can be described by having the net rate of change of the occupation numbers equal to zero, i.e.,

$$\frac{dm}{dt} = \frac{dn}{dt} = \frac{dn_c}{dt} = 0. \tag{6.38}$$

These equilibrium conditions associated with Equations (6.34) and (6.35) yield

$$\alpha I(M - m) = A_m mn_c, \tag{6.39}$$

$$\beta In = A_n(N - n)n_c. \tag{6.40}$$

Combining these two equations for $n_{eq} = m_{eq}$, the equilibrium values, one gets

$$n_{eq}^2(\beta A_m - \alpha A_n) + n_{eq}\alpha A_n(M + N) - \alpha A_n MN = 0. \tag{6.41}$$

The solution of this quadratic equation should yield the equilibrium value n_{eq} to which the total area under the TL glow peak should be proportional. If $\beta A_m = \alpha A_n$ which in a sense means that the recombination and trapping have the same efficiencies, Equation (6.41) reduces to a simple first-order expression in n_{eq}, the solution of which is

$$n_{eq} = MN/(M + N). \tag{6.42}$$

If $\beta A_m \neq \alpha A_n$, the solution of Equation (6.41) is

$$n_{eq} = \left\{ -\alpha A_n(M + N) \pm \left[\alpha^2 A_n^2(M + N)^2 + 4\alpha A_n MN(\beta A_m - \alpha A_n) \right]^{1/2} \right\}$$
$$/ \left[2(\beta A_m - \alpha A_n) \right]. \tag{6.43}$$

It can be easily shown that the expression under the square root is always positive. It is also obvious that the two possible solutions have opposite signs, and that only the positive one has physical significance. Thus, when $\beta A_m > \alpha A_n$ one should choose the plus sign, and when $\beta A_m < \alpha A_n$ the minus sign should be chosen.

6.4.2   Numerical Results

Chen *et al.* [175] solved the three simultaneous differential equations (6.34–6.36) numerically, using a Runge–Kutta sixth-order predictor-corrector program. For each set of chosen parameters, the program was run once starting from entirely empty electron traps and recombination centers, and again starting with saturated electron traps and recombination centers. It is to be noted that within the framework of the single electron trap and single recombination-center model, the maximum occupancy of electron traps and recombination

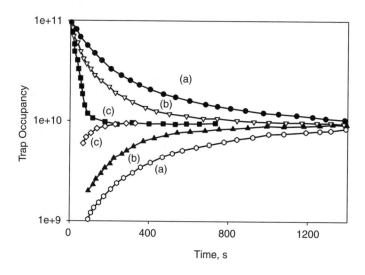

Figure 6.7 *Excitation and bleaching curves with the parameters given in the text. The ratio n_{eq}/n_0 is ~0.10. Reprinted from Chen, R., Hornyak, W.F. and Mathur, V.K., Competition between excitation and bleaching of thermoluminescence, J. Phys. D: Appl. Phys. 23, 724–728. Copyright (1990) with permission from Institute of Physics*

centers is $m_0 = n_0 = \min(M, N)$. The results of a number of runs are shown in Figures 6.7–6.9. In each case, the final equilibrium result has been compared with n_{eq} derived from Equation (6.41) with the same parameters. These results always agreed to better than 0.1%. It should be mentioned that a certain number of free electrons, n_c, remain in the conduction band at the end of the illumination period even if equilibrium has been reached. When the light is turned off, these decay partly into the recombination centers and partly into the electron traps, thus increasing n and decreasing m. This relaxation process (similar to the relaxation mentioned in Section 6.11) is simulated by allowing the program to continue for a certain period of time with $I = 0$ at the end of each period of illumination, until n_c practically vanishes. One thus ends up with $n = m$ and the subsequent total TL is usually proportional to this quantity.

Figure 6.7 depicts the results of the simulation for the following set of parameters; $\alpha = 10^{-15}\,\mathrm{s}^{-1}$; $\beta = 10^{-12}\,\mathrm{s}^{-1}$; $A_m = A_n = 10^{-9}\mathrm{cm}^3\,\mathrm{s}^{-1}$; $N = 10^{12}\mathrm{cm}^{-3}$; $M = 10^{11}\mathrm{cm}^{-3}$. As pointed out above, the units of αI and βI are s^{-1}. In curves (a) $I = 10^{10}$, in curve (b) $I = 2 \times 10^{10}$, and in curve (c) $I = 10^{11}$. The approach to the same equilibrium value is clearly evident and, obviously, the approach to this asymptote is faster with larger values of I. The value of n_{eq} found from Equation (6.43) is $9.47 \times 10^9\mathrm{cm}^{-3}$ for these given parameters, in excellent agreement with the numerical results. Note that for $I = 10^{10}$ the growth to one half of the equilibrium value takes ~530 s whereas the decay to one half the original value minus n_{eq} takes ~120s. With $I = 2 \times 10^{10}$ the corresponding numbers are ~260 s and ~40 s, respectively.

Figure 6.8 shows the results of a similar simulation with $\alpha = 10^{-17}\,\mathrm{s}^{-1}$ and $I = 2 \times 10^{12}$. Whereas in the previous case, the equilibrium value was about 10% of the saturated value (about 10^{11}, not shown in the graph), here it is about 1%. The value of n_{eq} calculated from

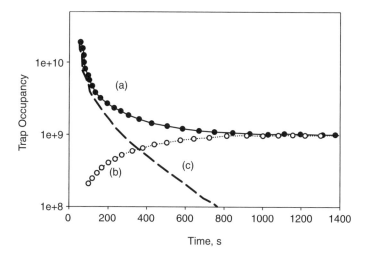

Figure 6.8 *Excitation and bleaching curves with the same parameters as in Figure 6.7 except $\alpha = 10^{-17}$. The value of I is 2×10^{12} for both bleaching (a) and excitation (b). The value of $n_{eq}/n_0 \simeq 0.01$. Curve (c) is for $n-n_{eq}$ for bleaching. Reprinted from Chen, R., Hornyak, W.F. and Mathur, V.K., Competition between excitation and bleaching of thermoluminescence, J. Phys. D: Appl. Phys. 23, 724–728. Copyright (1990) with permission from Institute of Physics*

Equation (6.43) is 9.94×10^8, again agreeing very well with the numerical results. Here the decay to half the maximum intensity takes $\sim$20 s, whereas the increase from zero to 50% of the equilibrium value takes $\sim$280 s.

Figure 6.9 shows the results with $\alpha = 10^{-13}$ s^{-1} and $I = 2 \times 10^{10}$, with the other parameters remaining the same as before. The value of n_{eq} from Equation (6.43) is 6.07×10^{10}, again in very good agreement with the numerical results. The rise time to 50% of the final equilibrium value is $\sim$190 s whereas the decay to half intensity is $\sim$140 s. Curves (c) in Figures 6.8 and 6.9 show the values of $n - n_{eq}$ as a function of time under bleaching conditions. As seen in Figure 6.8, this curve is not a straight line, which means that the bleaching is not simply of the form $n = n_0 e^{-\lambda t} + n_{eq}$. Note that in all cases examined, the equilibrium value of n_c was two to three orders of magnitude smaller than m or n.

As expected, the equilibrium levels reached are independent of the intensity I. This can be seen in Equations (6.41–6.43) and confirmed by the numerical simulation. The magnitude of I does change the time required to reach near-equilibrium values, e.g. as seen in Figure 6.7. In the expressions for equilibrium (6.41) and (6.43), it is seen that α is always associated with A_n (the excitation branch) and β with A_m (the de-excitation branch). Thus, while trying different parameters, only variations of αA_n on one hand and βA_m on the other, should be considered. In particular, if $\alpha A_n = \beta A_m$, the result is $n_{eq} = MN/(M + N)$. If $M \approx N$, this results in $n_{eq} = N/2$. On the other hand, if there is a substantial difference between N and M, n_{eq} will be equal to the smaller of N and M. Some further work along the same lines has been given by Dharamsi and Joshi [396] who found for given sets of trapping parameters the conditions under which an increase (excitation) or decrease (bleaching) of the electron density can be expected.

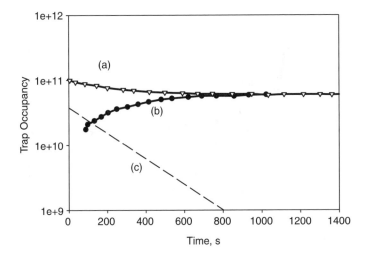

Figure 6.9   *Excitation and bleaching curves as in Figures 6.7 and 6.8 but with $\alpha = 10^{-13}\,s^{-1}$. The value of I is 2×10^{12} for both bleaching, (a) and excitation, (b). The value of n_{eq}/n_0 is ~0.60. Curve (c) is for $n - n_{eq}$ for bleaching. Reprinted from Chen, R., Hornyak, W.F. and Mathur, V.K., Competition between excitation and bleaching of thermoluminescence, J. Phys. D: Appl. Phys. 23, 724–728. Copyright (1990) with permission from Institute of Physics*

It should be noted that although these results bear some resemblance to those in Section 8.5, the main difference is that there, the discussion concerns the sensitivity of the sample, namely the response to a test dose as a function of the accumulated dose. However, here one deals with the trap occupancy as a function of the dose, to which the measured TL is often proportional.

6.5   A Model for Mid-term Fading in TL Dating; Continuum of Traps

The effect of anomalous fading of TL signals and its influence on the evaluated age of archaeological and geological samples has been described in Section 6.3. Mejdahl [397, 398] pointed out that when the ages to be determined are in the 10^5–10^6 year range, the loss of TL that occurs may be due to a long-term thermal decay rather than the mentioned anomalous fading. Thus, in the study of tertiary sands, he found TL ages in the region of 300–500 ka rather than 700–2000 ka predicted from the stratigraphy. Balescu *et al.* [399, 400] studied feldspar samples and reported a systematic underestimation of TL ages when compared with expected geological ages. However, the extent of the underestimation appeared to be dependent on the wavelengths selected by the optical filter used to observe the luminescence. Xie and Aitken [401] describe a model which explains the "mid-term" thermal fading during burial periods of the order of 10^5–10^6 years. They assume that the dating signal consists of a stable component and an unstable one, and that while the lifetime of the stable component is long compared with the oldest sample to be dated, that of the other component is mid-term, namely long compared with laboratory times but short compared with ancient ages. They assumed that the unstable component has the same growth characteristics as the stable one, both exhibiting saturating exponential functions of the dose, and that the fraction of the signal which is not stable is constant for all samples. Using these assumptions they showed that for relatively old samples, ages may be appreciably underestimated. In an example, a sample known to be 730 ka old appeared to be only 127 ka old based on TL evidence, presumably as a result of the mid-term fading effect.

Xie and Aitken [401] do not specify the nature of the fast and slower fading components. A plausible explanation suggested by Hornyak *et al.* [402] is that a distribution of trapping states exists, in which the deeper bound states are more stable than the shallower ones. The idea of a distribution of activation energies of trapping levels is as old as the theory of TL itself, as first suggested by Randall and Wilkins [6] who also gave evidence that such distributions exist in calcites and dolomites. These authors showed that if trapping levels are distributed uniformly, phosphorescence decay can be expected to behave like a $1/t$ function (see also Subsection 6.3.2). Kikuchi [403] showed that distributions of trapping level energies play an important role in the TL of glasses. Medlin [404] suggested that discrete levels may be broadened into a continuous band of levels as a result of local distortions in the crystal fields due to dislocations, vacancies and impurities. Further mention of the possibility of continuous distribution of traps was given by Bosacchi *et al.* [405, 406] who reported on TL results in "quasi-disordered" $ZnIn_2S_4$. These authors gave a theoretical explanation based on previous work by Simmons *et al.* [407] which has to do with p-n junctions, metal-oxide-semiconductor interfaces, and insulating or semiconducting thin films. Guzzi and Baldini [408] have also discussed the continuous distribution of traps in $ZnIn_2S_4$. A paper by Fleming and Pender [409] criticizes the conclusions of Bosacchi *et al.* suggesting that when electron retrapping occurs, the glow-curve profile may differ significantly from that associated only with the trap distribution. Another suggestion, namely that the distribution may be exponential has been given by Shousha [410] who compared the shapes of simulated glow curves associated with exponential distribution of traps and curves related to mono energetic traps. Srivastava and Supe [411] described trap distribution analysis in $CaSO_4$:Dy, and reported quasi-continuous distributions of traps with activation energies ranging from 1.15 to 2.30 eV in the 340–740 K glow-temperature range. In a later paper, the same authors [412] reported on TL in $Li_2B_4O_7$:Cu and explained the results as being due to the involvement of a number of trap groups with the activation energies in the range of 0.9–1.73 eV. Zahedifar *et al.* [413] discussed the expected width of glow peaks with continuous distribution of activation energies. Srivastava and Supe [414] reported on a distribution of frequency factors between 8×10^{12} and $2 \times 10^{16} \, s^{-1}$ in the same material and in the same temperature range. Sakurai and Gartia [415] also discussed trap distributions. Studying a Brown microcline they found unusually broad TL peaks, and a best-fit method showed an agreement with an exponential distribution of traps. In a later paper, Sakurai *et al.* [416] reported a similar exponential distribution in light-green glass.

Since both the local distortions and the traps may be randomly distributed throughout the crystal, a logical form for the distribution of activation energies would be a Gaussian of the usual form

$$\rho(E) = N\sqrt{\frac{a}{\pi}} \exp\left[-a(E - E_0)\right]^2 . \tag{6.44}$$

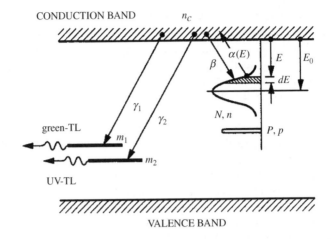

CONDUCTION BAND

VALENCE BAND

Figure 6.10   *A schematic representation of the charge trafficking giving rise to the observed TL emissions. A single active electron trap N possessing an activation energy distribution with central energy E_0, and two independent recombination centers interacting via the conduction band is assumed. A deep disconnected trap P is also taken into account. Reprinted from Hornyak, W.F., Chen, R. and Franklin, A., Thermoluminescence characteristics of the 375°C electron trap in quartz, Phys. Rev. B1, 46, 8036–8049. Copyright (1992) with permission from American Physical Society*

Hagekyriakou and Fleming [417, 418] studied the kinetic order of TL in the presence of a distribution of activation energies, and suggested a method for the determination of the retrapping/recombination ratio in TL. The actual filling of these trapping states may or may not result in a Gaussian charge distribution, depending on the circumstances. The effect of such a distribution on the TL characteristics has been studied in detail by Hornyak and Franklin [419]. Hornyak *et al.* [420] have shown that the TL characteristics of the 375 °C glow peak in quartz can be successfully explained using a model with such a Gaussian distribution of trapping states, and two recombination centers yielding green and UV emissions. They assume that electrons thermally elevated from traps are transported through the conduction band to recombination centers. Figure 6.10 is a schematic representation of the assumed charge trafficking giving rise to the observed TL emissions.

Central in the analysis is the presence of two glow peaks separated by 20 °C in temperature and emissions in the green and in the UV, respectively. Hornyak *et al.* [420] show that in this case, a single electron trapping state feeding two independent recombination centers produces two temperature-separated glow peaks when the light characteristic of the two centers is viewed with appropriate transmission filters. This phenomenon is a particularly simple version of interactive kinetics that generally embodies the coupling of several distinct electron and hole densities through electron trafficking in the conduction band. The possibility of such behavior has been suggested earlier by Bonfiglioli *et al.* [421]. Conforming to this model, only a single thermally active type of electron trap is assumed having a total concentration N, with n electrons present at any particular time. The Gaussian distribution given in Equation (6.44) is assumed and

$$\int_0^\infty \rho(E)dE = N.$$ (6.45)

In a manner adopted previously by Hornyak and Franklin [419], an electron occupancy number $f(E, t)$ is used such that an electron distribution $\eta(E, t)$ may be written

$$\eta(E, t) = f(E, t)\rho(E),$$ (6.46)

and

$$n = \int_0^\infty f(E, t)\rho(E)dE.$$ (6.47)

It is assumed that at the beginning of every TL process, i.e. at $t = 0$ or $T = T_0$, one has

$$\eta(E, 0) = f_0\rho(E),$$ (6.48)

with f_0 being a constant; that is, the original electron distribution is also Gaussian. The thermal release of electrons is assumed to be governed by an Arrhenius probability per electron, $\alpha(E) = s \cdot \exp(-E/kT)$ where the frequency factor s is assumed to be energy and temperature independent. The concentrations of trapped holes in the two recombination centers at any time are designated m_1 (for the green-TL center) and m_2 (for the UV-TL center). The concentration of electrons in the conduction band is labeled, as before, n_c. A temperature-independent retrapping factor β, and recombination transition probabilities γ_1 and γ_2 are assumed. The presence of a deep thermally disconnected electron trap with concentration P occupied by p electrons is also assumed in this model. The activation energy for this trap is assumed to be large enough to prevent its participation in the various TL processes. Also, in the context of sand-dune material discussed by Hornyak *et al.* [420], it can safely be assumed that the material has not been at an elevated temperature for a long geological period. It is therefore assumed that during this time, continual exposure to natural ionizing radiation resulted in this trap being saturated, i.e., $p = P$, so that no further trapping from the conduction band needs to be considered.

The governing charge trafficking integral-differential equations are [420]

$$\frac{d\eta(E, t)}{dt}dE = \rho(E)\frac{df(E, t)}{dt}dE = -\alpha(E)\rho(E)f(E, t)dE + \beta\left[1 - f(E, t)\right]\rho(E)n_c dE.$$ (6.49)

In view of Equation (6.47), one gets

$$n(t) = \int_0^\infty \eta(E, t)dE,$$ (6.50)

$$\frac{dn_c}{dt} = \int_0^\infty \alpha(E)\rho(E)f(E,t)dE - \beta n_c \int_0^\infty \left[1 - f(E,t)\right]\rho(E)dE - \gamma_1 m_1 n_c - \gamma_2 m_2 c_c,$$

$$(6.51)$$

$$\frac{dm_1}{dt} = -\gamma_1 n_c m_1, \tag{6.52}$$

$$\frac{dm_2}{dt} = -\gamma_2 n_c m_2, \tag{6.53}$$

and at all times

$$n + n_c + p = m_1 + m_2. \tag{6.54}$$

To allow for one overall intensity normalization that would include detection efficiency factors, etc., the value of N is not specified. If Equations (6.49–6.54) are all divided by N, then one more conveniently deals with relative concentrations and TL rates, namely n/N, $m_1/N,\ldots, d(n/N)/dt, d(m_1/N)/dt,\ldots$ and transition rates $\beta N, \gamma_1 N$, etc.

The method used to solve numerically Equations (6.49–6.54) was to subdivide the Gaussian distribution of N, $\rho(E)$, into 96 separate equal narrow sections. The electron population in each section was treated independently, each with its own release parameter $\alpha(E)$ and with a common retrapping parameter β. In addition to these equations embodied in Equation (6.49), the other four conditions Equations (6.51–6.54) were also programmed, thus 100 simultaneous differential equations were numerically solved by using a subroutine based on the sixth-order predictor-corrector Runge–Kutta method. Tests varying the number of sections into which N was subdivided showed no significant change in the results in using more than the 96 sections adopted.

As time or temperature increases, the distribution of electrons that first started out Gaussian in shape, gradually becomes asymmetric, generally favoring the deeper traps. This results from the fact that the electrons with the lowest activation energy are released earliest, while the retrapping probability is independent of energy, so that occupation of the deeper traps may actually be enhanced by retrapping.

Hornyak *et al.* [420] describe in detail the choice of parameters used in the simulation of the green and UV-TL peaks. These are $n_0/N = 0.330$, $m_{01}/N = 0.510$, $m_{02}/N = 0.017$, $p/N = 0.197$, $E_0 = 1.450\,\text{eV}$, $a = 80(\text{eV})^2$, $s = 5.1 \times 10^{11}\,\text{s}^{-1}$, $N\beta = 3.5$, $N\gamma_1 = 1.3$ and $N\gamma_2 = 3.5$. Figure 6.11 shows the experimental data points for the green-TL glow curve at a heating rate of $1\,^\circ\text{C}\,\text{s}^{-1}$ and the theoretically derived curve. The points and the corresponding theoretical curve for the UV-TL glow curve are also shown. Both simultaneous theoretical curves result from a single normalization $N = 1.05 \times 10^6$, in units relevant to actual detected counts. This value corresponds to an actual electron trap density of the order of 10^{14}cm^{-3}. The agreement on the peak maxima follows from the selection of m_{02}/m_{01} and the efficiency ratio of 8.3 for detecting the UV-TL relative to the green-TL. If the efficiency ratio were unity, the green-TL maximum would be 18.3 times as large as the UV-TL maximum.

Not only are the experimental full widths $\omega_1 = \omega_2 = 90.5\,^\circ\text{C}$ reproduced in Figure 6.11, but the correct shape parameters $\mu_1 = \mu_2 = 0.46$ are obtained as well. Finally, the peak

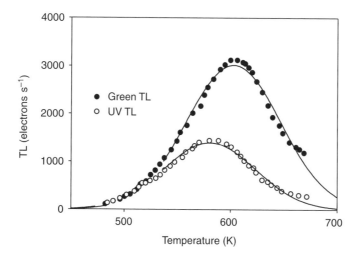

Figure 6.11 *The experimental data points (closed circles) for the green-TL glow curve at a heating rate of 1 °C s⁻¹. The curve is the simulation fit. The experimental data points (open circles) for the UV-TL glow curve at the same heating rate are also shown. Reprinted from Hornyak, W.F., Chen, R. and Franklin, A., Thermoluminescence characteristics of the 375 °C electron trap in quartz, Phys. Rev. B1, 46, 8036–8049. Copyright (1992) with permission from American Physical Society*

temperature difference between the two curves of 19.5 °C is also accounted for. It should be noted that the occurrence of the green peak here at 331 °C rather than 375 °C simply results from the relatively low heating rate of 1 °C s⁻¹ for both the experimental and the simulated results. The close agreement between the theory and the data suggests that in both cases the initial-rise plot of $\ln(-dm_1/dt)$ versus $1/T$ should also have the same slope and apparent value for an effective E. Indeed, the theoretical value arrived at is 0.95 eV, close to the data derived value of 0.92 eV. It is to be noted that these values are very much lower than the $E_0 = 1.450$ eV used in the theory. This result is in agreement with an earlier finding that showed that with a distribution of activation energies, the initial-rise effective value of E emphasizing as it does the initial release of the least bound electrons, will always be lower than the actual mean value of the distribution.

Hornyak *et al.* [402] studied the difference between dosing a sample in the laboratory time scale and in burial time scale when such a Gaussian distribution of traps is present [419, 422]. The differential equations governing the filling of traps and centers during long burial times were numerically solved and the results compared with those found when the same dose was imparted during laboratory times, say of the order of minutes. The distribution of trap activation energies given in Equation (6.44) is simulated with the same 96 segment histogram mentioned above. To avoid convergence problems and to achieve reasonably short computer running times, the earlier program was altered from a continuous dosing rate to that of doses administered in steps. For dosing during geological burial, the probability of thermal release per unit time for a given trapped electron, given by

$$\alpha = s \cdot \exp(-E/kT) \tag{6.55}$$

is used, while for laboratory dosing this probability is set to zero. The results for this latter calculation were confirmed [402] using the standard computer program with a total dose achieved in a 10 s period and a much higher continuous dose rate. In order to present the relevant behavior in its simplest form, a hypothetical model for charge trafficking during the dosing and subsequent glow periods was adopted, which consisted of a single type of active electron trap, but with a continuous distribution of activation energies, and a single type of recombination center. The physical parameters were taken from the above mentioned work [420], using only the green member of the pair of recombination centers; the relevant parameters are the same as above. The program assumes doses administered in steps of 10^7 s duration, delivering 1.54×10^{-4} Gy and generating the appearance of 0.96 electron–hole pairs in the conduction and valence bands. These values are based on the mentioned study [420]; note that the electron–hole pairs created when corrected for counting efficiency yield $10\,\text{cm}^{-3}\,\text{s}^{-1}$. The calculation for a given total dose results in the number and distribution of trapped electrons remaining at the end of dosing. This is the starting point for calculating the TL glow curves or isothermal decay curves.

To generate a point in a plot of true age versus apparent or laboratory age, the dosing calculation is done twice, each time resulting in the same number of remaining trapped electrons and therefore (in the absence of thermal quenching), the same area under the glow curve, assuming that this signal is used in the dating procedure. One calculation is done with $\alpha = 0$, representing laboratory dosing administered so rapidly that no thermal loss of trapped electrons occurs, and the time to reach the required number of trapped electrons is the apparent age. The other simulation is performed with α given by Equation (6.55) with T appropriate for the burial condition, thus turning on the thermal decay and allowing for the loss of trapped electrons. The time required to reach the same concentration of trapped electrons in spite of this loss is then the true age to be associated with the laboratory dosing signal or the resulting apparent age.

Hornyak *et al.* [402] defined the quantity Δ, the true age minus the apparent or laboratory age, and showed by numerical simulation that it is very strongly dependent on the burial temperature, in addition to the obvious dependence on the width parameter $a = 2.77(\text{eV})^{-2}$ appearing in Equation (6.44).

Figures 6.12 and 6.13 show these results for Δ as a function of the apparent age for some typical cases of burial temperature and FWHM (full width at half-maximum) values.

Figure 6.14 displays the same results showing the TL signal (area under the glow curve) in units of n/N as a function of total dose, both under burial and laboratory dosing conditions. As an example, the indicated TL signal corresponding to $n/N = 0.732$ would suggest a sample age of 300 ka based on laboratory dosing whereas the true burial age would be 90 ka older. Contrary to the model employed by Xie and Aitken [401], the ratio of the TL signal generated by a laboratory dose to that developed under burial conditions by the same dose is not a constant, but rather, the ratio increases sharply with increasing dose. For the case illustrated, samples less than 10 ka in age would involve negligible values of Δ.

Figure 6.15 shows the distributions of trapped electrons for the same value of n/N selected as the illustration in Figure 6.14. The electron density is per mg $\times$ eV (uncorrected for detection efficiency). Under laboratory dosing, the distribution has the same Gaussian shape as that for the trap activation energies. Under burial conditions for the same value of n/N, the concentration of the low energy electrons is severely reduced while that of the higher energy electrons is correspondingly increased as a consequence of retrapping. The

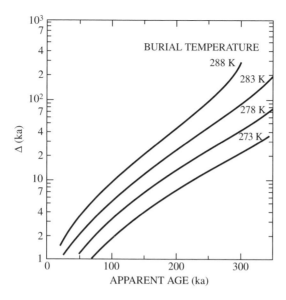

Figure 6.12 *The quantity* Δ = *true age – apparent age as a function of the burial temperature. The FWHM of the activation energy distribution is 0.186 eV and the central energy* $E_0 = 1.450$ *eV. Reprinted from Hornyak, W., Franklin, A. and Chen, R., A model for mid-term fading in TL dating, Ancient TL, 11, 21–26 (1993)*

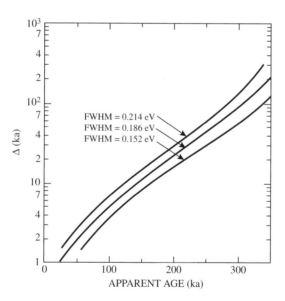

Figure 6.13 *The quantity* Δ *as a function of the apparent age for different values of the FWHM. In all cases, the central energy of the distribution is* $E_0 = 1.450$ *eV. The burial temperature is taken to be 283 K. Reprinted from Hornyak, W., Franklin, A. and Chen, R., A model for mid-term fading in TL dating, Ancient TL, 11, 21–26 (1993)*

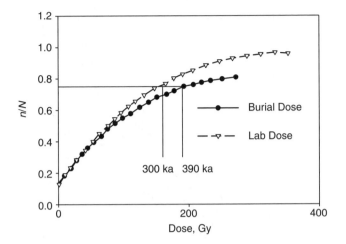

Figure 6.14 *The relative trapped electron population density n/N as a function of applied dose for both burial and laboratory dosing situations. The burial temperature is 283K and the FWHM = 0.186 eV with E_0 =1.450 eV. For the selected value, n/N=0.732, the corresponding TL signal (area under glow curve) would indicate the true burial age to be 390ka for the same TL signal. Reprinted from Hornyak, W., Franklin, A. and Chen, R., A model for mid-term fading in TL dating, Ancient TL, 11, 21–26 (1993)*

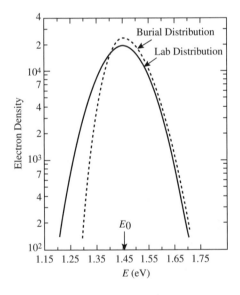

Figure 6.15 *The distribution of trapped electrons for the same values of n/N or TL signal as illustrated in Figure 6.14 contrasting the laboratory and burial dosing. The units for electron density refer to a total number of detectable electrons of 7.32×10^5 mg^{-1}. Reprinted from Hornyak, W., Franklin, A. and Chen, R., A model for mid-term fading in TL dating, Ancient TL, 11, 21–26 (1993)*

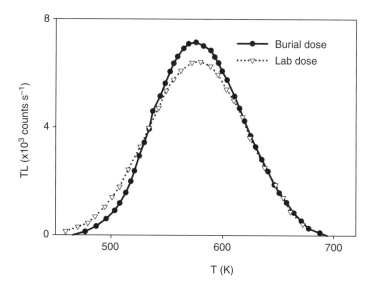

Figure 6.16   *The two glow curves corresponding to the electron distributions in Figure 6.15. Reprinted from Hornyak, W., Franklin, A. and Chen, R., A model for mid-term fading in TL dating, Ancient TL, 11, 21–26 (1993)*

areas under the two curves are equal. Note that the acquired dose is not the same in these two cases. For the same dose, the TL signal for burial dosing would give $n/N = 0.661$, the loss of $\Delta(n/N) = 0.071$ representing the thermal decay during burial.

Figure 6.16 shows the glow curves corresponding to the value of n/N previously selected as an illustration in Figure 6.14 and displayed as electron distributions in Figure 6.15. Here too, as required, the areas under the glow curves are equal. Although the peak position and FWHM of the two curves are almost the same, the effect of the reduced low energy electron population is clearly evident in the steeper initial rise for the burial situation.

Hornyak *et al.* [402] report that all these calculations were also repeated for the two-recombination-center model for quartz mentioned above (see also Hornyak *et al.* [420]). This study demonstrated that the 375°C peak in quartz involved a distribution of activation energies for the electron traps involved. Results very much like those reported in Figures 6.12–6.16 followed. One new feature reported was the substantially larger mid-term effect noted for the UV emission compared with that in the green. This behavior is attributed to the fact that the glow peak observed in the UV appears at a lower temperature by about 20°C, than the glow peak observed in the green, and hence is more sensitive to the loss of electrons with low activation energies under burial conditions.

6.6   Photo-transferred Thermoluminescence (PTTL)

In TL, the sample is always excited by irradiation or otherwise at a relatively low temperature, and the glow peaks occur as the sample is heated up. Another variant of the same effect is PTTL, also termed photo-stimulated TL, which can be performed in one of two ways. In both methods one can monitor carriers held in deep traps by transferring them optically into shallower traps, from which they can be released thermally during the heating of the sample yielding the PTTL peaks. In one version the sample is irradiated for example at RT, then cooled down for example to 80 K and illuminated by UV light. A subsequent heating of the sample may yield TL peaks that would not be observed following the UV exposure at 80 K alone (see e.g. Stoddard [423], Braner and Israeli [424], Kos and Nink [425]) . The explanation of the effect is that carriers captured by deep traps at RT are released by the UV light into the conduction/valence band and trapped at shallower traps which later yield PTTL peaks between 80 K and RT. Braner and Israeli [424] used this effect to distinguish between TL peaks associate with electron and hole trapping in some alkali halides as explained in Section 6.11. A similar method for distinguishing between electron and hole traps in CaF_2:Mn has been used by Wang *et al.* [426]. In the same way, Schlesinger [427] studied the electron and hole traps in quartz. The other version of the effect is that following, for instance, X-ray exposure at RT, the sample is heated up to 200 °C, possibly yielding TL peaks. The sample is then cooled back to RT and exposed, for example, to UV light. The UV light may transfer carriers from the deep into the shallower traps as explained above, and in a subsequent heating, TL peaks may reappear during heating from RT to 200 °C, peaks that are not excitable by UV only. The latter version can be used for dosimetry of deep traps, which are thermally more stable than the shallow traps. The main point here is that TL peaks associated directly with the deep traps may not be easily measurable, since they would occur at high temperature where the measurement is plagued by black-body radiation of the heater and sample. PTTL has been reported in several materials. For example, Crippa *et al.* [428] described the effect in KBr single crystals. Caldas and Mayhugh [429], Pradhan and Bhatt [430] and Goyet *et al.* [431] reported the effect in $CaSO_4$:Dy. A strongly sublinear dependence on the initial X-ray dose over several orders of magnitude of the excitation dose has been reported in this material (see Section 8.7). However, Goyet *et al.* [431] recommend this "high temperature dosimetry" associated with deep traps which yield directly a peak at 600 °C, by performing measurements at ∼240 °C. Puite [432] discovered the effect of UV transfer in γ-irradiated CaF_2:Mn and later, Allen and McKeever [141] reported on PTTL in CaF_2:Mn (TLD-400) and found that maximum transfer rate occurs at 280 nm, which corresponds to the wavelength of an absorption peak. The authors discuss possible models based on optical stimulation of electrons and diffusion of excited F-centers. Also, Lakshmanan and Vohra [433] reported on PTTL in Mg_2SiO_4.

PTTL has also been reported in the main dating material, quartz. Wheeler [434] reported on the possible use of PTTL in quartz for dating of geological samples. Milanovich-Reichhalter and Vana [435] described the PTTL from fine-grain samples. Three peaks at 100 °C, 150 °C and 195 °C were reported following α- or β-irradiation and after a first read-out up to 245 °C and UV illumination. The PTTL peaks were shown to be filled from the well known 325 °C peak. The PTTL emission was mainly at 315 and 275 nm. Alexander and McKeever [436] gave a theoretical account of PTTL, and demonstrated the conditions under which the signal is proportional to the electron concentration in the relevant trap at the end of the illumination period, as previously proposed by Wintle and Murray [437]. These authors give particular attention to the dependencies of the PTTL signal on the illumination time and the wavelength used for photo transfer. Comparison with experimental

PTTL from crystalline quartz is discussed. Alexander and McKeever use a model of two trapping states and two recombination centers and making some simplifying assumptions, they reach equations which yield the normalized maximum PTTL as a function of illumination time. These are compared with the numerical solutions of the original coupled rate equations describing the flow of carriers between the different levels. For a given set of trapping parameters, the numerically predicted curve and the analytically predicted one, both as a function of illumination time are in quite good agreement, yielding a maximum at the same illumination time.

PTTL has also been reported in LiF TLD-100. Sunta and Watanabe [438] were the first to report the effect in LiF TLD-100, first γ-irradiated and then annealed at different temperatures and then illuminated by UV light at RT. The intensity of the PTTL was found to depend on the temperature of annealing. Moharil and Kathuria [439] and Bhasin *et al.* [440] described the effect in γ-ray irradiated samples, annealed at 350 °C and exposed to UV at RT. Charge from the peak XII traps had been thought to be retrapped in shallower traps and subsequently yield a multi-peak glow curve. However, these authors suggest that these traps have only a catalytic role in the transfer process, and that the transfer is from unidentified centers. Osvay *et al.* [441] discussed the effect in TLD-100 as well as other dosimetric materials such as CaF_2:Mn and Al_2O_3:Mg, Y and compared the PTTL by UV light to photo-induced thermoluminescence (PITL), which is the TL directly excited by UV light in a fully annealed material. A later report on PTTL in LiF:Mg, Ti was given by Yazici *et al.* [442].

The effect was also described in calcite by Lima *et al.* [443]. The samples were first excited by γ-irradiation and then thermal bleaching was performed to erase the two lower temperature peaks. Subsequent UV exposure caused the reappearance of the erased peaks, at the expense of peaks related to the deeper traps.

Liu *et al.* [444] found the PTTL effect in $PbWO_4$(PWO). Here the samples were γ-irradiated at RT, annealed for 20 min at 300 °C, cooled back to RT and then illuminated by visible light. A number of TL peaks in PWO and PWO:Y reappeared due to photo transfer. The authors report that doping with Y^{3+} reduces the concentration of traps and subsequently the TL intensity, but enhances the stability of the light yield.

Iacconi *et al.* [445] described the effect in α-Al_2O_3. When an unirradiated sample is illuminated with 253.7 nm light at 77 K, only a very weak TL peak is observed at ~295 K. If it is X-irradiated at RT, the glow peak exhibits main peaks at 365, 408 and 531 K and a weak shoulder around 450 K. If the RT irradiation is followed by cooling to 77 K, very weak peaks appear at 113 and 238 K. If following RT X-irradiation the sample is cooled to 77 K and illuminated with 253.7 nm light, and then heated up, five main peaks are seen at 113, 238, 365, 408 and 531 K. The authors conclude that PTTL in α-Al_2O_3 qualifies for dosimetry use with two particular advantages. One is dispensing with the heating of the sample beyond ambient temperature, thus eliminating IR emission. The other is the possibility of repeated measurements of the dose following one exposure to the excitation irradiation. Some more details of TL and PTTL in α-Al_2O_3 have been reported by Lapraz *et al.* [446]. Kharita *et al.* [447] described the PTTL in LiF:Mg,Cu,P which occurred when γ-irradiated samples were later exposed to UV light either at RT or at LNT. Walker *et al.* [153] also reported on the PTTL effect in α-Al_2O_3. The optical fading of the main dosimetric TL signal from this material is accompanied by PTTL.

A somewhat similar effect where trapped carriers are transferred from deep to shallow traps by applying plastic deformation on previously X-irradiated KBr samples has been described by Panizza [448].

As described by Alexander *et al.* [449] (see also p. 230 in Chen and McKeever [33]) the simplest model to describe photo transfer includes one deep trap from which the electrons are transferred, one shallow trap into which the charge is transferred and one recombination center (see Figure 8.1). Let us denote by n_1 and n_2 the concentrations of electrons in the shallow and deep traps, respectively, and by m the concentration of holes in the centers. We can start the analysis of PTTL from the initial condition that $n_{10} = 0$ and $n_{20} = m_0$ after the irradiation and before the illumination. If the rate of optical stimulation is f, the following rate equations govern the process during the illumination

$$\frac{dn_2}{dt} = -n_2 f + n_c(N_2 - n_2)A_2, \tag{6.56}$$

$$\frac{dn_1}{dt} = n_c(N_1 - n_1)A_1, \tag{6.57}$$

$$\frac{dm}{dt} = -n_c m A_m = \frac{m}{\tau}, \tag{6.58}$$

where $\tau = (n_c A_m)^{-1}$ is the lifetime of captured holes in centers, which is assumed to be constant. With the additional quasi-equilibrium assumption and asserting that negligible retrapping takes place [$n_2 f \gg n_c(N_2 - n_2)A_2$], the solutions of these equations are

$$n_2(t) = n_{20} \exp(-ft), \tag{6.59}$$

$$n_1(t) = N_1 \left[1 - \exp(-Bt)\right], \tag{6.60}$$

$$m(t) = m_0 \exp(-t/\tau), \tag{6.61}$$

where $B = A_1 n_c$. Obviously, B and τ can be considered approximately constant only as long as the quasi-steady approximation holds. After an illumination time of $t = t^*$, the concentrations in the traps and centers will vary according to Equations (6.59–6.61). Following the optical stimulation, the situation resembles that of TL with competition during heating as discussed by Kristianpoller *et al.* [450]. The relevant equation given by these authors for TL is

$$S = m_0 - m_\infty \cong \left[A_m/A_2(N_2 - n_{20})\right] m_0 n_{10}, \tag{6.62}$$

where S is the area under the peak as discussed in detail in Section 8.3. [see Equation (8.34)]. Assuming a quasi-steady situation, $n_1(t^*) \ll N_2 - n_2(t^*)$, one can rewrite the equation for the present situation as

$$S(t^*) = \frac{Cm(t^*)n_1(t^*)}{[N_2 - n_2(t^*)]}, \tag{6.63}$$

where C is a constant, or

$$S(t^*) = \frac{Cm(t^*)n_1(t^*)}{[N_2 - n_2(t^*)]} = \frac{C \exp(-t^*/\tau) N_1[1 - \exp(-Bt^*)]}{\left[N_2/n_{20} - \exp(-ft^*)\right]}. \tag{6.64}$$

For more details, see page 231 in Chen and McKeever [33]. With the mentioned assumptions, Equation (6.64) describes the area under the PTTL peak as a function of the illumination time. Note that as $t^* \to \infty$, $n_2 \to 0$ and $m \to n_1$. S will start from 0 at $t^* = 0$ and will increase smoothly to a maximum of $S(\infty) = C n_{1\infty}/N_2$, where $n_{1\infty} = m_\infty$ are the final concentrations at $t^* = \infty$.

As opposed to the monotonic increase described above, experimental PTTL curves versus time often grow up to a maximum, and then decrease at longer illumination times. A possible explanation is that the illumination, while filling the shallow traps at the expense of the deeper ones, also empties them through an optical bleaching process. This process has been described in Al_2O_3 by Walker *et al.* [153]. Alexander *et al.* [449] described this effect in synthetic quartz and gave a detailed model to explain it. Assuming the same model with one center and two traps, they included the possibility that electrons be released at a rate of f_1 from n_1 and a rate of f_2 from n_2 for a given intensity of light incident on the sample. Including also the possibility that electrons are released thermally from n_1 during the optical stimulation, they wrote the coupled rate equations

$$\frac{dn_1}{dt} = n_c(N_1 - n_1)A_1 - n_1 s_1 \exp(-E/kT) - n_1 f_1, \tag{6.65}$$

$$\frac{dn_2}{dt} = -n_2 f_2 + n_c(N_2 - n_2)A_2, \tag{6.66}$$

$$\frac{dm}{dt} = -A_m m n_c, \tag{6.67}$$

$$\frac{dn_c}{dt} = \frac{dm}{dt} - \frac{dn_1}{dt} - \frac{dn_2}{dt}. \tag{6.68}$$

When the temperature is low enough, the term $n_1 s_1 \exp(-E/kT)$ can be ignored. For $f_1 = 0$, Alexander *et al.* report that the PTTL intensity as a function of temperature is always a monotonically increasing function of the time of illumination, up to a steady-state saturation level. On the other hand, if $f_1 > 0$, the numerical solutions of these simultaneous differential equations show that the PTTL intensity reaches a maximum and then declines and asymptotically decays to zero. An example is shown in Figure 6.17. The parameters used here were: $N_1 = N_2 = M = 10^{11} \, cm^{-3}$; $A_1 = 10^{-9} \, cm^3 \, s^{-1}$; $A_2 = 10^{-10} \, cm^3 \, s^{-1}$; $A_m = 10^{-7} \, cm^3 \, s^{-1}$; $s_1 = 5 \times 10^{11} \, s^{-1}$; $E_1 = 0.9 \, eV$; $T = 100 \, K$; and $f_2 = 0.1 \, s^{-1}$. These authors mention that many real examples exist in which the observed PTTL first grows with illumination time and then decreases but not to zero, which is difficult to explain within the framework of this model. A more complex model by Bøtter-Jensen *et al.* [451] explains successfully OSL and PTTL sensitivity changes in quartz. These authors add an additional deep trap and a nonradiative recombination center to the model. The extra deep trap is not optically active, but it does help in explaining sensitivity changes. The additional nonradiative recombination center is important in explaining the dose-dependence effect. It provides an additional competing channel and changes the charge neutrality condition which now is

$$n_c + n_1 + n_2 + n_3 = m_1 + m_2, \tag{6.69}$$

where n_3 is the concentration of electrons in the deep, and optically inactive electron trap, and m_1 and m_2 are the concentrations of holes in the radiative and non radiative recombination centers, respectively. As described by McKeever [452, 453], the reduction in the PTTL with

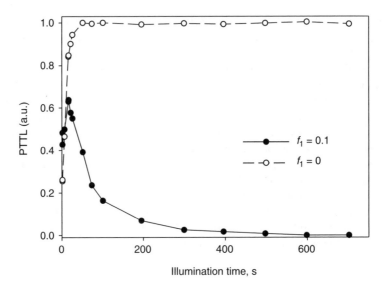

Figure 6.17 *Simulated curves of PTTL versus illumination time using the model described by Equations (6.65–6.68). The values of the parameters are given in the text and the values of f_1 are given in the figure. Reprinted from Alexander, C. S., Morris, M. F., and McKeever, S. W. S., The time and wavelength response of phototransferred thermoluminescence in natural and synthetic quartz , Radiat. Meas. 27, 153. Copyright (1997) with permission from Elsevier*

the illumination results from the removal of holes during the illumination period from the radiative recombination centers. This takes place by recombination and is also the source of the OSL signal observed during illumination. Note that in the model with one recombination center, n_1 cannot be larger than m because of the charge neutrality condition. Thus, if n_1 increases during illumination, there are always enough holes available so that the resulting PTTL follows n_1. However, in the more complex model, n_1 may be either smaller or larger than m_1, according to the new charge neutrality equation (6.69). This means that even if n_1 is increasing, for a given illumination time there may not be enough holes in the radiative center to accommodate for this increase. Obviously, the total number of available holes $m_1 + m_2$ is at least the same as the number of electrons in the shallow traps, but the PTTL will be dictated by the number of holes available in the radiative center, m_1.

At the start of illumination, the number of electrons in the shallow traps will be less than the number of holes in the radiative recombination center and one may expect the PTTL intensity to grow along with n_1. At longer times, n_1 may become larger than m_1 and the PTTL intensity may now be expected to follow m_1 and decrease as m_1 decreases. In other words, one might expect the PTTL intensity I_{PTTL} to be described by [see Equation (8.10) and discussion in Section 8.3]

$$I_{PTTL} = \min(n_1, m_1). \tag{6.70}$$

As pointed out by Alexander *et al.* [449], when several levels are involved and competition takes place, this equation may not hold and in order to understand the behavior for a certain

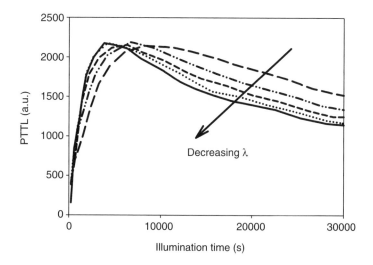

Figure 6.18 *Simulated curves of PTTL versus illumination time for several wavelengths using the complex model described in the text. The parameters used are given in the text. The wavelength varied from 300 to 440 nm. Reprinted from Alexander, C. S., Morris, M. F., and McKeever, S. W. S., The time and wavelength response of phototransferred thermoluminescence in natural and synthetic quartz, Radiat. Meas. 27, 153. Copyright (1997) with permission from Elsevier*

model and a given set of parameters, one has to solve the relevant set of equations in order to get the PTTL behavior. The parameters chosen were: $N_1 = N_2 = N_3 = M_1 = M_2 = 10^{11}\,cm^{-3}$; $A_1 = 10^{-9}\,cm^3\,s^{-1}$; $A_2 = A_3 = 10^{-10}\,cm^3\,s^{-1}$; $A_{m1} = 10^{-7}\,cm^3\,s^{-1}$; $A_{m2} = 10^{-8}\,cm^3\,s^{-1}$; $f_1 = 0$; $f_2 = 0.1\,s^{-1}$; $s_1 = 5 \times 10^{11}\,s^{-1}$; $E_1 = 0.9\,eV$; and $T = 200\,K$. Figure 6.18 shows the numerical results for this complex model. The characteristic behavior is of an increase in the intensity up to a maximum value and then a decrease toward a non-zero value at higher illumination times. At shorter wavelengths both the growth and decay are faster than at longer wavelength. The effect of wavelength was introduced by taking into account that the excitation rate is given by the product of the photoionization cross-section σ and the intensity of the illuminating light ϕ, i.e., $f = \sigma\phi$. The photoionization cross-section for the excitation of electrons from deep traps into the delocalized band is given by

$$\sigma = K(E_0)^{1/2} \left[(h\nu - E_0)^{3/2} / h\nu(h\nu - \gamma E_0)^2 \right], \tag{6.71}$$

where γ is a constant dependent on the electron effective mass. For quartz $\gamma = 0.559$. K is a constant, E_0 is the optical threshold energy for ionization, taken here to be 2.48 eV, and $h\nu$ is the photon energy (see Bäer [454]). Note that this expression is rather similar to Equation (5.32).

It is interesting to note that a decrease in the PTTL intensity, as shown in Figure 6.18, is obtained even if there is no optical excitation out of the shallow traps. The steady-state PTTL is not zero in the examples shown, a situation associated with the fact that there are no more electrons to be excited from the source traps, namely $n_2 \to 0$ as $t \to \infty$. Depending on the relative initial values of n_2 and m_1, one could have a non zero steady-state value of

m_1, and therefore the PTTL steady-state value must be non zero. A similar effect of photo transfer leading to OSL has been described by Umisedo *et al.* [455] who made a comparison between blue and green stimulated luminescence in Al_2O_3:C.

6.7 TL Response of Al_2O_3:C to UV Illumination

Yukihara *et al.* [456] carried out a comprehensive experimental study of the effect of deep traps on the TL of Al_2O_3:C by using both β-irradiation and UV illumination. The concentration of F- and F^+-centers in the samples were monitored by optical absorption measurements, and competing deep holes and deep electron traps were identified. The UV data showed no decrease of the TL signal at high UV fluences, in contrast to the drop in the corresponding TL signal observed at high β doses. Irradiation with UV (206 nm) excites electrons from the F-centers into the conduction band, and produces free electrons that can be captured by electron traps, at the same time causing a conversion of the F-centers into F^+-centers. The role of the same deep traps on the sensitivity of TLD-500 (Al_2O_3:C) has been further studied by Nikiforov and Kortov [457].

Pagonis *et al.* [39] extended their previous TL and OSL modeling work on Al_2O_3:C, to simulate the UV illumination experiments of Yukihara *et al.* [456]. Their simulations showed that the same model and kinetic parameters used to describe TL/OSL behavior, can also describe the variations of TL versus UV fluence, as well as the absorption coefficient K versus UV fluence. The model of Pagonis *et al.* [39] is identical to the one shown in Figure 8.21. Since no free holes are created during the UV experiments, one can set $dn_v/dt = n_v = 0$ and the set of simultaneous differential equations governing the UV excitation and heating stages of the simulation becomes

$$\frac{dm_1}{dt} = R\sigma(M_1 - m_1) - A_{m1}n_cm_1, \tag{6.72}$$

$$\frac{dm_2}{dt} = -A_{m2}n_cm_2, \tag{6.73}$$

$$\frac{dn_1}{dt} = A_{n1}n_c(N_1 - n_1) - s_1\exp(-E_1/kT)n_1, \tag{6.74}$$

$$\frac{dn_2}{dt} = A_{n2}n_c(N_2 - n_2), \tag{6.75}$$

$$\frac{dn_v}{dt} = 0, \tag{6.76}$$

$$\frac{dm_1}{dt} + \frac{dm_2}{dt} + \frac{dn_v}{dt} = \frac{dn_1}{dt} + \frac{dn_2}{dt} + \frac{dn_c}{dt}. \tag{6.77}$$

In Equation (6.72) R represents the photon flux in photons per cm^2 per s, and σ is the absorption cross-section in cm^2. The term $(M_1 - m_1)$ expresses the number of available F-centers (which the UV turns into F^+-centers), m_1 (cm^{-3}) is the instantaneous concentration of F^+-centers and $M_1(cm^{-3})$ is the total concentration of these radiative hole centers in the material. The product $R\sigma(M_1 - m_1)$ represents the rate of creation of the F^+-luminescence centers in the material by UV illumination and its units are centers per cm^3 per s. An estimate of the value of σ in Equation (6.72) was obtained from the experimental data of Yukihara *et al.* [456] by using the value of the F-center absorption coefficient for sample Chip101 $\left(45\,cm^{-1}\right)$, and the thickness of the samples (0.09 cm), which corresponds to four optical absorption lengths. The thickness of the samples introduces a spatial non uniformity in the UV illumination of the samples, with the UV fluence received by the front surface being much larger than the one received by the back surface. By using the experimental value of $M_1 = 10^{17}\,cm^{-3}$, one can estimate the value of $\sigma = (45\,cm^{-1})/(10^{17}\,cm^{-3}) = 4.5 \times 10^{-16}\,cm^2$. The values of the parameters in Equations (6.72–6.77) are the same as used previously to simulate the TL and OSL response of sample Chip101 to β-irradiation (see Table 1 in Pagonis *et al.* [38]).

The rest of the parameters are as follows: $M_2\,(cm^{-3})$ is the concentration of non radiative hole centers with instantaneous occupancy of $m_2\,(cm^{-3})$, $N_1(cm^{-3})$ is the concentration of the electron dosimetric trapping state with instantaneous occupancy of $n_1\,(cm^{-3})$ and $N_2\,(cm^{-3})$ is the concentration of the deep electron trapping state with instantaneous occupancy of $n_2\,(cm^{-3})$. n_c and n_v are the concentrations (cm^{-3}) of electrons and holes in the conduction and valence bands, respectively. A_{m1} and $A_{m2}\,(cm^3\,s^{-1})$ are the recombination coefficients for free electrons with holes in centers 1 and 2 and $A_{n1}\,(cm^3\,s^{-1})$ is the retrapping coefficient of free electrons into the dosimetric trapping state N_1. $A_{n2}\,(cm^3\,s^{-1})$ is the retrapping coefficient of free electrons into the competing trapping state N_2.

The simulation contains the irradiation stage for time t_D, a relaxation time of 1s, and the heating stage with a constant linear heating rate $\beta = 1\,K\,s^{-1}$. If the time of excitation is t_D, then $D = Rt_D$ represents the total UV fluence in photons per cm^2. The photon flux in Equation (6.72) is taken equal to some arbitrary value of $R = 1.1 \times 10^{16}$ photons $cm^{-2}\,s^{-1}$, and the irradiation time t_D was varied from 0.005 to 1s in steps of 0.05 s. This results in a range of calculated UV fluences between 5.5×10^{13} and 1.1×10^{16} photons cm^{-2}. The set of equations (6.72–6.77) is valid for both the irradiation and the heating stages of the simulation. During irradiation at room temperature the term $s_1 \exp(-E_1/kT)n_1$ in Equation (6.74) is negligible, while during the heating stage one sets $R = 0$ in Equation (6.72). The TL intensity $I(T)$ is associated with the electron–hole recombination in the recombination center m_1, and is given by

$$I(T) = A_{m1}m_1n_c\eta(T). \qquad (6.78)$$

The temperature-dependent factor $\eta(T)$ describes the thermal quenching of the TL intensity and is given by (see Section 2.9).

$$\eta(T) = \frac{1}{1 + C_1 \exp(-W_1/kT)}. \qquad (6.79)$$

The thermal quenching constants $W_1 = 1.1\,eV$ and $C_1 = 10^{11}$ for Al_2O_3 are known from experimental studies of thermal quenching effects in this material [78, 458].

Figure 6.19 shows the simulation results of Pagonis *et al.* [39] for the dependence of the TL signal on the UV fluence as solid lines, together with the experimental data in Yukihara *et al.* [456]. The TL signal is defined as the maximum TL intensity. The simulation also produced a satisfactory description of the corresponding variation of the concentration of holes in the luminescence center (m_1) as a function of the UV fluence for sample Chip101. The value of m_1 is proportional to the absorption coefficient K of this material. Subedi

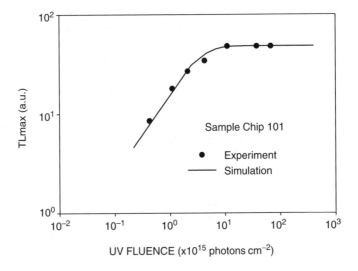

Figure 6.19 *Comparison of the experimental data for the TL intensity versus UV fluence for sample Chip101 by Yukihara et al. [456]. The solid curve indicates the calculated values using the kinetic model of Pagonis et al. [39]. Reprinted from Pagonis, V., et al., A quantitative kinetic model for Al₂O₃:C: TL response to UV-illumination, Radiat. Meas. 43, 175–179 Copyright (2008) with permission from Elsevier*

et al. [459] used numerical simulations to describe the effect of thermal quenching on TL. In particular, they have studied the shift of the quenched glow peak with heating rate and its effect on the various heating rate methods, the influence on the symmetry factor and the apparent kinetic order of the glow peak, and the effect of thermal quenching on the initial-rise and peak-shape methods.

6.8   Dependence of the TL Excitation on Absorption Coefficient

The experimental results of the dependence of TL excitation spectra and in particular, the effect in ranges of high optical absorption have been discussed in Section 3.2. In this section we discuss the interpretation of this behavior and its consequences concerning the excitation spectrum near an absorption edge using numerical simulation. As an example, let us concentrate on Figure 3.1 taken from Halperin and Nahum [460]. The behavior of this curve in wavelengths longer than 225 nm, where the absorption edge is located, is practically identical to the dependence of the absorption coefficient on the wavelength. This just means that both absorption and TL excitation depend on the ability of light to raise electrons from the valence band into the conduction band. The situation is entirely different at shorter wavelengths. Here, the absorption of light is very strong and remains so when the wavelength is further decreased. The fast decrease in the excitation efficiency is understood to be related to the fact that TL is a volume effect and that due to the very strong absorption, the light cannot penetrate the sample at the short wavelengths. Therefore, the effective volume that takes part in the process is very small, which accounts for the fact that below 200 nm, the absorption in the semiconducting diamond is very strong and the TL peak is hardly excitable. Chen *et al.* [461] studied theoretically the dependence of TL excitation in a range of rather high optical absorption and under the assumption that existing traps are being filled (rather than new traps and centers-being created by the irradiation). Let us consider the absorption coefficient μ to be given as a function of the excitation wavelength, i.e., $\mu = \mu(\lambda)$. Suppose that one has a sample with two parallel flat faces and that light is incident perpendicular to one of these faces. For a given wavelength, the intensity at a depth x in the sample is $I = I_0 \exp(-\mu x)$ where I_0 is the incident light intensity. We neglect the reflection from the first face of the sample; actually, I_0 should denote the amount of light which penetrates the sample. Let us also neglect the reflection from the second flat surface; in fact, for the ranges of μ which are of interest, hardly any light will reach the second surface. The concentration of free carriers n at a depth x is proportional to the number of absorbed photons, therefore

$$n = -\frac{dI}{dt} = I_0 \exp(-\mu x) \tag{6.80}$$

where the proportionality factor is set to unity. Let us assume that μ does not vary during the excitation, therefore, $n(x)$ does not depend on time during a constant photon flux. Suppose that the concentration of traps, which if filled by carriers would contribute to the measured glow, is M. Let $m(x, t)$ be the concentration of trapped carriers at a depth x after irradiation time t, and assume that the generated carriers may be trapped only in close vicinity to where they are formed, we get the following equation

$$\frac{\partial m}{\partial t} = \alpha n(M - m), \tag{6.81}$$

where α $(cm^3\ s^{-1})$ is a proportionality factor. This equation states that at any depth x, the rate of change of the concentration of trapped carriers is proportional to the local concentrations of free carriers and the still empty local trapping states. α here is similar in nature to the recombination or retrapping probability coefficients, namely, it can be described as the product of the thermal velocity of the free carriers and the relevant cross-section. Assuming $m(x, 0) = 0$ we can solve Equation (6.81) and insert the expression for n from Equation (6.80) to get

$$m(x, t) = M \left\{ 1 - \exp\left[-\alpha t I_0 \mu \exp(-\mu x)\right] \right\}. \tag{6.82}$$

$I_0 t$ is the incident radiation dose D. The total population M_t can be found by integration over x from 0 to the width of the sample, L,

$$M_t = \int_0^L m(x, y)dx = M \int_0^L \left\{ 1 - \exp\left[-\alpha D \exp(-\mu x)\right] \right\} dx. \tag{6.83}$$

We can usually assume that the TL intensity is proportional to M_t, which is related to the plausible assumption that the light emitted during heating is not absorbed by the sample itself. Thus, the behavior of M_t represents that of the glow intensity. From Equation (6.83) one can see that M_t is zero for $\mu = 0$, the case of no absorption (the long wavelength side, e.g. see Figure 3.1), and for $\mu = \infty$, where the effective volume is zero. From Equation

(6.83) it is also apparent that $M_t = 0$ for $D = 0$ and $M_t = ML$ for $D = \infty$, as might be expected.

For low doses, $\alpha D\mu \exp(-\mu x) \ll 1$ and Equation (6.83) reduces to

$$M_t = Ma\mu D \left[1 - \exp(-\mu L)\right],\tag{6.84}$$

which means that the dose dependence is linear, as is often found at low doses. By a change of variable, we find from Equation (6.83)

$$M_t/(ML) = 1 - (1/\mu L) \int_a^b \left[\exp(-y)/y\right] dy,\tag{6.85}$$

where $a = \alpha D\mu \exp(-\mu L)$ and $b = \alpha D\mu$.

The exponential integral on the right-hand side can be numerically evaluated or found in tables for any set of values of αD, μ and L. Note that the method for evaluating this integral as described in Chapter 15 is usually not applicable for the present case since the asymptotic series given for $\int_x^\infty (e^{-u}/u)du$ yields good results for $x \gtrsim 10$, which is not the case here.

Figure 6.20 shows the dependence of M_t/ML on μ. The curves (a–e) were found for different values of the excitation dose, varying by factors of $\sqrt{10}$. The maximum of the curve is seen to shift toward lower values of μ for higher doses. As could be expected, the curve in (e) flattens due to total saturation in a certain range of μ. The dependence of M_t/ML on the dose for a given sample and given excitation wavelength (namely, given μ)

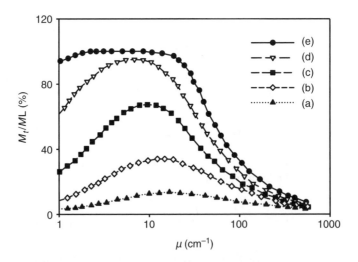

Figure 6.20   *Calculated dependence of M_t/ML on μ for $L = 0.1$ cm. The doses change from (a) to (e) by factors of $\sqrt{10}$. Reprinted from Chen, R., Israeli, M., Kristianpoller, N., Dependence of the excitation of glow curves on the absorption coefficient , Chem. Phys. Lett. 7, 171. Copyright (1970) with permission from Elsevier*

can be seen by examining the values of the intersection of each curve in Figure 6.20 with a vertical line at the desired value of μ.

As pointed out by Chen *et al.* [461], the dependence of M_t/ML on the wavelength rather than on the absorption coefficient can be found by combining the present results with given experimental results or theoretically acceptable functions of $\mu = \mu(\lambda)$. A possible theoretical dependence (see e.g. Johnson [462]) is $\mu \propto (h\nu - E_g)^k$ where E_g is the band-gap energy and k assumes the values of $1/2, 3/2, 2$ and 3. Another empirically known function is $\mu \propto \exp\left[(h\nu - E_g)/kT\right]$. It is quite obvious that the steeper the function $\mu = \mu(\lambda)$, the narrower the peak of $M_t = M_t(\lambda)$ which represents the excitation spectrum (see Figure 3.1).

6.9   TL Versus Impurity Concentration; Concentration Quenching

As pointed out above, the TL and OSL effects as well as other luminescence properties are directly related to the occurrence of imperfections, namely impurities and defects in the host crystal. It is quite obvious that one would be interested in finding correlations between the amount of the impurities and the TL and OSL intensities. A comprehensive account on defects and TL has been given in Chapter 5 of the book by McKeever [193]. Note that TL and OSL are always associated with the existence of at least two imperfection states which constitute the recombination center and the trap from which carriers are freed either thermally or optically. Establishing such correlations is not always straight-forward. It is not uncommon that a material includes a large amount of an impurity which does not influence the luminescent properties, whereas other impurities which occur in very small amounts are responsible for the luminescence emission. For example, in diamonds nitrogen impurity is very abundant (up to 1%), but it is not the source of the luminescence effects. On the other hand, small amounts of impurities like boron or aluminum may cause the occurrence of TL at concentrations of about 1 ppm and can cause the observable luminescence effects. Furthermore, one would expect that when a certain imperfection is responsible for the appearance of TL or OSL, higher impurity level would necessarily mean more emitted luminescence. This may not be the case if we consider the two mentioned entities responsible for TL or OSL. If one of these, say the luminescence center, has a given concentration and the concentration of the impurity responsible for the trapping state is increased, this may increase the amount of measured TL up to the point where the limiting factor is the amount of the impurity associated with the center. Furthermore, effects of impurity quenching and concentration quenching have been reported (see e.g. McKeever [193], pp. 150–152), which are described here in some detail.

The effect of concentration quenching of luminescence efficiency was discussed first by Johnson and Williams [463, 464]. Experimental results of the dependence of the luminescence efficiency on the concentration of Mn in ZnF_2 showed an increase with this concentration up to a certain fraction of the activator, a maximum at a certain concentration and a decline of the efficiency at higher concentrations. These authors suggest that on the basis of a random distribution of the activators (luminescence centers), and recognizing that only activators not adjacent to other activators are capable of luminescence, the efficiency of a phosphor can be quantitatively related to the activator concentration and to the capture cross-sections for the excitation energy of luminescent and non luminescence activators and of the host lattice. With these assumptions, the authors developed the following equation for the efficiency η,

$$\eta = \frac{c(1 - c)^z}{c + (\sigma/\sigma')(1 - c)},\tag{6.86}$$

where c is the total mole fraction of the activator and σ', σ are the capture cross-sections of radiative and non radiative centers, respectively, and z is the number of nearest cation neighbors. Equation (6.86) is a peak-shaped function which yields a maximum at a certain concentration, followed by a decrease. While trying to fit this equation to experimental results, most authors considered z and $\gamma = \sigma/\sigma'$ as adjustable parameters. Johnson and Williams [465] reported values of $z = 70$ for the 305 nm fluorescent emission of KCl:Tl and $z = 4000$ [464] for the fluorescent emission of ZnS:Cu. This value seems to be exceedingly large to represent the number of "nearest neighbors", and the authors suggest that this process involves transitions through the conduction band.

Ewles and Lee [466] developed another formula, based on the same assumptions and considering the effect of absorption,

$$E = \frac{K}{1 + \alpha c^{-1} \exp(nc)},\tag{6.87}$$

where E is the measured quantity which is proportional to the efficiency η, K is a normalizing constant, α another constant and $n = 1/c_{max}$ where c_{max} is the dopant concentration corresponding to η_{max}.

This nonmonotonic effect has been reported in several materials in addition to ZnF$_2$:Mn mentioned above. Schulman *et al.* [467] report on such a curve of the luminescence efficiency with a maximum at $\sim$0.1 mole% in KCl:Tl. Ewles and Lee [466] reported on this effect in yellow emission and UV emission in CaO:Bi and CaO:Pb. Van Uitert [468] describes the effect in CaWO$_4$:Tb and in CaWO$_4$:Eu. In the latter, the maximum occurs at different concentrations for different luminescence emission wavelengths. More results of the non-monotonic effect associated with concentration quenching in various materials have been given in a series of papers by Van Uitert *et al.* [469–473].

Similar effects of nonmonotonic dependence of TL on the concentration of the relevant impurity in various materials have been reported by different authors. Medlin [474] described the TL properties of calcite and reported on concentration quenching of the 350 K TL peak due to Mn^{2+} in calcite. He used the same formula, Equation (6.86) utilized before for fluorescence efficiency, and by curve fitting got values of $z = 75$ and $\gamma = 0.001$. For a 300 K peak in Pb^{2-} doped calcite he got $z = 700$ and $\gamma = 0.0001$ and for a 410 K peak in Pb^{2+}-doped calcite he got $z = 150$ and $\gamma = 0.0001$. In another work, Medlin [475] reported on similar results in dolomite. Figure 6.21 depicts an example of the 380 K TL peak in Mn^{2+}-doped dolomite. Similar shaped dependencies were found for the 330, 500 and 600 K peaks with maxima at different concentrations between 0.001 and 0.003 mole fraction of Mn^{2+}.

Some more publications on concentration quenching of TL in different materials will be mentioned briefly. Rossiter *et al.* [476] reported on the concentration dependence of peak 5 at 210 °C in LiF:Ti. They found a peak-shaped dependence on the Ti concentration with a maximum efficiency at $\sim$8 ppm Ti. Nambi *et al.* [477, 478] described the concentration

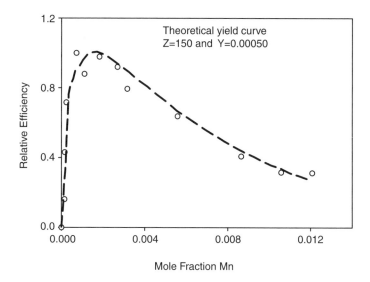

Figure 6.21   *Efficiency of TL as a function of activator concentration for the 380 K glow peak in dolomite due to Mn^{2+}. Reprinted from Medlin, W. L., Thermoluminescence in Dolomite, J. Chem. Phys. 34, 672. Copyright (1961) with permission from American Institute of Physics*

quenching effect in $CaSO_4$ with Dy and Tm impurities. In both cases, the maximum efficiency occurred at $\sim$0.1 wt% of the dopant. The authors fitted their experimental results to the expression given in Equation (6.87). Wachter [479] studied the dependence of the sensitivity of LiF:Mg, Ti (TLD-100) on the ratio of the two dopants, Mg/Ti, and found a peak-shaped dependence with a maximum at a ratio of $\sim$0.32. Lai *et al.* [480] investigated the TL of ZrO_2 doped with Yb_2O_3 and found concentration quenching with a maximum at 5 mol%, followed by a decrease at higher concentrations. Fitting the results to Equation (6.86) yielded values of the adjustable parameters of $z = 1$ and $\gamma = 0.003$. Tajika and Hashimoto [481] studied the blue TL in synthetic quartz with aluminum impurity. The characteristic quenching concentration behavior is observed with a maximum blue TL intensity at $\sim$10 ppm of aluminum. Vij *et al.* [482, 483] reported on the TL of UV-irradiated Ce-doped SrS nanostructures. The concentration quenching curve shows a maximum at a dopant concentration of $\sim$0.5 mol%. In a recent paper Sharma *et al.* [484] describe similar results in the TL of CaS:Ce nanophosphors which show a maximum intensity of TL with $\sim$0.4 mol% of cerium.

It should be noted that the attempts to fit the concentration dependence of TL to functions developed for the effect of concentration quenching of luminescence may not be always justified. Lawless *et al.* [485] have proposed an alternative model, based on the transitions of electrons and holes between trapping states, the conduction band and the valence band.

Figure 6.22 presents the proposed three trap-one recombination center (3T1C) energy-level model which includes three electron trapping states with concentrations of N_1, N_2 and N_3 and instantaneous occupancies of n_1, n_2 and n_3, respectively. The activation energies are E_1, E_2 and E_3 and the frequency factors are s_1, s_2 and s_3, respectively and the retrapping probability coefficients A_1, A_2 and A_3. Also shown is the recombination center with

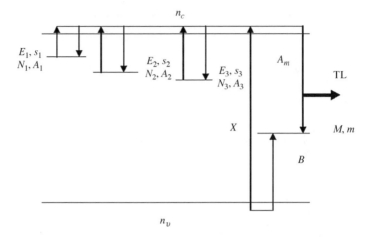

Figure 6.22   *Energy level diagram of the three trap-one recombination center (3T1C) model to explain concentration quenching of luminescence. Reprinted from Lawless, J.L., Pagonis, V. and Chen, R., Mater. Sci. Eng. Eurodim 2010, Pécs, Hungary, Book of Abstracts, 8.5*

concentration M and instantaneous occupancy of holes m. The probability coefficient for holes to be trapped in the center is B and the electron recombination probability coefficient is A_m. n_c and n_v denote the concentrations of free electrons and holes, respectively. X denotes the rate of production of free electron–hole pairs by the excitation dose, which is proportional to the dose rate. The concentration which is varied between different samples is assumed to be that of the recombination center M, whereas the other concentrations of the traps are assumed to remain constant. The excitation dose is also kept the same when simulating the excitation of the differently doped samples.

The set of simultaneous differential equations governing the process during excitation is

$$\frac{dn_1}{dt} = A_1(N_1 - n_1)n_c, \tag{6.88}$$

$$\frac{dn_2}{dt} = A_2(N_2 - n_2)n_c, \tag{6.89}$$

$$\frac{dn_3}{dt} = A_3(N_3 - n_3)n_c, \tag{6.90}$$

$$\frac{dm}{dt} = B(M - m)n_v - A_m m n_c, \tag{6.91}$$

$$\frac{dn_v}{dt} = X - B(M - m)n_v, \tag{6.92}$$

$$\frac{dn_c}{dt} = \frac{dm}{dt} + \frac{dn_v}{dt} - \frac{dn_1}{dt} - \frac{dn_2}{dt} - \frac{dn_3}{dt}. \tag{6.93}$$

In order to follow the experimental procedure, one has to solve this set of equations for a certain period of time t, and the dose applied is proportional to $X \cdot t$. The next stage of the simulation is relaxation, an additional period of time when the excitation is switched off, in which the free electrons and holes remaining in the conduction and valence bands decay into the respective trapping states, thus contributing to the final concentrations. Next, a third stage of TL read-out starts in which the sample is heated up and the emitted light denoted in Figure 6.22 by a thick arrow is recorded as a function of temperature. The governing equations for this stage are

$$\frac{dn_1}{dt} = A_1(N_1 - n_1)n_c - s_1 n_1 \exp(-E_1/kT), \tag{6.94}$$

$$\frac{dn_2}{dt} = A_2(N_2 - n_2)n_c - s_2 n_2 \exp(-E_2/kT), \tag{6.95}$$

$$\frac{dn_3}{dt} = A_3(N_3 - n_3)n_c - s_3 n_3 \exp(-E_3/kT), \tag{6.96}$$

$$\frac{dm}{dt} = -A_m m n_c, \tag{6.97}$$

$$\frac{dn_c}{dt} = \frac{dm}{dt} - \frac{dn_1}{dt} - \frac{dn_2}{dt} - \frac{dn_3}{dt}. \tag{6.98}$$

As seen in Figure 6.22, the TL emission is associated with the recombination of free electrons with holes in the center, thus, when the heating function $T(t)$ is known, the TL intensity can be written as

$$I(T) = -\frac{dm}{dt} = A_m m n_c. \tag{6.99}$$

An example of the numerical results reached in the simulation through the three mentioned stages is shown in Figure 6.23. The parameters used are given in the caption. It should be noted that if we start the excitation stage with empty centers, the non-monotonic concentration quenching effect is not seen. Instead, we assume that the initial occupancy is a given fraction of the concentration of the center, and we choose $m(t = 0) = m_0 = 0.1M$. When the heating stage is simulated, the area under each of the two peaks is recorded and then plotted against the varying concentration of the centers M. The results show that the two simulated glow-peaks areas vary nonmonotonically with the center concentration M, and that the maxima occur at different concentrations, which, as pointed out above, resembles the experimental results by Medlin [474, 475].

In another simulation, the details of which are not given here, it has been found that the concentration quenching effect for a single TL peak, which is an effect found in several materials as described above, can be simulated by a simpler model of two traps and one recombination center (2T1C). The authors [485] suggest that in future work one may check the assertion that the concentration quenching effect of four TL peaks, occurring at different concentrations, as described by Medlin [475] for Mn^{2+}-doped dolomite, can be simulated by a 5T1C model. It should be noted that the assumption made concerning the initial occupancy of the center not being zero has been made previously in relation to TL and OSL by Chen and Leung [486], Yukihara *et al.* [487] and Pagonis *et al.* [488]. As pointed out before (see e.g. Carter [489]), if the energy of the center is near the Fermi level, the center will be partially filled with electrons (and partially with holes), so that we can write $m_0 = \alpha M$, where α has some value determined by Fermi statistics. α can be assumed to be constant if the total center concentration M changes. The results do not seem to depend significantly

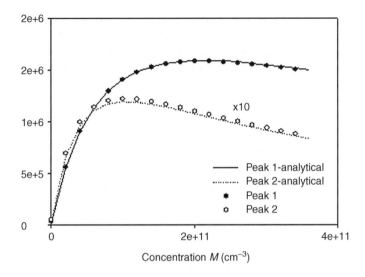

Figure 6.23 *Simulated results of the concentration dependence of the areas under the two peaks, at 91 °C and ~220 °C with: $N_1 = 3 \times 10^8 \, cm^{-3}$; $N_2 = 10^7 \, cm^{-3}$; $N_3 = 10^9 \, cm^{-3}$; $A_1 = 10^{-8} \, cm^3 \, s^{-1}$; $A_2 = 10^{-8} \, cm^3 \, s^{-1}$; $A_3 = 10^{-9} \, cm^3 \, s^{-1}$; $B = 10^{-9} \, cm^3 \, s^{-1}$; $E_1 = 0.97 \, eV$; $E_2 = 1.55 \, eV$; $E_3 = 1.8 \, eV$; $D = 5 \times 10^6 \, cm^{-3}$; $s_1 = 5 \times 10^{12} \, s^{-1}$; $s_2 = 5 \times 10^{12} \, s^{-1}$; and $s_3 = 5 \times 10^{13} \, s^{-1}$. Note that the analytical curves relate to Equations (6.104) and (6.109). Reprinted from Lawless, J.L., Pagonis, V. and Chen, R., A three trap one center model for explaining the concentration quenching of thermoluminescence (TL), Mater. Sci. Eng. Eurodim 2010, Pécs, Hungary, Book of Abstracts, 8.5*

on the strict constancy of α. The location of the Fermi level may change slightly with M. In order to check the sensitivity of the results to the values of α, the simulation with a constant initial occupancy, $m_0 = 10^{11} \, cm^{-3}$ has been run. The results were not the same as before, but the non monotonic concentration quenching effect was still clearly seen for the two simulated TL peaks with the maxima occurring at different concentrations. The main point in this respect remains that the initial occupancy of M is not zero.

Lawless *et al.* [485] have also presented an approximate analytical method of demonstrating the concentration quenching effect within the described models. Starting with the 2T1C model, one considers N_1 and N_2 to be fixed and the concentration of the recombination center M to be variable. For the initial conditions, Lawless *et al.* assumed $n_1(0) = n_2(0) = 0$ and also $m(0) = \alpha M$. Let us assume that the trapping states are far from saturation, $n_1 \ll N_1$ and $n_2 \ll N_2$ and also, n_1 and n_2 are much smaller than αM. Let us also make the quasi-equilibrium approximation for free electrons, namely $n_c \ll n_1, n_2$ and $|dn_c/dt| \ll |dn_1/dt|, |dn_2/dt|$. The population of the active trap n_1 after irradiation by dose $D = Xt$ is

$$n_1 = \frac{A_1 N_1}{A_1 N_1 + A_2 N_2 + A_m \alpha M} D. \tag{6.100}$$

This equation says that the concentration of this trap after irradiation is approximately the number of electron–hole pairs created, times the fraction of electrons which go into the first trap, as opposed to going into the second trap or recombining at the center. Going to the next stage, there is competition during read-out. The integrated TL intensity is determined approximately by the fraction of trapped electrons n_1 which recombine radiatively, as opposed to getting trapped at the deep trap,

$$I_{TL,int} = \frac{A_1 N_1}{A_1 N_1 + A_2 N_2 + A_m \alpha M} \frac{A_m \alpha M}{A_2 N_2 + A_m \alpha M} D. \tag{6.101}$$

Notice that the numerator is proportional to M. Consequently for small M, namely for $A_m \alpha M \ll A_2 N_2$, the TL intensity grows linearly with M. For large M, namely for $A_m \alpha M \gg A_1 N_1 + A_2 N_2$, the denominator grows as the square of M and as a result, the TL intensity declines like $1/M$. Thus, the integral intensity initially grows as M increases, then it levels off, and then declines with further increase in M, in a similar manner to any of the graphs in Figure 6.23.

Lawless *et al.* [485] have extended the approximate derivation to the 3T1C case. In full analogy with the previous case and under the same assumptions, one gets the concentration of trapped electrons at the end of excitation,

$$n_1 = \frac{A_1 N_1}{A_1 N_1 + A_2 N_2 + A_3 N_3 + A_m \alpha M} D, \tag{6.102}$$

$$n_2 = \frac{A_2 N_2}{A_1 N_1 + A_2 N_2 + A_3 N_3 + A_m \alpha M} D. \tag{6.103}$$

For the read-out stage, let us begin with the first TL peak. The electrons from the first trap are excited into the conduction band and subsequently trapped by N_2, N_3 or M. The fraction which is trapped by M and consequently cause TL emission is $A_m \alpha M/(A_2 N_2 + A_3 N_3 + A_m \alpha M)$. Thus, the integrated intensity for the first peak is approximately

$$I_{1,int} = \frac{A_m \alpha M}{A_2 N_2 + A_3 N_3 + A_m \alpha M} n_1 \tag{6.104}$$

$$= \frac{A_m \alpha M}{A_2 N_2 + A_3 N_3 + A_m \alpha M} \frac{A_1 N_1}{A_1 N_1 + A_2 N_2 + A_3 N_3 + A_m \alpha M} D.$$

Equation (6.104) is similar in form to (6.101), and therefore, the dependence on the concentration M is similar; it has a peak shape like each of the graphs in Figure 6.23. Similarly to the 2T1C case, $I_{1,int}$ rises linearly with M as long as M is small enough which in this case means

$$A_m \alpha M \ll A_2 N_2 + A_3 N_3. \tag{6.105}$$

$I_{1,int}$ drops with the inverse of M when M is large enough, i.e.,

$$A_m \alpha M \gg A_1 N_1 + A_2 N_2 + A_3 N_3. \tag{6.106}$$

During the thermal excitation of n_1, some of the freed electrons become trapped by N_2. This change in n_2, which will be called Δn_2, is the product of the number of electrons released n_1, and the fraction of those electrons which are captured by N_2,

$$\Delta n_2 = \frac{A_2 N_2}{A_2 N_2 + A_3 N_3 + A_m \alpha M} n_1 \tag{6.107}$$

$$= \frac{A_2 N_2}{A_2 N_2 + A_3 N_3 + A_m \alpha M} \frac{A_1 N_1}{A_1 N_1 + A_2 N_2 + A_3 N_3 + A_m \alpha M} D,$$

where Equation (6.102) was used. The total concentration, n'_2, after thermal excitation of n_1 is the sum of Equations (6.103) and (6.107). This yields immediately

$$n'_2 = \frac{A_2 N_2}{A_2 N_2 + A_3 N_3 + A_m \alpha M} \left(1 + \frac{A_1 N_1}{A_2 N_2 + A_3 N_3 + A_m \alpha M} \right) D \qquad (6.108)$$

$$= \frac{A_2 N_2}{A_2 N_2 + A_3 N_3 + A_m \alpha M} D.$$

Both concentrations n_3 and m change during this step, but under the low-dose assumption, one has no need to know the amount.

As temperature continues to rise during the read-out, the electrons are freed from trap 2. The total quantity of these electrons is given by Equation (6.108). These electrons are ultimately captured by either N_3 or M. The integrated intensity of the second peak is proportional to the fraction of electrons trapped by M, namely,

$$I_{2,int} = \frac{A_m \alpha M}{A_3 N_3 + A_m \alpha M} n'_2 = \frac{A_m \alpha M}{A_3 N_3 + A_m \alpha M} \frac{A_2 N_2}{A_2 N_2 + A_3 N_3 + A_m \alpha M} D. \qquad (6.109)$$

Note that Equation (6.109) is equivalent in form to Equations (6.101) and (6.104). $I_{2,int}$ rises linearly with M as long as M is small enough that

$$A_m \alpha M \ll A_3 N_3. \qquad (6.110)$$

$I_{2,int}$ drops with the inverse of M when M is large enough that

$$A_m \alpha M \gg A_2 N_2 + A_3 N_3. \qquad (6.111)$$

Comparing the relation (6.105) with (6.110) and (6.106) with (6.111) indicates that $I_{1,int}$ and $I_{2,int}$ will peak at different values of M. As M increases, the second peak reaches its maximum before the first one. How far before this occurs is controlled by the sizes of $A_1 N_1$ and $A_2 N_2$.

One should note that the underlying reason for the non monotonic dependence of the TL sensitivity on the concentration, as demonstrated in the simulation and Equations (6.104) and (6.109), is the competition in trapping between N_1, N_2 and N_3, both during excitation and read-out.

The associated effect of concentration quenching of OSL has also been mentioned in the literature. Inabe et al. [490] mentioned the occurrence of the effect in NaCl:Cu and NaCl:Ca, Cu crystals, X-irradiated at RT. Hackenschmied *et al.* [491] reported on the concentration quenching of OSL in $BaFBr:Eu^{2+}$ and $Ba_{0.82}Sr_{0.18}:Eu^{2+}$. Yawata *et al.* [492] described concentration quenching in blue TL, red TL and OSL in aluminum-doped quartz.

A somewhat related effect of "impurity quenching" has also been described in the literature (see e.g. McKeever [193], p. 152). By inserting "killer" impurities into a material, the intensity of luminescence may be reduced substantially. Medlin [493] mentions several metals which act as killers in a number of natural minerals and points out that the killing efficiency is related to the valency. Thus, iron is more effective in reducing the luminescence efficiency in the trivalent state than in the divalent state. Medlin [475] shows the reduction of TL in calcite with Mn^{2+} as a function of the concentrations of Co^{2+}, Fe^{2+} or Ni^{2+}. Similar effects of impurity quenching in $CaSO_4$ by Cu, Cr, Co and Fe were given by Schmidt *et al.* [494].

6.10   Creation and Stabilization of TL Traps During Irradiation

Most of the work in this book describes the filling of traps by irradiation and their thermal or optical emptying. However, when high energy radiation is used, new traps may be created which are associated with new defects in the irradiated samples. Claffy *et al.* [495] explained the effect of sensitization (see Sections 8.4 and 9.1), under certain circumstances, by defect creation. Attix [496] discussed the creation of new defects in connection with the track interaction model for supralinear response of TL dosimeters. Bull [497] suggested a mathematical model to describe the simultaneous creation and filling of TL traps during irradiation. In addition to the equations governing trap filling, recombinations both between free charges and between free and trapped charges as well as thermal detrapping, a simple trap creation term which is linear with the dose has been included. A simulation consisting of solving the system of differential equations has been performed. Bull [497] showed that thermal detrapping and band-to-band recombinations can introduce a dose-rate dependence into the accumulation of the trapped charges. Trap creation leads to the supralinear growth of charge trapped at certain defects with increasing dose.

More work on the creation of new traps or centers, associated with new point defects in crystals has been reported by Kristianpoller *et al.* [498–500]. These authors used a mechanism previously proposed by Pooley [501] and Hersh [502] and extended it to the case of defect creation by irradiation into a perturbed exciton state and excitation of TL. It is assumed that pairs of defects are created in the crystal by nonionizing UV irradiation and that the creation rate is proportional to the photon flux of the incident UV radiation. The constituents of such a pair which do not recombine immediately upon creation may be stabilized by existing traps in the crystal. It is assumed that one of the constituents is stabilized in a very short time, in a certain configuration specific to the type of the pair. This may be an F-center in the case of a Frenkel pair. The second constituent of the pair remains mobile for a longer period of time and may be trapped in an existing trap or recombine with a defect of the first type. Several types of traps denoted by T_i are presumed for the mobile constituent. These traps have generally different trapping probability coefficients β_i (cm^3 s^{-1}) and have different concentrations M_i (cm^{-3}). When more than one kind of trap exists, competition between the traps over the mobile constituents can be expected.

Based on these considerations, Israeli *et al.* [499] wrote the coupled rate equations governing the process as

$$\frac{dn_a}{dt} = X - \alpha n_a n_f,$$

(6.112)

$$\frac{dm_i}{dt} = \beta_i (M_i - m_i) n_f, \quad i = 1, \ldots$$

(6.113)

$$\frac{dn_f}{dt} = \frac{dn_a}{dt} - \sum_i \frac{dm_i}{dt},$$

(6.114)

where n_a (cm^{-3}) represents the concentration of the stabilized defects of the first type, n_f (cm^{-3}) is the concentration of the mobile defects which are free at time t and α (cm^3 s^{-1}) is the probability for the free defect to recombine with a stabilized defect. Equation (6.112) gives the time dependence of the density of the stabilized defects, X (cm^{-3} s^{-1}) is the rate of pair production associated with the dose rate, and the second term represents the rate of recombination. Equation (6.113) describes the change in the densities of the occupation of the traps at which the defects can be trapped, m_i being the concentration already trapped (out of M_i) at the instant t. Equation (6.114) gives the rate of change of the mobile defects in terms of the rates of change of the stabilized defects. Note that this presentation differs from that given by Israeli *et al.* [499] in the definition of m_i and in the assumption that the absorption coefficient is small.

An expression for the number of the defects created by the UV irradiation is derived as a function of the radiation dose. This expression takes into account the absorption coefficient of the crystal and the penetration depth of the exciting monochromatic light. Experimental results in *KBr* were found to be in good agreement with those predicted by the model. In a recent work, Chen *et al.* [503] studied further the stabilization of point defects by solving numerically Equations (6.112–6.114) for certain sets of the relevant parameters. In order to see the basic features of the relevant effect, the authors concentrate on the case with two kinds of traps, i.e., $i = 1$, 2. Also, in order to concentrate on the essence of the nonlinear dose dependence, they assume a homogeneous excitation of the whole sample, which means either relatively low absorption coefficient or, alternatively, a rather thin sample (for a discussion on the effect of the absorption coefficient on TL see Section 6.8). An example of the simulation results are shown in Figure 6.24; the chosen parameters are given in the caption. The TL intensity is assumed to be proportional to m_2, and

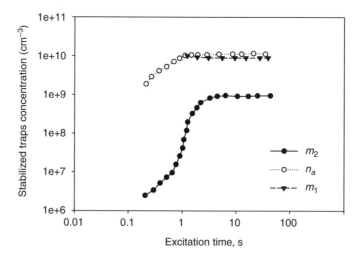

Figure 6.24 *Simulated results of the excitation-time dependence of the stabilized traps with the parameters: $M_1 = 10^{10}$cm^{-3}, $M_2 = 10^9$ cm^{-3}, $\alpha = 10^{-12}$ cm^3 s^{-1}, $\beta_1 = 10^{-10}$ cm^3 s^{-1}, $\beta_2 = 10^{-12}$ cm^3 s^{-1}, $X = 10^{10}$ cm^{-3} s^{-1}. Chen, R., Pagonis, V., and Lawless, J. L., Simulation of the Nonlinear Dose Dependence of Stabilized Point Defects, EURODIM 2010, Copyright (2010) with permission from Reuven Chen/Institute of Physics*

with the present set of parameters it is seen to be very strongly superlinear, as sometimes seen in the experiments. Note that this superlinear dependence bears some resemblance to curves discussed in Chapter 8, and in particular to Figure 8.15. These cases differ from the present situation in that transitions of charge carriers between existing impurity and defect states are considered there, rather than the creation of new ones as discussed here. The analogy between the two cases is related to the fact that competition between two or more trapping states is the main cause for the strong superlinearity. It seems that in "real life" cases, the superlinearity, seen as the very high slope in the log–log graph following the linear range, may not be that high due to the effect of finite absorption. If the excitation radiation is absorbed rather strongly in the sample, the front layers of the sample may approach saturation. Since the measured TL results from the whole sample, the slope of the dose-dependence curve may be mitigated as compared with the shown simulated curve. Chen *et al.* [503] have also shown that for other sets of parameters, where the probability for recombination α is significantly larger than the probabilities for retrapping β_i, a sublinear dose dependence with a close to $D^{0.5}$ dependence is found in the simulations, similar to some experimental results (see Section 8.7). This bears some analogy to cases discussed in Chapter 8 where the dose dependence within the OTOR model is discussed, in which case transitions of charge carriers between trapping states and the conduction band are considered.

Kristianpoller and Katz [504] described the generation of Z_1 centers in X-colored NaCl:Sr^{2+} and NaCl. Kristianpoller *et al.* [500] reported on the excitation of luminescence and the generation of point defects in NaF by monochromatic synchrotron radiation at liquid helium temperature. F-center generation as well as the excitation spectrum of the main emission bands showed a strong maximum at 123 nm, which coincides with the long-wavelength tail of the first exciton absorption band. These results agreed with the predictions of a previously proposed model. Israeli and Kristianpoller [505] explained the superlinear dose dependence of UV-excited 150 K peak TL in KBr as being due to a two-stage process. The quadratic behavior was ascribed to the F-center creation followed by the F'-center production from the F-centers. Further work along the same lines was performed by Israeli and Kristianpoller [147, 148] who also took into consideration the excitation by vacuum UV in regions of relatively high absorption, where different creation of centers can be expected at different depths of the sample. Nishidai *et al.* [506] also discussed the creation of new additional trapping sites by irradiation, which may in some cases be the reason for supralinearity. In some materials, previous exposure or pre-irradiation annealing could eliminate the supralinearity.

Another work on the production and destruction of centers in KBr:In has been reported by Popov [507]. His results show two TL peaks, one increasing linearly with the dose before approaching saturation and the other exhibiting a quadratic dose dependence in a rather broad range.

6.11 Duplicitous TL Peak due to Release of Electrons and Holes

TL glow curves usually consist of a number of peaks occurring at different temperatures. The accepted theory of TL considers charge carriers which are captured in the traps during excitation, and luminescence centers existing within the forbidden gap of the material being used. The model may include either electron traps and hole centers or hole traps and electron centers. In the former case, electrons are thermally raised into the conduction band during the heating stage, and recombine with holes in luminescence centers, yielding an emission of luminescence. In the latter, holes are thermally released from traps into the valence band and recombine with electrons in luminescence centers yielding the emission of TL. It has been shown that both kinds of peaks may take place in the same material. Thus, a thermoluminescent material may yield separate sets of peaks, one due to thermally released electrons recombining with stationary holes in centers and another set due to thermally released holes recombining with electrons in other luminescence centers (see e.g. Braner and Israeli [424]). Different TL peaks in a given material may yield different emission spectra due to the radiative recombinations in different recombination centers. The work by Braner and Israeli [424] distinguishes between electron and hole processes in each of four alkali halides, KBr, KI, NaCl and KCl. These authors report on the illumination with monochromatic light into the F⁻ or V⁻ absorption band of single crystals of these alkali halides at LNT after X-irradiation at room temperature, thus yielding PTTL peaks (see Section 6.6). In each of the samples, the UV exposure re-excites different glow peaks between LNT and RT. F illumination produces a set of electron TL peaks and V light produces a set of hole peaks. Both sets add up to the "normal" glow curve obtained after X-irradiation at LNT. These two sets of TL peaks can be expected to have different emission spectra. Panova *et al.* [508] have given a similar analysis concerning the TL peaks in CsI:Na and in CsI:Tl, and identified certain peaks in each of these materials as being associated with the thermal release of electrons from electron traps and other peaks with the thermal release of holes from hole traps. McKeever *et al.* [509] have shown that electrons and holes may be released simultaneously from their trapping states. The electron trapping states may be considered at the same time as electron traps and electron luminescence centers, while the hole trapping states can be considered as hole traps and hole luminescence centers. Under these circumstances, the observed TL may be the sum of two radiative transitions, namely free electron to bound hole and free hole to bound electron, which may take place simultaneously. More aspects of this rather complex process can be considered. For example, holes may be thermally released from a hole center, thus reducing the expected amount of luminescence from the center in question. Inversely, a hole trap may thermally release holes into the valence band, which can be captured in an active luminescence center, thus changing the subsequently measured luminescence from this center. This was the manner in which sensitization of the 110 °C peak in quartz was explained by Zimmerman [510]. Additional possibilities are implied in the work by Bailey [35] who studied in depth the TL as well as the OSL in quartz, a material which is broadly used for luminescence dating of archaeological and geological samples. His model includes five electron-trapping states and four hole levels. Depending on the relevant trapping parameters, electrons and holes may be released into the respective bands, and may recombine with opposite charge carriers in centers. The emitted luminescence, TL or OSL, is associated here with only one transition of free electrons with holes in one center. The resulting TL glow curve which is simulated by Bailey [35] consists of a number of peaks associated with electrons released from different electron traps and recombining with holes in one of the centers. The experimental justification for this is that in quartz, the different TL peaks have the same emission spectrum. The role of the other centers in the model is of being competitors which can capture free electrons. Also, the free holes may be captured by the active luminescence center, thus replenishing it with holes, which may contribute to the TL occurring at a higher temperature. Another related phenomenon is thermally stimulated electron emission (TSEE). Here, electrons previously thermally stimulated into the conduction band may be released from the surface of the sample and the resulting current may be measured as TSEE peaks as a function of temperature. Obviously, only electrons can be released from the surface, and therefore, TSEE has been used to distinguish between electron and hole transitions. However, an "Auger" TSEE was also reported in different materials in which the emission of electrons is associated with the prior thermal release of holes. This will be elaborated upon in Section 6.12. Chen *et al.* [511] focused on a situation in which in the temperature range of interest, electrons and holes are thermally released from their respective traps, whereas the radiative transition is assumed to be associated only with the recombination of electrons with holes in luminescence centers. The contribution of the hole trap is that it replenishes the hole center during the heating stage. As a result, one may see a duplicitous TL peak, the two components of which have the same emission spectrum since they are related to the same kind of transition. The behavior of the free electron and free hole curves as a function of temperature were also followed in this work; a new explanation was offered for a TSEE peak associated primarily with the thermal release of trapped holes.

The model given by Chen *et al.* [511] is shown in Figure 6.25. This includes one electron trapping state N_1, one hole trapping state N_2, and one hole recombination center N_3. This is the simplest model which enables the demonstration of the "duplicitous" peak, which results from the release of electrons and holes from trapping states, with recombinations taking place in one center. It should be noted that in most real materials, one may expect

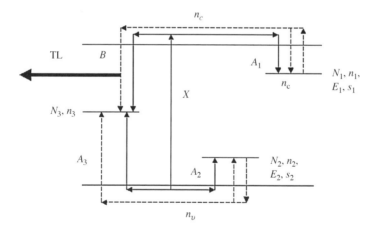

Figure 6.25   *The energy level diagram to explain the duplicitous TL peak phenomenon, consisting of one electron and one hole trapping states, and one kind of recombination center. Solid lines give transitions occurring during excitation and transitions taking place during read-out are shown by dashed lines. The TL emission is shown, schematically, by a thick arrow. Reprinted from Chen, R., Pagonis, V. and Lawless, J.L., Duplicitous thermoluminescence peak associated with a thermal release of electrons and holes from trapping states, Radiat. Meas. 43, 162–166. Copyright (2008) with permission from Elsevier*

more electron and hole trapping states, and possibly more recombination centers of different kinds. During excitation, the applied radiation raises electrons from the valence band into the conduction band. The free holes may get trapped at the hole trap N_2 or at the hole center N_3. At the same time, the free electron can be trapped in the electron trap N_1 or recombine with a hole in the hole center, provided it has trapped a hole at an earlier stage of the excitation. The assumption that N_3 first captures a hole, which only then can recombine with a free electron, defines it as being a "hole center".

The set of simultaneous differential equations governing the process during excitation is

$$\frac{dn_1}{dt} = A_1(N_1 - n_1)n_c, \tag{6.115}$$

$$\frac{dn_2}{dt} = A_2(N_2 - n_2)n_v, \tag{6.116}$$

$$\frac{dn_3}{dt} = A_3(N_3 - n_3)n_v - Bn_3n_c, \tag{6.117}$$

$$\frac{dn_c}{dt} = X - A_1(N_1 - n_1)n_c - Bn_3n_c, \tag{6.118}$$

$$\frac{dn_v}{dt} = \frac{dn_1}{dt} + \frac{dn_c}{dt} - \frac{dn_2}{dt} - \frac{dn_3}{dt}. \tag{6.119}$$

Here, N_1 (cm^{-3}) and n_1 (cm^{-3}) denote, respectively, the concentration and occupancy of electron traps, N_2 (cm^{-3}) and n_2 (cm^{-3}) the concentration and occupancy of hole traps and N_3 (cm^{-3}), n_3 (cm^{-3}) the concentration and occupancy of hole centers. n_c(cm^{-3}) is the instantaneous concentration of free electrons and n_v (cm^{-3}) the concentration of free holes. X (cm^{-3} s^{-1}) is the rate of production of electron–hole pairs by the applied radiation and is proportional to the applied dose rate. A_1 (cm^3 s^{-1}) is the trapping probability coefficient of free electrons from the conduction band into the electron trap, and A_2 (cm^3 s^{-1}) is the trapping probability coefficient of free holes from the valence band into N_2. A_3 (cm^3 s^{-1}) is the probability coefficient for free holes from the valence band to be captured into the luminescence center and B (cm^3 s^{-1}) the probability coefficient of free electrons recombining with holes in the center.

The set of equations (6.115–6.119) was solved numerically by Chen *et al.* [511] for given sets of the mentioned parameters and for the relevant value of X. If one denotes the time of excitation by t_D, $D = X \cdot t_D$ is the total concentration per unit volume of electrons and holes produced, which is proportional to the total dose imparted. In order to simulate the experimental situation properly, one has to consider a relaxation time following the excitation and prior to the heating stage, during which practically all the free carriers relax and end up in the traps and centers. It should be noted that the relaxation process is important since the carriers can build up to significant levels during excitation, and relaxation is needed to provide the time necessary for them to decay back toward zero. In the numerical simulation, one takes the final values of the five relevant concentrations as initial values for the next stage of relaxation, sets $X = 0$ and solves the equations for a period of time so that at the end, both n_c and n_v are practically zero.

The next stage is simulating the heating stage. Here, the relevant parameters for the thermal release of electrons are the activation energy E_1 (eV) and the frequency factor s_1 (s^{-1}), and for the release of holes these parameters are the activation energy E_2 (eV) and the frequency factor s_2 (s^{-1}). The set of five simultaneous differential equations governing the process is

$$\frac{dn_1}{dt} = A_1(N_1 - n_1)n_c - n_1 s_1 \exp(-E_1/kT), \tag{6.120}$$

$$\frac{dn_2}{dt} = A_2(N_2 - n_2)n_v - n_2 s_2 \exp(-E_2/kT), \tag{6.121}$$

$$\frac{dn_3}{dt} = A_3 n_v(N_3 - n_3) - Bn_3 n_c, \tag{6.122}$$

$$\frac{dn_c}{dt} = n_1 s_1 \exp(-E_1/kT) - A_1(N_1 - n_1)n_c - Bn_3 n_c, \tag{6.123}$$

$$\frac{dn_v}{dt} = \frac{dn_1}{dt} + \frac{dn_c}{dt} - \frac{dn_2}{dt} - \frac{dn_3}{dt}, \tag{6.124}$$

where T (K) is the temperature and k (eV K^{-1}) is the Boltzmann constant. One has to consider the heating function $T(t)$, and Chen *et al.* [511] took for simplicity a linear heating function, namely $T = T_0 + \beta t$ where β (K s^{-1}) is the constant heating rate. The emitted TL light, shown in Figure 6.25 as a thick arrow, is associated with the recombination of free electrons with holes in the recombination center and is given by

$$L(T) = Bn_3 n_c. \tag{6.125}$$

It should be pointed out that an important element of the occurrence of the duplicitous peak is the sizeable retrapping into the electron trapping state. Otherwise, it would not be plausible that a process associated with the release of electrons from traps, say 1.0 eV deep, will be governed by the supply of holes from the hole trapping state with activation energy of 1.3 eV.

Numerical-simulation results of the duplicitous TL peak will be discussed in Section 6.12. While making the following analytical considerations, we refer to these numerical results. In the example below, the high temperature TL peak, the $n_c(T)$ and the $n_v(T)$ peaks occur at nearly the same temperature, which appears to be rather surprising. It has been pointed out by Lawless *et al.* [512] that in the results of the simulations, while increasing the value of E_2 but keeping all the other parameters the same, not only the $n_v(T)$ peak shifts to higher temperature, but also the $n_c(T)$ and TL peaks shift by about the same amount. Following is a qualitative explanation for the fact that these three peaks almost coincide, with reference to the results of Figure 6.26 in Section 6.12. We note that in the temperature range of 400–500 K, $n_3(T)$ is slightly increasing due to the replenishment of holes from n_2 via the valence band, which appears to be slightly faster than the recombination process. Considering Equation (6.122), it seems obvious that dn_3/dt is significantly smaller than each of the two terms on the right-hand side, and therefore

$$A_3 n_v(N_3 - n_3) \cong Bn_3 n_c. \tag{6.126}$$

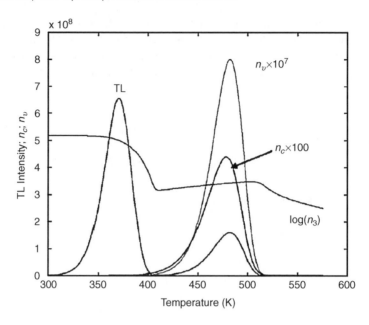

Figure 6.26   *The simulated TL curve for the set of parameters given in the text and for the model in Figure 6.25. Also shown are the free electron concentration n_c, and the free hole concentration n_v as well as the logarithm of n_3 along the temperature range of 300–600 K. Reprinted from Chen, R., Pagonis, V. and Lawless, J.L., Duplicitous thermoluminescence peak associated with a thermal release of electrons and holes from trapping states, Radiat. Meas. 43, 162–166. Copyright (2008) with permission from Elsevier*

With the parameters used in the simulation below, whereas $N_3 = 10^{14} \, \text{cm}^{-3}$, $n_3 \cong 10^6 \, \text{cm}^{-3}$, and therefore $N_3 - n_3 \cong 10^{14} \, \text{cm}^{-3}$ is practically constant. Using Equation (6.125), one therefore gets to a very good approximation

$$L(T) = Bn_3(T)n_c(T) \propto n_v(T), \tag{6.127}$$

which explains the close resemblance between $L(T)$ and $n_v(T)$. Since, as mentioned, $n_3(T)$ is nearly constant in this range (see Figure 6.27), one understands the similarity between the three curves of $L(T)$, $n_c(T)$ and $n_v(T)$. This also explains how, through the set of simultaneous equations (6.122–6.124), an increase in E_2 shifts all three curves to higher temperatures. A closer look at the figure shows that the $L(T)$ curve at $\sim$480 K is slightly shifted to higher temperatures as compared with $n_c(T)$. This is easily explained using Equation (6.125). $L(T)$ is proportional to $n_3(T)n_c(T)$, and since $n_3(T)$ is a slightly increasing function of temperature in this range, the peak shape of $L(T)$ is more or less similar to that of $n_c(T)$, but being the product of a peak-shaped function and an increasing function, it yields another peak-shaped function, which is slightly shifted to higher temperature as compared with $n_c(T)$.

It is possible to take the analysis of Equations (6.126) and (6.127) a bit further and use it to address some more issues. In the temperature range between the two peaks, n_c, n_v and n_3 are all small and the quasi-steady approach applies to all three. From Equation (6.123)

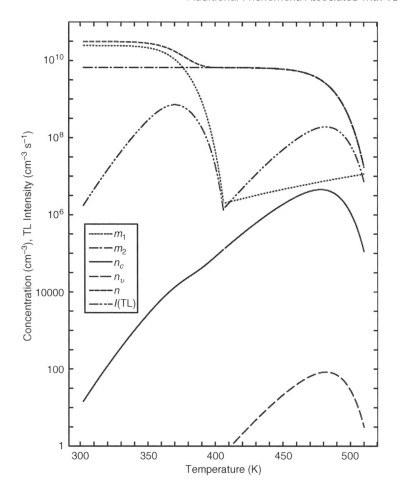

Figure 6.27 *For the same parameters as in Figure 6.26, the TL simulated curve and the concentrations of electrons and hole traps/centers, as well as free electrons and holes are shown as a function of temperature on a semi-log scale. Reprinted from Lawless, J.L., Chen, R. and Pagonis, V., On the theoretical basis for the duplicitous thermoluminescence peak, J. Phys. D: Appl. Phys. 42, 15, 155409. Copyright (2009) with permission from Institute of Physics*

and by setting $dn_c/dt = 0$, one finds

$$n_c = \frac{n_1 s_1 \exp(-E_1/kT)}{A_1(N_1 - n_1) + Bn_3}. \tag{6.128}$$

In conventional TL theory, n_1 drops to zero after a measurement of the TL peak. Here, by conservation of charge, n_1 cannot drop below n_2 (see also Figure 6.27). Furthermore, n_2 is fixed at its initial value until it starts to read out at whatever higher temperature is set by s_2 and E_2. Because n_1 is finite after the first peak but not zero, it follows that n_c cannot drop to zero either; it is determined by Equation (6.128) as above. This is the source of the anomalous n_c behavior described by Lawless *et al.* [513]. The equation (6.128) can

Table 6.1 *Parameters for evaluating the activation energies using the peak-shape method. From Lawless et al. [512]*

	T_1 (K)	T_m (K)	T_2 (K)	μ_g	E (eV)
1st TL peak	352.9	369.9	384.4	0.46	1.033
2nd TL peak	460.7	481.7	496.7	0.417	1.298
$n_c(T)$	453.6	477.9	494.5	0.406	1.06
$n_v(T)$	460.7	481.7	496.7	0.417	1.298

be simplified because in the simulations n_1 is far from saturation and it is clear that the retrapping term $A_1 \cdot N_1$ dominates over recombination $B \cdot n_3$ after the first peak, when n_3 is small. An additional consequence of n_3, n_c and n_v being small is that conservation of charge requires $n_1 = n_2$. Thus,

$$n_c = \frac{n_2 s_1}{A_1 N_1} \exp(-E_1/kT). \tag{6.129}$$

This shows that while n_c has an initial rise that reflects the activation energy E_1, its behavior at and after the second peak will be influenced by n_2, which is governed by the value of E_2. This would appear to be the reason why the initial-rise and various heating rate methods gave such different values for n_c in Table 6.1 taken from Lawless *et al.* [512]. One can find n_v similarly,

$$n_v = \frac{n_2 s_2 \exp(-E_2/kT)}{A_3(N_3 - n_3) + A_2(N_2 - n_2)}. \tag{6.130}$$

Thus, the initial rise of n_v should reflect E_2. This can be simplified by noting that $n_2 \ll N_2$ and $n_3 \ll N_3$ during the simulation, yielding

$$n_v = \frac{n_2 s_2 \exp(-E_2/kT)}{A_3 N_3 + A_2 N_2}. \tag{6.131}$$

For n_3, using a similar analysis with $n_3 \ll N_3$,

$$n_3 = \frac{A_3 N_3}{B} \frac{n_v}{n_c}. \tag{6.132}$$

As pointed out above, n_3 increases due to replenishment by holes. Equation (6.132) shows more specifically that n_3 reflects a competition between replenishment by holes and annihilation by electrons. To see why it is the holes that win, one can substitute n_c and n_v from Equations (6.129) and (6.131) to find

$$n_3 = \frac{A_1 N_1}{B} \frac{s_2}{s_1} \frac{A_3 N_3}{A_3 N_3 + A_2 N_2} \exp\left[-(E_2 - E_1)/kT\right]. \tag{6.133}$$

Thus, n_3 increases between the peaks because E_1 and E_2 were chosen (see Section 6.12) such that n_v rises faster than n_c with time. As a result, one can predict that if an initial rise analysis was done on n_3 between the peaks, it is expected to reflect $(E_2 - E_1)$. Lawless *et al.* [512] report that for the numerical results with the chosen parameters, the slope yields an effective energy value of $E_2 - E_1 = 0.3$ eV. The intensity of the second peak is given by

$$L(T) = Bn_3n_c = A_3N_3n_v,$$ (6.134)

where Equation (6.132) was used. This is why $L(T)$ and $n_v(T)$ have the same peak parameters in Table 6.1, accurate to all decimal places shown. Also, substituting from Equation (6.133) yields

$$L(T) = \frac{A_3N_3}{A_3N_3 + A_2N_2} n_2 s_2 \exp(-E_2/kT).$$ (6.135)

This equation shows that the second peak $L(T)$, like $n_v(T)$, has an initial rise determined by E_2.

It is quite interesting that the high temperature TL peak in the system (see Figure 6.26), along with the $n_v(T)$ peak (see Section 6.12 below) look like simple first-order curves as reflected by the shape factor μ_g being 0.417. Within the framework of results discussed by Lawless *et al.* [513], this can be explained as follows. Adding up Equations (6.120) and (6.123), and using Equation (6.122), one gets

$$L(T) = -\frac{dn_c}{dt} - \frac{dn_1}{dt}.$$ (6.136)

Using the conservation of charge, Equation (6.121), one gets

$$L(T) = -\frac{dn_3}{dt} - \frac{dn_2}{dt} - \frac{dn_v}{dt}.$$ (6.137)

As argued above, $n_3(T)$ is nearly constant and therefore, $\left|\frac{dn_3}{dt}\right| \ll \left|\frac{dn_2}{dt}\right|$. As for $n_v(T)$, it has a very similar shape to that of $n_2(T)$ in the range of occurrence of the second TL peak, but is several orders of magnitude lower. Therefore, in this range, $\left|\frac{dn_v}{dt}\right| \ll \left|\frac{dn_2}{dt}\right|$, and as a result,

$$L(T) = -\frac{dn_2}{dt}.$$ (6.138)

Using Equation (6.136), one can write

$$L(T) = -\frac{dn_2}{dt} = \frac{A_3N_3}{A_3N_3 + A_2N_2} n_2 s_2 \exp(-E_2/kT).$$ (6.139)

Thus, the high temperature duplicitous TL peak has a standard first-order shape with activation energy of E_2 and an effective frequency factor of

$$s_{eff} = \frac{A_3N_3}{A_3N_3 + A_2N_2} s_2.$$ (6.140)

6.12   Simulations of the Duplicitous TL Peak

Chen *et al.* [511] solved numerically the relevant sets of equations mentioned in Section 6.11. Equations (6.115–6.119) were first solved for a certain value of the dose rate X and for a certain length of time t_D, which together determine the dose $D = X \cdot t_D$. The solution of the same set of equations is continued for a further period of relaxation time but with $X = 0$. Finally, the coupled equations for the heating stage (6.120–6.124) were solved and Equation (6.125) gave the TL intensity as a function of temperature.

Figure 6.26 shows duplicitous TL peaks simulated using the following set of parameters (see Chen *et al.* [511]). $N_1 = 10^{12}\,\text{cm}^{-3}$, $N_2 = 10^{13}\,\text{cm}^{-3}$, $N_3 = 10^{14}\,\text{cm}^{-3}$, $A_1 = 10^{-7}\,\text{cm}^3\,\text{s}^{-1}$, $A_2 = 3 \times 10^{-8}\,\text{cm}^3\,\text{s}^{-1}$, $A_3 = 2 \times 10^{-8}\,\text{cm}^3\,\text{s}^{-1}$, $B = 5 \times 10^{-6}\,\text{cm}^3\,\text{s}^{-1}$, $X = 5 \times 10^7\,\text{cm}^{-3}\,\text{s}^{-1}$, $t_D = 1000\text{s}$, $E_1 = 1.0\,\text{eV}$, $E_2 = 1.3\,\text{eV}$, $s_1 = 5 \times 10^{12}\,\text{s}^{-1}$, $s_2 = 3 \times 10^{12}\,\text{s}^{-1}$ and $\beta = 1\,\text{K}\,\text{s}^{-1}$. The simulated TL curve is given in Figure 6.26, with two peaks at ~ 370 and $\sim 480\,\text{K}$. The $n_c(T)$ curve is shown, with a single maximum occurring below the higher temperature TL peak. In order to be able to show the $n_c(T)$ curve on the same graph, the values have been multiplied by 100. Also are shown the $n_v(T)$ values multiplied by 10^7, which also yields a maximum at $\sim 480\,\text{K}$, and the logarithm of $n_3(T)$.

The peak parameters for the two TL peaks and the $n_c(T)$ and $n_v(T)$ peaks are given in Table 6.1. Also shown in Table 6.1 are the shape factor μ_g and the evaluated activation energy found by Equation (5.8), along with (5.11) which determines the energy using the shape of the peak. It is readily seen that the second TL peak and the $n_v(T)$ peak look practically the same, namely in this temperature range, they are proportional to each other, and therefore their analysis yields the same activation energy. Using the values given here for A_2, N_2, A_3, N_3, and s_2 for the simulation, Lawless *et al.* [512] obtained from Equation (6.135) $s_{eff} = 2.6 \times 10^{12}\,\text{s}^{-1}$. Indeed, if one substitutes from Table 6.1 the values of $T_m = 481.7\,\text{K}$ and the resulting activation energy $E = 1.298\,\text{eV}$, in the expression for the maximum of a first-order TL peak, $s = (\beta E/kT_m^2)\exp(E/kT_m)$ [see Equation (5.12)], one obtains $s_{eff} = 2.47 \times 10^{12}\,\text{s}^{-1}$, in very good agreement with the expected value.

The evaluated activation energies as given in Tables 6.1 and 6.2 are of interest. The fact that the first TL peak yields 0.99 eV using the initial-rise method and 1.004 eV with the variable heating rate (VHR) method, agrees with its association with the release of electrons from a trap with a depth of 1.0 eV. The effective value of 1.033 eV found with the peak shape method is not too surprising, in particular since the shape factor of the first TL peak is $\mu_g \approx 0.46$. However, the second TL peak yields values of $E \sim 1.3\,\text{eV}$ using the three methods, which shows that although the recombinations are of free electrons with

Table 6.2   *Comparison of activation energies (eV) evaluated by three methods. From Lawless et al. [512]*

	Initial rise	Shape method	VHR
1st TL peak	0.99	1.033	1.004
2nd TL peak	1.29	1.298	1.337
$n_c(T)$	0.98	1.06	1.296
$n_v(T)$	1.29	1.298	1.337

holes in centers, the governing process here is the thermal release of holes and their trapping in recombination centers. The $n_v(T)$ curve yields consistently $E \approx 1.3\,\text{eV}$. The $n_c(T)$ curve, however, yields $E_{ir} \approx 1.0\,\text{eV}$ with the initial-rise method which, indeed, is a feature of the release of electrons. However, when the VHR method is used, a value of $E_{VHR} \approx 1.3\,\text{eV}$ is obtained, which is the feature of the hole trap. The peak shape method yields here a value of $1.06\,\text{eV}$, which, apparently has to do with the rather complex nature of the process under these circumstances. An important conclusion of the given results is that one should be rather cautious in ascribing specific transitions to the evaluated parameters when there is a possibility of having transitions of the described nature during the heating stage.

The same results as in Figure 6.26 are shown in Figure 6.27, on a semi-log scale, along with the concentrations. The curves of TL, n_c, and n_v are the same as in Figure 6.26, but on a semi-log scale. Note that following Lawless *et al.* [512], the notation in this figure is somewhat different. n_1, n_2 and n_3 are replaced by n, m_2 and m_1, respectively. n and m_2, the concentrations of electrons and holes in traps respectively, are shown by the dashed and dashed-dotted lines. The concentration of electrons in traps n decreases between $\sim 350\,\text{K}$ and $\sim 400\,\text{K}$, remains approximately constant up to $\sim 475\,\text{K}$ and drops rather quickly to zero around $\sim 530\,\text{K}$. The concentration of the hole center m_1 decreases up to $\sim 410\,\text{K}$, slowly increases up to $\sim 510\,\text{K}$ and then decreases similarly to n_2.

The occurrence of the high temperature TL and the $n_c(T)$ and $n_v(T)$ peaks at about the same temperature is somewhat surprising. In order to check whether this is merely a coincidence, Chen *et al.* [511] increased E_2 to higher values, while keeping all other parameters constant, expecting that the $n_v(T)$ peak will shift to higher temperature. This happened indeed, but at the same time the two other peaks shifted by about the same amount. This point has been mentioned in the theoretical discussion above, see Section 6.11.

TSEE has been mentioned in Section 6.11. As explained, thermally stimulated electrons from the conduction band may be released from the surface of the sample and the resulting current may be measured as TSEE peaks as a function of temperature. Only electrons can be released from the surface, and therefore, TSEE has been used to distinguish between electron and hole transitions. However, as pointed out above, an "Auger" TSEE was also reported in different materials (see e.g. Tomita *et al.* [514], Kuzakov [515], Bindi *et al.* [516], Rosenman *et al.* [517], Oster and Haddad [518] and Kappers [519]). Here, transitions which are identified as being associated with the release of holes, may produce a measurable TSEE peak. As explained by Tolpygo *et al.* [520] and Bindi *et al.* [516], the Auger-like model involves the recombination of an electron and a hole. The energy released when an electron and a hole recombine may be transferred to another electron localized in a nearby trapping state, which may thus acquire sufficient energy to overcome the potential barrier at the surface and be detected as an exoelectron. The mean energy of the exoelectron is of the order of $E_g - E_1 - E_2 - \chi$, where E_g is the band gap, χ the electron affinity, and E_1 and E_2 the ionization energies for the emitted electron. Thus, the occurrence of a TSEE peak associated with the initial release of holes into the valence band could be explained. Chen *et al.* [511] offered an alternative explanation based on the considerations presented in this section. As explained above, the released holes $n_v(T)$ indirectly induce a peak to form in the concentration of the free electrons $n_c(T)$, which may contribute to the TSEE. Further discussion on the simultaneous measurements of TL, TSC and TSEE is given in Chapter 11.

Chen *et al.* [511] point out that the behavior reported here, including the exhibited activation energy associated with either the electron or hole trap and the new interpretation of TSEE, are not necessarily general. However, they suggest the possibility that some TL peaks which appear to be associated with the release of holes from hole traps, are actually associated with the recombination of free electrons and holes in recombination centers. The same explanation may apply to the occurrence of TSEE peaks, which carry features related to the release of holes from traps.

7

Optically Stimulated Luminescence (OSL)

7.1 Basic Concepts of OSL

Some basic experimental evidence of OSL measurements has been discussed in Section 3.1. The theory of OSL is closely related to that of TL. To begin with, the excitation stage is exactly the same as in TL. As for the read-out, the thermal release of trapped carriers is replaced by a release by the stimulating light. Instead of the magnitudes of the activation energy E and the frequency factor s relevant to the release of carriers in TL, a parameter $f\,(\mathrm{s}^{-1})$ representing the intensity of the stimulating light is introduced. In fact, one can write $f = \sigma I$ where $\sigma\,(\mathrm{cm}^2)$ is the cross-section for a photon to cause the optical release of a trapped hole and $I(\mathrm{cm}^{-2}\,\mathrm{s}^{-1})$ is the flux of the stimulating photons. However, the activation energy and frequency factor are still of importance here since they determine the thermal stability of the OSL when the sample is held at ambient temperature. An important point to be mentioned is that there are several different versions of the OSL measurements. One can consider the OSL signal as being the area under the decaying curve during the stimulating light exposure. The integral over the emitted light intensity can be recorded for a relatively long time, until the light intensity is negligibly small, thus representing the total light emission. It can be alternatively taken over a shorter period of time, or even over a very short time when the stimulating light is delivered as a short pulse. The measured light can be taken during the optical pulse using wavelength discrimination, namely by removing the stimulating light by appropriate filters, or by time discrimination as suggested by McKeever et al. [521] and by McKeever and Akselrod [522]. These authors described a variant termed "pulsed optically stimulated luminescence (POSL)", in which the luminescence is only detected after the end of the stimulating light pulse. Although all these methods yield emission of luminescence associated with the occupancies of the same trapping states and recombination centers, the details of the process in the different variants are different and therefore, effects like the dose dependence of the signal may not be the same, as elaborated below.

Thermally and Optically Stimulated Luminescence: A Simulation Approach, First Edition. Reuven Chen and Vasilis Pagonis. © 2011 John Wiley & Sons, Ltd. Published 2011 by John Wiley & Sons, Ltd.

7.2 Dose Dependence of OSL; Basic Considerations

In nearly all the reports on OSL and its applications it is assumed, and sometimes shown [522], that the initial dose dependence at low doses is linear followed by an approach to saturation at high doses [523]. It is also assumed that there are no dose-rate effects and therefore, one can calibrate the sample at high dose rates and deduce the archaeological or geological dose which was imparted at a much lower dose rate. However, there are a few reports in the literature on superlinear dose dependence of OSL. In the study of OSL in quartz and mixed feldspars, Godfrey-Smith [524] found linear dose dependence of the unheated samples. However, following a preheat at 225 °C, the samples showed a clear superlinearity of the OSL signal at low excitation doses of γ-irradiation. Roberts *et al.* [525] have also found superlinearity of quartz OSL in several samples. For samples preheated at 160 °C, they reported a quadratic equation $aD^2 + bD + c$ describing the dose dependence, where D is the applied dose and a, b and c are positive magnitudes. Bøtter-Jensen *et al.* [451] also reported a superlinearity of the OSL signal in quartz samples extracted from bricks while performing retrospective dosimetry. They pointed out that linearity takes place in unannealed samples, whereas superlinearity occurs in annealed specimens.

Banerjee [526] discussed the superlinear dose dependence of OSL in quartz in analogy with this kind of dose dependence in TL. Following Kristianpoller *et al.* [450] and Chen and McKeever [194], he showed that in the presence of a competing trapping state, a strong superlinearity can be expected whereas in the presence of a competing non-radiative recombination center, either linear or slightly superlinear dose dependencies can take place. The superlinear dose dependence of OSL in Al_2O_3:C has been described by Jursinic [527]. A recent work by Chruścińska [311], explains the dependence of OSL decay on temperature, occurring in some materials, in terms of electron–phonon interactions. The author explains that the optical cross-section (OCS) of a trap depends on the experimental conditions, such as the energy of stimulation photons and the measurement temperature. Computer simulations have been carried out for different temperatures using one-trap and two-trap models.

A different kind of dose dependence of OSL on the dose in quartz samples extracted from sedimentary cores has been reported by Lowick *et al.* [528]. In addition to the expected simple saturating exponential function, an additional high dose component seems to exist in the dose response. Although often reported as linear, it appears that this is the early expression of a second saturating exponential. Similar behavior was reported for the UV and violet/blue emissions.

Chen and Leung [529] dealt with the problems of dose dependence and dose-rate dependence of OSL using simulations. The relevant sets of coupled differential equations governing the processes of excitation and optical stimulation have been solved numerically. The approach has been similar to that by McKeever *et al.* [530] except that these authors mainly followed the shape of the OSL decay curve with time, and did not study the dose and dose-rate dependencies.

The energy level scheme shown in Figure 7.1 is a simple OTOR model. One trapping state with a concentration of N (m^{-3}) electrons is assumed to be active, as well as one hole-center state with a concentration of M (m^{-3}). Electrons are assumed to be excited by the irradiation, the intensity of which is denoted by X (m^{-3} s^{-1}). The total dose imparted during a period of time t_D(s) is given in these units as $D = X \cdot t_D$ (m^{-3}) (regarding the units of dose and dose rate, see below). The irradiation produces holes in the valence band and electrons

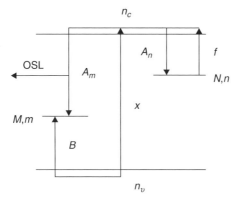

Figure 7.1   *Energy levels involved in the OSL phenomenon. A trapping state with a concentra-*
tion N (m^{-3}) and instantaneous occupancy n (m^{-3}) as well as a recombination center M (m^{-3})
with instantaneous occupancy of m (m^{-3}) are shown. B (m^{3}s^{-1}) is the probability coefficient for
holes in the valence band to be trapped in M. The trapping and recombination coefficients are
A$_n$ (m^{3}s^{-1}) and A$_m$ (m^{3}s^{-1}), respectively. X (m^{-3}s^{-1}) is proportional to the intensity (dose rate)
of radiation excitation of electrons from the valence band into the conduction band. f (s^{-1}) is
the intensity of the stimulating light. n$_v$ (m^{-3}) and n$_c$ (m^{-3}) are the instantaneous concentrations
of free holes and electrons, respectively. Reprinted from Chen, R. and Leung, P.L., Dose depen-
dence and dose-rate dependence of the optically stimulated luminescence signal, J. Appl. Phys.
89, 259–263. Copyright (2001) with permission of AIP

in the conduction band, the instantaneous concentrations of which are denoted by $n_v(m^{-3})$
and n_c (m^{-3}), respectively. The free electrons in the conduction band may be trapped in N
with a probability coefficient A_n (m^3 s^{-1}), whereas the free holes from the valence band
can be trapped in the centers with a probability coefficient of B (m^3 s^{-1}). The instantaneous
filling of electrons in traps and holes in centers during the excitation (and later, during the
optical stimulation as well) are denoted by n and m, respectively. After the excitation is
finished, a relaxation period is allowed for. This means that the excitation intensity is set
to zero, $X = 0$, and the carriers remaining in the conduction and valence bands relax to
the trapping states and centers, respectively. Note that once holes are accumulated in the
center, recombination may take place between these trapped holes and electrons from the
conduction band with a recombination coefficient probability of A_m (m^3 s^{-1}).

At the next stage of optical stimulation, we assume that the applied light releases electrons
from the trapping state at the rate of fn (m^{-3} s^{-1}), where the light intensity is proportional
to f (s^{-1}) as explained above. The released electrons may retrap in empty trapping states
with the same coefficient of trapping A_n (m^3 s^{-1}) mentioned earlier for the excitation phase,
but they can also recombine with trapped holes with the recombination coefficient A_m. The
OSL signal is assumed to result from this recombination as shown in Figure 7.1.

The set of simultaneous differential equations governing the process during the excitation
period is given by

$$dn_v/dt = X - B(M - m)n_v, \tag{7.1}$$

$$dm/dt = -A_m m n_c + B(M - m)n_v, \tag{7.2}$$

$$dn/dt = A_n(N - n)n_c, \tag{7.3}$$

$$dn_c/dt = dm/dt + dn_v/dt - dn/dt. \tag{7.4}$$

As pointed out earlier, the following stage of relaxation is simulated by taking the final values of n, m, n_c and n_v at the end of the excitation stage as initial values for the relaxation stage; setting X to zero and solving the set of equations for a further period of time until both n_c and n_v become negligibly small.

For the next stage of light stimulation, we take the final values of the functions n, m, n_c and n_v at the end of the relaxation as initial values, keep $X = 0$ and add the term associated with the optical stimulation. The set of equations to be solved now is

$$-dm/dt = A_m m n_c, \tag{7.5}$$

$$dn/dt = -fn + A_n(N - n)n_c, \tag{7.6}$$

$$dn_c/dt = dm/dt - dn/dt. \tag{7.7}$$

Since we associate the intensity of the OSL signal with the recombination rate, we can write the OSL intensity I as

$$I = -dm/dt. \tag{7.8}$$

In fact, a proportionality factor should have preceded the right-hand side of Equation (7.8). However, omitting it merely means that one measures the emission intensity in some different units. Chen and Leung [529] performed the simulation of the OSL process using these sets of equations for relatively short stimulating light pulses, in a similar way to what is done in experiments. They solved the set of equations (7.5–7.8) for 1s and recorded the final value of $I = -dm/dt$ as the OSL signal. This bypasses the question of what is happening during the first fraction of a second when the stimulating light is turned on. It also disregards the time dependence of the OSL during a long period of light stimulation, during which the emitted light intensity may be decaying with time when n and/or m are depleted.

Another point to be made is the dimensions of X appearing in Equation (7.1) and f in Equation (7.6). Basically, since both are proportional to irradiation intensity, one might expect that both have the same units. A closer look shows, however, that the equivalent of X in Equation (7.1) is fn in Equation (7.6), both having units of $m^{-3} s^{-1}$. However, whereas X is usually constant, fn varies when n varies, and f is constant.

7.3   Numerical Results of OSL Dose Dependence

As reported by Chen and Leung [529], it is quite easy to find sets of parameters within this basic energy level model that yields quadratic dependence on the dose of excitation while starting with empty traps and centers. As an example, they gave the following

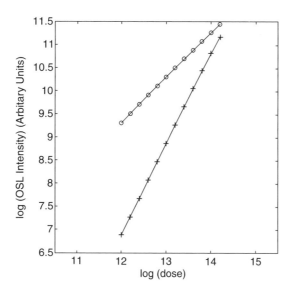

Figure 7.2   *Dose dependence of OSL on a log–log scale as calculated using the model shown in Figure 7.1; the parameters used are given in the text. The (+) symbols are for the case where n(0) = 0 at the beginning of the excitation whereas the (○) symbols for the case with n(0) = 0.9N. The straight line formed by the circles has a slope of 1, indicating linear dependence of the emitted intensity on the excitation dose. The straight line formed by the (+) points has a slope of 2, meaning a quadratic dependence of the OSL intensity on the excitation dose. Reprinted from Chen, R. and Leung, P.L., Nonlinear dose dependence and dose-rate dependence of optically stimulated luminescence and thermoluminescence, Radiat. Meas. 33, 475–481. Copyright (2001) with permission from Elsevier*

set: $A_m = 10^{-17} \, \text{m}^3 \, \text{s}^{-1}$; $B = 10^{-18} \, \text{m}^3 \, \text{s}^{-1}$; $N = 10^{17} \, \text{m}^{-3}$; $M = 10^{19} \, \text{m}^{-3}$; and $A_n = 10^{-19} \, \text{m}^3 \, \text{s}^{-1}$. The value of f was taken as $1 \, \text{s}^{-1}$ and X was varied between 10^{12} and $10^{14} \, \text{m}^{-3} \text{s}^{-1}$. They used the standard ode23 solver in the MATLAB package to solve Equations (7.1–7.4) which were solved with the given value of X, then the same set was solved with $X = 0$ for a further period of time for relaxation, and finally Equations (7.5–7.8) were solved with the given value of f. The results are shown in Figure 7.2. The (+) points in Figure 7.2 show on a log–log scale the dependence of the simulated OSL signal on the total applied dose; the applied dose was changed by changing the intensity of excitation X in the mentioned range while keeping the excitation time at 1 s. The straight line seen has a slope of 2, meaning a quadratic dose dependence. It is to be noted that very similar results were found if the same parameters and the same initial conditions of empty traps and centers were maintained, but the excitation dose was varied by changing the excitation time and keeping the excitation intensity constant.

The results discussed so far suggest that when one is dealing with empty traps and centers at the outset, usually associated with properly annealed samples, the quadratic dose dependence comes out as a natural result. Chen and Leung [529] have also studied the situation where one of the trapping states involved, either the electron trap or the hole center, is nearly full of carriers to begin with. From the theoretical point of view, if the quadratic dose dependence is associated with the product of occupancies of electrons in traps and holes in centers [see e.g. Equation (8.35), and the discussion in Section 8.3], having one of them practically constant might bring about a linear dependence of the OSL intensity on the dose if the other occupancy is linear with the dose. From the experimental results, the fact that superlinearity occurs at annealed samples whereas a linear dose dependence is observed in "as is" samples [524], points in the same direction. In order to simulate this situation, Chen and Leung [529] took $n_0 = 9 \times 10^{16}\,\mathrm{m}^{-3}$, which means that 90% of the traps are full with electrons at the beginning, kept all the parameters the same, and repeated the simulation cycle of excitation, relaxation and optical stimulation. The results are shown by the circles (○) in Figure 7.2. The straight line seen here has a slope of unity. This indicates that the dose dependence under these circumstances is linear. It is to be noted that the intensities of the points in the latter case of linear dependence are significantly higher (three to five orders of magnitude) than those calculated with the same parameters and doses when the trapping states start empty.

Chen and Leung [486] elaborate upon the similarities and dissimilarities between TL and OSL associated with the simple model shown in Figure 7.1. As far as the areas under the two curves are concerned, both are dependent on the quantity $\min(n_0, m_0)$ where n_0 and m_0 are the occupancies of the relevant trap and center at the end of irradiation. In fact, if this one trap-one recombination center is strictly considered, one actually has $n_0 = m_0$. As long as n_0 and m_0 are linear with the dose, which may very well be the case when this model holds, the dependence of the total areas under the TL curve and under the OSL decay curve on the dose is expected to be linear (for a possible exception, see Section 8.7). As far as TL is concerned, the maximum intensity of TL is usually found to be proportional or nearly proportional to the total area. Therefore, the maximum intensity of TL is usually proportional to the excitation dose in the simple case of one trapping state and one kind of recombination center. The situation appears to be different with OSL, as seen in Figure 7.2. The first point to remember in this concern is that in OSL, there is no obvious analog to the maximum of TL. The short pulse response of OSL is more similar to the intensity of TL in the initial-rise region. Here one can consider two extreme cases. If the kinetics order of the process is first, the intensity at each temperature in the TL curve is proportional to the initial occupancy of the traps, n_0. Therefore, if the dose dependence is examined at any temperature along the curve, it is expected to be linear. This is not the case in second-order kinetics. As pointed out by Chen *et al.* [531], in a second-order peak the dose dependence is expected to be quadratic within the initial-rise range. Although the exact conditions for such a quadratic behavior for cases beyond the strict second-order kinetics have not been specified, it is quite possible that the dose dependence in the initial-rise range is superlinear. It seems that there is a close analogy between the initial-rise range of TL and short pulsed OSL since in both we sample the specimen, depleting only a negligible fraction of the carriers in traps and centers. It seems that the quadratic dose dependence of pulsed OSL seen in Figure 7.2 is the correct analog of the quadratic dose dependence of the mentioned TL in the initial-rise range. Further results on the dose dependence of OSL are given in Chapter 8.

7.4 Simulation of the Dose-rate Dependence of OSL

Dose-rate effects of OSL in quartz have been discussed in Bailey *et al.* [532, 533], and they are mentioned in Section 8.6 along with the dose-rate effect of TL. It should be pointed out that generally speaking, for such complex processes taking place during the excitation and optical stimulation, there is no reason to assume that only the total dose is the determining factor. Chen and Leung [529] have simulated the results of this effect using the model in Figure 7.1 and solving numerically Equations (7.1–7.8). They took as an example the following set of parameters: $A_n = 3 \times 10^{-17}\,\text{m}^3\,\text{s}^{-1}$, $N = 10^{18}\,\text{m}^{-3}$, $M = 10^{19}\,\text{m}^{-3}$, $A_m = 10^{-17}\,\text{m}^3\,\text{s}^{-1}$, and $B = 10^{-17}\,\text{m}^3\,\text{s}^{-1}$. The dose rate X was varied from 10^{16} to $10^{23}\,\text{m}^{-3}\,\text{s}^{-1}$ whereas the excitation time was changed inversely, from 10^2 to $10^{-5}\,\text{s}$, so as to keep a constant dose. The stimulation intensity was set at $f = 10^{-5}\,\text{s}^{-1}$ and the stimulation time was taken to be 1s. This rather short stimulation time makes the results discussed here relevant to pulsed-OSL (see Section 7.5). The relevant sets of equations have been solved numerically by the ode23 solver and in the same sequence of stages as described above for the dose dependence case. The results are shown as circles (∘) in Figure 7.3, on a semilog dose rate scale. The OSL simulated intensity is seen to decrease by ∼20% while varying the dose rate from 10^{17} to $10^{20}\,\text{m}^{-3}\,\text{s}^{-1}$.

Chen and Leung [529] also tried to look for sets of parameters for which the increase of the dose rate brings about an increase in the OSL reading. The motivation has been that in those cases in which TL had been found to be dose-rate dependent, a decrease of the emitted intensity with the dose rate for a constant total dose had been seen in some cases and an increase in others. It turned out that by changing just one of the parameters in the set given above, namely if B was taken to be $10^{-18}\,\text{m}^3\,\text{s}^{-1}$ rather than $10^{-17}\,\text{m}^3\,\text{s}^{-1}$, such an increase was seen. The results are shown by the (+) symbols in Figure 7.3. In the same range of X varying between $10^{17}\,\text{m}^{-3}\,\text{s}^{-1}$ and $10^{20}\,\text{m}^{-3}\,\text{s}^{-1}$, the resulting OSL increased by ∼19%. It should be mentioned that the results shown in Figure 7.3 are related to a situation in which initially the trapping states and recombination centers are empty. This is associated with the superlinear dose dependence. The simulation was repeated when the

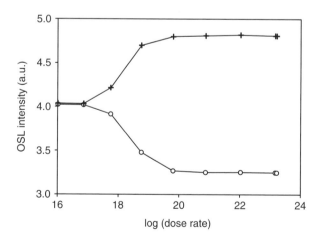

Figure 7.3 *Simulated dose-rate dependence of the emitted OSL signal with the total dose kept constant. The set of trapping parameters is given in the text. Within a range of three to four orders of magnitude in the doserate of excitation, the intensity goes down by ∼20% with one set of parameters, and goes up by ∼19% with another. Reprinted from Chen, R. and Leung, P.L., J. Appl. Phys. 89, 259–263. Copyright (2001) with permission of AIP*

initial filling of the trap was $0.9N$, which brought about linearity in the dose dependence. In this situation, practically no dose-rate dependence was seen in the simulated results. It should be emphasized that no general statement can be made concerning this dose-rate independence. It is possible that with other sets of parameters, dose dependence will be linear, whereas still some dose-rate effect can take place. Further discussion on the dose and dose-rate dependencies of OSL is given in Chapter 8.

7.5   The Role of Retrapping in the Dose Dependence of POSL

Chen and Leung [529] and Chen [534] have stressed the role of retrapping in the dose dependence of POSL. There are two possible versions of the experimental procedure. In the first one we detect the emitted light during the short-pulse illumination and in the second procedure the luminescence is only detected after the end of the simulating light pulse (see McKeever and Akselrod [522] and Akselrod and McKeever [535]). Chen [534] used simulations as well as an approximate analytical approach and showed that quadratic and more than quadratic dose dependence can be predicted for the initial dose range, mainly in annealed samples. The results of both the response to dose during the stimulating pulse and following it, were considered within the simplest possible model of one trapping state and one kind of recombination center shown in Figure 7.1. The conditions under which linear dose dependence can be expected were also discussed; the main point stressed was the effect of retrapping of optically freed carriers on the dose dependence. Let us examine again Equations (7.1–7.4) governing the processes of excitation and relaxation (with $X = 0$) and Equations (7.5–7.7) for the read-out phase, with Equations (7.8) yielding the OSL emission intensity. Similarly to the situation in TL, let us make the quasi-steady assumption [450], namely,

$$|dn_c/dt| \ll |dn/dt|, |dm/dt|; \ n_c \ll n, m, \tag{7.9}$$

then for OSL, in full analogy with TL [see Equation (4.9)], one obtains the simplified equation

$$I(t) = \frac{fA_m mn}{A_m m + A_n(N - n)}. \tag{7.10}$$

Let us consider first the restricted case of pure second order, where some of the conclusions can be reached in an analytical manner. The additional simplifying assumptions $N \gg n$ (trapping states are far from saturation), $n = m$ and $A_m m \ll A_n N$ (retrapping dominates) are now made. Equation (7.10) reduces into

$$I(t) = -\frac{dn}{dt} = \frac{fA_m}{NA_n}n^2. \tag{7.11}$$

Denoting the constant $A_m/(A_n N)$ by k, solving this differential equation for a given constant f, and reinserting n into Equation (7.11) results in

$$I(t) = \frac{fkn_0^2}{(1 + fkn_0 t)^2}. \tag{7.12}$$

For small values of t, $t \ll 1/(fkn_0)$, one has

$$I(t) = fkn_0^2. \tag{7.13}$$

As long as $n_0 \propto D$, which is usually the case for this simple situation of one trapping state and one kind of center in the initial range where $n \ll N$, one gets $I \propto D^2$ (for situations in which this is not the case, see Section 8.7). This is the situation for an instantaneous OSL intensity at a given time point in the short-time region. If one integrates over a certain time period within the short-time range, the dependence on the dose will still be quadratic. On the other hand, let us show that the total area under the curve depends linearly on the excitation dose by integrating over $I(t)$ in Equation (7.12) from 0 to infinity,

$$\int_0^\infty I dt = \int_0^\infty \frac{fkn_0^2}{(1 + fkn_0^2)} dt = \left[\frac{-n_0}{1 + fkn_0 t}\right]_0^\infty = n_0. \tag{7.14}$$

As long as n_0 depends linearly on the dose, the area under the whole OSL curve is also linear with the dose.

Returning to Equation (7.10) without additional simplifying assumptions, in the strict case of one trapping state and one kind of recombination center, the condition $n = m$ can be assumed to hold true since $n_c \ll n$. In the presence of additional trapping states and/or centers, the condition $n \neq m$ usually occurs. Banerjee [526] has shown that when competition takes place, the dependence of the area under the OSL curve may be superlinear. This may certainly influence the dose dependence of the pulsed signal discussed here. However, situations in which the effect of competition with other traps or centers is minimal and yet $n \neq m$ may hold. Even in this case, the conclusions reached in the pure second-order case do not change significantly. In the case of dominating recombination, $A_m m \gg A_n(N - n)$, one has first-order behavior,

$$I \cong fn, \tag{7.15}$$

and as long as n grows linearly with the dose, the response of OSL to a pulse of stimulating light (as well as the area under the decaying OSL curve) is linear. If however retrapping dominates, i.e. $A_n(N - n) \gg A_m m$, Equation (7.10) reduces to

$$I \cong \frac{fA_m mn}{A_n(N - n)}. \tag{7.16}$$

If, in addition, the trapping state is far from saturation, $N \gg n$, this reduces to

$$I \cong \frac{fA_m mn}{A_n N}, \tag{7.17}$$

and as long as m and n grow linearly with the excitation dose, the response of the OSL to a pulse of stimulating light will be quadratic with the dose. If, however, the occupancy of trapping states, n, approaches saturation, there will be situations in which a faster than quadratic dose dependence may take place. Here the term $N - n$ appearing in Equations (7.10) and (7.16) is a decreasing function of the dose. If both n and m are nearly linear with the dose, the additional term $N - n$ appearing in the denominator causes the whole expression to have a more than quadratic dose dependence.

For the response to a light pulse during the stimulation, Chen and Leung [529] have shown that starting with a nearly filled trapping state (e.g., $n_0 = 0.9N$), the dose dependence of OSL could be more or less linear. This agreed with the experimental fact that, at least for quartz samples, linearity was often observed in the unannealed samples and superlinearity in the annealed ones. It appears that the former has to do with $n_0 \neq 0$ and the latter with $n_0 \approx 0$. Some qualitative conclusions can be drawn from Equations (7.10), (7.16) and (7.17).

As pointed out by Chen [534], the case of dominating retrapping is of primary interest here since the situation of dominating recombination [Equation (7.15)] leads to the simple first-order case, in which linearity with the dose is expected. Equation (7.17) results from the assumption $N \gg n$ which does not agree with the condition $n_0 = 0.9N$ used in Chen and Leung [529]. As for Equation (7.16), taking $m_0 = 0$ and $n_0 = 0.9N$ may bring about a situation in which the variation in n with irradiation dose is relatively small, and therefore the dose dependence of OSL intensity is that of m. This in turn may be linear with the dose. The inverse possibility also exists, namely, $n_0 = 0$ and $m_0 \neq 0$, say, $m_0 = 0.9M$. Here, the variation of m with the dose may be small and the dose dependence of I is the same as that of n, which may be linear. As mentioned already, the situation changes at very high doses when n is close to N.

The approach that has been followed so far depended on making simplifying assumptions which enabled us to reach general conclusions in an analytical form. The drawback has been that the approximations made might have an influence on the final results. The alternative is choosing a set of reasonable trapping parameters and solving the relevant equations numerically. Thus, the results are accurate but limited, since they are valid only for the chosen set of parameters and therefore are basically a demonstration that certain dose dependence is possible within the simple model. However, if there is an agreement between the two routes taken, an amount of credibility is added to the theoretical results.

In order to follow the experimental procedure, Chen [534] first solved Equations (7.1–7.4) for a certain period of time t_D, for a given value of X, the rate of production of electrons and holes per unit volume. The total dose is represented by $D = X \cdot t_D$, which has dimensions of m^{-3} and actually is the total concentration of produced electrons and holes. This was followed by continuing the solution during the relaxation stage as described above, and finally, Equations (7.5–7.7) were numerically solved with the given optical stimulation rate f. Two different routes were taken which follow the two different experimental procedures in use. One is to study the response to a short optical stimulating pulse *during* the light exposure, and the other is the light emission *following* the exposure for a short period of the decay time.

An example of the calculated dose dependence of these two versions of the measurement is shown in Figure 7.4 on a log–log scale. The parameters chosen are $A_m = 10^{-10}\,\mathrm{m^3\,s^{-1}}$, $B = 10^{-15}\,\mathrm{m^3\,s^{-1}}$, $A_n = 10^{-9}\,\mathrm{m^3\,s^{-1}}$, $N = 10^{13}\,\mathrm{m^{-3}}$, $M = 10^{13}\,\mathrm{m^{-3}}$, $t_D = 0.1\,\mathrm{s}$, and $f = 10\,\mathrm{s^{-1}}$, X varies from 10^{12} to $10^{16}\,\mathrm{m^{-3}\,s^{-1}}$ and $n_0 = m_0 = n_{v0} = 0$. The points denoted by ($\times$) show the results simulated in a short period of time *during* the light stimulation and those by (o), the emission intensity *following* the stimulating pulse. The general appearance of the two curves is the same with slight differences in the details. The dose dependence starts quadratically, continues faster than quadratic in a narrow dose range, and then goes more or less linearly before approaching saturation. The apparent lack of similarity in the behavior of pulsed OSL and TL was explained by Chen [534] as follows. Had the total area under the OSL curve been taken, in this case of one trap-one center, linear

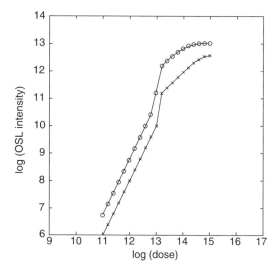

Figure 7.4 *Calculated dose dependence of pulsed OSL. (×) represents the OSL intensity in a short period of time during the simulation. (o) depicts the simulated light intensity following a short stimulating pulse. The values of the parameters are given in the text. Reprinted from Chen, R., Radiat. Prot. Dosim. 100, 71–74. Copyright (2002) with permission from Oxford University Press*

dependence could usually be expected (for an exception, see Section 8.7). Also, in the case of dominating recombination, the situation would be similar to a first-order TL peak in which each part of the curve grows linearly with the dose. However, this is not the case when retrapping dominates. The analog of the POSL is the intensity of TL in the initial-rise range. As shown by Chen *et al.* [531], in a second-order TL peak the dose dependence is expected to be quadratic in the initial-rise range. The case discussed here of dominating retrapping is broader than the relatively simple second order, but the main points of analogy hold.

7.6   Linear-modulation OSL (LM-OSL)

An important variation of the OSL measured under constant intensity of stimulating light was first proposed by Bulur [536]. Since the OSL measured during constant excitation intensity (CW-OSL) or following a pulse of stimulating light is a rather featureless decaying function, Bulur suggested using linear modulation of the stimulating light and the method used is termed LM-OSL. By using this technique, the OSL curves are observed in the form of peaks, and the peak maximum intensity which is easily determined, can be utilized as the measured OSL signal. Thus, the resolution of several OSL signals is significantly easier and the evaluation of the OSL parameters is much simpler. Bulur has written the following rate equation for first-order kinetics LM-OSL

$$\frac{dn(t)}{dt} = -\alpha(I_0/\theta)t \cdot n(t), \tag{7.18}$$

where $n(t)$ is the concentration of trapped carriers, with an initial value of n_0, I_0 is the final intensity of the stimulating light, α represents a proportionality constant, t is the stimulation time to which the stimulating light intensity is proportional and θ is the total length of time of the measurement. The solution is

$$L(t) = -\frac{dn(t)}{dt} = n_0\alpha(I_0/\theta)t \cdot \exp\left(-\frac{\alpha I_0}{2\theta}t^2\right) = L_0\frac{t}{\theta}\exp\left(-\frac{\alpha I_0}{2\theta}t^2\right), \qquad (7.19)$$

where $L(t)$ is the emitted LM-OSL intensity. This is a product of an increasing function t and a peak shaped Gaussian, and therefore $L(t)$ may have a maximum. Equating the derivative of Equation (7.19) with respect to time to zero yields

$$t_{max} = \sqrt{\frac{\theta}{\alpha I_0}}, \qquad (7.20)$$

and inserting this value into Equation (7.19) results in

$$L_{max} = n_0\sqrt{\frac{\alpha I_0}{\theta}}\exp\left(-\frac{1}{2}\right) = \frac{n_0}{t_{max}}\exp\left(-\frac{1}{2}\right). \qquad (7.21)$$

Bulur [536] also explored the LM-OSL governed by general-order kinetics. As pointed out above, general-order kinetics is a heuristic approach which enables an approximate treatment of cases that are neither of first- nor of second-order kinetics. The basic equation governing the process here is

$$\frac{dn(t)}{dt} = -\frac{\gamma I_0}{\theta}tn(t)^b, \qquad (7.22)$$

where b is a dimensionless positive "order", usually between 1 and 2, and γ a constant with dimensions of $cm^{3(b-1)}\,s^{-1}$. Bulur suggested that $\gamma = \alpha/n_0^{b-1}$, similar to the assumption by Wrzesinska [537] for the special second-order case ($b=2$) in TL, and as opposed to the assumption by Rasheedy [206] who used α/N^{b-1}, where N is the total number of the relevant traps. The difference between these two situations can be observed following different exposures, which result in different values of the initial population n_0, whereas N remains the same. Bulur and Göksu [538] opted for the latter possibility and amended the previous equations by Bulur [536],

$$L(t) = -dn/dt = \gamma I_0 n^b. \qquad (7.23)$$

Again, with the linear modulation of the stimulating light intensity, the equation changes into

$$L(t) = -dn/dt = (\gamma I_0/\theta)tn^b. \qquad (7.24)$$

The solution of this differential equation is

$$L(t) = n_0^b(\alpha I_0)(t/N^{b-1}\theta)\left[1 + (b-1)(\alpha I_0 n_0/2\theta N^{b-1})t^2\right]^{-b/(b-1)}. \qquad (7.25)$$

An important special case is that of second-order kinetics where $b = 2$, and the intensity is given by

$$L(t) = \frac{n_0^2 \left[\alpha I_0/(N\theta)\right] t}{\left\{1 + \left[\alpha I_0 n_0/(2N\theta)\right] t^2\right\}^2}. \tag{7.26}$$

Bulur and Göksu [538] have shown that for the case of general-order kinetics , by equating the derivative of Equation (7.25) to zero, the condition for the maximum of the LM-OSL signal is

$$t_{max} = \sqrt{\frac{2}{(b+1)}\frac{\theta}{\alpha I_0}\left(\frac{N}{n_0}\right)^{b-1}}. \tag{7.27}$$

For the special important case of second-order kinetics, this reduces to

$$t_{max} = \sqrt{\frac{2\theta}{3\alpha I_0}\frac{N}{n_0}}. \tag{7.28}$$

Inserting Equation (7.27) into (7.25) yields

$$L_{max} = \frac{2n_0}{(b+1)}\frac{1}{t_{max}}\left[\frac{2b}{b+1}\right]^{-b/(b-1)}. \tag{7.29}$$

In the special case of $b = 2$, this reduces to

$$L_{max} = \frac{3}{8}\frac{n_0}{t_{max}}. \tag{7.30}$$

It is quite obvious that since t_{max} depends on n_0, L_{max} is not proportional to n_0, but is rather superlinear with the initial carrier concentration. In ranges where n_0 is proportional to the dose, L_{max} will be expected to be superlinear with the dose. Combining Equations (7.27) and (7.29), one can immediately see that

$$L_{max} \propto n_0^{(b+1)/2}, \tag{7.31}$$

which, for the second-order case becomes

$$L_{max} \propto n_0^{3/2}. \tag{7.32}$$

Therefore, using L_{max} as the signal measuring the applied dose may be rather problematic. It should be noted here that if we consider as the measured signal the area under the LM-OSL curve taken from $t = 0$ to infinity, the situation is different. Since $L(t)$ is defined as $-dn/dt$, this integral must be n_0. The fact that the total area and the maximum intensity does not behave in the same way with n_0, and therefore with the dose, indicates that the shape of the peak distorts with increasing n_0.

In a recent work, Mishra *et al.* [539] suggest the use of non linear light modulation of OSL (NL-OSL). They mention the use of power dependence of the light intensity of $f \propto t^l$, which, where $l = 1$ corresponds to the LM-OSL discussed so far, and where $l > 1$ yields different shapes of the LM-OSL peak, depending on the order of kinetics. These authors suggest that the NL-OSL technique can be useful in analyzing OSL traps having closely lying values (see also the discussion on the work by Bos and Wallinga [540] in Section 7.12). Whitley and McKeever [541] described an analogous method which involves linearly modulated stimulation light while measuring photoconductivity. They term this technique linearly modulated photoconductivity (LM-PC) and apply it to Al_2O_3:C samples. They report two peaks of such conductivity which correspond to the same charge as in the 450 and 600 K thermally-stimulated-conductivity (TSC) peaks. While comparing the results of LM-PC to those of LM-OSL in the same samples, peak 1 appears in both measurements whereas peak 2 appeared only in LM-PC. The authors suggest that the carriers released from the deeper trap recombine with a different center, the transition into which produces light which is not measurable by the detection device being used.

Kitis *et al.* [542] described the computerized deconvolution of LM-OSL glow curves, in a way analogous to the deconvolution of TL glow curves (see Section 10.1). Singarayer and Bailey [302] reported on the use of LM-OSL in the study of OSL in quartz. Single aliquots of different quartz samples have been found to contain typically five or six common LM-OSL components when stimulated at 470 nm. More work on LM-OSL in quartz, along with TL, has been reported by Kitis *et al.* [543]. Bulur and Yeltik [544] used LM-OSL in BeO ceramics, and found two first-order components; they show that the properties of this material make it a candidate for dosimetry use.

Further discussion on the dose dependence and dose-rate dependence of LM-OSL signals is given in Chapter 8 .

7.7   Unified Presentation of TL, Phosphorescence and LM-OSL

In this section we digress from dealing with OSL only, and present a recently developed mathematical formalism that results in a unified presentation of TL, phosphorescence and LM-OSL. The exponential integrals appearing in Equations (4.11) and (4.19) are not elementary integrals; a number of methods have been offered in the literature for their approximate evaluation for given values of β, T_0 and T (see Chapter 15). However, one should note, that if one does not assume a specific heating function, the integral to be considered is

$$\mathrm{Int}(t) = \int_0^t \exp(-E/kT')dt', \tag{7.33}$$

in which the heating function $T(t)$ is implied. A number of authors   [545, 546] have suggested the use of a hyperbolic heating function where the relationship between time and temperature is $1/T = 1/T_0 - t/\beta'$, where β'(K s) is constant. One has here $dt = (\beta'/T^2)dT$ which when inserted into Equation (7.33) yields

$$\int_0^t \exp(-E/kT')dT' = \beta' \int_{T_0}^T (1/T'^2)\exp(-E/kT')dT'$$
$$= (k\beta'/E)\left[\exp(-E/kT) - \exp(-E/kT_0)\right]. \tag{7.34}$$

Since the function $\exp(-E/kT)$ is increasing very fast with temperature, the second term in the square brackets is negligible for T larger than T_0 even by a few degrees. In a sense, the fact that the TL curves obtained with the hyperbolic heating function are in their "natural" form will be shown below. Instead of Equation (4.11) we get now for first-order kinetics and a hyperbolic heating function

$$I(T) = sn_0 \exp(-E/kT) \exp\left[-(s\beta'k/E)\exp(-E/kT)\right]. \tag{7.35}$$

If we denote $-E/kT$ by x, we get

$$I(x) = sn_0 e^x \exp\left[-(s\beta'k/E)e^x\right], \tag{7.36}$$

which is practically the same as the expressions mentioned before by Kelly and Laubitz [547] and others. Note that as defined here, x is negative, a point discussed below while comparison is made with the presentation of two other luminescence phenomena. It should be mentioned that a slightly different presentation could be made, namely with $x = E/kT$. However, in the former presentation an increasing temperature T is associated with an increasing x and therefore we prefer it over the latter, where it is the other way around. In a similar way, we get from Equation (4.18) the solution for the hyperbolic heating rate and for general-order kinetics

$$I(T) = s''n_0 \exp(-E/kT) \left[1 + (b-1)(s''k\beta'/E)\exp(-E/kT)\right]^{-b/(b-1)}. \tag{7.37}$$

By using the same substitution $x = -E/kT$, we get

$$I(x) = s''n_0 \left[1 + (b-1)((s''k\beta'/E)e^x\right]^{-b/(b-1)}. \tag{7.38}$$

In the second-order case $b = 2$ and this reduces to

$$I(x) = s''n_0 e^x \left[1 + (s''k\beta/E)e^x\right]^{-2}. \tag{7.39}$$

As mentioned above, the phosphorescence decay curve is the emission of light from the irradiated sample while it is held at a fixed temperature T. From Equation (4.10), the decay function in the first-order case is

$$I(t) = sn_0 \exp(-E/kT) \exp\left[-s \cdot \exp(-E/kT) \cdot t\right]. \tag{7.40}$$

As pointed out by Randall and Wilkins [6], and later elaborated by Visocekas [214], this featureless decay curve can be transformed into a peak-shaped curve by plotting $t \cdot I(t)$ as a function of $\ln(t)$. We can write

$$y = t \cdot I(t) = \phi n_0 t e^{-\phi t}, \tag{7.41}$$

where $\phi = s \cdot \exp(-E/kT)$. If we denote $\ln(t)$ by x, we have $t = e^x$, and get

$$y = \phi n_0 e^x \exp(-\phi e^x). \tag{7.42}$$

Apart from the meaning of the constants, Equation (7.42) has exactly the same mathematical form as Equation (7.36) for the first-order TL with hyperbolic heating function. This TL-like presentation of phosphorescence can be easily extended to the general-order case, as shown by Chen and Kristianpoller [315]. Solving Equation (4.18) for a fixed temperature

T, we get

$$y = \phi n_0 e^x \left[1 + (b-1)\phi e^x\right]^{-b/(b-1)}, \tag{7.43}$$

where ϕ is the same as defined above. Obviously, the second-order function is the special case with $b = 2$, namely

$$y = n_0 e^x \left[1 + \phi e^x\right]^{-2}. \tag{7.44}$$

Chen and Kristianpoller have shown that by inserting the maximum condition derived from Equation (7.43) back into this equation, one gets for the general-order kinetics $y_m = n_0/b^{b/(b-1)}$, which for the special case of $b = 2$ yields $y_m = n_0/4$. In a similar way, using Equation (7.42) one gets for first-order kinetics $y_m = n_0/e$.

A new way of presenting the LM-OSL has been proposed by Chen and Pagonis [548]. Multiplying both sides of Equation (7.19) by t and taking $y = t \cdot L(t)$ and $t^2 = e^x$ or ($x = 2\ln t$) we obtain

$$y(x) = 2n_0 \delta e^x \exp(-\delta e^x), \tag{7.45}$$

where $\delta = \alpha I_0/(2\theta)$. As for the general-order situation, multiplying Equation (7.25) by t, taking the same x as before and defining $\delta' = \alpha I_0 n_0^{b-1}/(2\theta N^{b-1})$, we get

$$y = 2n_0 \delta' e^x \left[1 + (b-1)\delta' e^x\right]^{-b/(b-1)}. \tag{7.46}$$

Here too, the second-order expression is simply

$$y = 2n_0 \left[1 + \delta' e^x\right]^{-2}, \tag{7.47}$$

where $\delta' = \alpha I_0 n_0/(2\theta N)$. As shown by Chen and Pagonis [548], the advantage of this presentation of LM-OSL as compared with the original one, is that irrespective of the kinetics order, the maximum intensity of the curve in question is proportional to the initial concentration n_0 which, under the appropriate circumstances is proportional to the excitation dose. It is quite obvious that Equations (7.36), (7.42) and (7.45) for first-order kinetics in the three phenomena look exactly the same except for the meaning of the constants. The same is true for Equations (7.38), (7.43) and (7.46) for the general-order situation, and for Equations (7.39), (7.43) and (7.47) for the second-order case with $b = 2$. Chen and Pagonis [548] have also shown that in this new LM-OSL presentation, the total area under the curve is proportional to the concentration of the trapped carriers n_0, and this is the case for the similar presentations of TL and phosphorescence. As pointed out, this is of importance in dosimetric applications.

7.8   The New Presentation of LM-OSL Within the OTOR Model

The model shown in Figure 4.1 is the simplest possible OTOR model for explaining OSL, including LM-OSL. The same model was used by Chen *et al.* [549] to explain the possible sublinear dose dependence of the occupancy of trap n_0 on the excitation dose (see Section 8.7). The meaning of the different magnitudes is given in the text. The equations governing the TL process are Equations (4.5–4.7) and in direct analogy, the equations for OSL are (7.5–7.8). The linear modulation is achieved by increasing f linearly with time during read-out, namely $f = f_0 \cdot t$, where $f_0 \, (\mathrm{s}^{-2})$ is constant. In order to bypass the possible effect of sublinearity taking place during the excitation stage (see Section 8.7), these authors [549] suggested that in this very simple model, at the end of irradiation and relaxation one can expect an equal number of electrons and holes to be trapped in trapping states and recombination centers, respectively. Therefore they skipped the first steps of excitation and relaxation, assumed certain values of n_0, and solved the equations for the read-out stage for different values of n_0. The dependence of the maximum LM-OSL curve on n_0 was monitored for different sets of parameters, both in the original presentation by Bulur [536] and in the new presentation by Chen and Pagonis [548] (see Section 7.7). The possible dependencies of n_0 on the excitation dose within the OTOR model will be discussed in Section 8.7.

The rate equations governing the process are Equations (7.5–7.8) mentioned above. Chen *et al.* [549] discussed the sets of parameters to be chosen in order to get approximately first-order, second-order as well as intermediate situations. In all these cases, the quasi-steady condition mentioned above [Equation (7.9)] is assumed to hold. They pointed out that the properties of the solution depend on the relationship between A_n and A_m. However, for the quasi-steady condition to hold, both probability coefficients are to be large enough so that the values of n_c are significantly smaller than those of n. Halperin and Braner [30] gave the approximate equation [see Equation (7.10)] for OSL when quasi-steady situation holds. Let us repeat this equation,

$$L = -\frac{dm}{dt} = fn\frac{A_m m}{A_m m + A_n(N-n)}. \tag{7.48}$$

As pointed out above, in the OTOR case and when the quasi-steady condition holds, $n = m$. When $A_m m \gg A_n(N-n)$, the first-order equation results immediately in the following expression, in analogy with first-order TL in Equation (4.10)

$$L = -\frac{dn}{dt} = f \cdot n. \tag{7.49}$$

Note that $A_m \gg A_n$ is not a sufficient condition for Equation (7.49) to be a good approximation since $N - n$ may be significantly larger than m. The functional condition $A_m m \gg A_n(N-n)$ must hold for the first-order kinetics to take place. Moreover, since m and n decrease along the OSL measurement, one may start with $A_m m$ being much larger than $A_n(N-n)$, but the relationship between them may change, and a transition may occur from first- to second-order kinetics along the measurement. The analogous situation concerning TL has been discussed in Section 4.1. Note also that since $f = f_0 t$, Equation (7.49) is the same as Equation (7.18) where $f_0 = \alpha I_0/\theta$.

This is in full analogy to the discussion concerning TL in Section 4.1, Equations (4.12–4.14). If $A_m = A_n$, and since with this simple case of OTOR one has $m = n$, one gets

$$L(t) = -\frac{dn}{dt} = \frac{f_0}{N}t \cdot n^2. \tag{7.50}$$

The condition that the two probability coefficients be equal is not very likely to occur since the trapping state and recombination center are two independent entities, associated with two different impurities or defects in the sample. Another way of getting the second-order kinetics (see also Section 4.1) is assuming that retrapping dominates i.e. $A_n(N - n) \gg A_m m$, and that the trap is far from saturation, $N \gg n$. As in the first-order case, these are assumed relationship between variables rather than parameters. However, since m and n decrease during the measurement, if the conditions hold at the beginning, they hold all along the measurement. The governing equation here is

$$L(t) = -\frac{dn}{dt} = \frac{f_0 A_m}{N A_n} t \cdot n^2. \tag{7.51}$$

It appears more likely to get second-order kinetics with the dominating retrapping condition than with the equal probability coefficients case. Both Equations (7.50) and (7.51) are the same as Equation (7.22) with $b = 2$ (with substitution of the relevant constants, $\frac{\gamma I_0}{\theta} = \frac{f_0 A_m}{N A_n}$), and therefore Equation (7.26) is its solution.

Note that there are many intermediate cases in which $m A_m$ is of the same order of magnitude as $A_n(N - n)$. These cannot be precisely presented by the "general-order" equation. In order to check the dose dependence of the original LM-OSL as well as the revised presentation of LM-OSL curves, Chen *et al.* [549] chose sets of parameters, and solved numerically Equations (7.9–7.11) for different values of the initial traps and centers occupancies $n_0 = m_0$. They then changed the values of the initial occupancies, and observed the variations in the peak shape. This was done for cases of first order, second order of the two types mentioned and for intermediate cases. Along the simulations, they monitored the values of n_c and n in order to make sure that the quasi-equilibrium condition holds.

Figure 7.5 shows the results of the simulation for the set of parameters given in the caption. Here A_m is five orders of magnitude larger than A_n, and first-order kinetics may be expected. Indeed, these authors report that $m A_m$ was found to be three to four orders of magnitude larger than $A_n(N - n)$ in the simulations. The result is a peak of the LM-OSL

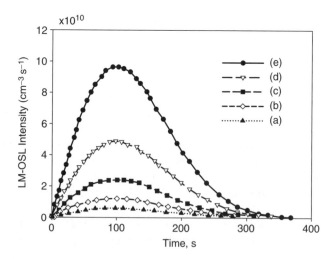

Figure 7.5 LM-OSL simulated with $A_m = 10^{-5} cm^3 s^{-1}$; $A_n = 10^{-10} cm^3 s^{-1}$; $B = 10^{-13} cm^3 s^{-1}$; $N = 10^{14} cm^{-3}$. The value of n_0 varies by factors of 2 in curves (a–e), being $10^{12} cm^{-3}$ in curve (a) Reprinted from Chen, R., Pagonis, V. and Lawless, J.L., A new look at linear-modulated optically stimulated luminescence (LM-OSL) as a tool for dating and dosimetry. Radiat. Meas. 44, 344–350. Copyright (2009) with permission from Elsevier

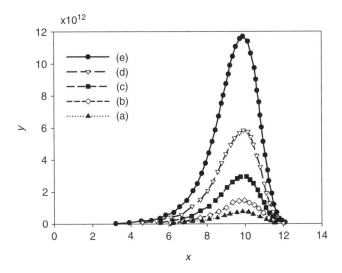

Figure 7.6 *The same results in Figure 7.5, plotted in the new presentation. Reprinted from Chen, R., Pagonis, V. and Lawless, J.L., A new look at linear-modulated optically stimulated luminescence (LM-OSL) as a tool for dating and dosimetry, Radiat. Meas. 44, 344–350. Copyright (2009) with permission from Elsevier*

curve which does not shift with the change of the initial value of n_0. As seen in the graph, the maximum intensity changes by a factor of 2 from one curve to the next, as expected when n_0 varies by factors of 2.

Figure 7.6 shows the same results in the new presentation suggested by Chen and Pagonis [548]. The peaks here are very similar in shape to first-order TL peaks. As expected, the maxima are linearly dependent on n_0.

Figure 7.7 shows the results of simulations with the parameters given in the caption. Here A_n is larger than A_m, and along with the fact that N is significantly larger than m, Chen *et al.* [549] found that $A_n(N - n) \gg A_m m$ and the second order results. As in the case of TL, the peak shifts to lower values of t for higher values of n_0. As for the maximum intensity, it is readily seen to increase superlinearly with n_0. For example, while moving from curve (a) to (e), n_0 increases by a factor of 16, and L_{max} by ~64 which is $16^{3/2}$, in agreement with Equation (7.32).

Figure 7.8 shows the same results in the new presentation. The peaks are symmetric, similar to second-order TL peaks. As expected from the work by Chen and Pagonis [548], the maximum intensity here is proportional to the value of n_0. Similar simulations were performed with $A_m = A_n$. As expected, irrespective of the relationship between $A_m m$ and $A_n(N - n)$, the second-order behavior was seen.

Figure 7.9 shows the LM-OSL results of an intermediate case, with the relevant parameters given in the caption. Here for the lowest value of n_0, $A_n(N - n)$ is reported to be an order of magnitude larger than $A_m m$, whereas for the highest value, the former is smaller than the latter. In this sense, this case is intermediate between first and second order. The peak shifts here with the value of n_0, pointing to a non-first-order behavior. The ratio of the L_{max} values between the largest and smallest n_0 is here ~48, which is rather superlinear, behaving like

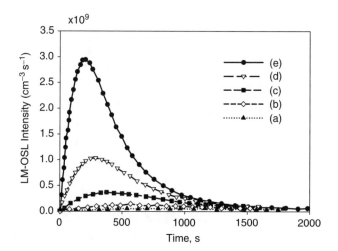

Figure 7.7 *LM-OSL simulated with:* $A_m = 10^{-12} cm^3 s^{-1}$; $A_n = 10^{-11} cm^3 s^{-1}$; $B = 10^{-13} cm^3 s^{-1}$; *and* $N = 10^{14} cm^{-3}$. *The value of* n_0 *varies by factors of 2 in curves (a–e), being* $10^{10} cm^{-3}$ *in curve (a). Reprinted from Chen, R., Pagonis, V. and Lawless, J.L., A new look at linear-modulated optically stimulated luminescence (LM-OSL) as a tool for dating and dosimetry, Radiat. Meas. 44, 344–350. Copyright (2009) with permission from Elsevier*

$\sim D^{1.4}$. If one considers the dependence mentioned in Equation (7.31) of $L_{max} \propto n_0^{(b+1)/2}$, this yields an effective order of $b = 1.8$. However, it should be noted that this approach is entirely empirical, and had the effective order been evaluated from another feature of the curve such as the symmetry factor, a different value of b could have been reached.

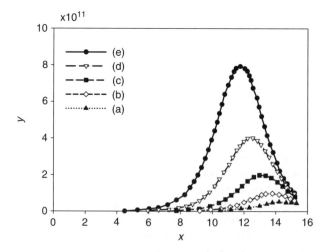

Figure 7.8 *The same results as in Fig. 7.7, plotted in the new presentation. Reprinted from Chen, R., Pagonis, V. and Lawless, J.L., A new look at linear-modulated optically stimulated luminescence (LM-OSL) as a tool for dating and dosimetry. Radiat. Meas. 44, 344–350. Copyright (2009) with permission from Elsevier*

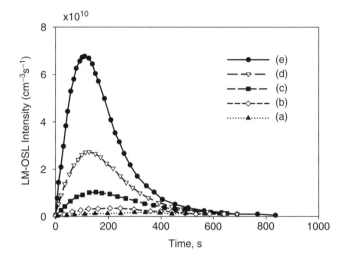

Figure 7.9 *LM-OSL simulated with:* $A_m = 10^{-9} cm^3 s^{-1}$; $A_n = 10^{-10} cm^3 s^{-1}$; $B = 10^{-13} cm^3 s^{-1}$; *and* $N = 10^{14} cm^{-3}$. *The value of* n_0 *varies by factors of 2 in curves (a–e), being* $10^{12} cm^{-3}$ *in curve (a). Reprinted from Chen, R., Pagonis, V. and Lawless, J.L., A new look at linear-modulated optically stimulated luminescence (LM-OSL) as a tool for dating and dosimetry, Radiat. Meas. 44, 344–350. Copyright (2009) with permission from Elsevier*

Figure 7.10 depicts the same results in the new presentation. Here, the dependence on n_0 is somewhat more than linear. The ratio between the maximum intensities is 18.38 for the change of n_0 by a factor of 16. The dependence on the initial occupancy behaves here like $\sim n_0^{1.05}$, a weak superlinearity as compared with the expected linearity for any value of b in this presentation, had the general-order kinetics Equation (4.18) been precisely the governing equation. One should not be surprised about this apparent discrepancy, since one cannot expect the result of the simulation to fit exactly into the general-order framework which is entirely heuristic in nature.

Instead of solving numerically Equations (7.5–7.8), one can proceed [550] by using Equation (7.48) along with the condition $n = m$ resulting from the OTOR model and $f = f_0 \cdot t$. This yields

$$L = -\frac{dn}{dt} = f_0 t \frac{A_m n^2}{A_m n + A_n (N - n)},\tag{7.52}$$

which is subject to the initial condition of $n = n_0$ at $t = 0$. This can be immediately integrated

$$\frac{A_m - A_n}{A_n} \ln\left(\frac{n_0}{n}\right) + \frac{A_n N}{A_m}\left(\frac{1}{n} - \frac{1}{n_0}\right) = \frac{1}{2} f_0 t^2.\tag{7.53}$$

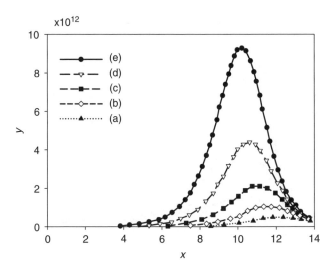

Figure 7.10 *The same as Figure 7.9, plotted in the new presentation. Reprinted from Chen, R., Pagonis, V. and Lawless, J.L., A new look at linear-modulated optically stimulated luminescence (LM-OSL) as a tool for dating and dosimetry, Radiat. Meas. 44, 344–350. Copyright (2009) with permission from Elsevier*

Equation (7.53) can be rearranged to solve for t as a function of n

$$t = \sqrt{\frac{2}{f_0}\left[\frac{A_m - A_n}{A_m}\ln\left(\frac{n_0}{n}\right) + \frac{A_n N}{A_m}\left(\frac{1}{n} - \frac{1}{n_0}\right)\right]}. \tag{7.54}$$

Combining Equations (7.52) and (7.54), the luminescence L can be found as a function of n alone,

$$L = \sqrt{2f_0\left[\frac{A_m - A_n}{A_m}\ln\left(\frac{n_0}{n}\right) + \frac{A_n N}{A_m}\left(\frac{1}{n} - \frac{1}{n_0}\right)\right]}\,\frac{A_m n^2}{A_m n + A_n(N - n)}. \tag{7.55}$$

One is also interested in the transformed luminescence intensity tL, which is found from Equations (7.54) and (7.55),

$$y = tL = 2\left[\frac{A_m - A_n}{A_m}\ln\left(\frac{n_0}{n}\right) + \frac{A_n N}{A_m}\left(\frac{1}{n} - \frac{1}{n_0}\right)\right]\frac{A_m n^2}{A_m n + A_n(N - n)}. \tag{7.56}$$

One now has t, L and tL, all as explicit functions of n in Equations (7.54-7.56). In order to create plots of L as a function of t and of Lt as a function of $2\ln t$, Chen *et al.* [549] chose a series of values of n in the range $0 \leq n \leq n_0$. For each such value of n, they used Equations (7.54) and (7.55) to compute the values of the time t and luminescence L which occur when the density reaches these values of n. This provides a point for use on a plot of L versus t. Similarly, by choosing various values of n, one can create the transformed plot of tL versus $2\ln t$. Chen *et al.* [549] report that exactly the same results were reached by this method and by the above mentioned simulations. Obviously, this has to do with the choice of the sets of parameters which ensured that the quasi-equilibrium condition holds.

Chen *et al.* [549] further show that Equations (7.54–7.56) can be simplified by defining the following non-dimensional quantities,

$$\hat{t} = t\sqrt{f_0}, \tag{7.57}$$

$$\hat{n} = n/N, \tag{7.58}$$

$$\hat{L} = \frac{L}{N\sqrt{f_0}}, \tag{7.59}$$

$$\hat{A} = \frac{A_n}{A_m}. \tag{7.60}$$

Using these variables, Equations (7.54–7.56) can be written as

$$\hat{t} = \sqrt{2\left[(1 - \hat{A})\ln\left(\frac{\hat{n}_0}{\hat{n}}\right) + \hat{A}\left(\frac{1}{\hat{n}} - \frac{1}{\hat{n}_0}\right)\right]}, \tag{7.61}$$

$$\hat{L} = \sqrt{2\left[(1 - \hat{A})\ln\left(\frac{\hat{n}_0}{\hat{n}}\right) + \hat{A}\left(\frac{1}{\hat{n}} - \frac{1}{\hat{n}_0}\right)\right]}\frac{\hat{n}^2}{\hat{n} + \hat{A}(1 - \hat{n})}, \tag{7.62}$$

$$\hat{t}\hat{L} = 2\left[(1 - \hat{A})\ln\left(\frac{\hat{n}_0}{\hat{n}}\right) + \hat{A}\left(\frac{1}{\hat{n}} - \frac{1}{\hat{n}_0}\right)\right]\frac{\hat{n}^2}{\hat{n} + \hat{A}(1 - \hat{n})}. \tag{7.63}$$

Whereas the right-hand sides of Equations (7.54–7.56) depend on six quantities, A_m, A_n, f_0, N, n_0 and n, in Equations (7.61–7.63) the right-hand sides depend on only three quantities, $\hat{n}$, $\hat{n}_0$ and $\hat{A}$. Let us consider the LM-OSL peak intensity using these quantities and equations. For fixed initial concentration $\hat{n}_0$ and fixed rate-constant ratio $\hat{A}$, we can find via Equation (7.62) the value of concentration $\hat{n}$ which maximizes the LM-OSL luminescence $\hat{L}$. Chen *et al.* [549] have done this using a numerical function maximizer. The value of the maximum luminescence $\hat{L}$ can be plotted as shown in Figure 7.11 as a function of n_0/N for different values of $\hat{A}$. This log–log plot has a slope of 1.5 for large values of A_n/A_m and the slope is more than unity even for smaller values, which shows that the response of the LM-OSL maximum is superlinear except when kinetics are first-order; the response is linear only when $\hat{n}_0 \gg \hat{A}$, i.e. $n_0 A_m \gg N A_n$.

The transformed LM-OSL can be studied similarly. One starts by finding the values of $\hat{n}$ which maximize Equation (7.63) for fixed values of $\hat{n}_0$ and $\hat{A}$. This was done again with the numerical maximizer. The values of $\hat{t}\hat{L}$ found from the maximizer are plotted against $\hat{n}_0$ on a log–log scale in Figure 7.12. Note that the peak of the transformed intensity $\hat{t}\hat{L}$, depends on only two parameters $\hat{n}_0$ and $\hat{A}$. Since Figure 7.11 shows the peak plotted against $\hat{n}_0$ for various values of $\hat{A}$, this plot includes the entire parameter space. From Figure 7.12, it is clear that the transformed LM-OSL peak intensity is mostly linear with n_0. Note that, unlike

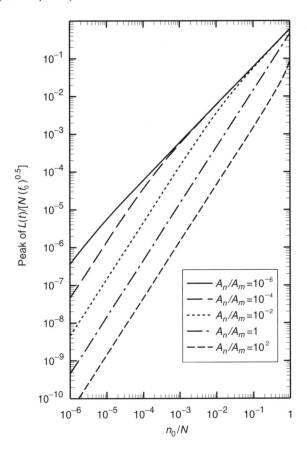

Figure 7.11   *The peak of LM-OSL as plotted against normalized initial concentration n_0/N for various values of $\hat{A} = A_n/A_m$. Reprinted from Chen, R., Pagonis, V. and Lawless, J.L., A new look at linear-modulated optically stimulated luminescence (LM-OSL) as a tool for dating and dosimetry, Radiat. Meas. 44, 344–350. Copyright (2009) with permission from Elsevier*

the case of usual LM-OSL, the peak intensity of the transformed LM-OSL is independent of the experimental value of f_0, though its position depends on $f_0^{-0.5}$, as can be seen in Equation (7.54).

In this work [549] single LM-OSL peaks are dealt with, but in dosimetry and dating it is common to work with a number of peaks. If there is some overlapping of components but the peaks are separated sufficiently, integration over certain interval may be used so as to maximize counts but at the same time to isolate as much as possible the different peaks. Chen *et al.* [549] have studied the dependence of such an area on n_0 in the simulations for the set of parameters leading to Figures 7.9 and 7.10, where the dependence on n_0 is strongly superlinear. They have taken the area from $t_m - \Delta t$ to $t_m + \Delta t$ where t_m is the time at the maximum of the LM-OSL signal and Δt a selected time interval. For each value of Δt, the value of x in the approximate dependence n_0^x has been determined. The results are shown in Table 7.1. The maximum intensity depends superlinearly on n_0 with sets of

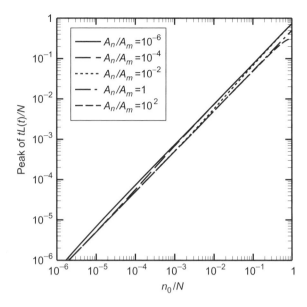

Figure 7.12 *The peak of normalized transformed luminescence, tL/N, plotted against the normalized initial concentration, n_0/N, for various values of $\hat{A} = A_n/A_m$. Reprinted from Chen, R., Pagonis, V. and Lawless, J.L., A new look at linear-modulated optically stimulated luminescence (LM-OSL) as a tool for dating and dosimetry, Radiat. Meas. 44, 344–350. Copyright (2009) with permission from Elsevier*

parameters leading to second-order and intermediate cases, whereas the total area is always linear with n_0. One might expect that the integral over a fixed limited time interval will depend superlinearly on n_0, though a "weaker" superlinearity can be anticipated. Indeed, the value of x in the approximate power dependence is seen in Table 7.1 to be smaller when the interval of integration gets larger. The value of x is still significantly larger than unity even when the integration goes from $t_m - 100$ to $t_m + 100$ (s).

As for the new presentation of LM-OSL with the same set of parameters, the dependence of the maximum intensity on the initial occupancy behaved like $\sim n_0^{1.05}$ in the example given, a very weak superlinearity as compared with the expected linearity for any value of b in the general-order presentation. The results of Figure 7.12 show that the near-linear dependence of the transformed LM-OSL on n_0 takes place generally within the OTOR situation. However, as pointed out above, one should not be too surprised that there are

Table 7.1 *Dependence of the power x of the dependence of the area under the LM-OSL peak on the time interval of integration. Integration was performed between $t_m - \Delta t$ and $t_m + \Delta t$. The parameters used for this simulation are those used for Figures 7.9 and 7.10. From Chen et al. [549]*

Δt (s)	0	25	50	75	100
x	1.395	1.390	1.375	1.350	1.316

slight deviations from linearity since one cannot expect the results of the simulation to fit exactly into the general-order framework.

In cases where the dependence of n_0 on the dose is linear, the dependence of the LM-OSL maximum on the dose is obviously the same as that of n_0, both in the original and revised presentations. In other cases, n_0 may depend non linearly on the dose, say if it is sublinear as described in Section 8.7, which may also result from the OTOR model. If, for example, $n_0(D)$ is a square-root function, this should be combined with the dependence of the LM-OSL signal on n_0. Thus, in cases where the latter depends linearly on the dose, the signal will vary like $D^{1/2}$, whereas if $n_0(D) \propto D^{3/2}$ the LM-OSL maximum may behave like $D^{3/4}$.

7.9   TL-like Presentation of CW-OSL in the OTOR Model

As pointed out in Section 7.7, the decaying phosphorescence curve can be presented as a TL-like peak-shaped function. One plots $y = Lt$ as a function of $x = \ln t$ and gets a peak identical to that of TL with hyperbolic heating function. As shown by Visocekas [214] and later discussed by Chen and Kristianpoller [315] for general-order kinetics, the symmetry properties of this TL-like curve are such that the symmetry factor μ_g varies between 0.4 and 0.5 for kinetics order between $b = 1$ and $b = 2$. As pointed out above, Chen and Kristianpoller have also shown that the maximum intensity of the TL-like function is proportional to n_0/e where n_0 is the initial occupancy of traps for first-order kinetics, and proportional to $n_0/b^{b/(b-1)}$ for general-order kinetics. For the specific second-order case with $b = 2$, the maximum intensity is proportional to $n_0/4$. In all these cases the maximum of the TL-like phosphorescence curve is proportional to the initial concentration, which is a desirable feature for dosimetry and dating. However, keeping in mind the heuristic nature of the general-order kinetics, we may wish to understand the dependence of the maximum signal on n_0 in the OTOR model which, as explained above, may be a better presentation of a realistic situation when one trapping state and one kind of recombination center are involved.

Let us discuss now the case of CW-OSL, the OSL measured following appropriate excitation and under continuous wave stimulating light. For such stimulating light intensity f, and assuming that the quasi-steady condition holds within the OTOR model, we can assume that at the end of excitation the concentrations of electrons in traps and holes in centers are equal, $m = n$. In analogy to Equation (7.52) (developed for LM-OSL), we get

$$L = -\frac{dn}{dt} = f \frac{A_m n^2}{A_m n + A_n(N - n)}. \tag{7.64}$$

Subject to the initial condition $n(0) = n_0$, we get, in analogy to Equation (7.53)

$$\frac{A_m - A_n}{A_n} \ln\left(\frac{n_0}{n}\right) + \frac{A_n N}{A_m}\left(\frac{1}{n} - \frac{1}{n_0}\right) = ft. \tag{7.65}$$

This can be rearranged in analogy to Equation (7.54) to yield

$$t = \frac{1}{f}\left[\frac{A_m - A_n}{A_n} \ln\left(\frac{n_0}{n}\right) + \frac{A_n N}{A_m}\left(\frac{1}{n} - \frac{1}{n_0}\right)\right]. \tag{7.66}$$

Finally, analogously to Equation (7.56), one gets from Equations (7.64) and (7.66)

$$y = Lt = \left[\frac{A_m - A_n}{A_n} \ln \left(\frac{n_0}{n} \right) + \frac{A_n N}{A_m} \left(\frac{1}{n} - \frac{1}{n_0} \right) \right] \frac{A_m n^2}{A_m n + A_n (N - n)}. \qquad (7.67)$$

This equation is not only analogous to Equation (7.56) it is identical to it excluding the factor 2 in the latter. In an analogous manner to that described for LM-OSL, one can proceed by defining $\hat{t} = ft$; $\hat{n} = n/N$; $\hat{L} = L/(fN)$; $\hat{A} = A_n/A_m$. However, this may not be needed. Since Equation (7.67) is practically the same as Equation (7.56), the conclusions concerning the n_0 dependence of the maximum hold for the present case. Therefore, the results in Figure 7.12 are valid also for the TL-like presentation of phosphorescence. Thus, it is obvious that in addition to the strict linearity of the signal in first- and second-order situations, in all the OTOR cases the n_0 dependence of the signal is nearly linear for all values of A_n/A_m. The practical advantage over the LM-OSL case is that the experiment is simpler, since there is no need for the linear modulation. One only has to manipulate the decaying OSL results and the conclusion of near linearity is the same for LM-OSL and the TL-like presentation.

Exactly the same conclusions can be reached for the decay of phosphorescence within the OTOR model. In Equation (7.64), one has to replace the constant stimulating light intensity f by $\gamma = s \cdot \exp(-E/kT)$ where s and E are the relevant frequency factor and activation, energy respectively, and T is the temperature. Since in the phosphorescence measurement the temperature is kept unchanged, this factor is constant. Thus, Equations (7.64–7.67) remain exactly the same for the case of phosphorescence, with γ replacing f everywhere.

7.10   Dependence of Luminescence on Initial Occupancy; OTOR Model

Chen *et al.* [551] have taken the discussion concerning Equations (7.52–7.56) one step further, considering different experimental OSL situations. The latter include a number of non-linearly modulated OSL (NL-OSL) cases suggested by Bos and Wallinga [540], namely exponentially increasing OSL (EM-OSL), hyperbolically increasing OSL (HM-OSL) and reciprocally increasing OSL (RM-OSL), in addition to the mentioned LM-OSL. A power-law dependence of the stimulating-light intensity is used of the form $f = f_0 t^k$, (where k is any power, an integer or a fraction), and f_0 has the appropriate units so that f has dimensions of s^{-1}. Equation (7.52) slightly changes now to

$$L = -\frac{dn}{dt} = f_0 t^k \frac{A_m n^2}{A_m n + A_n (N - n)}. \qquad (7.68)$$

This integrates immediately to yield

$$\frac{A_m - A_n}{A_n} \ln \left(\frac{n_0}{n} \right) + \frac{A_n N}{A_m} \left(\frac{1}{n} - \frac{1}{n_0} \right) = \frac{1}{k + 1} f_0 t^{k+1}. \qquad (7.69)$$

From Equation (7.68) one immediately gets

$$t \cdot L(t) = f_0 t^{k+1} \frac{A_m n^2}{A_m n + A_n (N - n)}, \qquad (7.70)$$

and by comparing Equations (7.69) and (7.70) one gets

$$y = tL = (k+1) \left[\frac{A_m - A_n}{A_m} \ln\left(\frac{n_0}{n}\right) + \frac{A_n N}{A_m} \left(\frac{1}{n} - \frac{1}{n_0}\right) \right] \frac{A_m n^2}{A_m n + A_n(N-n)}. \quad (7.71)$$

This is exactly the same as Equation (7.56), except for the factor $k+1$ replacing the factor 2. Clearly Equation (7.56) is a special case of Equation (7.71), applicable for the LM-OSL case where $k=1$, and Equation (7.67) is the special case for CW-OSL and phosphorescence within the general OTOR case with $k=0$.

Another simulation mode mentioned by Bos and Wallinga [540] with regard to the first-order NL-OSL, is the exponential mode which can be written as

$$f(t) = f_0 e^{\alpha t}, \quad (7.72)$$

for which Equation (7.64) can be changed into

$$L(t) = f_0 e^{\alpha t} \frac{A_m n^2}{A_m n + A_n(N-n)}. \quad (7.73)$$

By integration, one gets

$$\left[\frac{A_m - A_n}{A_m} \ln\left(\frac{n_0}{n}\right) + \frac{A_n N}{A_m} \left(\frac{1}{n} - \frac{1}{n_0}\right) \right] = \frac{1}{\alpha} f_0 e^{\alpha t}, \quad (7.74)$$

from which we have

$$L(t) = \frac{1}{\alpha} \left[\frac{A_m - A_n}{A_m} \ln\left(\frac{n_0}{n}\right) + \frac{A_n N}{A_m} \left(\frac{1}{n} - \frac{1}{n_0}\right) \right] \frac{A_m n^2}{A_m n + A_n(N-n)}, \quad (7.75)$$

where the right-hand side is the same as that in Equation (7.71) except that $1/\alpha$ replaces $(k+1)$. Note that on the left-hand side we have $L(t)$ and not $tL(t)$.

Going one step further, we can consider any monotonically non-decreasing stimulation mode $f(t)$. This may include the above mentioned cases, along with the hyperbolically increasing stimulation intensity (HM-OSL) and the reciprocal stimulation function mentioned by Bos and Wallinga [540], as well as other possibilities. In full analogy to the previous cases, one can get the general relationship

$$\frac{\int_0^t f(t')dt'}{f(t)} L(t) = \left[\frac{A_m - A_n}{A_m} \ln\left(\frac{n_0}{n}\right) + \frac{A_n N}{A_m} \left(\frac{1}{n} - \frac{1}{n_0}\right) \right] \frac{A_m n^2}{A_m n + A_n(N-n)}. \quad (7.76)$$

It can be readily seen that the case of constant stimulation intensity and phosphorescence, as well as all the mentioned monotonically increasing stimulation functions are included in Equation (7.76). In these cases the function of t in front of $L(t)$ can be easily evaluated. With other stimulation functions the evaluation of $\int_0^t f(t')dt'/f(t)$ may not be as easy. Note that the expression $\int_0^t f(t')dt'/f(t)$ has dimensions of time. While comparing Equation (7.76) on one hand and Equations (7.56), (7.67), (7.71) and (7.75) on the other hand, one can consider the expression $\int_0^t f(t')dt'/f(t)$ as a "generalized time" for the present purpose.

Chen *et al.* [551] conclude that in all these phenomena within the OTOR framework, namely LM-OSL, TL-like presentation of CW-OSL, NL-OSL with different stimulation modes and TL under hyperbolic and linear heating rates, presenting the luminescence in the appropriate manner yields the results shown in Figure 7.12. The maximum signal depends nearly linearly on the initial occupancy n_0. This can be compared with the "general-order" kinetics where for the revised presentation of the LM-OSL, it has been shown that the dependence is strictly linear. This, in turn, may be translated into a nearly linear dose dependence in cases where the initial occupancy (following the appropriate irradiation) is linear with the dose.

7.11   TL Expression Within the Unified Presentation

To complete the mathematical analogy between the different luminescence phenomena expressed in Section 7.7, let us discuss the expressions one can get for TL within the OTOR model. The basic equation here is (4.17). This can be re written as

$$-\frac{A_m n + A_n(N-n)}{A_m n^2} dn = s \cdot \exp(-E/kT) dt. \tag{7.77}$$

Upon integration, one gets

$$\left[\frac{A_m - A_n}{A_m} \ln\left(\frac{n_0}{n}\right) + \frac{A_n N}{A_m}\left(\frac{1}{n} - \frac{1}{n_0}\right)\right] = \int_0^t s \cdot \exp(-E/kT)(t') dt' \tag{7.78}$$

where t' is a variable of integration. Let us choose, for simplicity (see Section 7.7) a hyperbolic heating function,

$$\frac{1}{T(t)} = \frac{1}{T_0} - \frac{t}{\beta'} \tag{7.79}$$

where β' is a constant with units of K s. Equation (7.78) integrates exactly to

$$\left[\frac{A_m - A_n}{A_m} \ln\left(\frac{n_0}{n}\right) + \frac{A_n N}{A_m}\left(\frac{1}{n} - \frac{1}{n_0}\right)\right] = \frac{sk\beta'}{E}\left[e^{-E/kT} - e^{-E/kT_0}\right]. \tag{7.80}$$

Since the exponent is a very fast increasing function of T, for $T > T_0$ one gets $e^{-E/kT} \gg e^{-E/kT_0}$, and therefore the first exponent can replace the term in brackets on the right-hand side. Substituting into Equation (4.17) we get

$$L \approx \left[\frac{A_m - A_n}{A_m} \ln\left(\frac{n_0}{n}\right) + \frac{A_n N}{A_m}\left(\frac{1}{n} - \frac{1}{n_0}\right)\right] \frac{A_m n^2}{A_m n + A_n(N-n)}. \tag{7.81}$$

The right-hand side is exactly the same as in Equations (7.56) and (7.67) and therefore, the consequences concerning the dimensionless presentation and the nearly-linear dependence on n_0 are exactly the same.

The situation is more complicated when the more common linear heating function $T(t) = T_0 + \beta t$ is used. In this case Equation (7.78) integrates to

$$\left[\frac{A_m - A_n}{A_m} \ln\left(\frac{n_0}{n}\right) + \frac{A_n N}{A_m}\left(\frac{1}{n} - \frac{1}{n_0}\right)\right] = \frac{sE}{k\beta}\left[\Gamma(-1, E/kT) - \Gamma(-1, E/kT_0)\right],$$

(7.82)

where Γ is the incomplete gamma function, defined as

$$\Gamma(s, x) = \int_x^\infty t^{s-1} e^{-t} dt.$$

(7.83)

Note that since $\Gamma(-1, E/kT)$ is a very fast growing function of T, the term $\Gamma(-1, E/kT_0)$ can be neglected, since it is much smaller than $\Gamma(-1, E/kT)$ for $T_0 < T$. Suppose one defines a relationship

$$y = \Gamma[-1, -\ln(x)],$$

(7.84)

then an inverse function f can be defined such that

$$x = f(y).$$

(7.85)

Using f one can rearrange Equation (7.82) to find

$$e^{-E/kT} = f\left\{\frac{k\beta}{sE}\left[\frac{A_m - A_n}{A_m} \ln\left(\frac{n_0}{n}\right) + \frac{A_n N}{A_m}\left(\frac{1}{n} - \frac{1}{n_0}\right)\right]\right\}.$$

(7.86)

Substituting Equation (7.86) into Equation (4.17) , we get

$$L = sf\left\{\frac{k\beta}{sE}\left[\frac{A_m - A_n}{A_m} \ln\left(\frac{n_0}{n}\right) + \frac{A_n N}{A_m}\left(\frac{1}{n} - \frac{1}{n_0}\right)\right]\right\}\frac{A_m n^2}{A_m n + A_n(N - n)}.$$

(7.87)

The right-hand side here is a function of constants and n. The following non dimensionalized quantities can be introduced,

$$\hat{s} = \frac{sE}{k\beta},$$

(7.88)

$$\hat{n} = \frac{n}{N},$$

(7.89)

$$\hat{L} = \frac{EL}{Nk\beta},$$

(7.90)

$$\hat{A} = \frac{A_n}{A_m}.$$

(7.91)

Using these variables, the solution can be written as

$$\hat{L} = \hat{s}f\left\{\frac{1}{\hat{s}}\left[(1 - \hat{A})\ln\left(\frac{\hat{n}_0}{\hat{n}}\right) + \hat{A}\left(\frac{1}{\hat{n}} - \frac{1}{\hat{n}_0}\right)\right]\right\}\frac{\hat{n}^2}{\hat{n} + \hat{A}(1 - \hat{n})}.$$

(7.92)

Thus, the normalized luminescence $\hat{L}$ is a function of $\hat{n}$, $\hat{n}_0$, $\hat{A}$ and $\hat{s}$, namely, it depends on three parameters and one function, $\hat{n}$. Although the inverse function f cannot be given explicitly, the similarity with Equation (7.63) is quite obvious. Numerical results have shown that similarly to the hyperbolic heating case mentioned [see Equation 7.81], the maximum intensity of TL in the linear heating case is nearly linear with n_0 in the OTOR model [550].

7.12   Pseudo LM-OSL and OSL Signals under Various Stimulation Modes

Bulur [552] suggested a method by which OSL results measured under continuous wave stimulating light intensity can be manipulated, so that the results are mathematically similar to the LM-OSL situation. If the kinetic process is of first order, Bulur wrote the decay function of luminescence as

$$L(t) = n_0 b \exp(-bt), \tag{7.93}$$

where n_0 is the initial concentration of trapped electrons and b a constant describing the decay of the luminescence curve. In order to convert the CW-OSL to a pseudo LM-OSL curve, a new variable u is introduced defined as $u = \sqrt{2tP}$ where P is the measurement period and therefore, u has dimensions of time. Substituting into Equation (7.93), one gets

$$I(u) = n_0 \frac{b}{P} u \exp\left(-\frac{b}{2P}u^2\right). \tag{7.94}$$

This has the same form as Equation (7.19). Since, as pointed out above, in the first-order case the maximum intensity is proportional to the initial concentration of trapped carriers, the maximum intensity u_m can be used for evaluating the absorbed dose in cases where one is convinced that this concentration is proportional to the dose. Obviously, this may not be the case in non-first-order cases.

Bulur *et al.* [553] showed that instead of ramping the stimulation power received by the sample, one can use fast narrow-pulsed sources of excitation. Here, the off-time period between excitation pulses decreased from a maximum to a minimum, with the pulse intensity kept constant. This method of frequency-modulated pulsed stimulation (FMPS-OSL) was shown to exactly mimic the LM-OSL. Poolton *et al.* [554] show that there is a third simple method for directly recording LM-OSL type de-trapping effects, which is useful when employing sources where neither the intensity of the source can be altered, nor variable pulsed modes of operation are possible. This can be considered as a form of FMPS-OSL, but where all the dead-time of the stimulation "off" period is removed from consideration. The detection period needs to be modified accordingly, and the method works by ramping the sampling period (RSP) from zero to a maximum value. In this RSP-OSL method as demonstrated by Poolton *et al.* [554], the curves are plotted as a function of sampling channel, and not as functions of the real time. Like CW-OSL, but unlike LM-OSL and FMPS-OSL, luminescence is actually decaying during the RSP experiment. In this sense, it resembles the pseudo-LM-OSL mentioned above. Experimentally, in addition to the Risø reader, a second "slave" computer was used. A control trigger at the start of the CW-OSL measurement activated the slave computer to commence data gathering under the RSP timing sequences. Jain *et al.* [326] used IPT-ITL in an entirely analogous way to the pseudo LM-OSL, in their study of dose measurements beyond the OSL range in quartz.

As mentioned above, Bos and Wallinga [540] suggested the use of different monotonically increasing stimulation intensity functions, in addition to the mentioned linear modulation for the first-order kinetics OSL. These are hyperbolic (HM-OSL), exponential (EM-OSL) and reciprocal (RM-OSL) intensity functions. In full analogy to the pseudo LM-OSL, these authors also consider pseudo-HM-OSL, pseudo-EM-OSL and pseudo-RM-OSL transformations. Bos and Wallinga [540] conclude that, depending on the stimulation mode, the OSL signal can be monotonically increasing, monotonically decreasing, show a peak shape or be constant. The shape of the OSL signal is determined by the rate at which the stimulation intensity increases, and also by the decay-rate parameter. In the case of multicomponent OSL signals, the overlap of the components is identical for any stimulation mode. Thus, for the separation and identification of the different OSL components, there is no preference for a specific stimulation mode. It should be noted that the discussion in this section is limited to first-order kinetics OSL.

7.13   OSL Decay and Stretched-exponential Behavior

A subject very closely related to the phosphorescence decay (see Section 7.7) is the isothermal decay of OSL. It should be noted that neither the first-order decay nor the general-order (including second-order) behavior of OSL are common in experimental results. Huntley *et al.* [98] who first suggested the use of OSL for dating of sediments, reported that the decay of OSL during laser illumination of previously irradiated (during antiquity or in the laboratory) quartz grains, is non-exponential. Smith and Rhodes [555] elaborated on the decay of OSL and stated that the decrease of OSL emission with laser exposure in quartz does not follow a simple exponential as would be expected from first-order detrapping from a single trapping site. The authors explain this as being due to the presence of at least two sources of charge when OSL measurements are made at 17 °C. They tried to simplify the results by repeating the measurements at 220 °C, but still found a decay that was not a simple exponential. They interpret this as being due to photo transfer into a trapping state stable at 220 °C. These authors mention the role of retrapping in their sample (based on a previous work by Aitken and Smith [556]), but suggested that this retrapping produces only a minor modification to the exponential OSL decrease as long as the traps are not near saturation. Further work on the non-exponential decay of OSL has been done by Ankjærgaard and Jain [557]. These authors reported on time resolved optically stimulated luminescence (TR-OSL) from quartz covering over eight orders of magnitude from 50 ns to ~8 s. They observed an abnormal decay behavior seen as a sudden increase in the decay rate at about 2–3 s in OSL measured at 75 and 100 °C. This behavior is reproduced in a numerically solved kinetic model consisting of a three trap-one center model or a one trap-two center model involving localized transitions. Chen [558] has shown that substantial retrapping can cause a decay curve which is neither exponential nor of the second- or general-order form [Equations (6.4) and (6.5)]. This is opposed to the conclusion of Smith and Rhodes [555] that the released charge originates necessarily from trapping sites with more than one mean life under optical stimulation. Poolton *et al.* [559] studied the OSL decay of porcelain and found that the RT signal shows highly nonexponential decay characteristics which they ascribed to a range of defects with varying depth and/or capture cross-sections.

McKeever *et al.* [560] presented results of computer-simulated OSL based on a model of three trapping states and two recombination centers which explained a number of TL and OSL phenomena occurring in the single aliquot technique. They found a decay curve which is approximately exponential at short times, but deviates from exponential at longer times. They define the "OSL intensity" as the area under the OSL curve between 0 s and 100 s. Further theoretical work by McKeever *et al.* [561] describes the behavior of OSL as a function of temperature. The authors consider retrapping by shallow traps, thermally assisted optical stimulation, thermal quenching and localized donor–acceptor recombination. They emphasize the point that the decay is usually nonexponential, typically exhibiting a long "tail" at long illumination times. Bailey [562] reported a subexponential decay of OSL in quartz, which could be fitted to the stretched-exponential function discussed below. Markey *et al.* [152] introduced the POSL method where the laser stimulation source was pulsed with widths of 10 ns. The OSL emission of Al_2O_3:C was followed as a function of time after the stimulation pulse. The decay curve under these circumstances was found not to be a simple exponential, and in the analysis it was fitted to the sum of two exponentials. Akselrod *et al.* [563] further studied the TL and OSL properties of Al_2O_3:C and reported a non-exponential decay of the "delayed" OSL (following the stimulating pulse), and ascribed it to the occurrence of shallow trapping states. Similar results were reported by McKeever and Akselrod [522] who found that the decay after the light exposure shows multiple components. In a recent work, Ankjærgaard *et al.* [564] reported on a method for separation of quartz and feldspar OSL using pulsed excitation, which may improve the dating of geological and archaeological materials.

Chen and Leung [565] have shown that a possible decay function of OSL *during* as well as *following* the exposure to stimulating light might be closely approximated by the quite ubiquitous stretched-exponential function $\exp\left[-(t/\tau)^\beta\right]$ with $0 < \beta < 1$. As early as 1854, Kohlrausch [566] described mechanical creep by this function. Williams and Watts [567] who coined the term "stretched-exponential", showed that dielectric relaxation in polymers can be fitted to a stretched-exponential function. The "Brinkman Report" [568] stated that "there seems to be a universal function that slow relaxations obey. If a system is driven (or normally fluctuates) out of equilibrium, it returns according to the function $\exp\left[-(t/\tau)^\beta\right]$. Unfortunately, this is not a mathematical expression that is frequently encountered in physics, as little idea exists of what the underlying mechanisms are". In general, the parameters β and τ depend on the material and the specific phenomenon under consideration, and can be a function of external variables such as temperature, as shown by Klafter and Schlesinger [569]. For an account on the history of the stretched-exponential function see Cardona *et al.* [570].

A large number of papers have been published describing a stretched-exponential decay of luminescence in different materials and in different time-scales. Chen *et al.* [571] studied the decay of luminescence in porous silicon and in CdSe-ZnSe superlattice, and found that the stretched-exponential behavior

$$I = I_0 \exp\left[-(t/\tau)^\beta\right] \tag{7.95}$$

with $0 < \beta < 1$ describes it very well. Here, I_0 is the initial luminescence intensity following the excitation. For CdSe-ZnSe at 13 K, the decay time scale was $\sim$100 ns; for porous silicon at RT it was $\sim$100 μs and for porous silicon at 13 K, $\sim$10 ms. Chen *et al.* [571] reported that the fitting parameters β and τ were found to depend on excitation conditions, such as the excitation pulse width, intensity and photon energy. Many papers have been published in the 1990s describing the decay of luminescence as a stretched-exponential function. Much of the work reported results observed in porous silicon, but stretched-exponential decay of luminescence has been observed in other materials as well. Pavesi and Ceschini [572] suggested that a key role is played by disorder in the form of: (1) a wide distribution of the size of the Si nanocrystals which form a p-Si skeleton; (2) a random spatial arrangement of the nanocrystals; and (3) disorder in the structure of the nanocrystal surfaces. In conclusion, they state that the occurrence of the stretched-exponential decay of luminescence strongly points to the role of disorder. These authors report on two components in the decay of luminescence in porous silicon, a fast component in the time range of 10^{-9} s and a slow one in the time range of 10^{-6}–10^{-2} s, which can be modeled by a stretched exponential [Equation (7.95)] where τ is a lifetime and β is a dispersive factor. The authors suggest that this decay law, often encountered in disordered systems, can be considered a consequence of the dispersive diffusion of the photoexcited carriers. The diffusion of carriers among different spatial sites can be due to either the excitation of carriers from localized to extended states, or due to hopping between localized states. In the former, the localized states act as traps and the disorder causes a distribution of release rates and of trap energies. The diffusion arises from a multiple trapping detrapping (MTD) mechanism, where the parameter β is associated with the density of trap states and trapping release rates. In a later work, Pavesi [573] reiterated the importance of disorder in bringing about the stretched-exponential decay behavior. He stated that $\beta < 1$ corresponds to the existence of a broad distribution of lifetimes which describe the elementary relaxation processes, be it radiative or nonradiative.

Chen and Leung [565] have shown that stretched-exponential decay of OSL can be expected from a model of a crystal with one trapping state and one kind of recombination center, when transitions of the carriers are through the conduction band (see Figure 7.1), with no inherent necessity to have a disorder in the system. This is the case for both OSL *during* the application of the stimulating light or *following* it. The differences in the nature of the decay in these cases have been discussed. No analytical proof was given for this behavior, since it is not possible to solve analytically the relevant sets of non linear simultaneous equations, (7.1–7.4) and (7.5–7.8). However, these authors gave the results of the simulation consisting of the numerical solution of the equations for plausible choices of sets of trapping parameters, which demonstrate the result of stretched-exponential decay of OSL. Several runs of the mentioned sets of differential equations have been performed with different sets of the relevant parameters. As pointed out above, an important condition for getting a decay which is far from being a simple exponential is the occurrence of a relatively high retrapping coefficient. However, the conditions for getting different values of β depended on all the relevant trapping parameters. Anyway, in practically all the cases checked, a very good agreement with the stretched exponential function occurred with values of β ranging from 0.5 to 1.0. It has also been pointed out that the agreement with the stretched-exponential is usually better in the POSL where the emitted light is monitored following the stimulating light pulse, than in the case of decay of OSL during the exposure to the stimulating light.

Chen and Leung [565] used a best-fit procedure which consisted of minimizing the sum of squares of the differences between the simulated experimental points (I_i) and the relevant point on the stretched-exponential function

$$\Delta = \sum_{i=1}^{K} \left\{ I_i - I_0 \exp\left[-(t_i/\tau)^\beta\right] \right\}^2, \tag{7.96}$$

using the *fmins* minimization program in the MATLAB package, with the three variables I_0, τ and β. K is the number of points at which the luminescence intensities were evaluated. As the "figure of merit" (FOM) for the goodness of the fit, they took

$$\text{FOM} = (\Delta/K)^{1/2}/I_0. \tag{7.97}$$

The division by the number of points K, is in order to be able to compare the goodness of fit between cases with different number of points, and the square root is taken so that the numerator has dimensions of intensity. Thus, the FOM as defined in Equation (7.97) is dimensionless.

Chen and Leung [565] point out that within the discussed limited framework of one trapping state and one kind of recombination center, performing the numerical solution of Equations (7.1–7.4) is not needed. The reason is that in this case, at the end of the stages of excitation and relaxation, and if the latter is long enough, one ends up with equal final occupancies of the trap and center. These values are the initial occupancies for the next stage of optical stimulation, and therefore one can just choose some values of $n_0 = m_0$ and proceed. The relevant parameters to be chosen here for the simulation are the recombination and retrapping probability coefficients A_m and A_n, the total concentration of trapping states N, the initial filling of traps and centers following the excitation and relaxation $n_0 = m_0$, and the optical stimulation rate f. In the case of OSL *following* the light pulse, the length of the pulse t_f is also of interest.

Figure 7.13 shows the results of a simulation of CW-OSL with the chosen parameters: $A_m = 10^{-12} \text{ m}^3 \text{ s}^{-1}$; $A_n = 10^{-13} \text{ m}^3 \text{ s}^{-1}$; $N = 10^{17} \text{ m}^{-3}$; $n_0 = m_0 = 10^{16} \text{ m}^{-3}$; and $f = 1 \text{ s}^{-1}$. The results are shown on a semi-log scale. The solid line consists of 1000 simulated points at intervals of 0.01 s. The dashed line is the best-fitted curve to the logarithm of Equation (7.95), namely $\ln I_0 - (t/\tau)^\beta$. The simulated decay curve is seen to be sub-exponential; however, it is visually observed that the agreement is not very good. The effective parameters found are $\tau = 0.79 \text{ s}$ and $\beta = 0.57$. The value of the FOM defined in Eq. (7.97) is FOM= 4.9×10^{-3}.

Bearing in mind the possibility mentioned by McKeever *et al.* [560] that the decay curve may consist of two components, at short and long times, Chen and Leung [565] fitted the numerical OSL values separately for the first second and for the period of time from $t = 2$ to 10 s.

Figure 7.14 shows the latter. The fit looks significantly better with the fitting parameters $\beta = 0.35$ and $\tau = 0.057 \text{ s}$ with a the FOM = 1.7×10^{-4}, more than an order of magnitude better than the whole curve fit.

Figure 7.15 shows the best fit for the first second. Here, one gets $\beta = 0.94$ and $\tau = 1.47 \text{ s}$. The high value of β means that the initial decay is nearly exponential. Here the FOM = 3.3×10^{-5}, which is five times better than in Figure 7.15.

Figure 7.16 depicts the numerical results for a decay *following* the exposure to the stimulating light. The parameters chosen here are: $A_m = 10^{-20} \text{ m}^3 \text{ s}^{-1}$; $A_n = 10^{-16} \text{ m}^3 \text{ s}^{-1}$; $N = 10^{18} \text{ m}^{-3}$; $n_0 = m_0 = 9 \times 10^{17} \text{ m}^{-3}$; and $f = 10 \text{ s}^{-1}$. The length of the stimulation

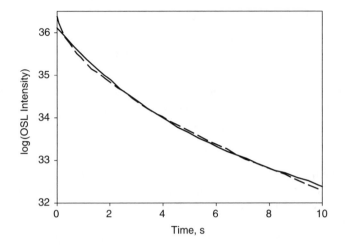

Figure 7.13 *Simulated decay curve (solid line) during optical simulation (CW-OSL), and best-fitted stretched-exponential function (dashed line) for a given set of parameters specified in the text and for the time range 0–10 s on a semi-log scale. Reprinted from Chen, R. and Leung, P.L., The decay of OSL signals as stretched-exponential functions, Radiat. Meas. 37, 519–526. Copyright (2003) with permission from Elsevier*

pulse is $t_f = 0.1$ s. One hundred numerically simulated points are shown and the solid line indicates the best-fitted curve. The fitting parameters here are $\beta = 0.75$ and $\tau = 0.041$ s. As opposed to Figures 7.13–7.15, the results here are on a linear (not logarithmic) scale in order to demonstrate the possibility of fitting the curve this way. Repeating the curve

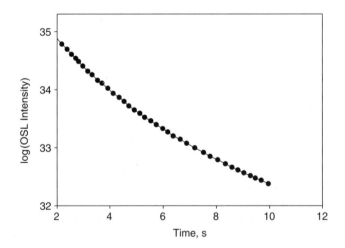

Figure 7.14 *Simulated decay curve and best-fitted stretched exponential on a semi-log scale. The results are the same as in Figure 7.13, but only in the time range of 2–10 s. Reprinted from Chen, R. and Leung, P.L., The decay of OSL signals as stretched-exponential functions, Radiat. Meas. 37, 519–526. Copyright (2003) with permission from Elsevier*

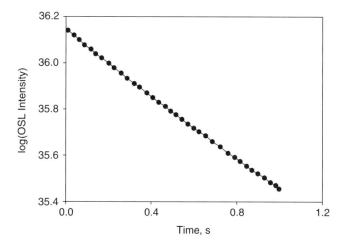

Figure 7.15 *The best-fitted curve for the same trapping parameters as in Figures 7.13 and 7.14, in the time range of 0–1 s. Reprinted from Chen, R. and Leung, P.L., The decay of OSL signals as stretched-exponential functions, Radiat. Meas. 37, 519–526. Copyright (2003) with permission from Elsevier*

fitting with the same data and on a logarithmic scale, resulted in practically the same β and τ values. The FOM here was 3.1×10^{-4}.

Chen and Leung [565] summed up their results, saying that for the case of OSL *following* a stimulating light pulse, the agreement of the decaying phosphorescence with the stretched-

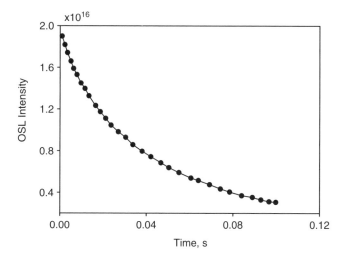

Figure 7.16 *Simulated decay curve (points) and best-fitted stretched-exponential function (solid line) following a light pulse. The set of chosen parameters and the resulting stretched-exponential parameters are given in the text. The results are shown here on a linear scale. Reprinted from Chen, R. and Leung, P.L., The decay of OSL signals as stretched-exponential functions, Radiat. Meas. 37, 519–526. Copyright (2003) with permission from Elsevier*

exponential function was quite good; more often than not, the decay curve was found to be nearly a simple exponential. Values of β significantly smaller than unity had been found either when the initial pulse had been very large (large values of f, meaning that the system was initially significantly out of equilibrium), or when the trapping state was initially close to saturation. As for the OSL decay during simulation (CW-OSL), the characteristic behavior found has been of an initial nearly exponential followed by a stretched-exponential decay. This is in good agreement with simulation based on a system with three trapping states and two kinds of recombination centers, previously reported by McKeever *et al.* [560]. These authors found a nearly exponential decay at short times and slower decay at longer times. The results by Chen and Leung [565] indicate that this behavior may be associated with the basic process occurring in the simple OTOR model, and not necessarily with the model by McKeever *et al.* which obviously includes many more parameters.

It is to be noted that even in the presently discussed simpler model, the stretched exponential is merely an approximation to the decay curve. We start with five parameters, namely $A_m, A_n, N, n_0 = m_0$ and f, and in the POSL also a sixth one, namely t_f (the length of time of the stimulating pulse), and fit the results to a function depending on three parameters, I_0, β and τ. Obviously, one cannot expect a perfect agreement. Chen and Leung [565] point out that it is encouraging to note that the decay of OSL may enter into a larger group of relaxation phenomena, which exhibit the stretched-exponential behavior. At the same time, the fact that with all these phenomena no analytical proof for this time dependence was found, is somewhat discouraging.

A common feature of the stretched-exponential functions for different values of β is that if one defines the "lifetime" as the time required to decay to e^{-1} of the initial intensity, one gets $t_{1/e} = \tau$ irrespective of β. As for the time needed to decay to half intensity, one gets $t_{1/2} = \tau (\ln 2)^{1/\beta}$. Thus, for $\beta = 1$, one gets the well-known relation $t_{1/2} = \tau \ln 2$, and for example for $\beta = 0.5$, one gets $t_{1/2} = \tau (\ln 2)^2$. In this respect, both in the stretched and simple exponential, the expression for half-life is simple and independent of the initial intensity I_0. This is opposed to other expressions such as the sum of two decaying exponentials, $A \exp\left[-(t/\tau_1)\right] + B \exp\left[-(t/\tau_2)\right]$ where the expression for half-life is significantly more complicated and depends on the parameters A and B as well as on τ_1 and τ_2.

It might be thought that in the systems discussed here the strong retrapping plays the role of disorder, usually quoted as being the source of the stretched-exponential behavior. One tends to believe that this is not the case. Even when the retrapping is rather strong, the distribution of trapping states is rather sparse, and the crystal under consideration basically remains unchanged. Furthermore, if one had a means to arrange the trapping states in a perfect orderly manner, the rate equations would still be the same, and therefore the consequences of having a stretched-exponential decay would remain unchanged. It has been suggested by Huber [574] that the real reason for the stretched-exponential behavior in the present case has to do with the nonlinearity of Equations (7.5) and (7.6), in which terms like $A_m m n_c$ and $A_n(N-n)n_c$ are included.

7.14   Optically Stimulated Exoelectron Emission

The study of exoelectron emission yields useful information on the trapped charge populations in solids. This phenomenon results from the emission of electrons which absorb enough excitation energy to overcome the work function of the material; these electrons

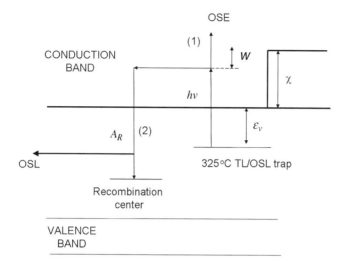

Figure 7.17   *A model describing the origins of OSE and OSL signals in quartz. Photon absorption gives rise to electron escape to the conduction band, followed either by transition (1) where electron emission from the surface is caused by thermal excitation (this is termed photo-thermostimulated emission) in Oster and Haddad [577]), or by transition (2) where electron recombination with a trapped hole gives rise to luminescence. Reprinted from Pagonis, V., Ankjærgaard, C., Murray, A.S. and Chen, R., Optically stimulated exoelectron emission processes in quartz: comparison of experiment and theory, J. Lumin. 129, 1003–1009. Copyright (2009) with permission of Elsevier*

may originate directly from traps, or from the conduction band. Ankjærgaard *et al.* [575] used a flow-through Geiger–Müller pancake electron detector to simultaneously measure optically stimulated electrons (OSE) and OSL from a sedimentary quartz sample. Their experiments also studied the thermal stability of the OSE and OSL signals, as well as their dependence on the stimulation temperature.

Pagonis *et al.* [576] used the general model of Oster and Haddad [577] which describes photostimulated exoelectron emission (PSEE) processes in solids, to describe the quartz experiments of Ankjærgaard *et al.* [575]. The quartz model of Pagonis *et al.* [576] is shown in Figure 7.17, where in transition (1) the electron release is caused by photons, while the emission from the surface is caused by thermal excitation. This is termed photo-thermostimulated emission in the more general model of Oster and Haddad [577], and is one example of several possible related phenomena which include photo-photostimulated emission (PPSEE), thermo-photostimulated emission (TPSEE), and thermo-thermostimulated emission. The model is developed by solving numerically the following set of differential equations

$$\frac{dn}{dt} = -P_v n = -\zeta E_s n, \tag{7.98}$$

$$\frac{dN}{dt} = P_v n - (A_R + \Omega_T)N, \tag{7.99}$$

$$A_R = A_m m. \tag{7.100}$$

Here A_R is the recombination probability (s^{-1}), n is the electron concentration in the traps (cm^{-3}), $P_v = \zeta E_s$ is the probability of electron ejection from the trap by photons (s^{-1}), Ω_T is the probability of electron emission from the conduction band by thermal stimulation (s^{-1}), ζ is the light absorption cross section for the OSL trap (in cm^2), E_s is the quantum intensity for the OSL trap (in eV). N is the electron concentration in the conduction band (cm^{-3}), and t is the time in s. Equation (7.100) states that the recombination probability A_R is given by the product of the instantaneous concentration of holes m in the recombination center (cm^{-3}) and the recombination coefficient A_m (in units of $cm^3\ s^{-1}$).

The probability of thermal electron emission from the conduction band is given by the following Boltzmann factor

$$\Omega_T = \Omega_o \exp\left(\frac{-\epsilon_v + \chi - h\nu}{kT}\right). \tag{7.101}$$

where Ω_0 is a dimensionless coefficient, χ is the work function (in eV), $h\nu$ is the energy of the stimulating photons (in eV), T is the temperature of the sample (in K), k is the Boltzmann constant and ϵ_v is the depth of the optically sensitive trap (in eV). These energy levels are shown schematically in Figure 7.17. By solving the above equations using the quasi-equilibrium assumption $dN/dt = 0$, Oster and Haddad [577] obtained the following expression for the instantaneous OSE intensity

$$I = B\exp\left(\frac{-\epsilon_v + \chi - h\nu}{kT}\right)\exp(-\zeta E_s t) = I_0\exp(-\zeta E_s t), \tag{7.102}$$

where B and I_0 are constants. From this equation it is clear that the initial emission intensity I_0 depends on the stimulation temperature T and also on the frequency of the stimulating light ν. The total intensity I decreases exponentially with time, with the rate of exponential decay equal to $-\zeta E_s$. This is the same as the time decay rate of the OSL signal, in agreement with the experimental findings of Ankjærgaard *et al.* [575]. In the case of transition (1) shown in Figure 7.17, the trapped electrons can be excited optically into the conduction band, and subsequently will be emitted from the surface via a thermally assisted process. In this case, the energy of the photon is such that

$$\epsilon_v < h\nu < \epsilon_v + \chi. \tag{7.103}$$

The experimental dependence of the OSE signal $I(T)$ on the stimulation temperature T is shown in Figure 7.18. According to Equation (7.102), a graph of $\ln I$ against $1/kT$ should yield a straight line at low temperatures T, with a negative slope of $W = h\nu + \epsilon_v - \chi$. By analysing the experimental data in this manner, Pagonis *et al.* [576] concluded that the OSE dependence on stimulation temperature is consistent with a thermal assistance process with an average thermal activation energy of $W = (0.29 \pm 0.02)\,eV$. By using the value $h\nu = 2.64\,eV$ for a photon of blue-stimulating light at 470nm, the value $W = 0.29$ eV and by substituting a typical value of $\epsilon_v = 1.75\,eV$ for the OSL trap at 320 °C for quartz, one obtains an estimate of the work function χ for quartz

$$\chi = h\nu - \epsilon_v - W = 2.64 - 1.75 + 0.29 = 1.2\,eV. \tag{7.104}$$

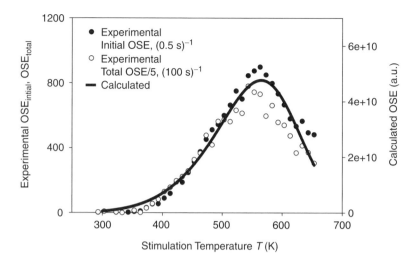

Figure 7.18 *Variation of the OSE signal with the stimulation temperature. The experimental data of Ankjærgaard et al. [575] are compared with the model of Oster and Haddad [577]. The solid line represents the product of the decreasing depletion factor D(T) for the OSL traps at 320 °C, and the increasing thermal assistance factor exp(−W/kT) as discussed in Pagonis et al. [576]. Reprinted from Pagonis, V., Ankjærgaard, C., Murray, A.S. and Chen, R., Optically stimulated exoelectron emission processes in quartz: comparison of experiment and theory, J. Lumin. 129, 1003–1009. Copyright (2009) with permission of Elsevier*

This value of χ was used by Pagonis *et al.* [576] to obtain a quantitative fit to the experimental data of OSE dependence on stimulation temperature, as shown in Figure 7.18. These authors showed that the peak-shaped behavior in Figure 7.18 can be described as the product of two terms, a thermal assistance factor of the form $\exp(-W/kT)$ and a term $D(T)$ representing the thermal depletion of the charge concentration in a single TL or OSL trap. The term $D(T)$ for a first-order kinetic process is of the form

$$D(T) = \exp[-s \cdot \exp(-E_1/kT)t_p], \qquad (7.105)$$

where $s \cdot \exp(-E_1/kT)$ represents the decay constant at temperature T (in s^{-1}), E_1 the energy depth of the trap below the conduction band (in eV), and s the frequency factor of the trap (in s^{-1}). The time t_p is the duration of heating of the sample at the temperature T, and k is the Boltzmann constant. Pagonis *et al.* [576] also allowed for a thermal broadening of the thermal depletion factor $D(T)$, with Figure 7.18 showing the final comparison of the experimental OSE data with the predictions of the model.

In a second experiment, Ankjærgaard *et al.* [575] studied the dependence of the OSE signal on the preheat temperature by using a single aliquot of sample WIDG8. The aliquot was irradiated with a dose of 330 Gy, preheated to a given preheat temperature for 10 s, and optically stimulated at 125 °C using blue LEDs. This cycle was repeated for different preheat temperatures using the same aliquot, and the results of the experiment are shown in Figure 7.19, where the OSE signal is shown to decrease continuously with the preheat

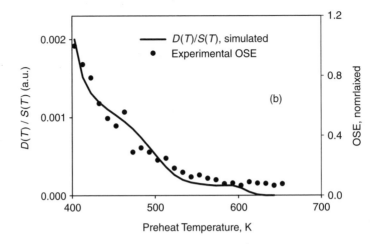

Figure 7.19　*Variation of the OSE signal with the preheat temperature. The experimental data of Ankjærgaard et al. [575] are compared with the energy scheme of Oster and Haddad [577]. The solid line represents the ratio of the decreasing depletion factor $D(T)$, and the changing sensitivity $S(T)$ of the sample, as discussed in Pagonis et al. [576]. Reprinted from Pagonis, V., Ankjærgaard, C., Murray, A.S. and Chen, R., Optically stimulated exoelectron emission processes in quartz: comparison of experiment and theory, J. Lumin. 129, 1003–1009. Copyright (2009) with permission of Elsevier*

temperature between $130\,°C$ and $400\,°C$. Pagonis *et al.* [576] explained this independent set of experimental data using the same photothermal mechanism shown in Figure 7.17. They showed that the OSE signal will be given by the expression

$$I = C\frac{n_o}{A_R}\exp\left(\frac{-\epsilon_v + \chi - h\nu}{kT_{stim}}\right)\exp(-\zeta E_s t), \qquad (7.106)$$

where C is a constant and $T_{stim} = 125\,°C$ is the temperature of the sample during the optical stimulation. The only two quantities in Equation (7.106) which are dependent on the preheat temperature T_p are the initial concentration of electrons in the trap n_0 at the beginning of the optical stimulation, and the recombination probability A_R. By using both simulation and available experimental data from Wintle and Murray [578], Pagonis *et al.* [576] showed that Equation (7.106) provides a good description of the experimental OSE dependence on the preheat temperature, as shown in Figure 7.19.

　　In the same set of experiments, Ankjærgaard *et al.* [575] measured the dependence of the *OSL signal* from the same sample, on the stimulation and preheat temperatures. In their protocol an aliquot of sample WIDG8 was irradiated with a dose of $330\,Gy$, preheated to a given preheat temperature for $10\,s$, and optically stimulated at $125\,°C$. This cycle was repeated for different preheat temperatures using the same aliquot. Pagonis *et al.* [576] used the comprehensive quartz model of Bailey [35] (see also Section 9.3) to simulate this protocol, with the results of both the simulation and the experimental data shown in Figure 7.20. However, it is important to realize that this type of simulation involves optical and thermal stimulation of electrons and holes through both the conduction and the valence band, and is of a very different nature from the OSE processes shown in Figure 7.17.

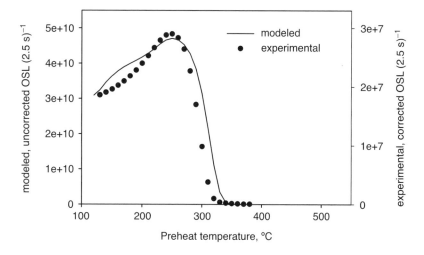

Figure 7.20 *Variation of the OSL signal with the preheat temperature. The experimental data of Ankjærgaard et al. [575] are compared with the energy scheme of Oster and Haddad [577] by using the quartz model of Bailey [35]. Reprinted from Pagonis, V., Ankjærgaard, C., Murray, A.S. and Chen, R., Optically stimulated exoelectron emission processes in quartz: comparison of experiment and theory, J. Lumin. 129, 1003–1009. Copyright (2009) with permission of Elsevier*

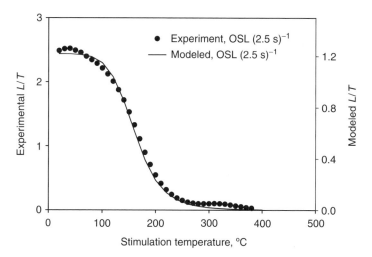

Figure 7.21 *Variation of the OSL signal with the preheat temperature. The experimental data of Ankjærgaard et al. [575] are compared with the model of Oster and Haddad [577] by using the quartz model by Bailey [35]. Reprinted from Pagonis, V., Ankjærgaard, C., Murray, A.S. and Chen, R., Optically stimulated exoelectron emission processes in quartz: comparison of experiment and theory, J. Lumin. 129, 1003–1009. Copyright (2009) with permission from Elsevier*

Pagonis *et al.* [576] also simulated the second experimental protocol of Ankjærgaard *et al.* [575] in which a variable stimulation temperature is used, by using the Bailey [35] model. The experimental data of Ankjærgaard *et al.* [575] are shown in Figure 7.21, together with the results of the simulation. These authors concluded that the energy scheme of Oster and Haddad [577] for OSE, and the Bailey [35] model for OSL are completely consistent with each other, and they elucidated the complex mechanisms involved in the charge movement in quartz samples.

7.15   Simulations of OSL Pulsed Annealing Techniques

During quartz dating procedures it is often necessary to apply short thermal treatments (such as 10 s at 260 °C) in order to isolate thermally stable signals. In addition, such short thermal treatments have been used to obtain information on the kinetic parameters of OSL traps and centers; such techniques are commonly referred to as pulsed annealing. During TL and OSL dating, the stability of the age-dependent signal is a primary concern. In particular, the evaluation of the activation energies and effective frequency factors of the relevant trapping states are of great practical significance. The method of pulse annealing has been developed for the evaluation of these kinetic parameters from OSL measurements. The method consists of obtaining and displaying the thermal stability by plotting the OSL signal measured at RT or somewhat elevated temperature (for example 50 °C) remaining after holding the sample for fixed time at various temperatures [74]. Rhodes [579] used 5 minute heat treatments at a range of temperatures up to 240 °C on naturally irradiated sedimentary quartz for the removal of the OSL component which has a low thermal stability. Duller and Wintle [580] studied TL and IRSL in K-feldspar separates. They described an experiment of heating the sample to 450 °C and monitoring the IRSL signal every 10 °C to see how the temperature of the sample affects the magnitude of the IRSL signal. Bailiff and Poolton [581] describe "pulse annealing" measurements in feldspars. In their measurements samples were subject to cycles of rapid heating to the selected temperature, cooled to RT and measurement of IRSL using a short duration stimulation was carried out, causing negligible depletion. Duller and Bøtter-Jensen [582] reported on the results of measurements of OSL stimulated by green and IR light in K-feldspars. The samples were heated at $10\,°C\,s^{-1}$ and the luminescence signal was measured for 0.1 s every 10 °C, both with and without optical stimulation, and the TL signal was measured for comparison. The OSL signal at any temperature was evaluated as the difference between total luminescence and TL. Duller [583] established the method of pulse annealing for the analysis of high precision data obtained using IRSL measurements. He used K-feldspars samples and measured the percentage IRSL signal remaining after annealing to successively increasing temperatures. For different experimental circumstances the IRSL was plotted as a function of temperature. The reported curves reach a maximum IRSL at 250 °C, which then gradually decreases to practically zero at 400 °C. The short IR light pulse only monitors the relevant concentration of trapped carriers without bleaching the sample appreciably. Thus, the recorded IRSL is a measure of the concentration of trapped carriers at different temperatures. Duller [583] plotted also the percentage IRSL signal lost per annealing phase of 10 °C. The results have a peak shape which resembles a TL peak. This is not too surprising because the procedure depicts in effect an approximation to the negative derivative of the relevant concentration, which under certain conditions should resemble a simple TL peak. Short and Tso [584] described a way of evaluating the activation energies from the descending OSL intensity vs. temperature curves. Their OSL curves versus temperature are computer generated using the GOT approximation, and their retrieved activation energies are in good agreement with the parameters used in the simulation. Huntley *et al.* [298] describe "thermal depopulation" in quartz, and from curve fitting of the descending OSL signal following subsequent heating to different temperatures up to 400 °C, obtained an activation energy of 1.03 eV and a frequency factor of 1.31×10^8 s^{-1}. Li *et al.* [585] pursued further the close analogy between the reduction rate in the OSL intensity of the pulse annealing steps and a TL peak. They performed pulse annealing measurements at different heating rates and assuming relatively simple kinetics, used the various heating rates method for evaluating the relevant activation energy from the shift of the reduction rate curve. For K-feldspar they found an activation energy of 1.72 eV, a frequency factor of 6.1×10^{13} s^{-1} and hence, a lifetime of 10^9 years at an ambient temperature of 10 °C. Bøtter-Jensen *et al.* [74] point out that the interpretation of the plots of remaining OSL after a fixed time at various temperatures is complicated by the sensitivity change brought about by thermal treatment. Wintle and Murray [578] used the response of the 110 °C TL peak to a test dose to correct for such sensitivity changes. Li and Chen [586] studied the pulse annealing results and the derived OSL reduction rate where, as in quartz, holes from a reservoir replenish the hole center during the heating stage, thus yielding a peak-shaped pulse-annealing curve. The reduction rate curve derived from this has a negative minimum and a positive maximum. Both of these shifted with the heating rate. Assuming first-order behavior, Li and Chen [586] evaluated the trapping parameters and from them estimated the lifetimes of the OSL trap and the reservoir at 20 °C. A further study of the thermal stability and pulse annealing in quartz was carried out by Li and Li [587]. Pulse annealing was also mentioned with regard to the linear modulation technique (LM-OSL) by Bulur *et al.* [552] and Singarayer and Bailey [302]. Pagonis and Chen [588] reported the results of simulations of the pulse-annealing effect within a framework of a model with a trapping state, a recombination center and a reservoir. They show that the numerical results were qualitatively similar to the experimental results in quartz given by Li and Chen [586]. However, the activation energies evaluated from simulation of the various heating rate measurements deviated in some cases from the values inserted into the simulation program.

Pagonis *et al.* [589] reviewed previous pulsed annealing studies in the literature and carried out simulations of the effect of pulsed annealing on OSL signals in quartz. These authors simulated the experimental pulse annealing protocol applied by Wintle and Murray [578] for both natural and laboratory irradiated samples. Furthermore, their simulation included an experimental OSL pulsed annealing procedure which uses different heating rates for both types of samples. Wintle and Murray [578] studied sensitivity changes for sample WIDG8, a 30 000-yr-old Australian sedimentary quartz sample. Their study used a natural sample, as well as a sample which had been optically bleached and subsequently irradiated in the laboratory. Pagonis *et al.* [589] used the comprehensive quartz model by Bailey [35] to simulate the protocol of Wintle and Murray [578]; the complete sequence of steps used in the simulation is as follows:

(1) Natural sample. All electron and hole concentrations set to zero during the crystallization process.

(2) Geological dose of 1000 Gy with a dose rate of 1 Gy s^{-1} at 20 °C.

(3) Heat to 350 °C (simulation of geological time).

(4) Illumination at 200 °C for 100 s, simulating repeated exposures to sunlight over a long period.

(5) Burial dose of 20 Gy at 0.01 βGy s^{-1} at 220 °C.

(6) Laboratory optical bleaching.

(7) Laboratory β dose of 10 Gy at Gy s^{-1}.

(8) Heat to preheat temperature $T_{PREHEAT}$ with a heating rate β (K s^{-1}) and keep at this temperature for 10 s.

(9) Apply short (0.1 s) pulse of blue OSL stimulation and record the integrated OSL signal.

(10) Laboratory test dose (0.1 Gy).

(11) Heat to 160 °C and record the 110 °C TL peak.

The first five steps in this simulation sequence are identical to those used by Bailey [35] to simulate the thermal and irradiation history of a natural quartz sample. Steps 6 and 7

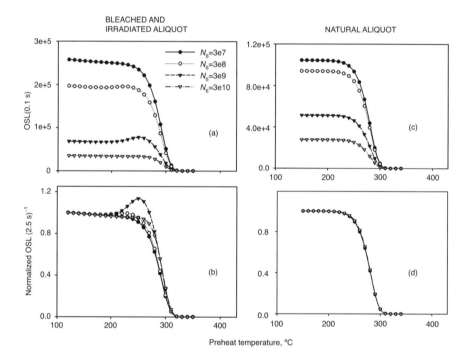

Figure 7.22 *The results of the OSL pulse annealing simulation obtained using different values of the parameter N_6 in the Bailey model, representing the available concentration of holes in the shallow reservoir R_1. (a) and (c) are for an aliquot that was bleached and irradiated with 56 Gy, and (b) and (d) are for a natural aliquot. (c) and (d) show the same data as (a) and (b) but normalized to the data point for the lowest preheat temperature. Reprinted from Pagonis, V., Wintle, A.G. and Chen, R., Simulations of the effect of pulse annealing on optically-stimulated luminescence of quartz, Radiat. Meas. 42, 1587–1599. Copyright (2007) with permission from Elsevier*

simulate the laboratory optical bleaching and irradiation of the quartz sample, and in step 8 the aliquot is heated to a preheat temperature $T_{PREHEAT}$ with a heating rate of K s^{-1} and is kept at this temperature for 10 s. In step 9, a short (0.1 s) pulse of blue OSL stimulation is applied, and the OSL signal is recorded. In steps 10–11, a test dose of 0.1 Gy is delivered, and the sample is heated to 160 °C to record the 110 °C TL peak. The sequence of steps 8–11 is repeated for progressively higher preheat temperatures up to 350 °C, in steps of 10 °C.

Figure 7.22 shows the results of the simulations for four different values of the parameter N_6 representing the total concentration of available holes in the shallow reservoir R_1 (level 6 in the Bailey model). The simulation is carried out for a natural aliquot [Fig. 7.22(c) and (d)],

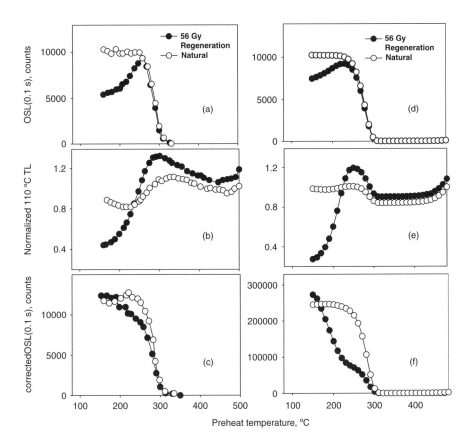

Figure 7.23 *A comparison of the experimental data of Wintle and Murray [578] (a–c) and the results of the simulations (d–f) carried out by Pagonis et al. [589]. The simulated OSL signals and sensitivity-corrected signals were multiplied by appropriate scaling factors to aid comparison. (a–c) Reprinted from Wintle, A.G. and Murray, A.S., Towards the development of a preheat procedure for OSL dating of quartz, Radiat. Meas. 29, 81–84. Copyright (1998) with permission from Elsevier. (d–f) Reprinted from Pagonis, V., Wintle, A.G. and Chen, R., Simulations of the effect of pulse annealing on optically-stimulated luminescence of quartz, Radiat. Meas. 42, 1587–1599. Copyright (2007) with permission from Elsevier*

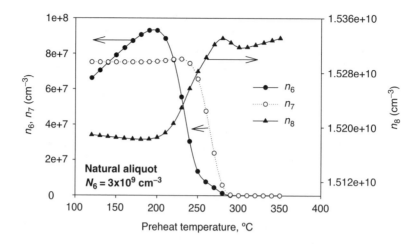

Figure 7.24   *The mechanism of hole transfer from the hole reservoirs R_1 and R_2 in quartz (levels 6 and 7 in the model), into the luminescence center L (level 8) during the experiment for the natural quartz aliquot. Reprinted from Pagonis, V., Wintle, A.G. and Chen, R., Simulations of the effect of pulse ennealing on optically-stimulated luminescence of quartz, Radiat. Meas. 42, 1587–1599. Copyright (2007) with permission from Elsevier*

and for a sample that was bleached and irradiated in the laboratory [Fig. 7.22(a) and (b)]. The simulated results in Figure 7.22 show that as the concentration of holes in the reservoir N_6 is increased, the OSL signal decreases. This was interpreted as due to direct competition between the reservoir R_1 (level 6) and the luminescence center L (level 8) for holes during irradiation of the sample. The behavior shown in Figure 7.22 is very similar to the simulation results presented by Pagonis and Chen [588] using a much simpler model; these authors found that the peak in the OSL versus preheat temperature graphs is associated with the presence of a hole reservoir. Figure 7.23 shows a direct side-by-side comparison of the simulation results in Pagonis *et al.* [589] with the experimental data of Wintle and Murray [578]. In Figure 7.23(a) and (d), the OSL signal for the bleached and irradiated aliquot (closed circles) increases up to 240 °C and subsequently decreases to zero at ∼310 °C. This increase of the OSL signal was attributed to the transfer of holes from the Zimmerman reservoirs R_1 and R_2 into the luminescence center, which leads to an apparent increase in the sensitivity as the preheat temperature increases. The corresponding data for the natural aliquot (open circles) do not show this increase, due to the fact that the shallow reservoir R_1 was already thermally activated during the burial history of the sample. Figure 7.24 shows the simulated mechanism of hole transfer from the hole reservoirs R_1 and R_2 into the luminescence center L during the pulse annealing experiments for the natural sample. As the preheat temperature is increased, the concentrations of holes in the reservoirs gradually decrease, while the corresponding concentration of holes in L increases. The results of the simulation show that the transfer of holes is almost 100% complete.

In the same work Pagonis *et al.* [589] simulated pulsed annealing experiments which use different heating rates, while similar results were presented by Pagonis and Chen [588]. A typical example of the results of such simulations is shown in Figure 7.25. The heating

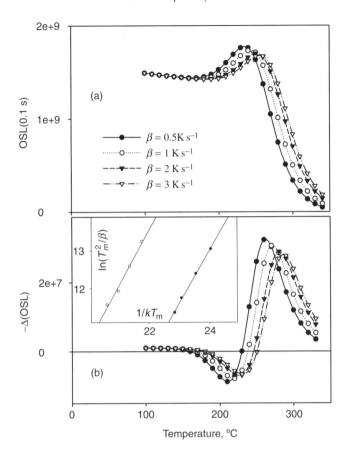

Figure 7.25 *Simulation results of the pulse annealing technique for obtaining the activation energy E for quartz samples. (a) Pulse annealing carried out using four different heating rates β. (b) Reduction of the OSL signal −Δ(OSL), i.e., the difference between consecutive measurements in (a) plotted as a function of the same preheat temperature. The inset shows Arrhenius plots used to obtain the activation energy of the electron trap and of the hole reservoir. Reprinted from Pagonis, V. and Chen, R., Geochronometria 30, 1–7. Copyright (2008) with permission from Wojewoda/Silesian University of Technology*

rates used were 0.5, 1, 2 and 3K s^{-1}, and the steps of the pulse annealing procedure were $\Delta T = 10\,°C$, as in the experiments of Li and Chen [586]. The reduction in the OSL signal was plotted as a function of preheat temperature, and yields peak-shaped plots which shift to higher temperatures with increasing heating rates, as shown in Figure 7.25(b). By assuming first-order kinetics the peak position T_m of the OSL-reduction plots will be related to the heating rate β according to the equation

$$\frac{\beta E}{kT_m^2} = s\exp(-\frac{E}{kT_m}). \qquad (7.107)$$

According to this equation, a plot of $\ln(T_m^2/\beta)$ versus $1/kT_m$ should yield a straight line as shown in the inset of Figure 7.25(b). The slopes of the two lines represent the activation energy E of the electron trap and of the hole reservoir, respectively. Furthermore, the corresponding frequency factors s can be calculated from the y-intercept of the fitted lines. Similar simulated results were obtained by Pagonis *et al.* [589] for the case of a bleached and irradiated quartz aliquot.

8

Analytical and Approximate Expressions of Dose Dependence of TL and OSL

8.1 General Considerations

The dose dependence of TL and OSL, be it the maximum intensity, the area under the curve or any other dose-dependent feature of the relevant effect, is of prime importance in radiation dosimetry and in dating applications in archaeology and geology. The main concern in these applications is to evaluate the absorbed dose from the measured luminescence. A desirable feature for these phenomena is that the dose dependence be linear in a broad dose range. This, however, is not a very common situation. Dose dependencies that begin superlinear at low doses, or begin linear and go superlinear at higher doses, are quite common. Also, sublinear dose dependence is sometimes found right from the lowest doses and sometimes takes place following a linear and/or superlinear range. Among other possibilities, one encounters the saturation effect where trapping states and recombination centers reach saturation, which normally results in a plateau in the dose dependence. One also finds in the literature several reports on non-monotonic dose dependence of TL and OSL, namely that the dose-dependent curve reaches a maximum at a certain dose and decreases at higher doses. If we consider only the very simple model depicted in Figure 4.1 where only one trapping state and one kind of recombination center exist, the linear-saturation behavior is expected. However, in real materials it is very common that several trapping states and recombination centers exist and take part in the relevant process. The explanation given to dose dependencies other than the linear-saturation case involves competition between active and competing levels in the forbidden gap of the crystal in question. In fact, the effects of competition can explain a wealth of different effects associated with TL and OSL. As far as superlinear dose dependence is concerned, it has been shown that competition during excitation can yield one kind of behavior, while competition during read-out yields another kind of dependence. Simulations have also shown the combined effect of both kinds of competition; this will be discussed in some detail below.

Thermally and Optically Stimulated Luminescence: A Simulation Approach, First Edition. Reuven Chen and Vasilis Pagonis. © 2011 John Wiley & Sons, Ltd. Published 2011 by John Wiley & Sons, Ltd.

Different definitions have been given to the term superlinearity (or supralinearity) which intuitively means a faster than linear dependence on the dose. If for example the TL intensity is measured by the maximum intensity I_{max}, one way of presenting this property (see e.g. Halperin and Chen [173]) where the superlinearity starts from the lowest doses, is by the expression

$$I_{max} = aD^k \qquad (8.1)$$

where D is the applied dose, a is a proportionality factor and k a constant. $k > 1$ means that the dependence is superlinear whereas $k < 1$ means sublinearity, and obviously $k = 1$ means a linear dependence. Halperin and Chen [173] reported on such strong superlinearity with $k \cong 3$, in UV-irradiated semiconducting diamonds. They ascribed it to a multi-stage transition, due to the subsequent absorption of a number of photons which raised electrons from the valence into the conduction band during the excitation. Neely *et al.* [590] reported on a somewhat similar effect in which TL induced by UV light appears to be due to a biphotonic ionization in benzophenone samples. Somewhat similar two-photon excitation of SiC by photons with energy lower than the band gap has been reported by Vakulenko and Shutov [591]; in this case, however, the dose dependence was not superlinear. Otaki *et al.* [592] reported on a similar strong superlinearity of TL in $CaF_2:Tb_4O_7$. When this behavior takes place in a certain dose range, a plot of I_{max} as a function of D on a log–log scale is expected to yield a straight line with a slope of k. Such superlinearity has been reported by Mikado *et al.* [593, 594] in electron–bombarded $Mg_2SiO_4:Tb$. Matrosov and Pogorelov [595] reported on a superlinear dose dependence of TL peaks at 180 and 305 °C in quartz. Superlinear dose dependence of TL in quartz was also described by Felix and Singhvi [596] who showed that the quartz extracted from samples from the Thar desert showed a superlinear dose-dependence range at exposures above $\sim$300 Gy.

A more general definition can be based on the fact that superlinearity means that the rate of increase is increasing with the dose. If $S(D)$ denotes the dependence of the measured quantity, (be it TL or OSL, maximum intensity or area under the curve), then superlinearity means $S''(D) > 0$, sublinearity $S''(D) < 0$ whereas $S''(D) = 0$ obviously means linearity. Note that these definitions are made for each point along the $S(D)$ curve, and this property may change along the dose axis for the same material. For example, a curve may start being super-linear, become linear and then sublinear before reaching saturation as is usually the case. The difficulty with this definition is that it is merely qualitative, since it has to do with the sign of the second derivative only. Chen and McKeever [194] proposed a normalized quantity which not only describes the qualitative feature of being super- or sub-linear, but is also a quantitative measure of these properties. They defined a magnitude which evaluates numerically the amount of superlinearity, and which would attain the value of k in the special case in Equation (8.1). They define the *superlinearity index* (or superlinearity factor) $g(D)$ as

$$g(D) = \frac{DS''(D)}{S'(D)} + 1. \qquad (8.2)$$

As long as one is dealing with a range of an increasing dose dependence function ($S' > 0$, which is nearly always the case), it is obvious that the condition $g(D) > 1$ indicates superlinearity since it is equivalent to the condition $S''(D) > 0$. Similarly $g(D) = 1$ means a range (or point) of linearity and $g(D) < 1$ signifies sublinearity. It can readily be seen that $DS''(D)/S'(D)$ is a dimensionless quantity, which makes the addition of unity legal; thus

$g(D)$ is a dimensionless function. It is immediately seen that for the expression in Equation (8.1) or even for $S(D) = aD^k + b$, one gets $g(D) = k$.

A number of materials including LiF:Mg,Ti have the property that at low doses the dose dependence is linear, followed by a nonlinear (i.e., superlinear) region before saturation effects set in (see e.g. Zimmerman [510]). Piesch *et al.* [597] report on the supralinearity of LiF dosimeters for neutron and γ exposures in a broad range of the excitation dose. Similar superlinearity with β excitation was reported by Wintersgill and Townsend [598]. Chen and McKeever [194] suggested the term supralinearity to describe this property of the measured quantity being above the continuation of the initial range. In fact, past authors starting with Cameron *et al.* [138] used the same term but without differentiating between the presently described attribute and superlinearity as defined above. Some authors (e.g. Waligórski and Katz [599], Horowitz [600] and Mische and McKeever [601]) quantified this property by defining the dimensionless quantity

$$f(D) = \frac{[S(D)/D]}{[S(D_1)/D_1]}, \tag{8.3}$$

termed by these authors the *supralinearity index* (*supralinearity factor*) or dose-response function. Here D_1 is a normalization dose in the initial linear range. This definition is obviously applicable only if such an initial range exists at the low dose range. It should be noted that in the literature, the terms superlinearity and supralinearity are used quite often interchangeably.

The mentioned behavior of linear-superlinear-saturation appears to be an indication of superlinearity due to competition during excitation. However, it has been suggested (see e.g. Nail *et al.* [602]), that if the superlinear dose dependence has to do with superlinear filling of the relevant trapping state, the optical absorption (OA) should be superlinear with the dose as well. This behavior of the OA has not been observed in the important material used for dosimetry LiF:Mg, Ti, in which superlinearity of the TL has been observed. In other alkali-halides such as NaCl:Mg and KCl, a superlinear dose dependence of OA has been observed (see e.g. Mitchell *et al.* [603]). Also, Wieser *et al.* [604] reported on rather strong superlinearity [with a 1.44 power in Equation (8.1)] in the dependence of OA and ESR, along with TL in fused silica, and associated it with the E_1' center. Results of the correlation between TL and OA of silica glasses have been reported by Gulamova *et al.* [605]. They show a TL peak at $\sim 350\,^\circ$C associated with a strong decrease in the optical absorption at 215 and 550 nm. These results and their relevance to TL have been discussed by Watanabe *et al.* [606]. Superlinear dose dependence in Mg_2SiO_4:Tb has been reported by Lakshmanan and Vohra [433]. The TL response of the virgin phosphor was found to be superlinear for both γ and UV radiations. The TL response of the sensitized phosphor (see Section 8.4) to γ radiation was found to be linear in a broad dose range. This was explained on the basis of the removal of competing deep traps. The TL response of the sensitized phosphor to UV was found to be sublinear. A debate on the nature of sensitization and superlinearity in LiF TLD-100 samples was discussed by McKeever [607]. As opposed to previous works by Stoebe and Watanabe [608] and Lakshmanan *et al.* [609], McKeever's analysis reveals that it is unnecessary for the competitors to be removed during irradiation. This conclusion supports the views of Rodine and Land [610] and Kristianpoller *et al.* [450] discussed below. Yet another discussion of the superlinear dose dependence of TL in TLD-100 was given by Nakajima [611]. Further work by Nakajima [612, 613] should also be mentioned. Another explanation of the effect of TL supralinearity due to competition during excitation in carbon-doped $Y_3Al_5O_{12}$ (YAG:C) has recently been given by Yang *et al.* [614].

8.2   Competition During Excitation

The first model explaining superlinearity has been given by Cameron and Zimmerman [615] and further discussed by Cameron *et al.* [138] (see pp. 168–174). This model proposes the creation of additional traps by radiation and hypothesizes a maximum trap density. A later model was given by Aitken *et al.* [10] and Suntharalingam and Cameron [616], and further elaborated upon by Chen and Bowman [617]. Basically, the assumption was made that the total number of different traps was unchanged, and the supralinearity was associated with the competition during excitation with some disconnected traps. The relevant energy level scheme is shown in Figure 8.1. These authors followed a much earlier work by Duboc [618] who explained superlinearity of photoluminescence and photoconductivity in phosphors by competition between luminescence centers and competing non radiative centers.

We begin with an intuitive explanation of the superlinear filling of the active trap N_1 under these circumstances. Suppose that the concentration N_2 of the competitor is smaller than that of the active trap N_1, but the trapping probability coefficient A_2 of the competitor is larger than that of the active trap A_1. At low doses the excitation fills both traps linearly. However, at a certain dose the competing trap saturates, hence more electrons are made available to the trap of interest N_1. This causes a faster, though linear, filling of this trap. However,

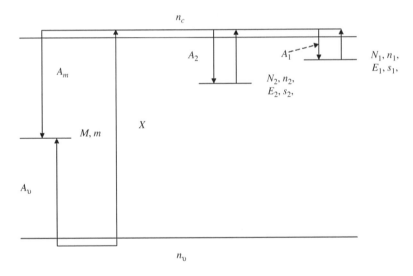

Figure 8.1   *Energy levels involved in the competition during excitation. M is the concentration of centers and m its instantaneous occupancy. N_1 and N_2 are the concentrations of active and competing traps, of which n_1 and n_2, respectively, are occupied. n_c is the instantaneous concentration of electrons in the conduction band and n_v is the instantaneous concentration of holes in the valence band. A_1, A_2 and A_m are the retrapping, competing trapping and recombination probability coefficients, respectively*

the transition region from one linear range to the other would appear to be superlinear since this transition obviously occurs continuously. Chen and Bowman [617] have studied the relationship between the applied dose and the concentration of electrons in the active trapping state, n_1. Using the relevant set of differential equations associated with the levels and transitions shown in Figure 8.1, and making the quasi-steady assumptions, they found the implicit function for $n_1(D)$,

$$D = N_2 \frac{A_m - A_2}{A_2} \left[\left(1 - \frac{n_1}{N_1}\right)^{\frac{A_2}{A_1}} - 1 \right] + \frac{A_1 - A_m}{A_1} n_1 - \frac{N_1 A_m}{A_1} \left(\frac{N_2}{N_1} + 1\right) \ln \left(\left(1 - \frac{n_1}{N_1}\right)\right).$$

$$(8.4)$$

For a given set of parameters one can solve numerically this implicit equation in n_1 for different values of the dose D and get a curve of $n_1(D)$. Chen and McKeever [33] have shown that a necessary condition for having a superlinear dose dependence is

$$A_2 > \max\left(A_m, A_1, \sqrt{A_1 A_m N_1 / N_2}\right),$$

$$(8.5)$$

whereas a sufficient condition for such behavior is

$$A_2 > 3 \max\left(A_m, A_1, \sqrt{A_1 A_m N_1 / N_2}\right).$$

$$(8.6)$$

An example of a simulated dose dependence of $n_1(D)$ is given in Figure 8.2. With a chosen set of parameters, Equation (8.4) was solved numerically for each value of D to yield the

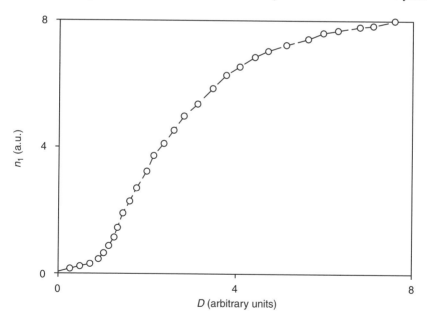

Figure 8.2 *Growth curve of the trapped carriers occupancy as a function of dose, calculated for $N_1 = N_2 = 10^{17}$ cm^{-3}, $n_{10} = n_{20} = 0$, $A_1 = A_m$, $A_2/A_1 = 30$. Reprinted from Bowman, S.G.E. and Chen, R., Superlinear filling of traps in crystals due to competition during irradiation, J. Lumin. 18/19, 345–348. Copyright (1979) with permission from Elsevier*

value of $n_1(D)$. The initial short linear range is followed by a range of superlinearity which, in turn, goes to sublinearity on its way to saturation. It should be noted that even if the active trapping state is filled superlinearly with the dose, the measured TL intensity will be superlinear with the dose only if certain conditions are fulfilled with regard to the processes taking place during the read-out stage. This will also be discussed below when the combined effect of excitation and heating is considered.

8.3   Competition During Heating

Rodine and Land [610] reported the dose dependence of some TL peaks in β-irradiated ThO$_2$ to be quadratic with the dose starting at the lowest doses, and explained the effect qualitatively as being the result of competition during heating. Let us start by giving an intuitive explanation, first for the circumstances leading to linear dose dependence and then for the mentioned quadratic dose behavior. Let us assume a situation in which the trap and center directly involved in the creation of a TL peak are linearly dependent on the dose, at least at low doses. Let the initial concentrations of filled traps and centers following the excitation be n_0 and m_0, respectively. One may think that since two entities, the trapping state and the recombination center are involved, then the emerging TL should be dependent on the product $n_0 m_0$. If we are dealing with a dose range in which both n_0 and m_0 are linearly dependent on the dose, one might expect a quadratic dependence of TL on the dose. As is well known, the normal dependence of TL on the dose is closer to linearity than being quadratic, which suggests that this explanation is incorrect in most cases. In fact, under most circumstances, if S denotes the measured TL signal, (be it the area under the peak or the maximum intensity), we may expect to have $S \propto \min(n_0, m_0)$ rather than $S \propto n_0 m_0$, and therefore a linear initial dependence is more likely to occur. The explanation for this is that the total number of recombinations should be determined by an integration over the TL intensity $I(T)$, which can also be written as $I(t)$ for a certain heating function $T = T(t)$. With the appropriate choice of units, we have $I(t) = -dm/dt$. For the present discussion let us assume the existence of only one TL peak. We can integrate over $I(t)$ from an initial time $t = 0$ to infinity and get

$$m_0 - m_\infty = \int_0^\infty (-dm/dt)dt = \int_0^\infty I(t)dt = S. \tag{8.7}$$

If $m_0 < n_0$, the peak terminates because the center (m) is exhausted, therefore $m_\infty = 0$ and thus

$$m_0 = \int_0^\infty I(t)dt = S, \tag{8.8}$$

and the right-hand side is the area S under the glow peak in the appropriate units. However if $m_0 > n_0$, the peak terminates because the trap (n) is exhausted, thus we are left with $m_\infty = m_0 - n_0$ holes in the centers and therefore,

$$S = \int_0^\infty I(t)dt = m_0 - m_\infty = n_0. \tag{8.9}$$

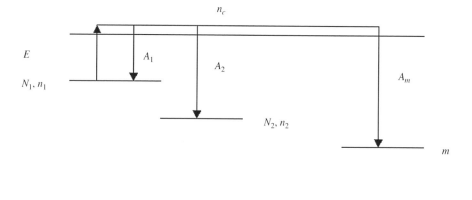

Figure 8.3 *Energy levels and transitions involved in competition during read-out. m is the concentration of holes in centers. N_1, N_2 are the concentrations of active TL traps and competing traps, respectively. A_1, A_2 represent the trapping probability coefficients into 1 and 2, respectively. A_m is the recombination probability coefficient, and n_c the concentration of free electrons in the conduction band. After Kristianpoller et al. [450]. Reprinted from Kristianpoller, N., Chen, R. and Israeli, M., Dose dependence of thermoluminescence peaks, J. Phys. D: Appl. Phys., 7, 1063–1072. Copyright (1974) with permission from Institute of Physics*

Equations (8.8) and (8.9) can be summarized as

$$S = \min(n_0, m_0). \tag{8.10}$$

Thus, if $n_0 \propto D$ and $m_0 \propto D$ we still have $S \propto D$, which agrees with the behavior often observed in experiments. As for the maximum intensity I_m, it is normally close to being proportional to the total area S, and therefore one gets $I_m \propto D$. This situation may change substantially in the presence of an active competitor.

Let us consider the energy level scheme of one kind of recombination center and two trapping states, this time concentrating on the transitions taking place during heating, as shown in Figure 8.3. We can slightly change the statement that $S \propto \min(n_0, m_0)$ by saying that the area under the peak should be proportional to the quantity n_0 or m_0 which expires first. When these two levels are the only active ones in existence (excluding disconnected traps) S should indeed be proportional to $\min(n_0, n_0)$. However, in the presence of a strong competitor this may not be the case anymore. We may have a situation in which we start with $n_0 > m_0$, but since during the heating some of the electrons are trapped in the competitor rather than recombining with holes in the centers, the peak may terminate when n_0 is exhausted whereas some holes still remain in the center. We therefore have $S \propto n_0$, where n_0 is assumed to depend linearly on the dose. In addition, if we consider the electrons in the conduction band during the heating, they will distribute so that some recombine with trapped holes m and others go into the competitor. Roughly speaking, we can say that if the concentration of empty competitors $N_2 - n_2$ does not depend strongly on the initial dose (that is, if N_2 is far from saturation, namely $N_2 \gg n_2$), then the larger the value of m, the more recombinations into the center are expected. Therefore, if m_0 is proportional to the dose, this is an additional dependence on the dose of the TL intensity, and therefore, under these circumstances, $S \propto n_0 m_0$ and therefore $S \propto D^2$. If the competitor is in the range approaching saturation, the dependence of the TL intensity on the excitation dose may be even "stronger" than D^2.

This hand-waving argument for explaining the superlinear dose dependence of TL under competition during heating has been given a more convincing mathematical formulation by Kristianpoller *et al.* [450]. With reference to Figure 8.3, the equations governing the process occurring during heating are

$$\frac{dn_1}{dt} = -\gamma n_1 + A_1 n_c (N_1 - n_1), \tag{8.11}$$

$$\frac{dn_2}{dt} = A_2 (N_2 - n_2) n_c, \tag{8.12}$$

$$I = -\frac{dm}{dt} = A_m m n_c, \tag{8.13}$$

$$\frac{dm}{dt} = \frac{dn_1}{dt} + \frac{dn_2}{dt} + \frac{dn_c}{dt}, \tag{8.14}$$

where $\gamma = s \cdot \exp(-E/kT)$, in which E is the trap depth and s is the frequency factor. From Equations (8.12) and (8.13) we get

$$n_c = \frac{1}{A_2} \frac{d}{dt} [\ln(N_2 - n_2)] = \frac{1}{A_m} \frac{d}{dt} [\ln(m)], \tag{8.15}$$

which can be integrated to yield

$$N_2 - n_2 = (N_2 - n_{20})(m/m_0)^{A_2/A_m}, \tag{8.16}$$

where m_0 and n_{20} are the initial values of m and n_2, respectively. From this point on we make some simplifying assumptions, which help us demonstrate that under certain conditions the quadratic and more than quadratic dose dependencies are a possible outcome of the set of equations (8.11–8.14). With the quasi-steady condition (4.8), Equation (8.14) can be written as

$$\frac{dm}{dt} \cong \frac{dn_1}{dt} + \frac{dn_2}{dt}. \tag{8.17}$$

By substituting Equations (8.11–8.14) into Equation (8.17), we get an explicit expression for n_c,

$$n_c = \gamma n_1 / [A_1(N_1 - n_1) + A_2(N_2 - n_2) + A_m m], \tag{8.18}$$

and by inserting Equation (8.18) into (8.13) we have

$$I(t) = -dm/dt = A_m \gamma n_1 m / [A_1(N_1 - n_1) + A_2(N_2 - n_2) + A_m m]. \tag{8.19}$$

The three terms appearing in the denominator represent the probabilities (s^{-1}) for retrapping at the original trap (level 1), trapping at the competitor (level 2) and for recombination, respectively. Integrating Equation (8.17) yields

$$m - m_0 = n_1 - n_{10} + n_2 - n_{20},$$ (8.20)

where n_{10}, n_{20} and m_0 are the initial values of n_1, n_2 and m, respectively. By substituting Equations (8.16) and (8.20) into Equation (8.19), we get the differential equation

$$I(t) = -dm/dt = \gamma A_{mm} m F(m)$$ (8.21)

with

$$F(m) = \frac{(n_{10} + n_{20} - m_0 - N_2) + m + (N_2 - n_{20})(m/m_0)^{A_2/Am}}{A_1(N_1 + N_2 - n_{10} + m_0) + (A_m - A_1)(N_2 - n_{20})(m/m_0)^{A_2/A_m}}.$$ (8.22)

This differential equation can be numerically solved for given sets of the parameters. However here, we would like to show semi-analytically the dose dependence when further simplifying assumptions are made. The quadratic dose dependence reported by Rodine and Land [610] can be deduced from Equation (8.21) if we assume that trapping in the competitor is much faster than both retrapping and recombination, namely

$$A_2(N_2 - n_2) \gg A_1(N_1 - n_1) + A_m m,$$ (8.23)

then Equation (8.19) becomes

$$I(t) = -dm/dt = A_m \gamma n_1 m / [A_2(N_2 - n_2)].$$ (8.24)

If we assume in addition that the retrapping into trapping state 1 is very small as compared with the rate of release of trapped carriers,

$$\gamma n_1 \gg A_1 n_c(N_1 - n_1),$$ (8.25)

Equation (8.11) becomes

$$dn_1/dt = -\gamma n_1,$$ (8.26)

the solution of which is simply

$$n_1 = n_{10} \exp\left(-\int_0^t \gamma d\tau\right).$$ (8.27)

Substituting Equations (8.16) and (8.27) into Equation (8.24) yields

$$I(t) = -dm/dt = A_m \gamma n_{10} \exp\left(-\int_0^t \gamma d\tau\right) m / \left[A_2(N_2 - n_{20})(m/m_0)^{A_2/A_m}\right].$$ (8.28)

Integration of this equation yields

$$\int_0^t \gamma n_{10} \exp\left(-\int_0^\tau \gamma d\tau'\right) d\tau = (N_2 - n_{20})\left[1 - (m/m_0)^{A_2/A_m}\right].$$ (8.29)

Since

$$\int_0^\infty \gamma \exp\left(-\int_0^t \gamma d\tau\right) dt = 1,$$ (8.30)

Equation (8.29) transforms into

$$n_{10} = (N_2 - n_{20}) \left[1 - (m_\infty/m_0)^{A_m/A_2} \right]. \tag{8.31}$$

Here m_∞ is the final value of the filling of the recombination center at $t = \infty$. From Equation (8.31) we immediately get

$$m_\infty = m_0 \left[1 - n_{10}/(N_2 - n_{20})^{A_m/A_2} \right]. \tag{8.32}$$

Assuming further that $n_{10} \ll N_2 - n_{20}$, we now get

$$m_\infty \cong m_0 \left[1 - (A_m/A_2)n_{10}/(N_2 - n_{20}) \right]. \tag{8.33}$$

As pointed out in Equation (8.7), the area S under the glow peak equals $m_0 - m_\infty$ which leads to

$$S = m_0 - m_\infty \cong \{A_m/[A_2(N_2 - n_{20})]\}\, m_0 n_{10}. \tag{8.34}$$

Following Kristianpoller *et al.* [450], we can add that if the competitor is far from saturation $N_2 \gg n_{20}$, we get

$$S = \left[A_m/(A_2 N_2) \right] m_0 n_{10}, \tag{8.35}$$

which yields the quadratic dose dependence reported by Rodine and Land [610] if both m_0 and n_{10} are in their linear dose dependence range. In addition, Chen *et al.* [620] pointed out that if n_{20} in Equation (8.33) is increasing with the dose, the overall dose dependence of S can be more than quadratic, which may be very significant in ranges where the competitor is coming close to saturation. It should be noted that although a number of simplifying assumptions have been made, the possibility of getting quadratic and even faster dose dependencies have been demonstrated. Similar considerations concerning the quadratic and more than quadratic dose dependence have been given by Savikhin [621], Zavt and Savikhin [622] and by Kantorovich *et al.* [189]. The quadratic dose dependence due to competition during heating was also discussed by Sunta *et al.* [623]. A further demonstration of superlinear dose dependence taking into account both processes occurring during excitation and read-out will be described in the simulations in Section 8.8.

A rather similar situation which at first sight may appear to lead to a similar kind of superlinearity, is a model with one trapping state with concentration n and two kinds of centers having concentrations of m_1 and m_2 as shown in Figure 8.4. As far as the filling of the radiative center is concerned, the general considerations made above are indeed expected to hold. This means that one of the centers, for example m_2, which we consider as being the radiationless competitor, starts growing linearly with the dose and tends to saturation when it is about to be filled to capacity. The center m_1 which is considered to be the active one (assuming that during the heating phase transitions into it will result in TL emission), starts linearly with the dose, gets superlinear in the range where m_2 is saturated, and finally goes also to saturation as it is close to being filled to capacity. The situation is different when the traffic of carriers during the heating phase is considered. Although our main concern here is the competition during heating, let us write first for the sake of completion the set

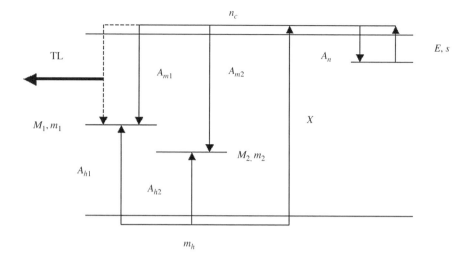

Figure 8.4 *An energy level scheme for the model of two competing centers and one trapping state. The meaning of the different parameters is given in the text. Transitions taking place during excitation and read-out are shown. Reprinted from Chen, R., Fogel, G. and Lee, C.K., Radiat. Prot. Dosim. 65, 63–68. Copyright (1996) with permission from Oxford*

of simultaneous differential equations governing the processes modeled in Figure 8.4. At a later stage we will discuss the combined effect of competition during both excitation and heating. We have again a set of five coupled equations during the excitation stage

$$dm_h/dt = X - A_{h1}n_v(M_1 - m_1) - A_{h2}n_v(M_2 - m_2), \qquad (8.36)$$

$$dm_1/dt = A_{h1}n_v(M_1 - m_1) - A_{m1}m_1n_c, \qquad (8.37)$$

$$dm_2/dt = A_{h2}n_v(M_2 - m_2) - A_{m2}m_2n_c, \qquad (8.38)$$

$$dn/dt = A_n(N - n)n_c, \qquad (8.39)$$

$$dm_1/dt + dm_2/dt + dn_v/dt = dn/dt + dn_c/dt. \qquad (8.40)$$

Here, M_1 and M_2 (cm^{-3}) are the concentrations of the two centers and m_1 and m_2 (cm^{-3}) are their instantaneous occupancies, respectively. A_{h1} and A_{h2} (cm^3 s^{-1}) are the trapping-probability coefficients for capturing free holes from the valence band, and A_{m1}, A_{m2} (cm^3 s^{-1}) are the recombination-probability coefficients of free electrons. X (cm^{-3} s^{-1}) is the rate of creation of free electrons and holes in the conduction and valence bands, respectively. n_c and n_v (cm^{-3}) are the concentrations of free electrons and holes. N (cm^{-3}) is the concentration of traps and n (cm^{-3}) is its instantaneous occupancy. A_n (cm^3 s^{-1}) is the trapping probability coefficient from the conduction band into the trap. The dose D is proportional to $X \cdot t_D$ where t_D is the duration of the excitation.

If we wish to follow analytically or numerically the whole procedure of excitation and heating, we have to consider a relaxation period in which X is set to zero, for a period of time such that n_c and m_h decay to negligible values. The transitions in the next stage of read-out are also shown in Figure 8.4. Here $E(\text{eV})$ and $s(\text{s}^{-1})$ are the trap depth and the frequency factor, respectively. The governing set of equations is now

$$I = -dm_1/dt = A_{m1}m_1n_c, \tag{8.41}$$

$$-dm_2/dt = A_{m2}m_2n_c, \tag{8.42}$$

$$-dn/dt = s \cdot n \cdot \exp\left(-E/kT\right) - A_n(N - n)n_c, \tag{8.43}$$

$$dn/dt + dn_c/dt = dm_1/dt + dm_2/dt. \tag{8.44}$$

The numerical solution of the simultaneous sets of equations (8.36–8.40) and subsequently (8.41–8.44) will be discussed below. However, approximate considerations similar to those made above in the competition between trapping states, show that competition between centers leads to entirely different results. Making the analogous assumptions leading to Equation (8.34) we get an apparently similar result with m_{20} replacing $(N_2 - n_{20})$, where m_{20} is the final filling of M_2 following the first stages of excitation and relaxation, i.e. the initial concentration for the read-out stage. The result is

$$S = \left[A_{m1}/(A_{m2}m_{20})\right] m_{10}n_0. \tag{8.45}$$

As much as this expression is analogous to Equation (8.34), the consequences concerning the dose dependence are entirely different. For example, if n_0 and m_{20} behave in a certain dose range in a similar manner, for instance both grow linearly with the dose, the measured TL dependence on the dose will behave in the same way as m_{10}. However, if due to competition during excitation m_{20} is sublinear with the dose, the superlinearity of TL will be stronger in this range. As opposed to the case of competition between traps, this kind of trap and center filling may account for a behavior that starts linearly, gets superlinear at higher doses and approaches saturation. This is opposed to the previous case where the initial dose dependence was expected to be quadratic, and at higher doses became even more superlinear before starting its approach to saturation. The reason for the fundamentally different behavior between the cases of competing traps and those of competing centers lies in the fact that the number of competing traps $(N_2 - n_2)$ decreases with the dose, whereas the number of competing centers m_2 increases with the dose.

Mady *et al.* [625] have also studied theoretically the dose dependence of TL when competition takes place during heating. They propose a dimensionless treatment when a number of traps are involved, and calculate the response of each peak by taking proper account of the transfer of carriers accompanying the successive peak read-outs. They report numerical results which show quadratic dose dependence in broad dose ranges, followed by more than quadratic dependence prior to the approach to saturation. One should also mention the work by Faïn and Monnin [626] and Faïn *et al.* [627] who explained the special behavior of a superlinear dose dependence following a linear range, based on a spatial correlation between charged sites. Their analysis does not require the assumption of saturating deep traps.

8.4   The Predose (Sensitization) Effect

The discussion so far has implicitly assumed that the response of a given sample to a certain excitation dose is a constant magnitude. Unfortunately, this may not always be the case. Annealing at high temperature may change the sensitivity of a sample, i.e. the response to a given "test-dose" may change by merely heating the sample to high temperature and cooling it back to, for example, RT. An even larger effect of increased sensitivity, sometimes rather dramatic, has been reported mainly in the 110 °C peak in quartz by the combined effect of β- or γ-irradiation followed by high temperature annealing. This effect of sensitization has been also termed the "predose" effect, and may cause difficulty in the applications of TL in dosimetry and dating . If one wishes to compare the effects of two doses, for example the archaeological dose and the laboratory calibration dose, a constant sensitivity is obviously preferable. Fleming [628], Aitken [10] and others showed that the change in sensitivity could, by itself, be used as a measure of the applied dose. These authors used β test-doses of ~0.01 Gy and sensitizing doses of ~1 Gy and measured the emission of the quartz 110 °C peak at ~380 nm. Zimmerman [510] reported on the "UV reversal" effect. If the quartz sample, thus sensitized so as to yield high response to the test-dose was illuminated by UV light in the range of 230–250 nm, the sensitivity was reduced substantially, nearly to the original sensitivity prior to irradiation plus annealing. Yet another piece of experimental information was reported by Fleming and Thompson [629]. If following a given irradiation of a sample the annealing is performed to different temperatures, different responses to a test-dose were recorded. In the case of the 110 °C peak in quartz, a typical behavior was that the sensitivity increased up to 500 °C and then decreased at higher temperature. A further discussion of this thermal activation curve (TAC) and its relation to the relevant model will be discussed in Section 9.2 along with the results of the simulations. Further work on the predose effect in quartz has been carried out by Martini *et al.* [630] who gave a tentative proposal as to the identity of the relevant impurities taking part in the sensitization of the 110 °C TL peak in quartz. Lee *et al.* [631] have studied theoretically the predose effect of quartz, based on the modified Zimmerman model. A sublinearity of the sensitivity on the predose is found, and is ascribed to competition of electrons in both the excitation and read-out stages. The possible impact on the evaluation of palaeodoses has been discussed. For further discussion of the predose effect in quartz and its applications in archaeological and geological dating see Section 9.1.

A similar effect of TL sensitization has also been found by Suntharalingam and Cameron [616], Petralia [632] and Wintersgill and Townsend [598] in LiF. High radiation dose followed by annealing at 280 °C enhances the TL response to a subsequent exposure. Models describing sensitization and supralinearity of TL in TLD-100 have been given by Lakshmanan *et al.* [633] and by Moharil [634]. The former has to do with the possible removal of competition of nonluminescent centers and the latter with track interaction (see Section 14.3). Nail *et al.* [635] reported on sensitization of peak 5 in LiF TLD-100 and explained it using the Unified Interaction Model (UNIM) (see chapter 14). The sensitization of TL in LiF by prior exposure to γ-rays has been reported by Tournon [637]. The sensitization of the dosimetric material $CaSO_4$:Dy has been reported by Srivastava and Supe [636], who ascribed the effect to radiation damage.

Kirsh *et al.* [638] reported on a sensitization effect in the TL of silica optical fibers. In addition to the initial emission at $\sim$400 nm, the combined effect of irradiation and heating generates a new band at $\sim$520 nm. Besides the broad distribution of activation energies typical of amorphous silica, the authors report on several monoenergetic traps. They suggest that the recombination with O_2^- hole centers produces the 400 nm emission, while the 520 nm band is associated with a Nd^{3+} hole center. Gd^{4+} is assumed to be the dominant electron trap. Murthy *et al.* [639] also reported on a similar effect of sensitization in BeO. Here, both TL and TSEE (see Section 11.6) were sensitized significantly by the combined effect of γ-irradiation and 325 °C annealing. A similar effect of sensitization in Mg_2SiO_4:Tb has been reported by Lakshmanan and Vohra [433]. After high γ-exposure and a post-annealing treatment at 300 °C, the TL sensitivity to both γ and UV excitation was significantly enhanced.

A similar effect of predose sensitization of OSL in porcelain has been reported by Hübner and Göksu [640]. While using conventional OSL in porcelain it was found that the material was insensitive to radiation below 2 Gy if the measurement was taken a few days after irradiation. Thus, direct measurements of the OSL signal could not be used for the assessment of accidental doses. However, the enhanced OSL signal after thermal activation was found to be sensitive enough to enable the measurement of doses well below 2 Gy up to several years after irradiation. Thus, absorbed doses as low as 0.04 Gy could be reconstructed after thermal activation in laboratory controlled samples. It should be noted that the reported TAC in porcelain is significantly different than the TAC associated with the sensitization of the 110 °C TL peak. The OSL measured following irradiation and heating increases moderately with the activation temperature for activations between 300 °C and 600 °C, and increases much faster for temperatures from 600 to 700 °C. The thermally activated OSL signal is found to be linear with the size of the test-dose up to $\sim$500 mGy. A strong correlation is reported between the thermally activated 230 °C TL peak and the OSL signal following thermal activation. The predose effect of OSL in quartz has been described by Koul and Chougaonkar [641, 642] who compared the sensitization of the 110 °C TL peak and the OSL signal.

A different type of pre dose sensitization of UV excited TL in semiconducting diamonds has been reported by Chen and Halperin [643]. They found that the TL behavior changes if the sample is illuminated by longer wavelength light, e.g., 700 nm prior to the UV (300–400 nm) excitation. Although the 700 nm light does not excite TL by itself, the TL observed after the combination of 700 nm followed by 360 nm is much higher than that after 360 nm alone as shown in Figure 8.5. Curve (a) shows on the log–log scale the strong superlinearity mentioned in Section 8.1. The points on curves (b) and (c) are significantly higher than those in curve (a). Also, the slope of curve (b) is less steep in (a) and even less steep in (c), which is practically linear with the dose (slope=1). This effect was explained with the model of excitation of TL in this material by lower than band gap photons, by a multistage transition mentioned in Section 6.3.5 and in Section 8.1. The 700 nm light seems to raise the electrons into the mentioned intermediate states. This explains both the higher intensity for the same UV dose, and the linearity of the UV response following a high 700 nm illumination.

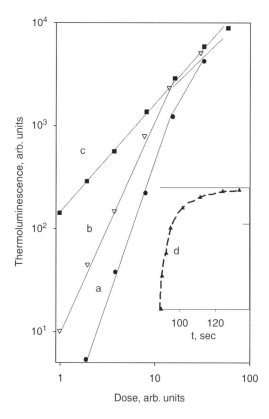

Figure 8.5 *The effect of red (700 nm) pre-excitation prior to the UV excitation (360 nm) of the 250 K peak in semiconducting diamond. (a) Excitation by 360 nm. (b) Same 360 nm doses, each preceded by a low 700 nm dose. (c) Same as (b) but with 200 times higher 700 nm dose. (d) The 250 K peak intensity as a function of the 700 nm pre-excitation dose. After Chen and Halperin [643]. Reprinted from Chen, R., and Halperin, A., Superlinear Excitation of TL in Diamonds, Proc. Int. Conf. Lumin. 1414–1417 (1966)*

8.5 Sensitization and De-sensitization in Quartz

The mentioned sensitization of the 110 °C peak in quartz discussed above was associated with the 380 nm emission in this material; the change in sensitivity had to do with an increase in the concentration of the holes in the relevant center. However, the same peak has another component at 460 nm resulting from recombination in another center, the sensitivity of which changes in an entirely different manner. The results are shown in Figure 8.6. Chen *et al.* [644] reported on the sensitization and de-sensitization of this 460 nm component, when significantly higher test-doses ($\sim$8 Gy) and sensitizing doses ($\sim$100 Gy) were utilized. Cycles of β-irradiation followed by annealing were employed. Starting with a pristine powder sample and an applied test-dose of 8 Gy, followed by annealing at 300 °C, caused an increase in the sensitivity up to an equilibrium level as seen in curve (b_1).

Similar results are seen in (a_1) and (c_1), but with 4 and 12 Gy test-doses, respectively. However, if the same sets of cycles are performed on a previously sensitized sample by,

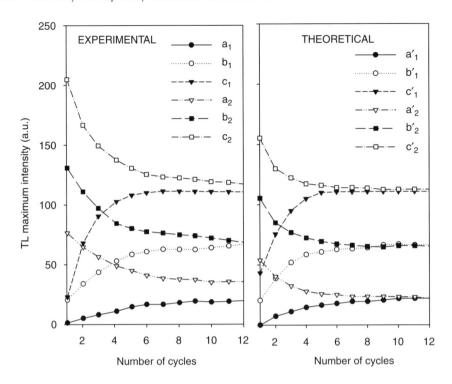

Figure 8.6 *Experimental results and theoretical results of sensitization and de-sensitization of the 460 nm emission from the 110 °C TL peak in synthetic quartz. The experimental details are given in the text. The theoretical results show the simulation of the same situation using the model described in Figure 8.1, with the parameters mentioned in the text. Reprinted from Chen, R., Fogel, G. and Kristianpoller, N., Theoretical account of the sensitization and de-sensitization in quartz, Radiat. Meas. 23, 277–280. Copyright (1994) with permission from Elsevier*

for example, an initial β dose of 100 Gy followed by annealing at 300 °C, the repeated test-dose application plus 300 °C annealing caused a decrease in the sensitivity. In this case the sensitivity went down to equilibrium values (a_2, b_2, c_2) located rather close to those reached by the sensitization curves with the same test-doses.

Chen *et al.* [645] showed that the same kind of results can be numerically simulated using a simple energy level model with one kind of recombination center, one active trapping state and one kind of competing trap (see e.g. Figure 8.1). The energy scheme shown in Figure 8.1 can be used here. The two trapping states are N_1 and N_2 (denoting also the total concentration of the two trapping states, in m^{-3}) with instantaneous occupancies of n_1 (m^{-3}) and n_2 (m^{-3}). E_1 and E_2 are the activation energies (eV), s_1 and s_2 (s^{-1}) are the frequency factors, and A_1, A_2 (m^3 s^{-1}) are the retrapping probability coefficients. n_c and n_v (m^{-3}) are the concentrations of electrons in the conduction band and holes in the valence band, respectively. X is the rate of production of electron–hole pairs (m^{-3} s^{-1}) by the radiation, which is proportional to the dose rate. A_m and A_v are the recombination probability coefficient of electrons and trapping probability coefficient of electrons and holes into the hole centers, respectively. M and m (m^{-3}) are the hole center concentration and occupancy. The differential equations governing the process during the excitation phase are

$$\frac{dn_1}{dt} = A_1 n_c (N_1 - n_1) - s_1 n_1 \exp(-E_1/kT), \tag{8.46}$$

$$\frac{dn_2}{dt} = A_2 n_c (N_2 - n_2) - s_2 n_2 \exp(-E_2/kT), \tag{8.47}$$

$$-\frac{dm}{dt} = A_m m n_c - A_v n_v (M - m), \tag{8.48}$$

$$\frac{dn_v}{dt} = X - A_v n_v (M - m), \tag{8.49}$$

$$\frac{dn_c}{dt} + \frac{dn_1}{dt} + \frac{dn_2}{dt} = \frac{dn_v}{dt} + \frac{dm}{dt}. \tag{8.50}$$

A relaxation period is being simulated, in a similar way to that described before, in which the temperature of the sample is kept constant for a certain period of time after the excitation has stopped ($X = 0$). During this time n_c and n_v decay down to negligible values. The final concentrations reached at this stage serve as initial values for the read-out phase (heating). A constant heating rate is assumed such that $T = T_0 + \beta t$; the hole center is assumed to be far enough from the valence band so that in this stage $n_v = dn_v/dt = 0$. The emitted light is assumed to be proportional to the rate of recombination, so that Equation (8.48) changes in the final stage into [see e.g. Equation (8.13)]

$$I = -\frac{dm}{dt} = A_m m n_c. \tag{8.51}$$

Chen *et al.* [645] chose the following set of parameters for simulating the sensitization and de-sensitization effects: $E_1 = 0.9$ eV; $E_2 = 1.4$ eV; $s_1 = s_2 = 10^{12}$ s^{-1}; $A_v = A_m = 2 \times 10^{-20}$ m^3 s^{-1}; $N_1 = 1.99 \times 10^{20}$ m^{-3}; $N_2 = 10^{18}$ m^{-3}; M$=2 \times 10^{20}$ m^{-3}; $\beta = 1$ °C s^{-1}; $A_1 = A_2 = 10^{-20}$ m^3 s^{-1}; $X = 10^{18}$ m^3 s^{-1} in (a'); $X = 2 \times 10^{18}$ m^3 s^{-1} in (b'); $X = 3 \times 10^{18}$ m^3 s^{-1}. The simultaneous differential equations were solved using the sixth-order Runge–Kutta algorithm. The theoretical results shown in Figure 8.6 are of the maximum intensity of the 110 °C peak as a function of the number of preceding cycles of *test- dose* applications plus heating to ~300 °C. As can be seen, the theoretical results resemble the experimental ones for the three values of X taken. In each case, the TL yield reached the same equilibrium level starting from either low or high sensitivity. The final level reached was slightly more than linearly dependent on the dose rate X, again in a similar manner to the experimental results.

Chen *et al.* [645] note that the effects of superlinearity and sensitization are intrinsically connected. One way of stating that superlinearity occurs at a given range of doses, is to say that once a unit of dose has been imparted, the sensitivity of the sample to the next unit of dose is higher. However, the effect discussed here is more complicated since the stage of annealing takes place between two consecutive excitations. Nevertheless it is encouraging to see that exactly the same energy-level model results in a behavior which agrees so well with the experiments. Note that both the experimental and theoretical curves are given in arbitrary units (see Figure 8.6). One adjustable parameter has been set so that the equilibrium response in (c_1) and (c'_1) is the same. The results in curves (a) and (b) are seen to be very similar though not identical with those in (a′) and (b′). The reasons for this are probably that more elements are influencing the experimental results, and that the choice of parameters is not unique; it stands to reason that another choice of parameters might have been even more successful. Bearing these reservations in mind, the agreement between experimental and theoretical results is surprisingly good.

It should be reiterated that in this model the reservoir, which is essential in the explanation of the predose effect of the 380 nm emission (see Section 8.4), plays no role. Although further experiments along these lines are required, Chen *et al.* [645] state that it can be predicted with some caution that the two effects are unique, each to its respective emission range; the main reason is that the two effects are seen in the same kind of sample, but at two different dose ranges. However, the important common factor in the two models is the central part played by a third trapping state.

A related effect of de-sensitization of TL in LiF has been reported by Marrone and Attix [646] and further discussed by Cameron *et al.* [138] (pp. 158–165). The response of LiF noticeably decreases if the phosphor has been exposed previously to high dose radiation. The conclusion was that large exposure produces permanent radiation damage to the crystal, and thus desensitizes it.

8.6   Dose-rate Dependence

In the discussion so far, we have addressed the question of dose dependence of TL, and later we will discuss the dose dependence of OSL as well. In principle, the rate at which the dose is applied and the length of time during which it is administered are two independent variables. In the literature, there are a number of accounts showing that in some cases, the emitted TL is not the same following the same given dose, if it is applied at different rates. In most cases this effect is not very large, but it deserves some attention. Early investigators such as Karzmark *et al.* [647] and Tochilin and Goldstein [648] reported that no dose-rate effects were observed in LiF for dose rates ranging from 100 to 10^9 Gy s^{-1}, and for a number of years, it was accepted that no such effect could be expected [10, 138]. It should be noted that when archaeological and geological dating are concerned, the natural dose rate may be as low as 10^{-3} Gy yr^{-1} (3×10^{-11} Gy s^{-1}), whereas the laboratory dose rate applied for calibration may be as high as several Gy s^{-1}. One has to distinguish between a dose-rate effect such as that reported by Facey [649], which has to do with a thermal decay during excitation, and "real" dose rate effect where the TL peak in question is stable such that no thermal decay is expected at ambient temperature. Groom *et al.* [650] reported a significant effect where no thermal decay takes place. Here, a decrease of TL by up to a factor of 5 was observed with increased dose rate in powdered samples of Brazilian crystalline quartz when irradiated by ^{60}Co γ-rays at dose rates ranging from 1.4×10^{-3} to 3.3 Gy s^{-1}. A smaller effect of the same sort was reported by Hsu and Weng [651]. Urbina *et al.* [652] described a rather unusual dose rate effect in calcite. If a calcite sample has been γ-irradiated for a long time at a high dose rate, a following β-irradiation exhibits a dose-rate effect. Shlukov *et al.*

[653] suggested that in archaeological and geological dating based on the measurement of TL in quartz, the fact that there is a difference of eight to nine orders of magnitude in the dose rate between natural and calibration irradiations, may cause a serious error in the age evaluation. In a later paper, Shlukov *et al.* [654] suggests a dating technique for Quaternary sediments that tries to bypass this problem since it does not require laboratory irradiation. The method seems to be rather limited since it is based on a comparison with another sample of known age which may have different thermoluminescent properties. Also, these authors limit their analysis to the two possibilities of first- and second-order kinetics whereas intermediate cases are obviously possible.

Additional experimental evidence of a similar nature was given by Wintersgill and Townsend [598] who described a dose-rate effect in the excitation of TL in LiF. An opposite effect of higher measured TL for larger dose rates was reported by Kvasnička [655] who found the effect in Brazilian and milky quartz excited by ^{60}Co γ-rays, using dose rates of $2 \times 10^{-5}\,\mathrm{Gy\,s^{-1}}$ $-2 \times 10^{-2}\,\mathrm{Gy\,s^{-1}}$. The apparent discrepancy between the two sets of results may be due to the different ranges of dose being used. A very interesting dose-rate effect has been reported by Valladas and Ferreira [656] . They distinguish between three components in the emission of TL in quartz, namely UV, blue and green. Applying the same dose of excitation at two dose rates which are three orders of magnitude apart, they found different behaviors for the three components. The UV component was nearly twice as large with the high dose rate than with the low one. As for the green component, the low dose rate yielded about 10% less than with the high one. However, with the blue component, the low dose rate yielded about 50% more TL than the high one. Valladas and Valladas [657] gave some further details concerning TL results in detrital quartz following annealing at different temperatures, as well as variations of the response of the samples irradiated at the high dose rate ($0.16\,\mathrm{Gy\,s^{-1}}$) as a function of temperature during irradiation. In a recent paper, Gastélum *et al.* [658] have reported on the dose-rate effect of TL in microwave-assisted chemical vapor deposition (MWCVD) diamonds. Using γ-irradiation they found that for fixed irradiation dose the TL efficiency increased as the dose rate increased in these CVD diamond films. Dose-rate effects of OSL have also been mentioned in the literature. Bailey [532] and Bailey *et al.* [533] have reported that dose-rate effects affected the dose response of quartz OSL.

8.7   Sublinear Dose Dependence of TL and LM-OSL in the OTOR System

Cases of sublinear dose dependence of TL and OSL at relatively low doses, namely ranges where $d^2 S/dD^2 < 0$, where S is the relevant luminescence intensity and D is the dose, have also been reported. Rodine and Land [610] reported on the dose dependence of TL in ThO$_2$. Different peaks showed different dependencies on the dose, some were superlinear whereas others were sublinear. These authors explained the results in terms of competition over carriers during the excitation and heating stages. Boustead and Charlesby [659] studied TL in squalane and reported superlinearity in some peaks and sublinearity in others. Goldstein [660, 661], Durand *et al.* [662], Farge [663] and Israeli *et al.* [499] studied the TL resulting from the production of point defects in UV-irradiated KI, LiF and other alkali halides. The situation here is different from the mentioned case of the filling of existing trapping states, however a close analogy can be found between the sets of simultaneous differential equations governing the two processes. These authors showed that a $D^{1/2}$ dependence can be expected under certain conditions (see section 6.10). A similar square-root dependence in TL of SiC was reported by Vakulenko and Shutov [591]. A material in which sublinearity was found as of very low doses is $CaSO_4$:Dy. Caldas and Mayhugh [429] reported on photo-transferred TL (PTTL, see Section 6.6) which behaves like $D^{0.55}$ where the initial dose varies by five orders of magnitude. Lakshmanan *et al.* [664] presented results of sublinear γ-ray dose response of the high temperature peaks in $CaSO_4$:Dy. In a recent paper, Lawless *et al.* [513] have pointed to the possibility that even with the simplest system of one trapping state and one kind of recombination center, and at relatively low doses, when both the electron trapping state and the recombination center are far from saturation, the dose dependence of the trapping states may be significantly sublinear. Thus, depending on the details of the process, the dose dependence of the measured luminescence signal, be it TL or OSL (see below) may be sublinear. The model used is that shown in Figure 4.1 and the governing equations are (4.1–4.4). It should be mentioned that the second term on the right-hand side of Equation (4.1), $s \cdot n \cdot \exp(-E/kT)$ can be omitted if the sample is excited at a low enough temperature such that practically no electrons are released thermally during the excitation. In order to follow the mentioned feature of having sublinear dose dependence far from saturation, Lawless *et al.* [513] rewrote these equations in a slightly different way,

$$\frac{dn}{dt} = A_n(N - n)n_c, \tag{8.52}$$

$$\frac{dn_c}{dt} = X - A_n(N - n)n_c - A_m m n_c, \tag{8.53}$$

$$\frac{dm}{dt} = Bn_v(M - m) - A_m m n_c, \tag{8.54}$$

$$\frac{dn_v}{dt} = X - Bn_v(M - m). \tag{8.55}$$

Assuming physically realistic rate constants and not exceedingly large dose rates, one can make the usual quasi-equilibrium assumption $dn_c/dt \approx dn_v/dt \approx 0$ which results in

$$n_v = \frac{X}{B(M - m)} \tag{8.56}$$

and

$$n_c = \frac{X}{A_n(N - n) + A_m m}. \tag{8.57}$$

Substituting Equation (8.57) into Equation (8.52) yields

$$\frac{dn}{dt} = \frac{dm}{dt} = \frac{A_n(N - n)}{A_n(N - n) + A_m m} X. \tag{8.58}$$

Under the present assumptions of OTOR, and assuming that n_c and n_v are significantly smaller than n and m, one has $m = n$. With this assumption Equation (8.58) can be rewritten as

$$\frac{dn}{dt} = \frac{A_n(N-n)}{A_n N + A_m n} X, \tag{8.59}$$

which can be rearranged

$$X dt = \frac{A_n(N-n) + A_m n}{A_n(N-n)} dn = \left[1 + \frac{A_m n}{A_n(N-n)}\right] dn, \tag{8.60}$$

and integrating brings about

$$Xt = n + \frac{A_m}{A_n}\left[-N \cdot \ln\left(1 - \frac{n}{N}\right) - n\right]. \tag{8.61}$$

For a small argument $x \ll 1$, one can use the first two terms in the relevant series expansion,

$$\ln(1-x) = -x - \frac{1}{2}x^2, \tag{8.62}$$

which yields

$$Xt = n + \frac{A_m}{2A_n}\frac{n^2}{N}. \tag{8.63}$$

This, in turn, results in

$$A_n N n + A_m n^2/2 = A_n N X t. \tag{8.64}$$

Expression (8.64) is a quadratic equation in n which actually gives a relationship between n and t (or the dose $D = X \cdot t$). The solution for n yields

$$n = \frac{-A_n N \pm \sqrt{(A_n N)^2 + 2A_m A_n N X t}}{A_m}. \tag{8.65}$$

Obviously, in order to get a positive value for n one has to choose the plus sign, which results in

$$n = \frac{A_n N}{A_m}\left(\sqrt{1 + \frac{2A_m}{A_n N}Xt} - 1\right). \tag{8.66}$$

If following the excitation one allows for relaxation, i.e., that the free electrons and holes get either recombined or trapped following the excitation, one ends up with $n_c = n_v = 0$. Lawless *et al.* [513] explained that the parameters chosen for a certain system must be such that n_c and n_v are much smaller than n and m, respectively, and therefore the additional contribution to n and m during relaxation is relatively small. By conservation of charge one must have $n = m$, hence Equation (8.66) can be considered as a good approximation to the dependence of n on t or D under the mentioned assumptions. The square root sign shows that the expression is sublinear. Two limiting cases can be considered:

(1) Small doses. For $Xt \ll A_n N/(2A_m)$, Equation (8.66) simplifies to

$$n \approx Xt. \tag{8.67}$$

This means that during excitation and for small doses, the trap captures the available free electrons with 100% efficiency.

(2) Large doses. For $Xt \gg A_n N/(2A_m)$, Equation (8.66) simplifies to

$$n \approx \sqrt{\frac{2A_n N}{A_m} Xt}. \tag{8.68}$$

This relation indicates that as doses grow larger, the trap captures an ever smaller fraction of the available free electrons. This is because m grows with dose and, as it grows, the recombination rate $A_m mn_c$ becomes stronger relative to the rate for capture by traps $A_n Nn_c$. This explains the sublinearity of n.

As in other situations, one may wonder whether the approximations made cannot influence substantially the results. One can get some support for the conclusions by solving numerically the mentioned set of coupled differential equations with no approximations but for certain sets of parameters, and comparing the results to the analytical results mentioned here.

An example of such simulations by Lawless *et al.* [513] is given in Figure 8.7. The parameters chosen (see Figure 4.1 and Equations (4.1–4.4)) were: $A_m = 10^{-5}\,\mathrm{cm^3\,s^{-1}}$; $A_n = 10^{-11}\,\mathrm{cm^3\,s^{-1}}$; $B = 10^{-10}\,\mathrm{cm^3\,s^{-1}}$; $N = 10^{14}\,\mathrm{cm^{-3}}$; $M = 10^{15}\,\mathrm{cm^{-3}}$ and $X = 10^{14}\,\mathrm{cm^{-3}\,s^{-1}}$. Concerning the recombination-probability coefficients, the values of B and A_n chosen for the simulations were in the range typical of the repulsive type of interaction between carriers and A_m is of the attractive type, whereas B and A_n are consistent with the neutral type (see Section 4.1.1). Note that assuming that A_m is of the attractive type agrees with the fact that recombination takes place only after a hole has been captured in the center. The results of the simulations are shown by the solid line. The ($+$) signs indicate the results of Equation (8.66) and the ($\times$) signs those of Equation (8.68). The agreement between the simulated results and the approximations looks very good with this set of parameters, and therefore the possibility of having a square-root dependence on the dose

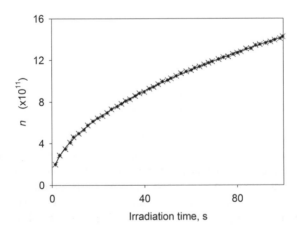

Figure 8.7 *Simulated concentration of carriers (solid line), results of Equation (8.66) ($+$) and of Equation (8.68) ($\times$), the chosen values of the relevant parameters are given in the text. After Lawless et al. [513]. Reprinted from Lawless, J.L., Chen, R. and Pagonis, V., Sublinear dose dependence of thermoluminescence and optically stimulated luminescence prior to the approach to saturation level, Radiat. Meas. 44, 606–610. Copyright (2009) with permission from Elsevier*

within the framework of the OTOR model is established. The reason for the sublinearity appears to be the competition between the trap and the recombination center for free electrons during irradiation. These results resemble those by Goldstein [660, 661], Durand *et al.* [662], Farge [663] and Israeli *et al.* [499] mentioned above, although these authors dealt with the production of new F-centers rather than the filling of existing trapping states. The similarity between the two cases takes place in spite of the fact that in the present case there is a limitation on the number of existing trapping states which is not present in the case of production of defects. However, the interesting range here is much below saturation even with relatively high doses, which under the appropriate conditions results in the $D^{1/2}$ dependence similarly to the defect production case. It should be noted that the TL and OSL may depend linearly or non linearly on the concentration following the irradiation. These dependencies should be combined with the presently discussed dependence of the concentration of carriers on the dose, to yield the total dependence of the luminescence intensities on the excitation dose.

One may be interested in the dose dependence all the way to saturation. Figure 8.8 gives an example based on the numerical solution of the implicit function given in Equation (8.61). As shown in the figure, three dose dependencies are shown on a log–log scale of n/N as a function of the normalized dose Xt/N for different values of the relative recombination to retrapping probabilities. The solid line depicts the results for $A_m/A_n = 10^6$, the dashed-dotted line for $A_m/A_n = 10^3$ and the dashed line for $A_m/A_n = 1$. One should note that the scales of the x- and y-axes are not the same and the solid line starts with a slope of 0.5, meaning a dose dependence of $D^{1/2}$ nearly all the way to saturation. The other two curves start with a slope of unity, but the dashed line behaves linearly for a long range of doses

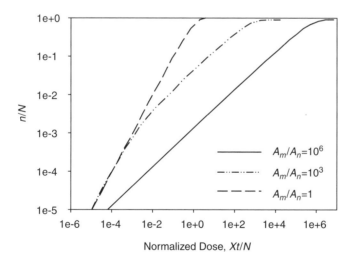

Figure 8.8 *Dose dependence of the trap occupancy calculated by solving Equation (8.61) numerically for different values of the dose Xt. The solid line depicts the results for $A_m/A_n = 10^6$, the dashed-dotted line for $A_m/A_n = 10^3$ and the dashed line for $A_m/A_n = 1$. Note the different scale of the two axes. After Lawless et al. [513]. Reprinted from Lawless, J.L., Chen, R. and Pagonis, V., Sublinear dose dependence of thermoluminescence and optically stimulated luminescence prior to the approach to saturation level, Radiat. Meas. 44, 606–610. Copyright (2009) with permission from Elsevier*

before approaching saturation, whereas the dashed-dotted line changes from linearity to an approximate $D^{1/2}$ dependence before going to saturation.

With respect to Figure 8.8 it is interesting to note that as long as the condition leading to Equation (8.61) is valid, namely that the quasi-steady condition holds, the figure shows the dose dependence all the way to saturation with different ratios of the recombination to retrapping probabilities. Taking into consideration the different scale of the x- and y-axes, the solid line with $A_m/A_n = 10^6$ has a slope of 0.5. One can consider the cases of small and large doses leading to Equations (8.67) and (8.68), respectively. The condition bringing about Equation (8.68), $Xt \gg A_n N/(2A_m)$, is much easier to be fulfilled for this large ratio of 10^6, and one can expect a broader range of a $D^{1/2}$ dependence which is seen in the figure almost all the way to saturation. As for the opposite condition $Xt \ll A_n N/(2A_m)$, this can be fulfilled in a broad range of doses for a much smaller ratio of recombination to retrapping probabilities, and for the example given of $A_m/A_n = 1$ (dashed line in Figure 8.8), a linear dependence on the dose is seen, in agreement with Equation (8.67). In the intermediate case (dashed-dotted line) with $A_m/A_n = 10^3$ the behavior is more complex. The dose dependence starts being linear, it then goes sublinear at a relatively low dose far from saturation, remains approximately $\propto D^{1/2}$ for a few orders of magnitude change of the dose, and finally goes to saturation. In conclusion, a long range of $D^{1/2}$ occurs if the ratio A_m/A_n is large and becomes longer as this ratio increases. By contrast, if the ratio is not large then no such region occurs regardless of the values chosen for N and M. Changing N has no effect on the data once it is normalized as in Figure 8.8. It should be mentioned that the model requires $M > N$ in order to avoid the possibility of accumulation of electrons in the conduction band at the very large doses. As long as this condition is obeyed and the quasi-steady condition is valid, the value of M has no effect on the dose-scaling results.

It should be noted that although the luminescence signal (be it TL, OSL or LM-OSL) may represent the occupancy of trapping states following excitation by irradiation, one does not always expect the signal and the occupancy to be proportional to each other. This was quite obvious when we discussed the dose dependence in multi-level materials where competition with other traps or centers during excitation and read-out caused superlinear, sublinear and even non-monotonic dose behavior of both TL and OSL (see Sections 8.2 and 8.3). Lawless *et al.* [513] have studied the possible nonlinear dose dependencies of TL and LM-OSL within the simple OTOR model, emphasizing the conditions under which these effects are not directly proportional to the trap concentration at the end of irradiation. In this case it is obvious that the only competition one can think of is that of the recombination center competing with the trap for electrons during excitation, and retrapping competing with the recombination center during read-out (see e.g. Figure 4.1).

In order to follow the full process taking place during excitation, relaxation and heating, one first solves the set of Equations (4.1–4.4) for the excitation stage, then sets $X = 0$ and solves the same set and finally, for the heating of the sample one has to solve numerically the relevant set of coupled rate equations, which are

$$\frac{dn}{dt} = A_n(N - n)n_c - sn \exp(-E/kT), \tag{8.69}$$

$$\frac{dn_c}{dt} = -\frac{dn}{dt} - A_m m n_c, \tag{8.70}$$

$$\frac{dm}{dt} = -A_m m n_c, \tag{8.71}$$

where E (eV) is the activation energy and s (s^{-1}) the frequency factor. The intensity of the emitted TL is proportional to minus the rate of change of the concentration of holes in centers, and choosing the proportionality factor to be unity simply means a certain choice of the light emission units. Using some heating function $T(t)$, the linear heating function $T = T_0 + \beta t$ being a popular choice, one can write

$$I(T) = -\frac{dm}{dt} = A_m m n_c, \tag{8.72}$$

where $I(T)$ denotes the TL emitted light intensity as a function of temperature.

In order to show the dose dependence of the trapped electrons, Lawless *et al.* [513] chose sets of trapping parameters and solved numerically Equations (4.1–4.4) using the MATLAB ode23s solver for a certain period of time t. The simulated excitation dose was therefore proportional to $D = Xt$. It should be noted that since the materials in question are insulators, even under irradiation the concentrations of free electrons and holes are expected to be only a fraction of the concentrations of trapped electrons and holes. Lawless *et al.* [513] and Chen *et al.* [550] have monitored these concentrations and made sure that for the chosen sets of parameters, n_c and n_v are significantly smaller than n and m. In the second stage of the simulation, the temperature has been raised at a constant rate of $1\,°\mathrm{C\,s}^{-1}$, and Equations (8.69–8.71) were solved using the same ode23s solver. When the maximum intensity was reached, it was registered as the TL signal.

Figure 8.9 depicts the values of $m = n$ as the dashed line, and the TL maximum intensity as the solid line for the set of parameters given in the caption. Had the TL curve

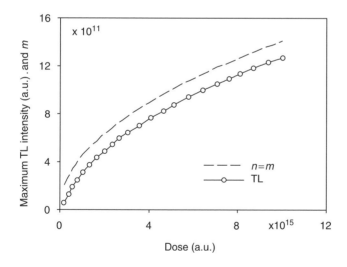

Figure 8.9 *Maximum TL intensity (solid line) and concentration of electrons in traps and holes in centers ($n = m$) as simulated by solving the sets of coupled equations for $A_m = 10^{-5}\,\mathrm{cm^3\,s^{-1}}$; $A_n = 10^{-11}\,\mathrm{cm^3\,s^{-1}}$; $B = 10^{-10}\,\mathrm{cm^3\,s^{-1}}$; $N = 10^{14}\,\mathrm{cm^{-3}}$; $M = 10^{15}\,\mathrm{cm^{-3}}$; and $X = 10^{14}\,\mathrm{cm^{-3}\,s^{-1}}$. The irradiation time varies between 2 s and 100 s. For the two lines to be seen separately, the dashed line is shifted upward. Reprinted from Chen, R., Pagonis, V. and Lawless, J.L., Nonlinear dose dependence of TL and LM-OSL within the one trap-one center model, Radiat. Meas. 45, 277–280. Copyright (2010) with permission from Elsevier*

been normalized so as to have it coincide with the concentration curve at one point, they would overlap all along. The TL curve has been shifted by a constant amount downwards so that the parallel two curves can be shown separately. The concentration curve here is the same as in Figure 8.7 for the same set of parameters, thus both the occupancy curve and the TL curve are very closely square-root functions. The temperature of occurrence of the TL peak was constant for these results at $T_m = 447$ K. As pointed out by Lawless *et al.* [513], the dependence of the concentrations is exactly of $D^{1/2}$ and as shown here, this is also the case for the maximum TL intensity I_m. In the results, $n = m$ vary between 2×10^{11} cm^{-3} and 1.4×10^{12} cm^{-3} with $A_m = 10^{-5}$ cm^3 s^{-1}; this yields $mA_m = (2 \times 10^6 - 1.4 \times 10^7)$ s^{-1}. With $A_n = 10^{-11}$ cm^3 s^{-1} and $N = 10^{14}$ cm^{-3}, we get $A_n(N - n) \approx 10^{-11} \times 10^{14} = 10^3$ s^{-1}. It is obvious that $mA_m \gg A_n(N - n)$ and the first-order condition prevails. This agrees very well with the fact that I_m is proportional to n (and m). Also, the fact that the maximum temperature does not vary with the excitation dose is compatible with first-order kinetics. With $X = 10^{14}$ cm^{-3} s^{-1} the dose varies here between 2×10^{14} cm^{-3} and 10^{16} cm^{-3}. With the given parameters, $A_n N/(2A_m) = 5 \times 10^7$ and, obviously $D \gg A_n N/(2A_m)$, which clearly indicates a $D^{1/2}$ dose dependence of the trap occupancy at the end of irradiation.

Figure 8.10 shows the results of the maximum TL and $n = m$ for the set of parameters given in the caption. These are the same as in Figure 8.9 except that the retrapping-probability coefficient is three orders of magnitude smaller, $A_n = 10^{-14}$ cm^3 s^{-1}. The concentration $n = m$ and maximum TL curves are both sublinear. X here is 10^4 cm^{-3} s^{-1} and the simulated

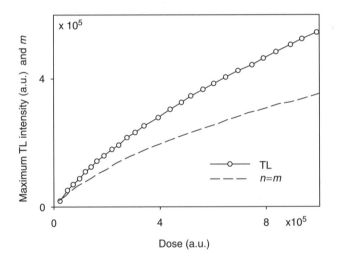

Figure 8.10 *Maximum TL intensity (solid line) and concentration of electrons in traps and holes in centers ($n = m$) (dashed line) as simulated by solving the mentioned sets of coupled differential equations for: $A_m = 10^{-5}$ cm^3 s^{-1}; $A_n = 10^{-14}$ cm^3 s^{-1}; $B = 10^{-10}$ cm^3 s^{-1}; $N = 10^{14}$ cm^{-3}; $M = 10^{15}$ cm^{-3}; and $X = 10^4$ cm^{-3} s^{-1}. The irradiation time varies between 2 s and 100 s. The TL intensities were normalized so that the two curves coincide in the lowest dose. Reprinted from Chen, R., Pagonis, V. and Lawless, J.L., Nonlinear dose dependence of TL and LM-OSL within the one trap-one center model, Radiat. Meas. 45, 277–280. Copyright (2010) with permission from Elsevier*

irradiation times are the same as previously. Therefore, the dose D varies between $2 \times 10^4 \, \text{cm}^{-3}$ and $10^6 \, \text{cm}^{-3}$. With the parameters quoted in the caption, $A_n N/(2A_m) = 5 \times 10^4 \, \text{cm}^{-3}$ and therefore we have $D < A_n N/(2A_m)$ at low doses, but $D > A_n N/(2A_m)$ at higher doses. Obviously, one has here a transition from one behavior to another as the dose grows. Neither of the two sublinear curves is expected to behave like $D^{1/2}$ with the dose. The values of $m = n$ are seen to be between $2 \times 10^4 \, \text{cm}^{-3}$ and $3.5 \times 10^5 \, \text{cm}^{-3}$. Therefore mA_m varies from 0.2 to 3.5 s^{-1} whereas $A_n(N-n) = 1 \, \text{s}^{-1}$. The kinetics here is obviously intermediate in a sense between first and second order. The TL curve is normalized so that the two curves coincide at the lowest dose and are seen to behave in a different manner at higher doses. The maximum TL temperature varied with the dose, changing gradually from 479 to 452 K as the dose increases.

Figure 8.11 depicts on a log–log scale the results with another set of parameters given in the caption. The maximum temperature here varied between 640 K and 534 K with increasing dose. The slope of the $n = m$ curve is nearly unity whereas that of the maximum TL is slightly superlinear with a slope of 1.07. The results are given on a logarithmic (base 10)

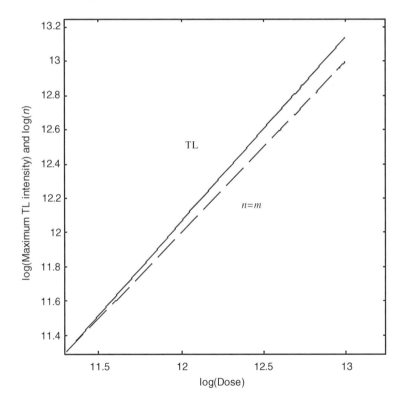

Figure 8.11 *Maximum TL intensity (solid line) and concentration of electrons in traps and holes in centers ($n = m$) (dashed line) as simulated by solving the mentioned sets of coupled differential equations for: $A_m = 10^{-6} \, \text{cm}^3 \, \text{s}^{-1}$; $A_n = 10^{-5} \, \text{cm}^3 \, \text{s}^{-1}$; $B = 10^{-10} \, \text{cm}^3 \, \text{s}^{-1}$; $N = 10^{14} \, \text{cm}^3$; $M = 10^{15} \, \text{cm}^{-3}$; and $X = 10^{11} \, \text{cm}^{-3} \, \text{s}^{-1}$. The irradiation time varies here too between 2 s and 100 s. The results are shown on a log–log scale*

scale. As shown, m varies between $\sim 2.5 \times 10^{11}$ and $10^{13}\,\mathrm{cm}^{-3}$, therefore mA_m changes between $2.5 \times 10^5\,\mathrm{cm}^{-3}$ and $10^7\,\mathrm{cm}^{-3}$ whereas $A_n(N-n) \approx 10^9\,\mathrm{cm}^{-3}$. This, along with the fact that with this simple model $n \approx m$ and n is significantly smaller than N, results in nearly exact second-order kinetics. The fact that the temperature of the TL maximum decreases significantly with increasing dose agrees with the second-order property. The value of the simulated dose $D = Xt$ varies between $\sim 2.5 \times 10^{11}\,\mathrm{cm}^{-3}$ and $10^{13}\,\mathrm{cm}^{-3}$ and $NA_n/(2A_m) = 5 \times 10^{14}\,\mathrm{cm}^{-3}$. The latter is significantly higher than the former and therefore, as mentioned above [see Equation (8.68)], the filling of the traps and centers is expected to be linear with the dose. Indeed, in Figure 8.11 the occupancy is seen to be linear with the excitation dose, whereas the maximum TL behaves slightly superlinearly, yielding a slope of 1.07 on the log–log scale. This result is commensurate with those by Chen *et al.* [531] who showed that one can expect a dose power of 1.06–1.08 of the maximum TL in pure second-order peaks (see also Rasheedy and Amry [665]). Note that Guzzi and Baldini [408] reported a power of 1.09 ± 0.06 in the dose dependence of TL in $\mathrm{ZnIn_2S_4}$. It should be mentioned that the area under the TL glow peak must be proportional to the carrier concentration at the end of the excitation, and the slight superlinearity mentioned here has to do with a distortion of the shape of the second-order peak with the excitation dose, which is also expressed in the change of the maximum temperature.

As for the simulation of LM-OSL, exactly the same equations (4.1–4.4) are relevant for the excitation and relaxation stage (with $X = 0$), whereas for the read-out phase Equations (8.69–8.72) hold with f (s^{-1}) denoting the stimulating light intensity replacing $s\exp(-E/kT)$ in Equation (8.69). For f being linearly dependent on time, one takes $f = f_0 t$ where f_0 (s^{-2}) is constant.

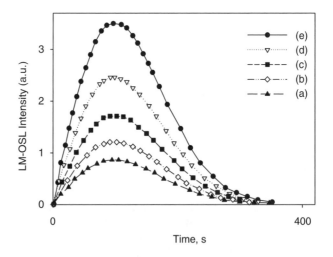

Figure 8.12 *Simulated LM-OSL curves for the same set of trapping parameters as in Figure 8.9. Excitation times are 1, 2, 4, 8 and 16 s in curves (a–e), respectively. After Chen et al. [549]. Reprinted from Chen, R., Pagonis, V. and Lawless, J.L., A new look at the linear-modulated optically stimulated luminescence (LM-OSL) as a tool for dating and dosimetry, Radiat. Meas. 44, 344–350. Copyright (2009) with permission from Elsevier*

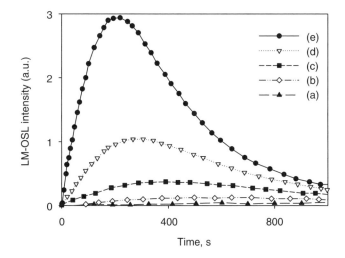

Figure 8.13 *Simulated LM-OSL curves for the same set of trapping parameters as in Figure 8.11. Excitation times are 1, 2, 4, 8 and 16 s in curves (a–e), respectively. After Chen et al. [549]. Reprinted from Chen, R., Pagonis, V. and Lawless, J.L., A new look at the linear-modulated optically stimulated luminescence (LM-OSL) as a tool for dating and dosimetry, Radiat. Meas. 44, 344–350. Copyright (2009) with permission from Elsevier*

Figure 8.12 shows the LM-OSL curves with the same parameters as in Figure 8.9 for TL. The simulation was performed using the MATLAB ode23s solver [549]. As pointed out above for TL with the same set of trapping parameters, the occupancy of the trapping states behaves like $D^{1/2}$, and as seen here, the same is true for the maximum LM-OSL signal. The peaks occur at the same time t_m for all five doses.

Figure 8.13 depicts the LM-OSL curves with the same set of parameters as in Figure 8.11. The behavior is of second order with a strong dependence of t_m on the excitation dose and a $D^{3/2}$ dose dependence of the maximum intensity. It should be mentioned that Figure 8.13 is rather similar to Figure 7.7, except that there, the concentrations at the beginning of the optical stimulation are varied by a factor of 2 from one curve to the next, whereas here the excitation dose is varied by the same factor of 2.

Figure 8.14 shows, similarly to the TL shown in Figure 8.10, the situation under a kinetics which is intermediate in some sense. As pointed out above, mA_m varies here from 0.2 to $3.5\,s^{-1}$ and the kinetics order is intermediate between first and second order. Comparing the maximum intensity of the simulated LM-OSL for curves (d) and (e) shows that for a factor of 2 in the dose, the maximum intensity changes by a factor of 2.51. This is more than 2 and therefore represents superlinearity, but less than $2^{3/2} = 2.83$ characterizing the second-order kinetics. As for the maxima of curves (a) and (b), their ratio is 1.90 for a dose ratio of 2 meaning that the dose dependence is sublinear, but the ratio is more than the factor of $\sqrt{2}$ described in Equation (8.68). Here, t_m varies with the dose, but much more moderately than in the pure second-order case. The situation here is evidently intermediate between first and second order, though not in the sense of general-order kinetics.

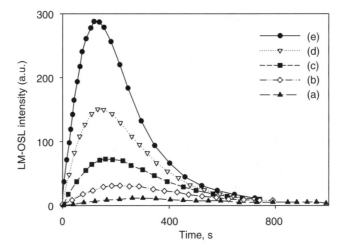

Figure 8.14 *Simulated LM-OSL curves for the same set of trapping parameters as in Figure 8.10. Excitation times are 1, 2, 4, 8 and 16 s in curves (a–e), respectively. After Chen et al. [549]. Reprinted from Chen, R., Pagonis, V. and Lawless, J.L., A new look at the linear-modulated optically stimulated luminescence (LM-OSL) as a tool for dating and dosimetry, Radiat. Meas. 44, 344–350. Copyright (2009) with permission from Elsevier*

8.8   Dose-dependence and Dose-rate Behaviors by Simulations

The discussion above concerning the non linear dose dependence of TL and OSL and in particular the superlinear dependence of these signals, dealt separately with competition during excitation and during heating and included simplifying assertions such as the quasi-equilibrium assumption. One may fear that the assumptions may cause some cumulative error, and it is of great interest to follow the theoretically expected dose dependence without making these assumptions. Moreover, in the experiments one cannot distinguish in most cases between effects taking place during the excitation and during the read-out. As pointed out above, the alternative is to use simulations, namely to solve the relevant sets of simultaneous differential equations along the three stages of the experiment, excitation, relaxation and heating (or optical read-out in OSL). With the available packages for solving differential equations, this can be done to practically any desired accuracy with the disadvantage that each "numerical experiment" can be carried out only for a certain set of trapping parameters. Thus, in most cases certain behaviors such as superlinear dose dependence or nonmonotonic dose dependence (see below) can be shown to be feasible and compatible with a certain energy level scheme, but definitely not as an analytical function of dose dependence. In this section we deal with dose dependence as well as dose rate dependence in both TL and OSL.

8.8.1   Superlinearity of TL

Chen and Fogel [666] performed such calculations for the model of two trapping states and one kind of recombination center shown in Figures 8.1 and 8.3. For a certain choice of parameters and of the intensity of excitation X they solved the simultaneous equations for

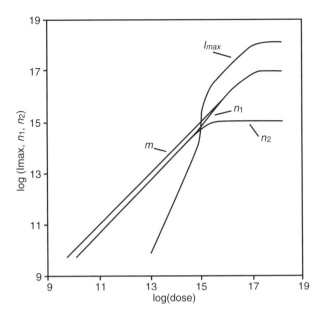

Figure 8.15 *An example of results of the numerical calculations for a model with two traps and one center. n_1, n_2 and m are values obtained for the given doses of irradiation following the first two stages of excitation and relaxation. I_{max} is the resulting maximum of TL, simulated during the heating stage. Reprinted from Chen, R. and Fogel, G.M., Supralinearity in thermoluminescence revisited, Radiat. Prot. Dosim. 47, 23–26. Copyright (1993) with permission from Oxford University Press*

the excitation stage. They then took the final occupancies of the traps, centers and valence and conduction bands and set them as the initial values for the next stage of relaxation. The same equations were now solved for a further period of time until the concentrations of free electrons and holes became practically nil. The final occupancies of this stage were now used as initial values for the read-out stage of TL, and the equations (8.11–8.14) were now solved step by step for increasing temperature until the maximum intensity was reached. This maximum I_{max} was then taken as the signal for the given dose. The dose was then changed and the signal plotted as a function of the dose. An example of the results of such stimulation are shown in Figure 8.15.

The parameters chosen (see the energy-level diagram in Figure 8.1) were: $N_1 = 10^{23}$ m^{-3}; $N_2 = 10^{21}$ m^{-3}; $A_1 = A_m = A_n = 10^{-21}$ m^3 s^{-1}; and $A_2 = 10^{-19}$ m^3 s^{-1}. As can be seen Figure 8.15, n_1 and n_2 start linearly with the dose and at a certain dose where n_2 approaches saturation, n_1 goes slightly superlinear before approaching saturation. As shown in Figure 8.15, the emitted TL (as measured by I_{max}) starts quadratically, as demonstrated above in the semi-analytical way [Equation (8.35)] . In the following rather narrow dose range where n_2 approaches saturation during the excitation, the dose dependence is much "stronger" than quadratic, which is compatible with the discussion related to Equation (8.34). The dose dependence curve becomes linear at higher doses and finally approaches saturation. These results resemble those by Kristianpoller *et al.* [450] and point to the fact that

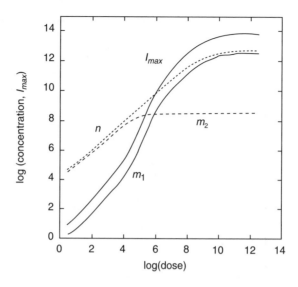

Figure 8.16 *A sample result of the TL, as well as the carriers concentrations in the trapping state and recombination centers, simulated by the model with two centers and one trap. The parameters chosen are given in the text. Reprinted from Chen, R., Fogel, G. and Lee, C.K., A new look at the models of the superlinear dose dependence of thermoluminescence, Radiat. Prot. Dosim. 65, 63–68. Copyright (1996) with permission from Oxford University Press*

the dominant effect is that of competition during heating. The given simulation results are representative in the sense that when parameters were changed within the same energy-level model, the results remained basically the same with minor changes in the details.

An alternative effect associated with a model of two competing centers and one trapping state was studied by Chen *et al.* [624]. The energy level model is seen in Figure 8.4, and the relevant sets of equations are Equations (8.36–8.40) for the excitation stage, the same equations with X set to zero for the relaxation stage and Equations (8.41–8.44) for the heating stage. A representative simulated result is shown in Figure 8.16. The set of parameters chosen here is: $E = 1.0\,\mathrm{eV}$; $s = 10^{13}\,\mathrm{s^{-1}}$; $X = 10^{21}\,\mathrm{m^{-3}\,s^{-1}}$; $A_{m1} = A_{m2} = 10^{-21}\,\mathrm{m^3\,s^{-1}}$, $A_n = 2 \times 10^{-21}\,\mathrm{m^3\,s^{-1}}$; $M_1 = 9 \times 10^{23}\,\mathrm{m^{-3}}$; $M_2 = 10^{23}\,\mathrm{m^{-3}}$; $A_{h1} = 10^{-21}\,\mathrm{m^3\,s^{-1}}$; $A_{h2} = 10^{-20}\,\mathrm{m^3\,s^{-1}}$ and $N = 10^{24}\,\mathrm{m^{-3}}$. The x-axis shows the logarithm of $X \cdot t_D$ where t_D is the duration of the irradiation, and the product is proportional to the dose. The curves show the dose dependencies of m_{10}, m_{20}, n and I_{max}. Here the active center m_1 grows linearly as long as m_2 is far from saturation. It then goes up superlinearly in the range where m_2 approaches saturation at higher doses. During the heating stage, transitions into m_2 are considered to be radiationless, or to emit light in a range to which the light detecting device is blind. The trap filling of n is seen to grow linearly until it approaches saturation at about the same dose level as m_1. The resulting TL is seen to follow nearly exactly the dose dependence of the active trap m_1. It is obvious that the dominating effect here is that of competition during excitation, whereas competition during heating adds only a minor contribution. Here too, varying the relevant parameters does not seem to change the main features of the results.

To conclude this section, some points are to be made. In general, it is not sufficient to consider separately the phenomena of competition during excitation and during heating. Both possibilities are to be considered together since TL is always the result of both, and the observed superlinearity is an outcome of both. Since there is no analytical way to follow all three stages of excitation, relaxation and heating, the considerations given above can only give a general idea about the expected outcome, and the use of simulations is indispensable. Moreover, the superlinear effect due to competition during heating by itself has two elements. One has to do with the final result being proportional to the concentration of charge carriers in both traps and centers in the presence of a strong competitor. Thus, if each of them is proportional to the dose, the dose dependence is quadratic. In addition, in the dose range where the competitor approaches saturation, the reduction in competition causes some extra superlinearity, namely more than quadratic behavior.

The numerical solutions of the relevant sets of simultaneous differential equations enable us to bypass the possible problems of artifacts due to the approximations made. If a certain dose dependence is observed in the simulation, one can be sure that it is the real effect of the model under examination and therefore can be compared with the relevant experimental results. On the other hand, the numerical results are associated with the specific sets of parameters chosen, and it is rather difficult to draw general conclusions. However, one feels that it is of great importance to be able to demonstrate that the experimentally observed dose dependencies can indeed be associated with these models. It is also important to mention the approach by Lee and Chen [667] who used the opposite direction, namely they reached some conclusions concerning superlinearity by making approximations, thus getting semi-analytical results.

As opposed to what might have been thought, it has been found both from analytical considerations and numerical results that there is a significant difference between a situation involving a competition of two trapping states and a competition between two centers. In the former, a typical behavior is an initially quadratic dose dependence followed by a steeper superlinearity before saturation effects set in. In the latter, a typical dependence is an initial linear range followed by moderate superlinearity, and in turn going back to linearity and saturation. It appears that the main difference between a trap competitor and a center competitor is that in the former case, the concentration $N_2 - n_{20}$ (which competes with the concentration m_{10} for released electrons), decreases with the dose. In the latter case, m_{20} which is the concentration of the relevant competitor, increases with the dose.

The apparent discrepancy between the theoretical dependence which starts quadratically and gets more superlinear at higher doses, and the more than cubic initial dose dependence of the 110 °C peak in synthetic quartz [620] can be explained quite easily. It is plausible that the initial quadratic dependence cannot be observed due to an insufficient sensitivity of the measuring system in the low dose range. Therefore as the TL intensity grows, the threshold of the measurement sensitivity may very well be within the "super-quadratic" dose range.

8.9   Simulations of the Dose-rate Effect of TL

Section 8.6 described the dose rate effect of excitation of TL. In the present section we discuss an energy-level model given by Chen and Leung [668] which explains, by the use of simulation, the occurrence of this effect. The model shown in Figure 8.17 is practically

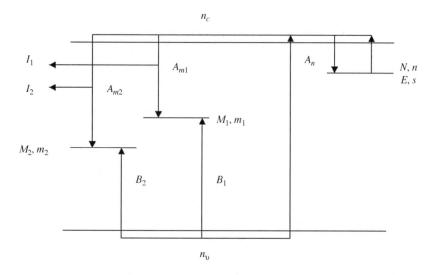

Figure 8.17   *The energy level scheme of a solid with one trapping state and two recombination centers. The transitions taking place both during excitation and heating are shown. The meaning of the parameters and the values chosen for them are given in the text. After Chen and Leung [668]. Reprinted from Chen, R. and Leung, P.L., A model for dose-rate dependence of thermoluminescence intensity, J. Phys. D: Appl. Phys. 33, 846–850. Copyright (2000) with permission from Institute of Physics*

the same as that shown in Figure 8.4, except that here both transitions into the two competing centers are taken into consideration. As pointed out above, in some experimental results TL increases with the dose rate, whereas in other materials the measured TL decreases with the increasing dose rate when the total dose is kept constant. However, the results given by Valladas and Ferreira [656] show that two spectral components in quartz behave in opposite ways, namely one increases and the other decreases with the dose rate. The work by Chen and Leung [668] simulates this situation by solving the relevant equations associated with the given model. The sets of equations to be solved numerically are Equations (8.36–8.40) for the excitation stage and for the relaxation stage (with $X = 0$), and Equations (8.41–8.44) for the read-out. One important addition is that the two emission components are $I_1 = -dm_1/dt$ and $I_2 = -dm_2/dt$.

The parameters chosen by Chen and Leung for the simulation demonstrating the effect are: $E = 1.38\,\text{eV}$; $s = 10^{10}\,\text{s}^{-1}$; $A_{m1} = 10^{-18}\,\text{m}^3\,\text{s}^{-1}$; $A_{m2} = 10^{-19}\,\text{m}^3\,\text{s}^{-1}$; $B_1 = 10^{-19}\,\text{m}^3\,\text{s}^{-1}$; $B_2 = 1.5 \times 10^{-21}\,\text{m}^3\,\text{s}^{-1}$; $N = 10^{20}\,\text{m}^{-3}$, $M_1 = 10^{20}\,\text{m}^{-3}$; $M_2 = 10^{21}\,\text{m}^{-3}$; $A_n = 3 \times 10^{-20}\,\text{m}^3\,\text{s}^{-1}$.

Figure 8.18 depicts the results of the solution of the equations in the sequence of excitation–relaxation–heating with this set of chosen parameters and with $X = 10^{18}\,\text{m}^{-3}\,\text{s}^{-1}$ and time of excitation $t_D = 10^3\,\text{s}$. The curve shown by a dotted line indicates the intensity as a function of temperature $I_1(T)$, whereas the solid line gives $I_2(T)$. Note that the shape of the peaks shows the effect of competition in the range of overlap; $I_1(T)$ is narrow at the fall-off part whereas $I_2(T)$ is narrow in the low-temperature half.

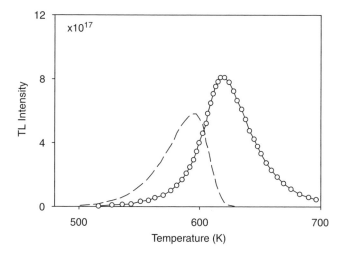

Figure 8.18 *Two TL peaks as simulated using the model of two recombination centers and one trapping state (see Figure 8.17). The set of parameters chosen is given in the text and the dose rate is $10^{18} m^{-3} s^{-1}$. The dashed line represents I_1, whereas the solid line represents I_2. After Chen and Leung [668]. Reprinted from Chen, R. and Leung, P.L., A model for dose-rate dependence of thermoluminescence intensity, J. Phys. D: Appl. Phys. 33, 846–850. Copyright (2000) with permission from Institute of Physics*

Figure 8.19 shows the results of the intensities for these same two peaks following an excitation with $X = 10^{22}$ m^{-3} s^{-1} for $t_D = 0.1$ s. The total dose is the same in the two cases, with X being 10^4 times higher than before and t_D being 10^4 times lower. It is readily seen that I_1 increases by a factor of nearly 2 whereas I_2 decreases by ~20%.

Figure 8.20 depicts the variation of the maximum intensity of I_1 and I_2 with the dose rate, varying gradually within this range of four orders of magnitude in the dose rate, both in the increasing and decreasing glow peak. In order to understand the source of this variation of the two peaks with the dose rate, Chen and Leung [668] have also monitored the filling of the trap and center at the end of the relaxation period which follows the excitation. The results with $X = 10^{18}$ m^{-3} s^{-1} and $t_D = 1000$ s were $m_1 = 2.67 \times 10^{19}$ m^{-3}, $m_2 = 4.61 \times 10^{19}$ m^{-3} and $n = 7.28 \times 10^{19}$ m^{-3}. With $X = 10^{22}$ m^{-3} s^{-1} and $t_D = 0.1$ s, they found $m_1 = 5.65 \times 10^{19}$ m^{-3}, $m_2 = 3.37 \times 10^{19}$ m^{-3} and $n = 9.02 \times 10^{19}$ m^{-3}. The ~24% increase in the final n value can be explained by suggesting that fewer annihilations take place in the short excitation time (0.1 s) through the recombination centers. It is much more difficult to follow the different competing transitions so as to give an intuitive explanation to the dose-rate effect. The transition of electrons associated with A_{m1} competes with the retrapping associated with A_{m2} and both compete with that associated with A_n. Also, the transition of holes associated with B_1 competes with that associated with B_2. The electron and hole transitions taking place during excitation are coupled since m_1 and m_2 participate in both kinds of transitions.

The values of m_1 and m_2 at the end of the excitation plus relaxation and prior to the heating are qualitatively commensurate with the intensities of peak 1 and peak 2 shown

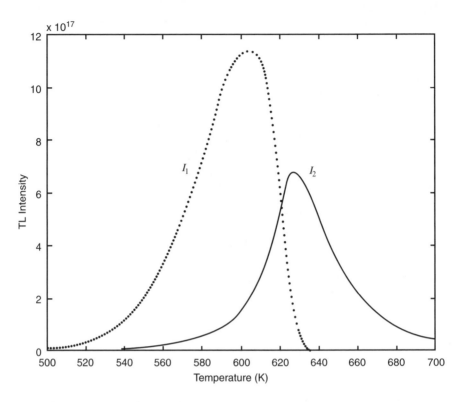

Figure 8.19   *Same as Figure 8.18, but the dose rate is $10^{22}m^{-3}s^{-1}$. The total dose is $10^{21}m^{-3}$, the same as in Figure 8.18. After Chen and Leung [668]. Reprinted from Chen, R. and Leung, P.L., A model for dose-rate dependence of thermoluminescence intensity, J. Phys. D: Appl. Phys. 33, 846–850. Copyright (2000) with permission from Institute of Physics*

in Figures 8.18 and 8.19 with the low and high dose rates. This indicates the importance of the competition during the irradiation in the excitation stage. However, the fact that these concentrations are not really proportional to the calculated TL intensities shows the importance of competition during heating in determining the final results.

The simulated results of the variations with the dose rate as given by Chen and Leung [668] are not identical with the experimental results reported by Valladas and Ferreira [656] for the UV and blue components of quartz. This is simply because the model is quite obviously oversimplified, and in addition it is not claimed that the real relevant parameters are known. However, the behavior resembles the experimental results qualitatively, with even the small shift in the maximum temperature occurring in the same way with the variation of the dose rate.

The same model demonstrates also the possibility which was described in the literature of having only one TL peak that increases or decreases with the dose rate. This is the case when one of the transitions described in the present discussion is either radiationless, or radiative but not measurable with the detecting device used in the experiment.

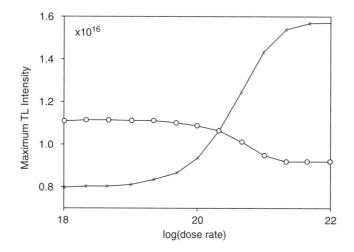

Figure 8.20 *The dose rate dependence of the TL maximum values of $I_1(T)$(crosses) and of $I_2(T)$ (circles) for the set of parameters given in the text. The strong increase in the maximum of I_1 and the decrease in the maximum of I_2 are clearly seen at a constant total dose. After Chen and Leung [668]. Reprinted from Chen, R. and Leung, P.L., A model for dose-rate dependence of thermoluminescence intensity, J. Phys. D: Appl. Phys. 33, 846–850. Copyright (2000) with permission from Institute of Physics*

8.9.1 Dose and Dose Rate Units

We digress here to give the range of dose rates and doses associated with the X values chosen, in the conventional units of Gy s^{-1} and Gy correspondingly. Let us consider LiF which has a specific gravity of 2.6 or a density of 2600kg m^{-3} (the calculation for other materials can be done in an analogous way). According to Avila *et al.* [669], an average of 36 eV is required for heavy charged particles to produce one electron – hole pair, and $\sim$34 eV for γ-rays (this is about three times the band gap). Since 1 Gy equals 1 J kg^{-1} and 1 J equals 6×10^{18} eV, the number of pairs produced per kg when 1 Gy is applied is about 1.7×10^{17}. Therefore the number of pairs produced per m^3 is 4.4×10^{20}. Thus, $X = 10^{18}$ m^{-3} s^{-1} is equivalent to 2.3×10^{-3} Gy s^{-1}. The transition of the TL level shown in Figure 8.20 around 10^{20} m^{-3} s^{-1} for the given set of parameters occurs at $\sim$0.23 Gy s^{-1}. This is within the range given by Groom *et al.* [650]. The total dose of 10^{21} m^{-3} is in fact 2.3 Gy. The situation with quartz should be more or less similar to that with LiF.

8.10 Nonmonotonic Dose Dependence of TL and OSL

As pointed out above, the dose-dependence function of TL is normally an increasing function. However, it has been reported in the literature that in several materials the intensity of TL peaks reached a maximum as a function of the excitation dose and decreased at higher doses. The effect was first reported by Charlesby and Partridge [670]. They describe a decline of the maximum TL intensity in γ-irradiated polyethylene at 10^4 Gy and postulate that the cause of the effect is radiation damage in the material. Halperin and Chen [173] reported on the UV excited TL in semiconducting diamonds. The secondary peak at 150 K increased linearly with the dose at low doses, reached a maximum at a certain dose and decreased at higher doses. It is quite obvious that in diamonds photons with an energy of 3–5.5 eV cannot cause radiation damage. Cameron *et al.* [138] described the non-monotonic dose dependence in LiF:Mg,Ti as a function of ^{60}Co γ-ray excitation dose. In this material, which has been serving for many decades as one of the main dosimetric materials, they report on a rather broad range of linear dependence, followed by a superlinear range, after which a maximum value and a slight decline are observed. Jain *et al.* [671] describe a significant decrease of the TL output of peak V in LiF, by a factor of 2.5 from the maximum, and ascribe it to radiation damage. Their graphs show that at very high doses, the dose-dependence curves tend to level off following a range of significant decrease in the TL intensity. Similar non-monotonic results but following X-ray excitation of LiF:Mg,Ti were reported by Horowitz *et al.* [672]. The effect of nonmonotonic dose dependence has also been seen in quartz, the main material used for archaeological and geological dating. Ichikawa [673] found that in γ-irradiated natural quartz, the peak at $\sim$200 °C reached a maximum at 6×10^4 Gy and decreased at higher doses by a factor of 2.5. Damm and Opyrchal [674] reported on TL of γ-irradiated KCl crystals, and described a non-monotonic dose dependence of some of the pure and Sr- and Pb-doped samples. David *et al.* [675] showed the dose dependence of some TL peaks in γ-irradiated pink quartz, which revealed a decline following a maximum at 10^3–10^4 Gy. A number of authors reported on the non-monotonic effect in the important dosimetric material Al_2O_3:C. For example, Yukihara *et al.* [456] described a somewhat superlinear dependence up to 30 Gy of β-irradiation in the 450 K peak in some of the samples. The peak reached a maximum value and declined at higher doses. These authors explain the observed effects in carbon-doped aluminum oxide (Al_2O_3:C) using a model based on the occurrence of F/F$^+$-centers as well as on the other trapping states and recombination centers. Also, a nonmonotonic dependence of TL of γ-irradiated silica glass has been found by Gulamova *et al.* [605]. This was associated with the behavior of optical absorption (OA) at the 550 nm band, namely that at $\sim$10^5 Gy the OA reached a maximum and declined following higher doses. Vij *et al.* [482] have reported such non-monotonic dose dependence of TL in SrS:Ce excited by UV light.

In a number of papers, a similar nonmonotonic dose dependence of OSL has been reported. Schulman *et al.* [676] described the changes in photoluminescence due to prior γ-excitation in organic solids. In naphthalene, the dependence of 464 nm emission stimulated by 365 nm UV light depended non linearly on the γ-excitation dose, reaching a maximum at $\sim$10^5 Gy and decreasing at higher doses. Freytag [677] and later Tesch [678], and Böhm *et al.* [679] described the γ-dose dependence of silver-activated phosphor glass, which is used for dose measurements. The visible orange light stimulated by 365 nm increased nearly linearly with the γ-dose between 10^{-2} Gy and 10^2 Gy, reached a maximum at $\sim$3$\times$10^3 Gy and declined at higher doses up to 10^8 Gy. The signal size was then reduced by nearly three orders of magnitude in the range between 3×10^3 Gy and 10^8 Gy. The strict linearity with the dose between 10^{-2} Gy and 50 Gy enabled the evaluation of the dose in this region. Freytag [677] has pointed out that the region of decreasing "radio-photoluminescence (RPL)" (see Chapter 13) could also be used for dose measurements, provided a well-determined calibration curve is available. Zeneli *et al.* [680] have further studied the behavior of the RPL glasses, which are being used for high-level dosimetry around high-energy particle accelerators. These authors study the light emission by 365 nm stimulation following γ-irradiation. The γ-dose dependence was a nearly linear function between 10^{-1} Gy and 10^2 Gy, reached a maximum at $\sim 10^3$ Gy and declined substantially at doses up to 10^6 Gy at all temperatures, namely 4.6, 77 and 300 K. The authors ascribed the decline to the self-absorption of the luminescence light, previously discussed by other authors in association with TL (see e.g. Horowitz *et al.* [681] and Mukherjee and Vana [682]). Bloom *et al.* [683] studied POSL in Al_2O_3 single crystals and reported a small decline at β doses above ~ 10 Gy. Yukihara *et al.* [487, 684] reported the effect in β-irradiated Al_2O_3:C. Both integrated CW-OSL and initial CW-OSL were found to reach a maximum at ~ 30 Gy of β excitation, the effect being somewhat stronger in the latter than in the former. In parallel, TL measurements revealed a maximum in the dose dependence at about the same dose.

It should be mentioned in this context that Hughes and Pooley [685] described an effect of nonmonotonic optical absorption in proton-irradiated alkali halides, an effect resembling the non monotonic luminescence effects described here. They suggest that at low temperatures, the limiting process is the formation of substitutional hydrogen ions (U-centers), as protons are captured by F-centers. The observation of IR absorption associated with substitutional hydrogen ions confirms that this combination of protons with F-centers does occur, and may contribute to the decrease in F-center concentration which follows saturation.

8.11 Nonmonotonic Dose Dependence of TL; Simulations

Chen *et al.* [686] suggested a model to explain the phenomenon of nonmonotonic dose dependence of TL mentioned in Section 8.10. The model includes two trapping states N_1 and N_2 and two kinds of hole recombination centers M_1 and M_2 as shown in Figure 8.21. The trapping level N_1 is considered to be active in the sense that it releases electrons thermally into the conduction band with the relevant parameters being the activation energy E_1, and the frequency factor s_1. The instantaneous occupancy of N_1 is denoted by n_1. N_2 is considered to be thermally and optically disconnected; practically no electrons are released from N_2 into the conduction band and therefore this trapping state serves only as a competitor both during the excitation and the heating stages. The instantaneous occupancy of this level is denoted by n_2. The retrapping probability coefficients are denoted by A_{n1} and A_{n2}. The instantaneous concentrations of electrons and holes in the conduction and valence bands are n_c and n_v, respectively. The rate of production of electrons and holes by the irradiation per unit volume per second is denoted by X, which is proportional to the dose rate. If the radiation is imparted at a constant rate for a period of time t_D, the total concentration of electrons and holes produced by the irradiation is $X \cdot t_D$, which is proportional to the dose D. The concentration of the radiative centers is denoted by M_1 and that of the competitor non radiative center by M_2. The instantaneous occupancies of these centers are m_1 and m_2, respectively. The trapping coefficients of free holes into M_1 and M_2 during the excitation are B_1 and B_2. The recombination coefficients of free electrons with trapped holes in M_1 and M_2 both during excitation and heating, are denoted by A_{m1} and A_{m2}. The rate of recombination into M_1 during heating is associated with the TL emission, as shown in Figure 8.21.

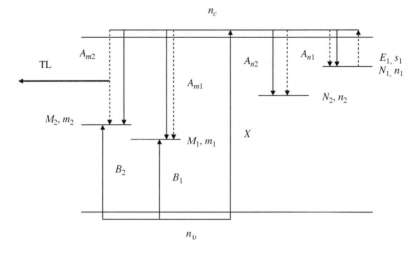

Figure 8.21 *The energy level scheme of two trapping levels and two kinds of recombination centers. Transitions occurring during the excitation are given by solid lines and transitions taking place during the heating by dashed lines. Reprinted from Chen, R., Lo, D. and Lawless, J.L., Non-monotonic dose dependence of thermoluminescence, Radiat. Prot. Dosim. 119, 33–36. Copyright (2006) with permission from Oxford University Press*

The set of coupled differential equations governing the process during the excitation is

$$-\frac{dm_1}{dt} = A_{m1}n_c m_1 - B_1 n_v (M_1 - m_1), \qquad (8.73)$$

$$-\frac{dm_2}{dt} = A_{m2}n_c m_2 - B_2 n_v (M_2 - m_2), \qquad (8.74)$$

$$\frac{dn_1}{dt} = A_{n1}n_c(N_1 - n_1) - s_1 \exp\left(-E_1/kT\right)n_1, \qquad (8.75)$$

$$\frac{dn_2}{dt} = A_{n2}n_c(N_2 - n_2), \qquad (8.76)$$

$$\frac{dn_v}{dt} = X - B_1 n_v(M_1 - m_1) - B_2 n_v(M_2 - m_2), \qquad (8.77)$$

$$\frac{dm_1}{dt} + \frac{dm_2}{dt} + \frac{dn_v}{dt} = \frac{dn_1}{dt} + \frac{dn_2}{dt} + \frac{dn_c}{dt}. \qquad (8.78)$$

Note that usually the last term in Equation (8.75) can be ignored since the irradiation is performed at a low enough temperature so that practically no electrons are thermally released during the excitation. Obviously a similar term has been omitted from Equation (8.76) since the E_2 and s_2 values are such that no electrons and holes reside in the conduction and valence bands, respectively. Since the materials in question are insulators, it is rather obvious that the concentration of the free carriers in the bands should be significantly smaller than those of electrons and holes in traps and centers.

To follow the experimental procedure a period of relaxation time is considered. This is done by taking the final values of all the occupancy functions as initial values for the relaxation stage, setting $X = 0$ in Equation (8.78) and solving numerically the coupled equations for a further period of time until n_c and n_v are negligibly small. In the final stage the following set of simultaneous differential equations is to be numerically solved (see the dashed-line transitions in Figure 8.21), with the final values at the relaxation stage serving as initial values for the heating stage.

$$\frac{dn_1}{dt} = -s_1 n_1 \exp(-E_1/kT) + A_{n1}(N_1 - n_1)n_c, \tag{8.79}$$

$$\frac{dn_2}{dt} = A_{n2}(N_2 - n_2)n_c, \tag{8.80}$$

$$\frac{dm_1}{dt} = -A_{m1}m_1 n_c, \tag{8.81}$$

$$\frac{dm_2}{dt} = -A_{m2}m_2 n_c, \tag{8.82}$$

$$\frac{dn_1}{dt} + \frac{dn_2}{dt} + \frac{dn_c}{dt} = \frac{dm_1}{dt} + \frac{dm_2}{dt}. \tag{8.83}$$

One has to specify the heating function, which can be conventionally chosen to be linear according to $T = T_0 + \beta t$, where β is the constant heating rate. The TL emission is associated with the recombination into m_2, therefore the intensity is

$$I(T) = A_{m2}m_2 n_c. \tag{8.84}$$

Chen et al. [686] chose appropriate sets of parameters and solved the simultaneous sets of equations using the MATLAB ode23 solver. Equations (8.73–8.78) were first solved for a certain value of the dose rate X, and then solved with $X = 0$, for a further period of relaxation time. Subsequently, the coupled set of equations (8.79–8.83) was solved with the appropriate heating rate using the same solver; the values of $I(T)$ were determined from Equation (8.84), and the maximum value was recorded. It is quite clear that the measured TL following excitation for a given set of trapping parameters depends on the competition processes during both the excitation and the heating of the sample. However, one can identify situations where the effects of competition during excitation are dominating and others where competition during heating is more important. Figures 8.22 and 8.23 show the simulated results of the maximum TL as a function of the dose on a logarithmic scale.

The parameters chosen by Chen et al. [686] for Figure 8.22 are as follows: $M_1 = 3 \times 10^{21}\,\mathrm{m}^{-3}$; $M_2 = 1 \times 10^{18}\,\mathrm{m}^{-3}$; $A_{n1} = 3 \times 10^{-17}\,\mathrm{m}^3\,\mathrm{s}^{-1}$; $A_{n2} = 3 \times 10^{-17}\,\mathrm{m}^3\,\mathrm{s}^{-1}$; $E_1 = 1.0\,\mathrm{eV}$; $s_1 = 1 \times 10^{12}\,\mathrm{s}^{-1}$; $A_{m1} = 1 \times 10^{-17}\,\mathrm{m}^3\,\mathrm{s}^{-1}$; $A_{m2} = 1 \times 10^{-16}\,\mathrm{m}^3\,\mathrm{s}^{-1}$; $N_1 = 1 \times 10^{19}\,\mathrm{m}^{-3}$; $N_2 = 1 \times 10^{21}\,\mathrm{m}^{-3}$; and $B_1 = 1.5 \times 10^{-17}\,\mathrm{m}^3\,\mathrm{s}^{-1}$; and $B_2 = 1 \times 10^{-17}\,\mathrm{m}^3\,\mathrm{s}^{-1}$. The heating rate taken for all the simulated glow curves was $\beta = 1\,^\circ\mathrm{C}\,\mathrm{s}^{-1}$. The results show an increase of the TL maximum (solid line) with the dose up to a maximum

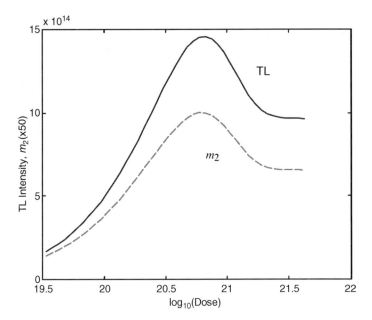

Figure 8.22 *Simulated dose dependence of the maximum TL (solid line), and the radiative center concentration at the end of irradiation m_2 (dashed line), when competition during excitation dominates. The relevant set of parameters is given in the text. Reprinted from Chen, R., Lo, D. and Lawless, J.L., Non-monotonic dose dependence of thermoluminescence, Radiat. Prot. Dosim. 119, 33–36. Copyright (2006) with permission from Oxford University Press*

at a 'dose' of $\sim 7 \times 10^{20}$ m^{-3}, followed by a decrease of $\sim 35\%$ after which the maximum TL intensity tends to level off at higher doses. This behavior is very similar to experimental results reported in some materials [671]. A similar behavior is seen in the plot of m_2 (at the end of the relaxation period) as a function of the dose (dashed line), and the leveling off at higher doses is also similar.

The parameters chosen for Figure 8.23 are as follows: $M_1 = 3 \times 10^{21}$ m^{-3}; $M_2 = 1 \times 10^{18}$ m^{-3}; $A_{n1} = 3 \times 10^{-17}$ m^3 s^{-1}; $A_{n2} = 3 \times 10^{-20}$ m^3 s^{-1}; $E_1 = 1.0$ eV; $s_1 = 1 \times 10^{12}$ s^{-1}; $A_{m1} = 1 \times 10^{-17}$ m^3 s^{-1}; $A_{m2} = 1 \times 10^{-18}$ m^3 s^{-1}; $N_1 = 1 \times 10^{21}$ m^{-3}; $N_2 = 1 \times 10^{19}$ m^{-3}; and $B_1 = 1.5 \times 10^{-19}$ m^3 s^{-1} and $B_2 = 1 \times 10^{-17}$ m^3 s^{-1}. The results show an increase of the maximum TL intensity (solid line) up to a simulated dose of 7.5×10^{19} m^{-3}, and a decline from there on; this time no leveling off is observed at high doses. This resembles at least qualitatively the experimental results in quartz by Ichikawa [673]. The dashed line shows the dependence of m_2 as a function of the dose, which is an increasing function. Chen *et al.* [686] report that checking the dose dependence of the concentrations of the two trapping states and the recombination center shows that all of them are increasing functions of the dose all along, including the dose at which the TL peak is maximal as a function of the dose. It is quite obvious that in this case, the competition during excitation does not have the main role in producing the non monotonic dose dependence. It is the mutual action between the trapping states and recombination centers which results in

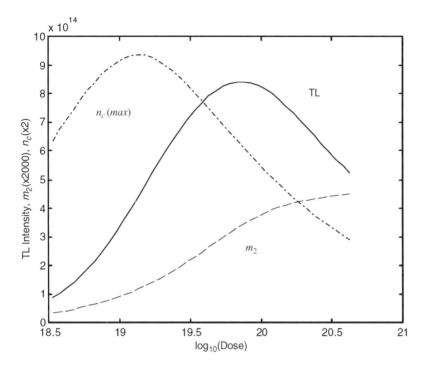

Figure 8.23 *Simulated dose dependence of the maximum TL (solid line), the radiative center concentration at the end of irradiation, m_2 (dashed line), and the values of n_c(max) (dashed-dotted line) when competition during heating dominates. The relevant set of parameters is given in the text. Reprinted from Chen, R., Lo, D. and Lawless, J.L., Non-monotonic dose dependence of thermoluminescence, Radiat. Prot. Dosim. 119, 33–36. Copyright (2006) with permission from Oxford University Press*

a strong decrease in the value of the concentration of free electrons, n_c, at the TL maximum as a function of the dose in this range. The product of increasing m_2 and decreasing n_c yields the emitted TL [Equation 8.84], reaches a maximum at a certain dose as seen in Figure 8.23 and decreases at higher doses due to the fast decrease of n_c. Chen *et al.* [686] conclude that the success of their work is in demonstrating that the non monotonic dose dependence effect is not necessarily a result of a destruction of trapping states or recombination centers as previously suggested, but could be due to competition between existing traps and centers.

In later work Pagonis *et al.* [38] used the same model as Chen *et al.* [686] to simulate the TL dose-response data and optical absorption experiments of three different samples of Al_2O_3:C. The three samples termed D320, Chip101 and B1040 were studied extensively in the experimental work of Yukihara *et al.* [456] using both beta-irradiation and UV-illumination, as well as by employing step-annealing techniques. Yukihara *et al.* [456] concluded in their paper that the observed differences in sensitization, desensitization and TL dose–response between the three samples were due to: (1) different concentrations of deep electron and deep hole traps; and (2) due to different initial concentrations of the recom-

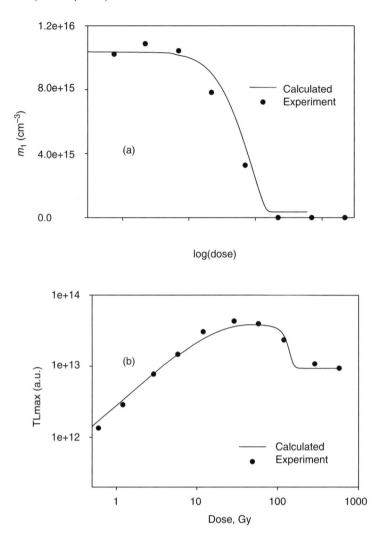

Figure 8.24 (a) The simulated variation of radiative centers with the β dose for sample Chip101 from Pagonis et al. [38] is compared with the experimental data of Yukihara et al. [456]. Reprinted from Yukihara, E.G., et al., The effects of deep trap population on the thermoluminescence of Al_2O_3:C, Radiat. Meas. 37, 627–638. Copyright (2003) with permission from Elsevier. (b) The corresponding experimental and simulated dose response of this sample. Reprinted from Pagonis, V., Chen, R. and Lawless, J.L., A quantative kinetic model for Al_2O_3:3 TL response to ionzing radiation, Radiat. Meas. 42, 198–204. Copyright (2007) with permission from Elsevier

bination centers (F^+-centers). This hypothesis was verified by the set of simulations carried out by Pagonis *et al.* [38], who suggested keeping the trapping and recombination probabilities constant for all three samples within the model, while the total and initial concentrations of the traps and centers were varied to fit the experimental data. One notable feature of these

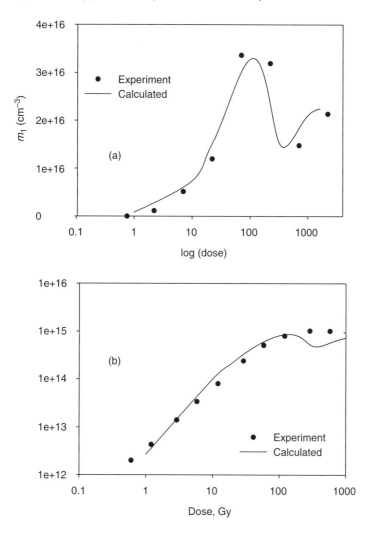

Figure 8.25   *(a) The simulated variation of radiative centers with the beta dose for sample B1040 from Pagonis et al. [38] is compared with the experimental data of Yukihara et al. [456]. (b) The corresponding experimental and simulated dose response of this sample. Reprinted from Yukihara, E.G., et al., The effects of deep trap population on the thermoluminescence of Al₂O₃:C, Radiat. Meas. 37, 627–638. Copyright (2003) with permission from Elsevier*

simulations is the inclusion of a conversion factor between the experimentally measured dose rate (in Gy s^{-1}), and the excitation rate X in the model (in electron–hole pairs cm^{-3} s^{-1}). The conversion factor for a dose rate $X = 1$ Gy s^{-1} was found to be equal to $X = 1.7 \times 10^{15}$ electron–hole pairs cm^{-3} s^{-1}. As a result of introducing this conversion factor, the experimental data could be shown on the same graph as the simulated curves of TL versus dose and of m_1 versus dose. The results of the simulation work of Pagonis *et al.* [38] are shown in Figure 8.24 and 8.25 for samples Chip101 and B1040, respectively.

Of particular importance in the model are the initial occupancies of the luminescence centers (m_{10}) for the samples studied, with the value of this parameter causing a different behavior in the two samples. The initial concentration of recombination centers is $m_{10} = 0$ for sample B1040, while sample Chip101 had a non-zero value of m_{10}. The solid lines in Figure 8.24 and 8.25 represent the calculated dependence of the TL signal and of the occupancy of luminescence centers (m_1) on the excitation dose; good agreement can be seen between simulation and experiment, with the nonmonotonic effect of the appropriate magnitude appearing at the correct experimental dose.

There have been several other notable efforts to produce a model for luminescence phenomena in Al_2O_3:C. Akselrod *et al.* [78] used time-resolved photoluminescence techniques to measure the lifetime of F-centers in Al_2O_3:C and found that the kinetic parameters W and C for thermal quenching are independent of the trap filling. They interpreted their result as indicating a classic Mott–Seitz mechanism for thermal quenching (see Section 2.10), and supported their conclusions with simulations carried out using simple models with and without thermal quenching effects. Milman *et al.* [94] proposed a different mechanism for this material, based on a temperature-dependent competition process between the radiative recombination centers and the deep traps during the TL read-out process. Kitis [458] demonstrated experimentally the influence of thermal quenching on the initial-rise method of analysis and found that this method underestimates the values of the activation energy E for the main dosimetric trap. He applied a correction method previously suggested by Petrov and Bailiff [687] to obtain the correct E values.

Nikiforov *et al.* [37] described a model which includes thermal and optical ionization of excited states of F-centers. The model provided an explanation for several experimental results on the effect of the filling of deep traps on TSEE and TSC. In a later work, Kortov *et al.* [688, 689] showed that this model explained experiments on radioluminescence, photoluminescence and thermoluminescence, as well as TSC and TSEE phenomena, and also the observed changes in sensitivity of TLD-500 detectors. Their experimental work showed that the nonlinearity of the dose dependence of the TL output depends on the heating rate used during measurement of TL, and this effect was explained using the same comprehensive interaction model between the dosimetric and deep traps.

Agersap Larsen *et al.* [690] reported on simultaneous TL and TSC measurements in Al_2O_3:C crystals, and found that their data was consistent with the presence of several overlapping first-order TL-TSC peaks. They carried out their analysis using a one-dimensional Fredholm equation assuming a constant s-factor and a distribution of activation energies E. Vincellér *et al.* [691] summarized the two prominent theories of thermal quenching in Al_2O_3:C, namely the Mott–Seitz mechanism and the alternative charge-transfer mechanism. In the former, the decrease of the luminescence efficiency is due to non-radiative transitions whose probability increases with the sample temperature. This mechanism received strong experimental support by Akselrod *et al.* [78]. In the alternative charge-transfer mechanism proposed by Kortov *et al.* [688], and Nikiforov *et al.* [37], thermal quenching is associated with a competition mechanism between recombination at F-centers and charge trapping at deep traps. Thermal and optical ionization of F-centers is temperature dependent and plays a key role in this alternative thermal quenching mechanism. By examining the dependence of the F^+-center emission on the heating rate, Vincellér *et al.* [691] proposed an energy band scheme which involves three possible competitive pathways for thermally released charge carriers in this material. Molnár *et al.* [692] investigated the influence of irradiation temperature on the TL peaks of Al_2O_3:C and confirmed the temperature dependence of the photoionization of F-centers. They noted that the Mott–Seitz mechanism as well as the alternative charge transfer mechanism lead to the same equation of the temperature dependence of the luminescence efficiency, and summarized the prevalent debate about which mechanism is dominant in this dosimetric material. Berkane-Krachai *et al.* [693] presented a rather complex model for simultaneously analyzing TL and TSEE curves for the main dosimetric peak of Al_2O_3:C. A multilayer sample structure was used to account for TL being a bulk phenomenon, while TSEE is mostly a surface process. Their model uses 11 free parameters and experimental data are fitted using a minimization procedure with the MINUIT program. Good agreement was found between calculated and experimental TL/TSEE curves, and with other available experimental parameters for alumina crystals.

In their comprehensive experimental work, Yukihara *et al.* [456] studied the effect of deep-electron and deep-hole traps on the sensitization, desensitization and TL dose–response properties of Al_2O_3:C. They suggested a band diagram for this material containing the main dosimetric trap, the deep-electron trap, the deep-hole trap, and the F and F^+-centers. Suggested possible electron and hole transitions included electron–hole pair creation, electron capture by the main dosimetric trap and by the deep-electron trap, electron–hole recombination at F^+-centers, hole capture by F-centers, hole capture by the deep–hole traps and electron–hole recombination at deep hole traps. Polf *et al.* [694] studied the real-time luminescence signal from Al_2O_3:C single crystals attached to fibers, using both experiments and numerical simulations. They simulated OSL and radioluminescence signals by using both a simple OTOR model, as well as a more complex model containing also shallow- and deep-electron traps. These authors presented experimental and simulation results for the time dependence of the luminescence signals, and also how these signals are affected by the presence of additional traps in the samples. An overview of OSL properties of Al_2O_3:C and its uses in medical dosimetry, including relevant modeling work on this material is given in Akselrod *et al.* [695].

8.12 Nonmonotonic Effect of OSL; Results of Simulations

Pagonis *et al.* [696] and Chen *et al.* [697] have simulated the observed non-monotonic effect in OSL mentioned in Section 8.11. The energy level diagram assumed is that shown in Figure 8.21, with the slight difference that instead of the thermal stimulation associated with *s* and *E* as shown in the figure, a quantity f (s^{-1}) proportional to the stimulating light intensity is considered. Obviously the set of governing equations for the excitation stage is exactly the same as in TL, given by Equations (8.73–8.78). As for the read-out stage, Equations (8.80–8.84) remain unchanged whereas Equation (8.79) is replaced by

$$\frac{dn_1}{dt} = -fn_1 + A_{n1}(N_1 - n_1)n_c. \tag{8.85}$$

Whereas in the study of TL the maximum intensity was usually taken as the TL signal, for OSL Pagonis *et al.* [696] follow the experimental practice in which one usually considers the integral over the decaying curve over a certain period of time. For the simulation, this time interval was chosen long enough so as to enable the depletion of the trap concentration to zero. In order to evaluate the area under the simulated OSL decaying curve one should usually perform a numerical integration. In order to bypass this step, Chen *et al.* [697] used an alternative method. If it is assumed that m_1 is the hole concentration in the luminescent center, obviously the OSL intensity is $I(t) = -dm_1/dt$. The integral over time between zero and the chosen final time t_f yields

$$\int_0^{t_f} I(t)dt = m_{10} - m_{1f}. \tag{8.86}$$

Therefore, one can perform the simulation along the stimulation time, but instead of integrating over $I(t)$, one can just take the difference between the initial and final value of m_1. This method, which saves computation time, also enables a better understanding of the results as discussed below. Pagonis *et al.* [696] and Chen *et al.* [697] used the following set of parameters relevant in particular to the Al_2O_3:C samples, based in part on a previous work by Yukihara *et al.* [456]: $M_1 = 10^{17}\,cm^{-3}$; $M_2 = 2.4 \times 10^{16}\,cm^{-3}$; $N_1 = 2 \times 10^{15}\,cm^{-3}$; $N_2 = 2 \times 10^{15}\,cm^{-3}$; $A_{m1} = 4 \times 10^{-8}\,cm^3\,s^{-1}$; $A_{m2} = 5 \times 10^{-11}\,cm^3\,s^{-1}$; $B_1 = 10^{-8}\,cm^3\,s^{-1}$; $B_2 = 4 \times 10^{-9}\,cm^3\,s^{-1}$; $A_{n1} = 2 \times 10^{-8}\,cm^3\,s^{-1}$ and $m_{10} = 9.4 \times 10^{15}\,cm^{-3}$. The rest of the initial carrier concentrations in the model are taken to be equal to zero. The chosen values of M_1 and m_{10} are in agreement with independent measurements inferred from optical absorption data [456]. The selected rate-constant values were all chosen to be within the range of commonly observed electron capture in solids (see e.g., Lax [57]).

Different lengths of excitation time were used and n_1, n_2, m_1 and m_2 were recorded at the end of excitation plus relaxation. The value of m_1 following the next optical stimulation stage was found for each dose, as well as the value of the integral OSL as described in Equation (8.86). Figure 8.26 depicts in curve (a) the dose dependence of the occupancy of

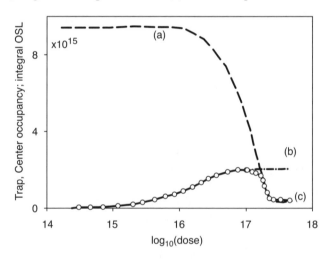

Figure 8.26 *Simulated dose dependence of the occupancy of the radiative center following excitation and relaxation is shown in curve (a), and that of the active trap in (b). Curve (c) shows the simulated integral OSL in the same range of doses. After Chen et al. [697]. Reprinted from Chen, R. et al., The nonmonotonic dose dependence of optically stimulated luminescence in Al_2O_3:C: Analytical and numerical simulation results, J. Appl. Phys. 99, 033511. Copyright (2006) with permission from American Institute of Physics*

the radiative center m_1 at the end of the relaxation period. Curve (b) shows the increase of the active trap concentration n_1 at the end of relaxation with the dose, from an initial value of zero to full saturation at 2×10^{15} cm^{-3}. n_1 reaches saturation at a "dose" of $\sim 10^{17}$ cm^{-3}. Not shown in the graph is the approach to saturation of the competitor at high doses; n_2 reaches a saturation level ($n_2 \approx N_2$) of 2×10^{15} cm^{-3} at a simulated dose of 2.5×10^{17} cm^{-3}. Curve (c) represents the area under the OSL curve as a function of the dose, as calculated by Equation (8.86). The relationship between these three quantities is discussed below.

Although this work by Chen *et al.* [697] dealt mainly with the behavior of Al$_2$O$_3$:C samples, more simulations were performed which helped in obtaining a better insight into the processes taking place and the inter-relationship between trapping states and recombination centers. Figure 8.27 shows the dose dependence of n_1 and m_1 as a function of the dose in the same system with the same trapping parameters, except that the initial value of m_1 is set to zero. This is similar to the case of sample B1040 reported by Yukihara *et al.* [456]. In curve (a), m_1 gets to a maximum value at $\sim 1.6 \times 10^{16}$ cm^{-3} and then declines and remains at a plateau level of $\sim 6.13 \times 10^{13}$ cm^{-3} at high doses. In curve (b), n_1 is seen to increase from zero to its saturation value of 2×10^{15} cm^{-3}. Curve (c) represents the integral OSL which roughly speaking has a similar peak shape to m_1, but slightly shifted to higher doses. The OSL also reaches the same plateau level of $\sim 6.13 \times 10^{13}$ cm^{-3}. These relationships are discussed below.

Figure 8.28 shows the same magnitudes as Figures. 8.26 and 8.27, but the capacity of the competitor center is increased by two orders of magnitude to be 2.4×10^{18} cm^{-3}; the initial value of m_1 is 9.4×10^{15} cm^{-3} as in Figure 8.26. Here too, the behavior at low

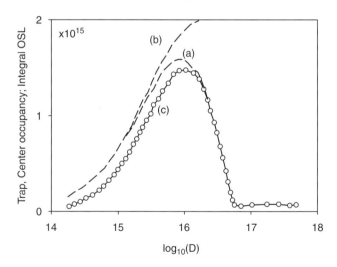

Figure 8.27   *The same as Figure 8.26, but with initial occupancy of $m_{10} = 0$. Curves (a) and (b) show, respectively, the occupancies of m_1 and n_1 at the end of relaxation, and (c) depicts the simulated integral OSL. After Chen et al. [697]. Reprinted from Chen, R. et al., The non-monotonic dose dependence of optically stimulated luminescence in Al$_2$O$_3$:C: Analytical and numerical simulation results, J. Appl. Phys. 99, 033511. Copyright (2006) with permission from American Institute of Physics*

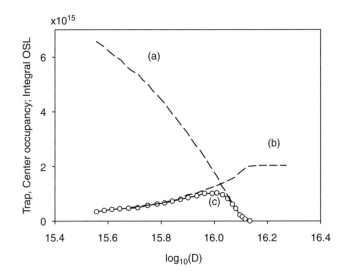

Figure 8.28 *This figure is similar to Figure 8.26, but the total concentration of competing centers is set to be $M_2 = 2.4 \times 10^{18} cm^{-3}$. After Chen et al. [697]. Reprinted from Chen, R. et al., The nonmonotonic dose dependence of optically stimulated luminescence in Al_2O_3:C: Analytical and numerical simulation results, J. Appl. Phys. 99, 033511. Copyright (2006) with permission from American Institute of Physics*

doses resembles that of the dose dependence of n_1, whereas the OSL signal at high doses is significantly smaller; its simulated value is $1.75 \times 10^{12}\,cm^{-3}$, nearly three orders of magnitude smaller than the maximum occurring at the dose of $\sim 10^{16}\,cm^{-3}$. This behavior of a significant decline in the OSL intensity resembles that reported by Freytag [677], Tesch [678] and Böhm *et al.* [679].

8.12.1   Discussion of the Simulation Results of the OSL Nonmonotonic Effect

The nonmonotonic dose dependence of the integral OSL has been associated with the competition during either excitation or read-out (or, perhaps both) [696]. In the simulation described here (Figure 8.26), m_1 is decreasing with the dose whereas n_1 is increasing, and therefore competition during excitation is not the leading factor for the nonmonotonic dependence. Competition during read-out may have an important role, and the apparent effect may be explained as follows. Equation (8.10) in Section 8.3 shows that the TL area under the glow peak depends on the quantity $\min(n_0, m_0)$, where n_0 and m_0 are, respectively, the occupancies of active traps and centers at the end of irradiation and prior to heating. The same considerations hold for n_1 and m_1 at the end of excitation in the present case of integral OSL. Considering Figure 8.26 at relatively low doses $n_1 < m_1$, and therefore the simulated OSL behaves like n_1 [curve (b)], i.e. it is increasing with the dose. In fact, since OSL is given in the same units as the concentration, the two curves practically coincide at low doses. At high doses one has $n_1 > m_1$, and therefore the OSL signal [curve (c)] behaves like m_1 [curve (a)], namely decreases with the dose, and at a high enough doses the two curves coincide. One might expect that the maximum will occur at the intersection where

$n_1 = n_2 \approx 2 \times 10^{15}$ cm^{-3}. In fact, the maximum signal occurs at a somewhat smaller dose of $\sim 10^{17}$, whereas the intersection takes place at $\sim 1.6 \times 10^{17}$. This slight discrepancy should be attributed to nontrivial dynamic effects that have to do with the variations which take place in n_2 and m_2 in the same dose range.

As for the results of Figure 8.27 where the concentration of the radiative recombination center is initially nil, curve (a) for the concentration of these centers and curve (c) for the OSL are similar peak-shaped functions, but not identical. The similarity between the curves points to an effect of competition during excitation. The slight shift of the OSL peak to higher doses than that of m_1 can be explained by taking into consideration the n_1 curve (b) which is an increasing function at the same dose range. Generally speaking, the integrated OSL can be considered as the product of n_1 and m_1 when competition takes place during read-out [see Equation (8.35)]. When the peak-shaped function is multiplied by an increasing function, the result is usually a peak-shaped function, the maximum of which occurs at higher values of the argument.

Chen *et al.* [697] discussed the nonzero plateau values of OSL which are not saturation values, observed at high doses both in some experimental results and in the simulation Figures 8.26 and 8.27. Considering the set of parameters yielding Figure 8.26, the simulation gave an equilibrium value of 3.7×10^{14} cm^{-3} for m_1, from which the same value was found for OSL, and a value of 1.3×10^{16} cm^{-3} for m_2. Chen *et al.* [697] have shown how these values are associated with the set of parameters chosen for the simulation. The fact that initially $m_{10} = 9.4 \times 10^{15}$ cm^{-3} necessarily means that there is another entity in the sample, an entirely disconnected trapping state that does not participate in the exchange of carriers during excitation or read-out, but should be assumed to exist due to neutrality considerations. This entity holds 9.4×10^{15} cm^{-3} electrons. During a given excitation, the same number of electrons is accumulated in N_1 and N_2 as holes in M_1 and M_2, and those accumulated in M_1 and M_2 are in addition to the initial concentration m_{10} mentioned above; therefore, if one considers points in time following excitation, one has the neutrality condition

$$n_1 + n_2 + 9.4 \times 10^{15}\,\text{cm}^{-3} = m_1 + m_2. \tag{8.87}$$

If one considers a high enough dose such that both N_1 and N_2 are saturated, a situation that can take place only if $N_1 + N_2$ is smaller than $M_1 + M_2$, then one can write for this range

$$N_1 + N_2 + 9.4 \times 10^{15}\,\text{cm}^{-3} = m_1 + m_2. \tag{8.88}$$

The above mentioned nonzero plateau results from a state of dynamic equilibrium in which the net change of both m_1 and m_2 is nil. In other words, the number of holes trapped per second in each center is equal to the number of holes recombining with electrons in the same center per second, i.e.,

$$A_{m1}m_1 n_c = B_1(M_1 - m_1)n_v, \tag{8.89}$$

$$A_{m2}m_2 n_c = B_2(M_2 - m_2)n_v. \tag{8.90}$$

Dividing the two equations by one another, one obtains.

$$\frac{A_{m1}m_1}{A_{m2}m_2} = \frac{B_1(M_1 - m_1)}{B_2(M_2 - m_2)}. \tag{8.91}$$

With the given set of parameters, Equations (8.88) and (8.91) are two equations in the two unknown parameters m_{1e} and m_{2e}, the equilibrium values of the centers concentrations, which can be solved by inserting m_{2e} from Equation (8.88) into (8.91), thus yielding a quadratic equation in m_{1e}. Out of the two possible solutions, one is found to be negative, and therefore non-physical. The positive equilibrium value is $\sim 3.7 \times 10^{14}\,\mathrm{cm^{-3}}$, which is exactly the plateau level seen on the right-hand side of curves (a) and (c) of Figure 8.26. The equilibrium value of the competing center m_{2e} can immediately be found from that of m_{1e} using Equation (8.88) and the value found is $m_{2e} = 1.3 \times 10^{16}\,\mathrm{cm^{-3}}$. It should be noted that this consideration applies to the equilibrium *during* excitation, and that the relevant occupancies may be somewhat different at the end of relaxation. However, in the examples given by Chen *et al.* [697] it has been found that the additional contribution to the concentration of the traps and centers is minor and less than 1%, and therefore the evaluated concentrations are valid.

A similar calculation was made for the situation in which the initial filling of the radiative center is nil, as shown in Figure 8.27. The left-hand side of Equation (8.88) now includes only $N_1 + N_2$ which is equal to $4 \times 10^{15}\,\mathrm{cm^{-3}}$ with the given parameters. Solving the analog of Equation (8.88) with this value along with Equation (8.91) yields $m_{1e} = 6.13 \times 10^{13}\,\mathrm{cm^{-3}}$, which is the plateau level shown in curves (a) and (c) of Figure 8.27. The resulting value for m_{2e} is $3.94 \times 10^{15}\,\mathrm{cm^{-3}}$. A similar calculation performed for the parameters yielding Figure 8.28, namely the same as in Figure 8.26 but with $M_2 = 2.4 \times 10^{18}\,\mathrm{cm^{-3}}$, yielded $m_{1e} = 1.75 \times 10^{12}\,\mathrm{cm^{-3}}$, which is indeed the value reached for high doses in Figure 8.28. This value is more than three orders of magnitude smaller than the maximum of the scale, and therefore looks like zero on the figure. Chen *et al.* [697] conclude that the main features seen in the dose dependence of integral OSL in some materials, namely the occurrence of a peak-shaped curve followed by an equilibrium plateau at high doses, have been demonstrated using simulations as well as intuitive physical considerations, within the framework of a model with two trapping states and two kinds of recombination centers. The relationship between the concentrations of the active center at different stages of the process and the OSL integral has been discussed. Also, the direct connection between the relevant parameters chosen for simulation and the high dose equilibrium value has been established.

A related effect of thermal bleaching of the OSL signal along with TL, has been numerically simulated by Chruścińska [698–701]. The author explains the complexity of these luminescence phenomena within a model of three traps and two recombination centers. Complications arising in the interpretation of the results associated with bleaching are described in detail.

9

Simulations of TL and OSL in Dating Procedures

9.1 The Predose Effect in Quartz

The predose technique, based on the predose effect mentioned in Section 8.4 above, is a well-established experimental method for determining the total cumulative dose from natural radiation sources, for accident and retrospective dosimetry [702–706] and for archaeological authenticity testing ([704, 707–709], and references therein). The method is based on the observed change of sensitivity of the 110 °C TL peak of quartz, caused by a combination of irradiation and high temperature annealing [707].

During the past 20 years several researchers have shown that natural materials containing quartz can be used as environmental dosemeters. The range of doses that can be estimated using the predose technique has been shown to range from low values of a few tens of mGy, to larger doses of the order of several Gy, with the upper limit restricted by the early onset of saturation. Bailiff and Haskell [702] measured the thermal characteristics of Japanese tile and showed that the predose technique can be used to measure low doses in the region of 10 mGy. Stoneham [710] applied several predose techniques to artificially irradiated porcelain fragments, and showed that porcelain would be a good predose dosimeter for both low doses (44 mGy) and for high doses (440 mGy). Haskell *et al.* [703] applied the predose TL technique to measure the cumulative dose to bricks exposed to fallout radiation during atmospheric testing at the Nevada site, and estimated doses as low as 0.20 Gy. Haskell *et al.* [706] measured the cumulative dose to quartz crystals embedded in housing bricks, which were used for construction shortly before the fallout incidents in Utah. They obtained an average dose of 38 ± 15 mGy. Hütt *et al.* [705] used the predose, fine grain and quartz inclusion TL techniques to analyze the cumulative dose to a variety of environmental materials in regions downwind from Chernobyl. These authors obtained dose estimates in the range of 0.1–2 Gy. In recent work, it was shown that the 210 °C peak of quartz and porcelain exhibited similar characteristics to the 110 °C TL peak, and that the combined

Thermally and Optically Stimulated Luminescence: A Simulation Approach, First Edition. Reuven Chen and Vasilis Pagonis.   © 2011 John Wiley & Sons, Ltd. Published 2011 by John Wiley & Sons, Ltd.

use of the two peaks can be used to estimate small accrued doses in several types of materials [702, 710–715].

While the predose technique is well established experimentally, many of the experimental protocols used in predose TL dosimetry are of an empirical nature, necessitating a better theoretical understanding and modeling of the predose phenomenon. Zimmerman [510] first proposed a predose model consisting of one electron trapping state T, a luminescence center L and a hole reservoir R. It should be noted that the UV-reversal effect mentioned in Section 8.4 is usually ascribed to the optical release of holes from the recombination center and retrapping in the reservoir. An alternative explanation has been proposed by McKeever [452] who suggested that the illumination releases trapped electrons from deep traps, which in turn will recombine with trapped holes and thereby reduce the recombination center concentration. Chen [716] pointed out that in the Zimmerman model, the population of electrons in T must be smaller than that of holes in L in order that the signal be linear with the test-dose, as indeed happens in the experiments. On the other hand, in order for the measured TL to be dependent on the population of L, its occupancy should be smaller than that of T [see Equation (8.10)]. As explained with reference to Equation (8.10), one has to assume the existence of a competing trap, in which case the TL intensity will depend on both the concentrations of trapped electrons and trapped holes, as required. Chen [716] proposed a modification of the Zimmerman model by adding the extra electron level S which competes for electrons during the heating stage. By using a modification of the Zimmerman theory, Chen and Leung [717, 718] developed a mathematical model for the predose effect on quartz, which is based on two electron and two hole trapping states. These authors simulated a typical sequence of experimental actions taken during the predose experimental technique, consisting of a sequence of irradiations and annealings. By using physical arguments concerning the observed experimental behavior of quartz samples, these authors were able to arrive at a "good" set of parameters for their model, and successfully explained several experimental results associated with the predose effect in quartz. Pagonis and Carty [719] showed that the model of Chen and Leung [717, 718] can also be used to simulate successfully the complete sequence of experimental steps taken during the additive dose variation of the predose technique. By solving the kinetic differential equations in the model, these authors demonstrated the mechanism of hole transfer from the reservoir (R) to the luminescence center (L), caused by heating to the activation temperature. Their simulation results showed that the additive dose technique can reproduce accurately the accumulated dose or paleodose (PD) received by the sample with an accuracy of 1–5% over several orders of magnitude of the PD. Pagonis *et al.* [720] simulated the complete sequence of experimental steps taken during the predose procedure, using the comprehensive quartz model developed by Bailey [35]. These simulations extended the previous work of Pagonis and Carty [719] to include a description of the multiple activation version of the predose technique, as well as a comparison with the popular single aliquot regenerative (SAR-OSL) technique, which is based on OSL signals. These authors also demonstrated the phenomenon of radiation quenching and its effects on the results of the predose technique. Several alternative explanations of the predose effect have been proposed which are based on specific impurities in quartz crystals [721–723]. Yang and McKeever [723] proposed that the predose effect might be due to the movement of hydrogen ions between defects during the irradiation and annealing steps of the predose technique. A second physical model based on quartz impurities was suggested by Itoh *et al.* [721, 724]. Their model is based on known impurities and defect processes in quartz, as well as on the introduction of an additional ionic process. The model of Itoh *et al.* [721] provides important evidence for a link between the predose effect and known impurities and defect centers in quartz. However, it does not explain the effect of radiation quenching described in Section 9.4.2.

Chen and Pagonis [725] used the model of Chen and Leung [717, 718] to simulate the sensitization of quartz samples as a function of the annealing temperature, yielding the so-called *thermal activation characteristics* (TACs) of quartz samples. Their results will be presented in some detail in Section 9.2. A variation of the model of Chen and Leung [717, 718] has also been used to describe the experimentally observed superlinearity of predosed and annealed quartz samples [726, 727]. Kitis *et al.* [267, 268] presented a broad study of the predose effect and the TACs of three quartz samples of different origin. The TACs were measured as a function of several experimental parameters using both multiple-aliquot and single-aliquot techniques, and complete TL versus dose and sensitivity *S* versus predose curves were obtained for the dose range of $0.1 < D < 400$ Gy. The results of this experimental study for all three quartz samples were interpreted by using a simple modified Zimmerman model for quartz consisting of two electron traps, a luminescence center and three hole reservoirs. Pagonis *et al.* [728] simulated the sensitization of quartz samples as a function of the annealing temperature, yielding the TAC. These authors also used a modified Zimmerman model to simulate multiple TACs and the effect of UV radiation on the TL sensitivity of quartz; these phenomena are important analytical and diagnostic tools during applications of the predose technique. Adamiec *et al.* [729, 730] presented a comprehensive kinetic model for the predose effects in quartz, consisting of three nonradiative recombination centers and one radiative center, as well as two electron traps and a reservoir electron trap. The model successfully simulated the experimentally observed TACs and isothermal sensitization of quartz samples.

In Section 9.2 we shall present the simple model of Chen and Leung [717, 718] and how it has been used to simulate the TACs in quartz. Section 9.3 contains the more comprehensive quartz model developed by Bailey [35] and its use in modeling TACs. In Section 9.4 several simulations will be presented of the experimental protocols used during application of the predose technique. Simulations of the very successful SAR-OSL protocol are demonstrated next, followed by a discussion of thermally transferred OSL (TT-OSL) signals in quartz.

9.2 Simulation of Thermal Activation Characteristics in Quartz

The sensitization of various TL materials by *β*- or *γ*- irradiations followed by activation at high temperature is a quite well known phenomenon (see e.g., Cameron *et al.* [138], Zimmerman [510], Aitken [97], Chen [716], Yang and McKeever [731], Bailiff [708] and Pagonis *et al.* [726]). By sensitization one means the change of sensitivity, namely the response to a given (usually small) test-dose, caused by a heavier irradiation followed by a thermal activation. This effect was reported by Fleming and Thompson [629] and by others in quartz, and therefore it is of importance when dealing with the dating of archaeological pottery [628] and with retrospective dosimetry [708]. The sensitization effect, also called the predose effect of the $110\,^{\circ}\mathrm{C}$ peak in quartz was reported by Zimmerman [510] who also gave the first model to account for this phenomenon. The Zimmerman model deals with an electron-trapping state T and two hole centers R and L, and is shown schematically in

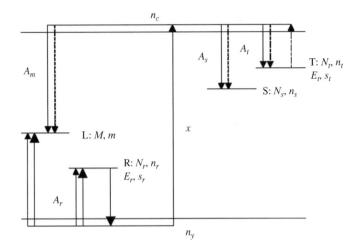

Figure 9.1 *An energy-level diagram including the electron-trapping state, T, the competitor, S, the hole reservoir, R and the luminescence center L. n_c and n_v are the free electron and hole concentrations, respectively, and X is the rate of production of free electrons and holes. The transitions taking place both during excitation and heating are shown. (After Chen and Leung [717]). Reprinted from Chen, R. and Leung, P.L., Processes of sensitization of thermoluminescence in insulators, J. Phys. D: Appl. Phys. 31, 2628–2635. Copyright (1998) with permission from Institute of Physics*

Figure 9.1. The cross-section for trapping holes is assumed to be much larger for R than for L, and therefore during irradiation practically all the free holes accumulate in R, whereas some of the generated electrons congregate in T. However, this trap is rather shallow (yielding a peak at $\sim$110 °C at a heating rate of 5 °C s^{-1}), and is emptied at RT within hours, or may be emptied by heating to, say, 150 °C. It is assumed that R is not a very deep state in the sense that it is close enough to the valence band, so that heating to $\sim$500 °C would release the holes from R. Although the probability of their trapping in L is rather low, at this high temperature the favorable direction of the holes is to go from R into L. This is so since L is assumed to be much farther from the valence band, so that once a hole is captured at L, it cannot be thermally released back into the valence band. In this sense, R is a "reservoir" which holds holes following the irradiation and prior to the thermal activation. The increase of the concentration of holes in L represents according to Zimmerman [510] an increase in the sensitivity, since L is the luminescence center.

Chen [716] suggested an amendment to the Zimmerman model. He argued that since the measured response of the TL to a test-dose is a monotonically increasing function of the concentration of trapped holes in L following the "large" excitation plus thermal activation on one hand, and also a monotonically increasing function (sometimes linear) of the size of the test-dose on the other hand, the existence of a second trapping state for electrons is to be assumed. This has to be a deeper trap, and should act as a competitor to the released electrons during the "read-out" stage, in which the sample is heated following the application of the test-dose, and the emission is recorded at $\sim$110 °C. Let us denote from

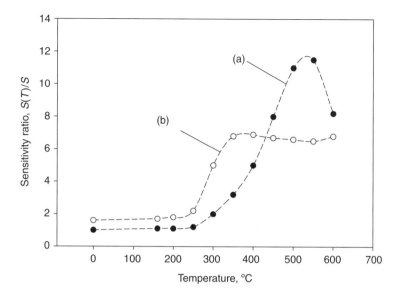

Figure 9.2   *The experimental data of Fleming and Thompson [629] redrawn here from Chen and Pagonis [725]. Chen, R. and Pagonis, V., Modelling thermal activation characteristics of the sensitization of thermoluminescence in quartz, J. Phys. D: Appl. Phys. 37, 159–164. Copyright (2004) with permission from Institute of Physics*

here on the shallower trap by T and the competitor by S (see Figure 9.1). For more details see Chen [716] and Chen and McKeever ([33], p. 208). Although the optimal activation temperature is often quoted in the literature as 500 °C, there are significant variations of the TAC depending on the type of quartz and the temperature and duration of firing of archaeological quartz in antiquity [707]. The TACs are measured by heating the sample to successively increasing temperatures from 200 °C upwards, and by measuring the sensitivity of the sample to a small β test-dose after each heating. Aitken [707] (see his Figure 3.9) describes in detail a graph previously presented by Fleming and Thompson [629]. This graph is redrawn in Figure 9.2 and shows two typical examples of different TAC behaviors; curve (b) shows "early activation" at ~350 °C and curve (a) a "late activation" at ~520 °C. Aitken assumed a distribution of hole traps and suggested that the form of the TAC reflects this distribution in a manner that traps close to the valence band have their holes transferred to the L centers at a lower temperature than traps that are not as close. An important feature seen in Figure 9.2 is that the sensitivity reaches a maximum and falls off when higher temperatures are used. This is referred to as *thermal deactivation* which, according to Aitken [707] is due to thermal eviction of holes from L centers into the valence band. Aitken pointed out that the temperature at which the sensitivity maximum of the TAC is reached depends on the time spent at the high temperature. If the heating rate is slow, and more particularly, if the maximum temperature is held for, say, a minute before cut off, then the maximum will shift downwards in temperature. Obviously, when different samples are compared, it is essential that exactly the same thermal treatment is maintained in all activations.

Other researchers of the predose sensitization include David and Sunta [732] who reported that the optimum temperature of activation is dependent on the growth conditions of the quartz crystal, and therefore may be correlated with the formation temperature of the natural quartz. Haskell and Bailiff [733] discussed the retrospective dosimetry of bricks using the sensitivity changes in quartz. They describe the details of the TACs which, generally speaking, show a significant increase in sensitivity starting with activation at $\sim$250 °C, reaching maximum at 500–550 °C and decreasing to rather low values (<10% of the maximum) following activation at $\sim$700 °C . G. Chen *et al.* [734] compared the TACs for the 110 °C TL peak and for the related phenomenon of OSL in quartz, and found rather similar results. Figel and Goedicke [735] gave a model with two electron-trapping states and three hole centers, the reservoir R and the centers L_1 and L_2, associated with the two kinds of TL emission in quartz at 380 and 460 nm. Apart from the addition of a second center, the model is very similar to that by Chen [716], with one significant difference, namely that Figel and Goedicke considered the competitor denoted above by S to be thermally connected. They assumed that the electrons in S may be thermally evicted into the conduction band, which may bring about the reduction of sensitivity at temperatures above $\sim$550 °C due to the recombination of the released electrons with trapped holes.

Chen and Leung [717, 718], using the original model by Chen [716], elaborated upon the dependence of the sensitivity on several dose increments and showed that, as previously found experimentally, the behavior could be presented as an exponential dependence on the accumulated dose D. Chen and Leung did not explain the TAC behavior within the framework of their model. Chen and Pagonis [725] offered an explanation different from that given by Figel and Goedicke for the thermal activation characteristic of the 110 °C peak in quartz. These authors demonstrated by numerical simulation and also explained on an intuitive basis that the different kinds of measured TACs can be reached by using the two trap-two center model, and one needs not resort to assuming a distribution of R centers as suggested by Aitken, or to assume the thermal release of competitor traps as suggested by Figel and Goedicke [735]. Furthermore, the de-activation occurring at higher temperatures is shown to be a direct result of this simpler model, rather than assuming that de-activation is due to thermal eviction of holes from L [33].

The energy level scheme given by Chen and Leung [717, 718] is shown in Figure 9.1. The transitions shown are during the excitation and during the read-out stages. T is the active trapping state having a concentration of N_t (m^{-3}) and an instantaneous occupancy n_t (m^{-3}); the activation energy is E_t (eV) and the frequency factor s_t (s^{-1}). S is the thermally disconnected trapping state with concentration N_s (m^{-3}) and occupancy of n_s (m^{-3}). A_t (m^3 s^{-1}) and A_s (m^3 s^{-1}) are the trapping coefficients into T and S, respectively. L is the luminescence center with concentration M (m^{-3}) and instantaneous occupancy of m (m^{-3}). The transition coefficient of the free holes from the valence band into L is A_l (m^3 s^{-1}) and the recombination coefficient of free holes is A_m (m^3 s^{-1}). R is the reservoir having a concentration of N_r (m^{-3}) and instantaneous occupancy of n_r (m^{-3}); E_r (eV) is the activation energy of freeing holes into the valence band and s_r (s^{-1}) the relevant frequency factor. The rate at which electron–hole pairs are produced by the irradiation is X (m^{-3} s^{-1}) which is proportional to the dose rate imparted on the sample. Thus, if the irradiation time is t_D, the total dose given to the sample is proportional to $X \cdot t_D$ which, in fact, is the number of electrons and holes produced by the irradiation per unit volume.

The set of equations governing the process during the excitation stage is:

$$\frac{dn_t}{dt} = A_t n_c (N_t - n_t),$$ (9.1)

$$\frac{dn_s}{dt} = A_s n_c (N_s - n_s),$$ (9.2)

$$\frac{dn_v}{dt} = X - A_l n_v (M - m) - A_r n_v (N_r - n_r),$$ (9.3)

$$\frac{dm}{dt} = A_l n_v (M - m) - A_m m n_c,$$ (9.4)

$$\frac{dn_r}{dt} = A_r n_v (N_r - n_r),$$ (9.5)

$$\frac{dn_c}{dt} = \frac{dm}{dt} + \frac{dn_r}{dt} + \frac{dn_v}{dt} - \frac{dn_t}{dt} - \frac{dn_s}{dt}.$$ (9.6)

Chen and Pagonis [725] solved this set of equations for certain sets of parameters using the ode23s MATLAB package solver, as well as the Mathematica differential equations solver. These particular solvers are designed to deal with "stiff" sets of equations and the above set is such, in particular since in Equations (9.3), (9.4) and (9.6) the derivative on the left-hand side is given as a small difference between two large functions. The results found from using these two packages were practically the same. The procedure described here, which takes care of the process of excitation, is also being used for both the test-dose and high-dose excitations.

A second part of the procedure deals with the transitions taking place during the heating of the sample, also shown in Figure 9.1. In order to simulate more accurately the experimental procedure, one has to add an intermediate stage of "relaxation" in which the irradiation has ceased, and the set of equations is solved with $X = 0$ for a period of time so as to bring n_c and n_v down to negligible values. The final values of n_t, n_s, m and n_r at the end of the relaxation time are taken as the initial values for the heating stage. It is assumed that during the high temperature heating, thermal release of holes is possible from the reservoir R (this is at the basis of the Zimmerman model), which may be trapped preferably in L, and that holes from L cannot be thermally released into the valence band in the relevant temperature range. The rate equations governing the read-out process are:

$$\frac{dn_t}{dt} = A_t n_c (N_t - n_t) - s_t n_t \exp(-E_t/kT),$$ (9.7)

$$\frac{dn_s}{dt} = A_s n_c (N_s - n_s),$$ (9.8)

$$\frac{dn_v}{dt} = n_r s_r \exp(-E_r/kT) - A_l n_v (M - m) - A_r n_v (N_r - n_r),$$ (9.9)

$$\frac{dm}{dt} = A_l n_v (M - m) - A_m m n_c,$$ (9.10)

$$\frac{dn_r}{dt} = A_r n_v (N_r - n_r) - n_r s_r \exp(-E_r/kT), \tag{9.11}$$

$$\frac{dn_c}{dt} = \frac{dm}{dt} + \frac{dn_r}{dt} + \frac{dn_v}{dt} - \frac{dn_t}{dt} - \frac{dn_s}{dt}. \tag{9.12}$$

The intensity of the emitted light is assumed to be the result of recombination of free electrons with trapped holes in the center L. Therefore, it is given by

$$I(T) = A_m m n_c. \tag{9.13}$$

In the simulation a conventional linear heating function is used, $T(t) = T_0 + \beta t$ where β is the constant heating rate chosen to be $5\,°\text{C}\,\text{s}^{-1}$. Additional parameters appearing in Equations (9.7–9.12) are E_t and s_t, the activation energy (eV) and frequency factor (s^{-1}), respectively, of the electron-trapping states, and E_r and s_r, the activation energy and frequency factor of the reservoir, respectively.

The complete sequence of steps simulated by Chen and Pagonis [725] are as follows:

(1) All traps and centers are assumed to be empty; this is usually considered to be the case in samples previously annealed at high enough temperature.
(2) The sample is given a large "natural dose", similar to that it would have received in nature.
(3) The sample is heated to a varying temperature T, starting at $\sim 200\,°\text{C}$.
(4) The sample is given a small β test-dose (typically 0.01 Gy).
(5) The sample is heated to $150\,°\text{C}$ to measure the TL peak at $110\,°\text{C}$. The area under this peak (or, alternatively, the maximum height) represents the sensitivity S of the sample.
(6) Steps 3–5 above are repeated several times, by heating to a slightly higher temperature T each time. In this way the sensitivity S of the sample to the test-dose is obtained as a function of the activation temperature T, to yield the TAC.

It is noted that these simulations refer to the "multi-aliquot" case, namely that for the measurement of $S(T)$ the sample is heated to temperature T, and when the sensitivity is to be measured at another value of T, a different aliquot of the same weight is given the same irradiation, activated at the new temperature and then given the test-dose to find the value of $S(T)$ as a function of the temperature.

The values of the parameters in the model were adjusted to demonstrate TAC behaviors similar to the graph by Fleming and Thompson reproduced in Figure 9.2. Figure 9.3 resembles the TAC curve (b) in Figure 9.2, which reaches a maximum at $\sim 350\,°\text{C}$, followed by a slight decrease, and then by a plateau for activation temperatures above $400\,°\text{C}$. The parameters used were similar to those chosen by Chen and Leung [717, 718] with some changes. The main difference as compared with the Chen–Leung parameters is the inversion of the values of A_s and A_m, and a small change in the value of E_r. The values used were: $s_t = 10^{13}\,\text{s}^{-1}$; $E_t = 1.0\,\text{eV}$; $s_r = 10^{13}\,\text{s}^{-1}$; $E_r = 1.6\,\text{eV}$; $A_t = 10^{-12}\,\text{m}^3\,\text{s}^{-1}$; $A_r = 10^{-10}\,\text{m}^3\,\text{s}^{-1}$; $A_s = 10^{-12}\,\text{m}^3\,\text{s}^{-1}$; $A_m = 10^{-11}\,\text{m}^3\,\text{s}^{-1}$; $A_l = 10^{-12}\,\text{m}^3\,\text{s}^{-1}$; $N_t = 10^{13}\,\text{m}^{-3}$; $N_s = 10^{13}\,\text{m}^{-3}$; $N_r = 10^{13}\,\text{m}^{-3}$; $M = 10^{14}\,\text{m}^{-3}$; $X = 10^9\,\text{m}^{-3}\,\text{s}^{-1}$; $t_D = 3000\,\text{s}$ for the "long" excitation; and $t = 1\,\text{s}$ for the test-dose.

Figure 9.4 was obtained using a different set of parameters in the model, and the TAC in Figure 9.4(c) is very similar to the sensitivity shown in curve (a) of Figure 9.2. The parameters here are the same as in Figure 9.3, except that E_r was changed to $1.8\,\text{eV}$; s_r to

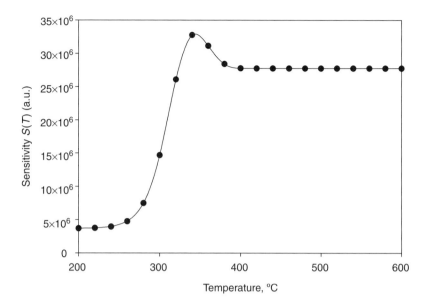

Figure 9.3 *A simulated TAC from Chen and Pagonis [725], in which the curve reaches a maximum at ∼350°C, decreases slightly and reaches a plateau at higher temperatures. The parameters used are given in the text. Chen, R. and Pagonis, V., Modelling thermal activation characteristics of the sensitization of thermoluminescence in quartz, J. Phys. D: Appl. Phys. 37, 159–164. Copyright (2004) with permission from Institute of Physics*

10^{10} s^{-1} and t_D to 7000 s. The sensitivity increases with the activation temperature, reaches a maximum at 520 °C, and finally reduces significantly at 650 °C. Figure 9.4(a) and (b) gives the dependencies of n_t and of m on the activation temperature, and provides an explanation for curve Figure 9.4(c), as will be elaborated below.

Chen and Pagonis [725] explained the peak-shaped TAC in Figure 9.4(c) by using an intuitive explanation as follows. In order to understand the peak shape shown in Figure 9.4, one needs to consider the curves of n_t and m as a function of the activation temperature, as shown in Figure 9.4(a) and (b). The temperature dependence of the concentration $m(T)$ in Figure 9.4(b) shows that heating the sample transfers a large number of holes from the hole reservoir R to the luminescence center L, which is exactly the Zimmerman hole transfer mechanism in the predose effect. The dependence of this $n_t(T)$ is a result of the dependence of m on the activation temperature. When the test-dose is imparted, the larger the value of m, the more electrons are "lost" through the recombination which takes place during this "short" irradiation, and therefore at the end of the irradiation less electrons are trapped at the active trapping state N_t. This explains the decreasing function $n_t(T)$. During the following "short" heating to 150 °C, fewer electrons reach the conduction band, therefore the concentration of free electrons n_c will be smaller at higher activation temperatures. For a qualitative description, let us assume that n_c as a function of T behaves more or less like $n_t(T)$. Finally, the TL intensity following the test-dose excitation is given by Equation (9.13), namely, proportional to m and n_c. As long as n_c behaves like $n_t(T)$, $S(T)$ should look like the product of the two curves, the decreasing function in Figure 9.4(a) and the increasing

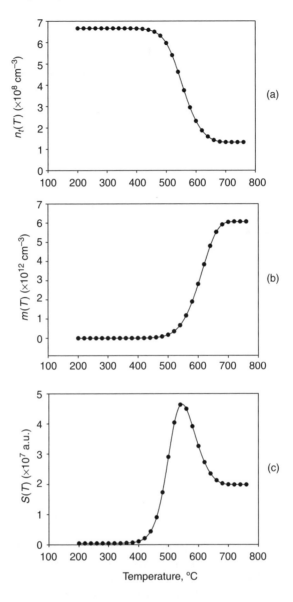

Figure 9.4 *Simulated mechanism of TACs by Chen and Pagonis [725]. For a chosen set of parameters, n_t at the end of the test-dose excitation as a function of the activation temperature T is shown in (a); the occupancy of the recombination centers m as a function of T is shown in (b) and the resulting sensitivity S(T) is given in (c). The parameters are given in the text. Chen, R. and Pagonis, V., Modelling thermal activation characteristics of the sensitization of thermoluminescence in quartz, J. Phys. D: Appl. Phys. 37, 159–164. Copyright (2004) with permission from Institute of Physics*

one in Figure 9.4(b). The product function yields a peak-shaped curve like the one shown in Figure 9.4(c). Another point to be mentioned is that in Figure 9.4(c), $S(T)$ drops to less than 50% of the maximum value. In some experimental cases, the reduction was even more dramatic. With the present model such a decrease can easily be reached with another choice of the set of parameters, but Figure 9.4(c) is sufficient to demonstrate the effect.

Figure 9.5 shows the results of the simulation of the test-dose dependence of the measured $110\,^{\circ}\mathrm{C}$ TL peak [Figure 9.5(a)] and of the natural-simulated "large" dose [Figure 9.5(b)],

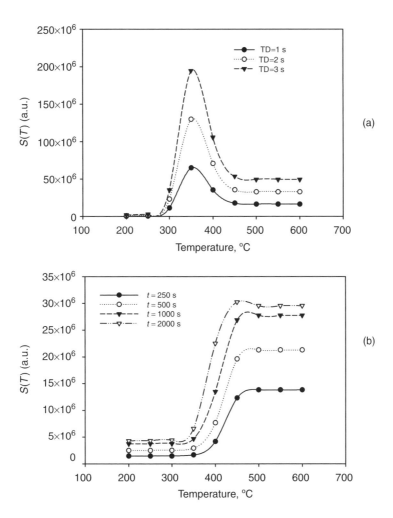

Figure 9.5 *The dependence of the simulated intensity as a function of the activation temperature is shown in (a) for three test-doses with relative values of 1:2:3. In (b), the curves shown are of the TACs for simulated-natural doses with relative values of 1:2:3:4. The set of parameters used is the same as in Figure 9.4. From Chen and Pagonis [725]. Chen, R. and Pagonis, V., Modelling thermal activation characteristics of the sensitization of thermoluminescence in quartz, J. Phys. D: Appl. Phys. 37, 159–164. Copyright (2004) with permission from Institute of Physics*

as a function of the activation temperature. In Figure 9.5(a) the maximum intensity both at the optimal activation temperature and at lower and higher temperatures is seen to be practically linear with the test-dose. In Figure 9.5(b), it is seen that the whole TAC goes up with increasing natural-simulated dose monotonically though sub linearly, when the size of the test-dose is kept constant. The sublinearity appears to be a result of the filling of the recombination center M. Figure 9.5 demonstrates the capability of the model to explain the experimental results of the approximately linear dependence of the TL intensity on the size of the test-dose, and also of the monotonic dependence of the emitted TL on the size of the natural dose received by the sample.

Chen and Pagonis [725] demonstrated that the common kinds of behavior of the TACs can be explained in the framework of this relatively simple model consisting of one recombination center, one reservoir, one active trap and one disconnected trap which competes for free electrons. The thermal deactivation effect in TACs (namely, that it reaches a maximum at 520 °C and decreases substantially at higher activation temperatures), is demonstrated within the model without the need to assume neither the possibility that holes are thermally evicted into the valence band as suggested by Aitken [707], nor that the competing trapping state can thermally release electrons as suggested by Figel and Goedicke [735]. Concerning the Aitken assertion, it should be mentioned that the possibility of releasing thermally holes from the centers to the valence band is not very reasonable, since the band gap of quartz is ∼8.5 eV [736] and the emission photons resulting from transition from the conduction band into the center are of ∼3.3 eV (corresponding to the wavelength of 380 nm). This puts the hole centers >5 eV above the valence band, and hence they cannot be thermally released at the temperature range in question. It is noted that in the real material there are perturbations such as additional trapping states and centers that may change the details of the measured phenomena. The behavior seen in other samples in which the TAC reaches a maximum at a certain temperature (e.g. 350 °C), reduces slightly and remains constant for higher activation temperatures, was also demonstrated and shown to result directly from the model, by using different sets of parameters. Finally, the nearly linear dose-rate dependence of TL as well as the monotonically increasing dependence on the natural dose were also explained within this rather simple model.

In Section 9.3 we present a more comprehensive model for quartz, and how it has been applied to modeling of the TACs in quartz.

9.3   The Bailey Model for Quartz

The TAC simulations presented in this section were carried out by Pagonis *et al.* [720] using the comprehensive quartz model developed by Bailey [35]. Figure 9.6 shows the energy level diagram for this model. The parameters used in the model were arrived at on the basis of empirical data from several different types of quartz, and are presented in Table 9.1. The model has been successful in simulating several TL and OSL phenomena in quartz [35, 532]. Level 1 in the model represents the 110 °C TL shallow electron trap, which gives rise to a TL peak at 110 °C when measured with a heating rate of 5K s^{-1}. This TL peak has been the subject of numerous studies because of its importance in predose dating [708] and retrospective dosimetry [709], as well as for its use in measuring the luminescence sensitivity [737–739] . Within the model, the 110 °C TL level is assigned

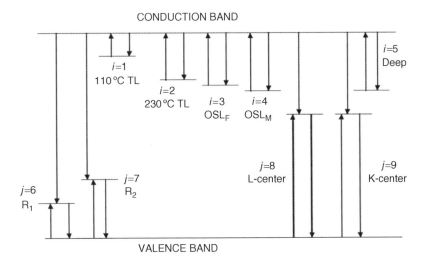

Figure 9.6 *Schematic diagram of the Bailey model, consisting of a total of nine energy levels. The arrows indicate possible transitions. Reprinted from Bailey, R.M., Towards a general kinetic model for optically and thermally stimulated luminescence of quartz, Radiat. Meas. 33, 17–45. Copyright (2001) with permission from Elsevier*

a photostimulation probability since it has been shown to be light sensitive [437]. Level 2 represents a generic "230 °C TL level", typical of such TL peaks found in many sedimentary quartz samples. It is assumed that this TL trap is not light sensitive and thus it is not assigned a photostimulation probability. Levels 3 and 4 are usually termed the fast and medium OSL components [740] and they yield TL peaks at 330 °C as well as giving rise to OSL signals. The photostimulation rates for these levels are discussed in some detail in the original paper by Bailey [35]. The model does not contain any of the slow OSL components which are known to be present in quartz [302, 303], and which were incorporated in later versions of the model [532].

Table 9.1 *The Qtz-A1 parameters of the original Bailey [35] model. The photo-eviction constants are θ_{0i} at $T = \infty$, and the thermal assistance energies are E_{thi}*

	Description	N_i (cm^{-3})	E_i (eV)	s_i (s^{-1})	A_i	B_i (cm^3 s^{-1})	θ_{0i} (s^{-1})	E_{thi} (eV)
1	110 °C TL	1.5×10^7	0.97	5×10^{12}	1×10^{-8}		0.75	0.1
2	230 °C TL	1×10^7	1.55	5×10^{14}	1×10^{-8}			
3	OSL$_F$	1×10^9	1.7	5×10^{13}	1×10^{-9}		6	0.1
4	OSL$_M$	2.5×10^8	1.72	5×10^{14}	5×10^{-10}		4.5	0.13
5	Deep	5×10^{10}	2	1×10^{10}	1×10^{-10}			
6	R$_1$-center	3×10^8	1.43	5×10^{13}	5×10^{-7}	5×10^{-9}		
7	R$_2$-center	1×10^{10}	1.75	5×10^{14}	1×10^{-9}	5×10^{-10}		
8	L-center	1×10^{11}	5	1×10^{13}	1×10^{-9}	1×10^{-10}		
9	K-center	5×10^9	5	1×10^{13}	1×10^{-10}	1×10^{-10}		

Level 5 is a deep, thermally disconnected electron center. Such a level is known to be necessary in order to explain several TL and OSL phenomena based on competition between energy levels. The model contains also four hole-trapping centers which act as recombination centers for optically or thermally released electrons.

Levels 6 and 7 are thermally unstable, nonradiative recombination centers, similar to the "hole reservoirs" first introduced by Zimmerman [510] in order to explain the predose sensitization phenomenon in quartz. Level 8 is a thermally stable, radiative recombination center termed the "luminescence center" (L in the Zimmerman model). Holes can be thermally transferred from the two hole reservoirs (levels 6 and 7) into the luminescence center via the valence band. Level 9 is a thermally stable, nonradiative recombination center termed a "killer" center (K in the Zimmerman model). The computer code in Pagonis *et al.* [720] was written in Mathematica, and was tested for consistency by successfully reproducing several of the simulation results in Bailey [35]. The parameters are defined as follows; N_i are the concentrations of electron traps or hole centers (cm^{-3}), n_i are the concentrations of trapped electrons or holes (cm^{-3}), s_i are the frequency factors (s^{-1}), E_i are the electron trap depths below the conduction band or hole center energy levels above the valence band (eV). A_i ($i = 1, \ldots, 5$) are the conduction band to electron trap transition probability coefficients (cm^3 s^{-1}), A_j($j = 6, \ldots, 9$) are the valence band to hole trap transition probability coefficients (cm^3 s^{-1}) and B_j ($j = 6, \ldots, 9$) are the conduction band to hole center transition probability coefficients (cm^3 s^{-1}).

The equations to be solved in the Bailey model are as follows:

$$\frac{dn_i}{dt} = n_c(N_i - n_i)A_i - n_i P\theta_{0i}e^{(-\frac{E_i^{th}}{k_B T})} - n_i s_i e^{(-\frac{E_i}{k_B T})} \quad (i = 1, \ldots, 5), \qquad (9.14)$$

$$\frac{dn_j}{dt} = n_v(N_j - n_j)A_j - n_j s_j e^{(-\frac{E_j}{k_B T})} - n_c n_j B_j \quad (j = 6, \ldots, 9), \qquad (9.15)$$

$$\frac{dn_c}{dt} = R - \sum_{i=1}^{5} \left(\frac{dn_i}{dt}\right) - \sum_{j=6}^{9} n_c n_j B_j, \qquad (9.16)$$

$$\frac{dn_v}{dt} = \frac{dn_c}{dt} + \sum_{i=1}^{5} \left(\frac{dn_i}{dt}\right) - \sum_{j=6}^{9} \left(\frac{dn_j}{dt}\right), \qquad (9.17)$$

and the luminescence is defined as

$$L = n_c n_8 B_8 \eta(T), \qquad (9.18)$$

with $\eta(T)$ representing the luminescence efficiency, and R denoting the pair production rate [35]. The underlying assumption of Equation (9.18) is that the TL/OSL emission uses one and the same radiative recombination center.

The luminescence efficiency is affected by the phenomenon of thermal quenching in TL, which is discussed in some detail in Bailey [35] and by Nanjundaswamy *et al.* [741] (see Section 2.9). These authors suggest a Mott–Seitz mechanism, based on the existence of an activation energy W interpreted as the energy barrier which must be overcome for an excited-state electron to transition non-radiatively to the ground state, with the emission of phonons. The thermal-quenching phenomenon was first described by Wintle [75, 742] and further discussed by Petrov and Bailiff [687, 743] and several other authors. The luminescence efficiency is given by the expression:

$$\eta(T) = \frac{1}{1 + K \exp(-W/kT)}, \tag{9.19}$$

where K is a dimensionless constant equal to 2.8×10^7 and W is the activation energy with a value of 0.64 eV. It should be noted that when the luminescence efficiency is given by Equation (9.19), the initial-rise method provides an evaluation not of E, but of $E - W$ [687]. Further work along the same lines, on the thermal quenching of luminescence in Al_2O_3:C was given by Akselrod *et al.* [78].

The exact steps in the simulation of a natural sedimentary quartz sample are given in Bailey [35] as follows:

(1) Natural sample. All electron and hole concentrations set to zero during the crystallization process.
(2) Geological dose of 1000 Gy with a dose rate of 1 Gy s^{-1} at 20 °C.
(3) Heat to 350 °C (simulation of geological time).
(4) Illumination at 200 °C for 100 s, simulating repeated exposures to sunlight over a long period.
(5) Burial dose of 20 Gy at 0.01 Gy s^{-1} at 220 °C.
(6) Laboratory β dose of 10 Gy at 1 Gy s^{-1}.

The main purpose of this section is to show that the Bailey model can be used to simulate the experimental protocols used during the predose dating technique. We start with the work of Pagonis *et al.* [720] , who simulated a typical TAC measurement in quartz using the Bailey model. The simulation steps were as follows:

(1) Natural sample. All electron and hole concentrations set to zero.
(2) Natural irradiation with a paleodose of 1 Gy with the slow natural dose rate of 10^{-11}Gy s^{-1}.
(3) The sample is given a β dose of 10 Gy in the laboratory.
(4) The sample is heated to the activation temperature (in the range 20–600 °C).
(5) Give test-dose TD $= 0.1$ Gy. Measure TL sensitivity S_N by heating to 150 °C.
(6) Repeat steps 3–6 using a higher activation temperature, to obtain the sensitivity S_N as a function of the activation temperature (TAC).

In step 1 above, the concentrations of all traps are set to zero, while in step 2 the sample is given a paleodose of 1 Gy. In order to simulate the natural dose rate as close as possible, a dose rate of 10^{-11} Gy s^{-1} is used during the natural irradiation of the sample in step 2 [532]. In step 3 the sample is given a β dose of 10 Gy in the laboratory with a dose rate of 1 Gy s^{-1}, and in step 4 it is heated to the activation temperature (in the range 20–600 °C). In step 5 the sample is given a small β test-dose (TD) of 0.1 Gy and is heated to 150 °C to measure the sensitivity S_N, which is represented by the maximum of the TL peak at 110 °C. In step 6, steps 3–6 are repeated by re-irradiating the sample with 10 Gy and using a higher activation

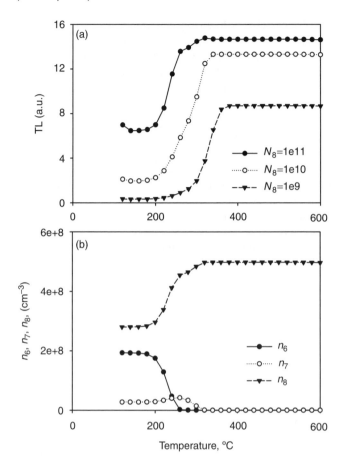

Figure 9.7   *(a) Simulated TACs for quartz samples, from Pagonis et al. [720], for three different values of the total concentration of holes N_8 in the luminescence center. The value of N_8 is varied over three orders of magnitude. (b) The variation of the concentrations of holes $n_8(T)$ in the luminescence center L and $n_6(T)$ and $n_7(T)$ in the hole reservoirs R_1 and R_2, respectively, during the TAC simulation, demonstrating the Zimmerman hole transfer mechanism. Reprinted from Pagonis, V., Balsamo, E., Barnold, C., Duling, K. and McCole, S., Simulations of the predose technique for retrospective dosimetry and authenticity testing, Radiat. Meas. 43, 1343–1353. Copyright (2008) with permission from Elsevier*

temperature each time, thus obtaining the variation of the sensitivity S_N with the activation temperature. The results of simulating such a TAC measurement sequence are shown in Figure 9.7(a) for three different values of the total concentration of holes N_8 in the luminescence center. The value of N_8 is varied over three orders of magnitude, from the original large value of $N_8 = 10^{11}$ cm^{-3} in the Bailey model, to a much reduced value of $N_8 = 10^9$ cm^{-3}.

For a 5 mg sample of quartz, a typical increase of the sensitivity S_N is around 2–10 times for a predose of 1 Gy ([707], p. 155). In practical applications it is desirable to have a value of the RT sensitivity S_0 many times smaller than the thermally activated sensitivity S_N.

This is known as the requirement of "low-S_0" values for accurate application of the predose technique. The top curve in Figure 9.7(a) represents the TAC simulated using the original Bailey parameters (with $N_8 = 10^{11}$ cm^{-3}), and shows an increase of the sensitivity from a value of $S_0 \sim 6$ at low temperatures, up to the thermally activated value of $S_N \sim 15$. The activation ratio in this case is only $S_N/S_0 \sim 2.5$ and this can be considered representative of a sample exhibiting a "high-S_0" value. The bottom cruve in Figure 9.7(a) represents the TAC simulated using the much smaller value of $N_8 = 10^9$ cm^{-3} and shows an increase of the sensitivity from a value of $S_0 \sim 0.4$ at low temperatures, up to the thermally activated value of $S_N \sim 8.5$. The activation ratio in this case is much larger, $S_N/S_0 \sim 21$ and this can be considered representative of a sample exhibiting a "low-S_0" value.

Figure 9.7(b) shows the simulated concentrations of holes $n_8(T)$ in the luminescence center L and $n_6(T)$ and $n_7(T)$ in the hole reservoirs R$_1$ and R$_2$, respectively as a function of the annealing temperature T during the TAC simulation. The results of Figure 9.7(b) show that the model successfully describes the predose activation process, which consists of the hole transfer from R$_1$ and R$_2$ to L during the heating of the sample to 400–600 °C. As the temperature T increases, the concentration of holes $n_8(T)$ in the luminescence center L is increased, while the corresponding concentrations of holes $n_6(T)$ and $n_7(T)$ in the Zimmerman hole reservoirs R$_1$ and R$_2$ are decreased by the same total amount. The results of Figure 9.7 also show that an activation temperature in the region 400–600 °C is sufficiently high to transfer all holes from the Zimmerman hole reservoirs R$_1$ and R$_2$ into the luminescence center L. The exact shape of the TAC and the optimal value of the activation temperature depend on the experimental details of the measurement [725].

9.4 Simulation of the Predose Dating Technique

Two main variations of the predose technique exist in TL, known as the *multiple thermal activation technique* and the *additive dose technique*. Pagonis and Carty [719] simulated the complete sequence of events in the additive dose version of the predose technique, and obtained quantitative results for the effect of several experimental parameters on the estimated accrued dose. Their numerical results showed that the predose technique can reproduce the paleodose (PD) with an accuracy of 1–5% or better. The accuracy of the method remains within the ±5% limit when the PD was varied over at least an order of magnitude. Quantitative results were also obtained for the effect of the test dose (TD) and of the calibration β dose on the accuracy of the predose method in retrospective dosimetry. The disadvantage of the model of Pagonis and Carty [719] is the presentation of the radiation dose in units of cm^{-3} s^{-1}, instead of the commonly used experimental dose units of Gy. Pagonis *et al.* [720] lifted this disadvantage in a later work in which they simulated both versions of the predose technique using the Bailey model. The results of their simulations will be presented in the next two subsections.

9.4.1 Simulation of the Additive Dose Version of the Predose Technique

The basic sequence of measurements during the predose technique is as follows ([707], pp. 153–168). There are two commonly used versions of the predose technique; in the first version only two aliquots of the quartz sample are used, while in the second version multiple aliquots are used for better accuracy.

In the simplest version, two portions (aliquots) of the sample are used. Using these two aliquots, the following steps are implemented in order to measure the TL sensitivities S_0, S_N and $S_{N+\beta}$ of the material to a small TD:

(1) Give a TD (usually 0.01 Gy) and measure the 110 °C peak response, denoted by S_0.
(2) Heat to the activation temperature (typically in the range of 400–600 °C), as in the course of a normal TL glow curve.
(3) Give the same TD and measure the activated 110 °C response, denoted by S_N. Using the second portion of the material, the TL sensitivities S_0 and $S_{N+\beta}$ are measured as follows:
(4) Repeat a measurement of the 110 °C peak response S_0, as in step 1 above. This is done for sample normalization purposes between the two aliquots.
(5) Give a laboratory calibrating dose β, usually several Gy.
(6) Heat again to the activation temperature of 500 °C as in step 2 above.
(7) Give the same TD and measure the new response, denoted by $S_{N+\beta}$.

The response to the TD is measured by heating the sample to 150 °C, just above the 110 °C TL peak. The additive dose method avoids multiple thermal activation of the material, which can cause changes to its predose characteristics. The equations used in the additive dose technique are based on the assumption of a linear response of the sensitivity of the sample. By assuming that the change in sensitivity $S_N - S_0$ between steps 1 and 3 above is proportional to the accrued dose during the lifetime of the material, and that the further increase $S_{N+\beta} - S_0$ between steps 4 and 7 is proportional to the calibrating dose β, the accrued dose (AD) can be calculated using the equation:

$$\text{AD} = \frac{S_N - S_0}{S_{N+\beta} - S_0}\beta. \tag{9.20}$$

An example of simulated TL glow curves during application of the additive dose variation of the predose technique is shown in Figure 9.8, indicating the three different sensitivities S_0, S_N and $S_{N+\beta}$.

Pagonis *et al.* [720] simulated also the second version of the additive dose technique, in which several aliquots are used to measure the TL sensitivities of the sample. Their simulation steps for the additive dose technique using the Bailey model are shown in Table 9.2. Typical results from these simulations are shown in Figure 9.9, demonstrating how the accrued dose (AD) is obtained by linear extrapolation. Figure 9.10 shows the simulated variation of the accuracy of the calculated AD with the PD, for three different values of the TD and for a calibration dose β equal to the PD ($\beta = \text{PD}$).

9.4.2   Simulation of the Multiple Activation Version of the Predose Technique

The basic sequence of measurements during the multiple activation technique is shown in Table 9.3, and is based on the use of a single aliquot of the sample ([707], pp. 153–168). The TL sensitivities S_0 and S_N of the material to a small TD are measured as shown in steps 3–5 of Table 9.3. The thermally activated sensitivities $S_{N+\beta}$ and $S_{N'}$ are measured using the same aliquot as shown in the rest of Table 9.3. Experiments using the multiple activation

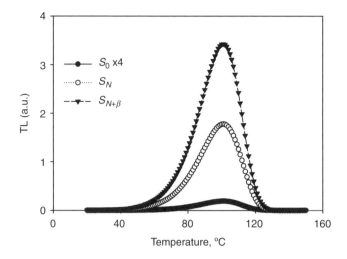

Figure 9.8 *Simulated TL glow curves during application of the additive dose variation of the predose technique, from Pagonis et al. [720], demonstrating the TL sensitivities S_0, S_N and $S_{N+\beta}$ measured during the predose dating technique. The TD used in the simulation is TD $= 0.01$ Gy, the PD accrued by the sample is PD $= 4$ Gy, a calibration β dose of 4 Gy is used, as well as an activation temperature of 500 °C. Reprinted from Pagonis, V., Balsamo, E., Barnold, C., Duling, K. and McCole, S., Simulations of the predose technique for retrospective dosimetry and authenticity testing, Radiat. Meas. 43, 1343–1353. Copyright (2008) with permission from Elsevier*

variation exhibit the phenomenon of *radiation quenching*. This phenomenon consists of the sensitivity $S_{N'}$ in step 8 being usually lower than the sensitivity S_N in step 6, due to the β irradiation in step 5 of Table 9.3. In addition, in this technique the aliquot undergoes a multiple thermal activation, which can cause changes to its pre dose characteristics. The

Table 9.2 *The simulation steps for the additive dose version of the predose technique, as given by Pagonis et al. [720]*

Step	Description
1	Natural sample. All electron and hole concentrations are set to zero
2	Natural irradiation with a paleodose with the slow natural dose rate of 10^{-11} Gy s^{-1}
3	Give test dose, usually 0.01 Gy. Measure initial sensitivity S_0 using the 110 °C TL peak
4	Heat to 500 °C
5	Give test dose. Measure S_N
6	Use a new aliquot; give test dose. Measure S_0
7	Give calibration dose β
8	Heat to 500 °C
9	Give test-dose. Measure $S_{N+\beta}$
10	Repeat steps 6–10 using doses of 2β, 3β, etc. in step 7, to obtain the sensitivities $S_{N+2\beta}$, $S_{N+3\beta}$ etc. Obtain the graph of the sensitivity S versus added dose, as shown in Figure 9.9

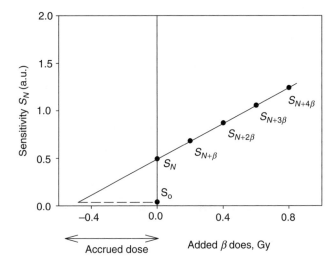

Figure 9.9 *Simulation of the additive dose technique; the TL sensitivities are measured as a function of the added dose using several aliquots. The AD is obtained by linear extrapolation on the dose axis, as shown. Reprinted from Pagonis, V., Balsamo, E., Barnold, C., Duling, K. and McCole, S., Simulations of the predose technique for retrospective dosimetry and authenticity testing, Radiat. Meas. 43, 1343–1353. Copyright (2008) with permission from Elsevier*

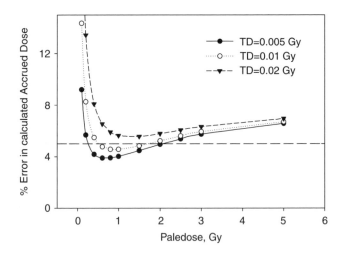

Figure 9.10 *The variation of the accuracy of the calculated AD with the PD as obtained by Pagonis et al. [720], for three different values of the TD. The calibration dose β is taken to be equal to the PD (β = PD). Reprinted from Pagonis, V., Balsamo, E., Barnold, C., Duling, K. and McCole, S., Simulations of the predose technique for retrospective dosimetry and authenticity testing, Radiat. Meas. 43, 1343–1353. Copyright (2008) with permission from Elsevier*

Table 9.3 *The simulation steps for the multiple activation version of the predose technique, as given by Pagonis et al. [720]. A single aliquot is used for all measurements*

Step	Description
1	Natural sample. All electron and hole concentrations set to zero
2	Natural irradiation with a paleodose β with the slow natural dose rate of 10^{-11} Gy s^{-1}
3	Give test-dose. Measure S_0
4	Heat to 500 °C
5	Give calibration dose β
6	Give test-dose. Measure S_N
7	Heat to 150 °C to empty the traps
8	Give test-dose. Measure $S_{N'}$
9	Heat to 500 °C
10	Give test-dose. Measure $S_{N+\beta}$
11	Calculate accrued dose using S_0, S_N, $S_{N+\beta}$ and $S_{N'}$ in Equation (9.21) or (9.22)
12	Find percent difference of paleodose and accrued dose

equations used in the multiple activation technique are also based on the assumption of a linear response of the sensitivity of the sample. The AD can be calculated using the equation

$$AD = \frac{S_N - S_0}{S_{N+\beta} - S_N}\beta. \tag{9.21}$$

When the effect of radiation quenching is taken into account, Equation (9.21) is modified by using the quenched sensitivity $S_{N'}$ instead of the sensitivity S_N, to obtain a corrected estimate of the AD using the equation [707]:

$$AD = \frac{S_N - S_0}{S_{N+\beta} - S_{N'}}\beta. \tag{9.22}$$

Typical results from these simulations are shown in Figure 9.11, demonstrating the variation of the accuracy of the calculated AD with the PD and for three different values of the calibration dose β. A fixed TD $= 0.015$ Gy is used in this simulation.

The simulations of Pagonis *et al.* [720] provide also a quantitative mathematical description of the phenomena of radiation quenching and its opposite phenomenon, *radiation enhancement* of TL sensitivity. Experimentally it is found that some types of quartz exhibit enhancement instead of reduction of the sensitivity during application of the predose technique, i.e., the ratio $S_{N'}/S_N > 1$. Figure 9.12 shows the simulated glow curves during the multiple activation method, for two different values of the concentration of holes N_8 in the sample, namely for the original value of $N_8 = 10^{11}$ cm^{-3} in the Bailey model, and for a much smaller value of $N_8 = 10^9$ cm^{-3}. The data in Figure 9.12(a) show that in the case of "low-S_0" value (corresponding to $N_8 = 10^9$ cm^{-3}), the ratio $S_{N'}/S_N \sim 0.90 < 1$ and the quartz sample exhibits radiation quenching. On the other hand, Figure 9.12(b) shows that in the case of "high-S_0" value (corresponding to $N_8 = 10^{11}$ cm^{-3}) the ratio $S_{N'}/S_N \sim 1.20 > 1$ and the quartz sample exhibits radiation-enhanced sensitivity. Close examination of the concentrations of the nine energy levels in the Bailey model during the simulations provides the following explanations of these opposite behaviors. In the case

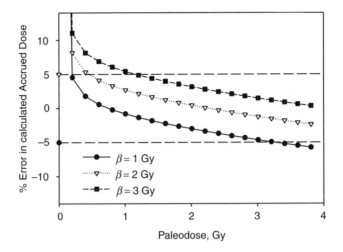

Figure 9.11 *The effect of the calibration beta dose β on the accuracy of the multiple activation technique, as given in the simulations of Pagonis et al. [720]. A fixed TD=0.015 Gy is used in this simulation. Reprinted from Pagonis, V., Balsamo, E., Barnold, C., Duling, K. and McCole, S., Simulations of the predose technique for retrospective dosimetry and authenticity testing, Radiat. Meas. 43, 1343–1353. Copyright (2008) with permission from Elsevier*

of a "low-S_0" sample, electrons released from the 110 °C TL trap during measurement of $S_{N'}$ are captured in the luminescence center L, resulting in a reduced instantaneous concentration n_8 of holes in L. A lower value of n_8 leads to a lower value of the observed TL signal according to Equation (9.18), and hence to a reduced (quenched) activated sensitivity $S_{N'}$. In the case of the "high-S_0" value sample of Figure 9.12(b), the higher total concentration of holes in the luminescence center (N_8) leads to more holes being captured in the luminescence center during the natural irradiation of the sample. This leads to a higher concentration of holes n_8 during the measurement of $S_{N'}$, and hence to a higher value of the observed TL signal according to Equation (9.18). It is noted that two slightly different mechanisms for radiation quenching and radiation enhanced sensitivity than the ones suggested by Pagonis *et al.* [720], were proposed previously by Aitken [707] on a purely empirical basis.

9.5   The Single Aliquot Regenerative Dose (SAR) Technique

During the past decade, the technique of OSL dating has been established as one of the most accurate and precise methods for dating of sedimentary quartz deposits. The goal of OSL dating is the determination of the radiation dose received by the samples since the last time they were exposed to sunlight. This dose is known as the equivalent dose (D_e). Perhaps the most powerful, yet relatively simple tool employed during OSL dating is the measurement sequence known as the SAR protocol. This measurement technique has been reviewed in detail recently by Wintle and Murray [26].

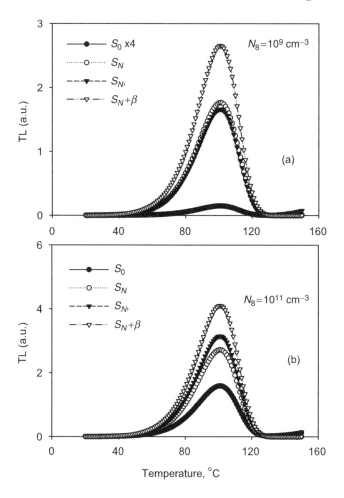

Figure 9.12 *Simulated TL glow curves during the multiple activation method of the predose technique, for two different values of the concentration of holes in the luminescence center. In (a), the sample exhibits radiation quenching, while in (b) it exhibits radiation sensitization. Reprinted from Pagonis, V., Balsamo, E., Barnold, C., Duling, K. and McCole, S., Simulations of the predose technique for retrospective dosimetry and authenticity testing, Radiat. Meas. 43, 1343–1353. Copyright (2008) with permission from Elsevier*

Before describing the SAR protocol in some detail, we will first present the TL and OSL dose response of a typical quartz sample, together with a relevant discussion of the various continuous-wave OSL (CW-OSL) signal components . This is followed by a description of the luminescence sensitivity changes occurring during routine heating and/or irradiation of quartz samples. The assumptions underlying the SAR procedure will be discussed in some detail next, together with several tests of the applicability of the SAR protocol. Finally, various aspects of the SAR protocol itself will be discussed in some detail.

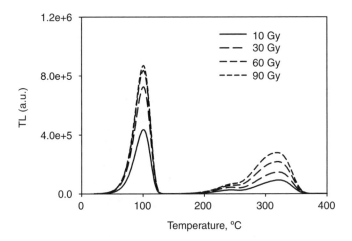

Figure 9.13 *Simulated TL glow curves using the Bailey model [35]. The natural sample was given various doses between 10 Gy and 90 Gy, and the heating rate was 5 K s^{-1}. (Simulations and drawing by the authors). From Bailey, R.M., Towards a general kinetic model for optically and thermally stimulated luminescence of quartz, Radiat. Meas. 33, 17–45. Copyright (2001) with permission from Elsevier*

9.5.1   The TL and OSL Dose Response in Quartz

Figure 9.13 shows simulated TL glow curves which were obtained using the Bailey model presented in Section 9.3. The natural sample was given various doses between 10 Gy and 90 Gy. As may be expected, the height of the TL peaks in Figure 9.13 increases with the β dose. The TL peak at $\sim$100 °C saturates at an earlier dose than the higher temperature TL peaks at $\sim$240 and $\sim$330 °C shown in Figure 9.13. This simulation assumes the use of multiple aliquots, namely a different aliquot for each dose given to the sample. It is also an additive-dose simulation, i.e., the laboratory dose is added to the natural dose of the sample. The natural sample in this simulation is assumed to have received a burial dose of 20 Gy in nature at a temperature of $\sim$220 °C, as discussed in [35].

Figure 9.14 shows the maximum TL intensity of the TL peak at $\sim$100 °C as a function of the dose given to the natural sample, as obtained from the data in Figure 9.13. The natural sample (which had received a burial dose of 20 Gy) is given the additional β dose in the range 0–300 Gy in the laboratory. The uncorrected TL data in Figure 9.14 can not be fitted with a single saturating exponential (SSE), but it can be fitted with a SSE plus a linear function, as is common practice in laboratory measurements.

The TL intensity in Figure 9.14 can be corrected for sensitivity changes occurring due to the sample irradiation. This is the phenomenon of *radiation quenching* commonly observed in experimental work on quartz. The sensitivity of the samples is monitored by using the response of the 110 °C TL peak to a small test-dose of 0.1 Gy [35]. The result of dividing the measured TL data by the sensitivity of the sample is also shown in Figure 9.14. Figure 9.15 shows the sensitivity of the quartz sample as a function of the dose given to the natural sample. For small doses the sensitivity of the sample exhibits *radiation quenching*, while for doses> 25 Gy it exhibits *radiation sensitization*.

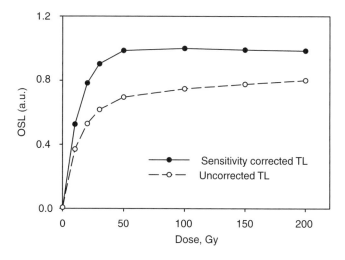

Figure 9.14   *The maximum TL intensity as a function of the dose given to the natural sample, simulated using the Bailey model [35]. The TL intensity is corrected for sensitivity changes as described in the text. This simulation assumes that multiple aliquots of the quartz sample are used, namely a different aliquot for each dose D given to the sample. Reprinted from Bailey, R.M., Towards a general kinetic model for optically and thermally stimulated luminescence of quartz, Radiat. Meas. 33, 17–45. Copyright (2001) with permission from Elsevier*

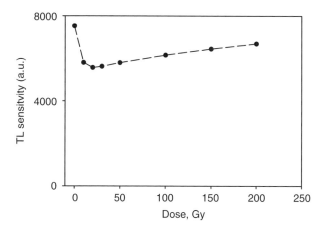

Figure 9.15   *The sensitivity of the quartz sample as a function of the dose added to the natural sample, for the data shown in Figure 9.14. For small doses the sensitivity of the sample exhibits radiation quenching, while for doses > 25 Gy it shows radiation sensitization. (Simulation and drawing by the authors)*

The sensitivity-corrected TL data can be fitted to either an SSE, or to an SSE plus a function of the form

$$TL = A + Be^{-C \times D} \tag{9.23}$$

where C is a constant and A represents the TL signal for the unirradiated sample. Alternatively, the behavior may be an SSE plus a linear component

$$TL = A + A' \cdot D + Be^{-C \times D}. \tag{9.24}$$

The rate of growth of the TL with the dose D is commonly described using the value D_0, which is equal to the dose where the slope of the response is $1/e$ of the initial slope. Commonly found values for D_0 in quartz are usually in the range 10–30 Gy [35]. Note that $D_0 = 1/C$. In the case of the uncorrected simulated TL data in Figure 9.14, the value obtained for D_0 was ~ 14 Gy. When fitting the sensitivity-corrected data, one obtains a value of ~ 17 Gy.

Figure 9.16 shows the simulated OSL signal versus the dose added to the natural sample, as obtained using the Bailey model. This simulation also assumes the use of multiple aliquots. An example of a dose-response curve obtained using a single aliquot technique is found in the next section. The OSL data in Figure 9.16 can be fitted with an SSE yielding a value of $D_0 \sim 140$ Gy. Alternatively, these data can also be fitted with a double SSE, corresponding to the two OSL components in the Bailey model, namely the fast and medium OSL components.

Analysis of the experimental CW-OSL decay curves for quartz has shown that these curves contain contributions from five different components, usually termed fast, medium, and three slow components. Bailey [532] developed a more comprehensive quartz model

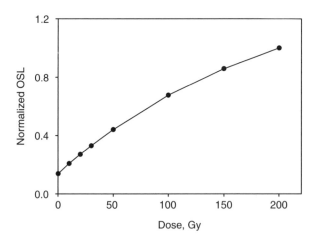

Figure 9.16   *The OSL signal versus the dose added to the natural sample, simulated using the Bailey model [35]. This simulation assumes the use of multiple aliquots. These data can be fitted using an SSE. Reprinted from Bailey, R.M., Towards a general kinetic model for optically and thermally stimulated luminescence of quartz, Radiat. Meas. 33, 17–45. Copyright (2001) with permission from Elsevier*

Table 9.4 *Example of a typical SAR protocol*

Step		Description
1	Laboratory irradiation: dose D_i at 1 Gy s^{-1} at 20 °C.	$D_i = 0$ for natural sample
2	Preheat the sample at a preheat temperature (e.g. 260 °C) for 10 s.	Remove electrons from the shallow thermally unstable TL traps
3	Blue stimulation at 125 °C for 100 s, *record the OSL signal (L)*.	OSL signal due to dose D_i
4	Give a small β test dose (e.g. 0.1 Gy)	Sensitivity correction
5	Heat the sample to 220 °C for 10 s (known as the *cutheat*)	Remove electrons from shallow, thermally unstable TL traps
6	Blue stimulation at 125 °C for 100 s, *record the test dose OSL signal (T)* Return to step 1	OSL signal due to test-dose

briefly mentioned above, which contains these experimentally observed five OSL components. A recent summary of the properties of these components and the possibility of using them for quartz dating have been presented by Wintle and Murray [26].

9.5.2   Simulations of the SAR Protocol

A typical series of measurements used during the SAR experimental protocol is shown in Table 9.4. This experimental protocol was developed over the course of a few years, initially by Murray and Roberts [739], and more completely by Murray and Wintle [744]. The most important part of the SAR technique is the correction of the measured OSL signals by using a measurement of the sensitivity of the sample to a small test-dose. Murray and Roberts used the signal from the 110 °C TL peak for monitoring these sensitivity changes, while an OSL measurement is more commonly used in later versions of the protocol. During the SAR protocol, the measured "sensitivity-corrected" *natural* OSL signal is compared with additional sensitivity-corrected signals which are *regenerated* by irradiating the samples in the laboratory. A typical sequence of steps undertaken during the SAR protocol is shown in Table 9.4.

In step 1 the sample is given a laboratory dose D_i, known as the regenerative dose. In step 2 the sample is heated to a preheat temperature, typically for 10 s at 260 °C, in order to empty the shallow thermally unstable TL traps (at $\sim$ 110 and $\sim$230 °C). In step 3 the sample is optically stimulated for 100 s using blue light (typically 470 nm), and the resulting OSL signal (L) is recorded. The optical stimulation at step 3 is carried out at an elevated temperature of 125 °C, in order to avoid complications due to the optically sensitive TL trap at $\sim$110 °C. In step 4 the sample is given a small test-dose of 0.1 Gy, and in step 5 it is heated for 10 s at a lower temperature of 220 °C (known as the *cutheat*), to again remove electrons from shallow TL traps. Finally in step 6 the sample is again optically stimulated for 100 s, in order to measure the OSL signal (T), which is used to carry out the sensitivity correction for the OSL signal measured in step 3. Steps 1–6 in Table 9.4 are repeated for a sequence of doses D_i, with the first dose D_i taken to be zero in order to measure the OSL signal in the natural sample. Figure 9.17 shows a simulation of the SAR protocol using the original Bailey parameters [35]. The natural sample is bleached in the simulation, and then

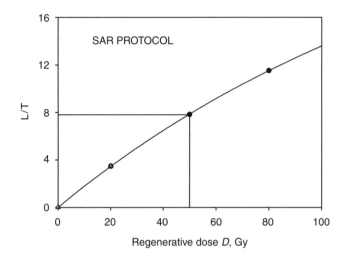

Figure 9.17   *Simulation of the SAR protocol using the original Qz-A1 Bailey parameters [35]. The natural sample is bleached and then given a sequence of regenerative doses (50, 20, 50, 80, 50, 0 Gy). The preheat temperature in this example is 10 s at 260 °C, the cutheat temperature is 10 s at 220 °C, and the test dose is 5 Gy. (Simulation and drawing by the authors)*

given a sequence of regenerative doses (50, 20, 50, 80, 50, 0 Gy). The preheat temperature in this example is 10 s at 260 °C, the cutheat temperature is 10 s at 220 °C, and the test-dose is 5 Gy.

The optimal choices of preheat temperature, cutheat temperature, test-dose and pre-heat plateau test, etc., for the SAR protocol are discussed in some detail by Wintle and Murray [26].

9.6   Thermally Transferred OSL (TT-OSL)

Several experimental studies have shown that thermal transfer of charge in quartz can lead to inaccurate determination of the equivalent dose D_e during application of the SAR protocol, especially for young quartz samples [739, 745–753].

Wintle and Murray [26] suggested using plots of measured D_e as a function of preheat temperature, as a test of whether thermal transfer of charge is a problem during application of the SAR protocol. For younger samples preheat temperatures above 220 °C can cause charge from thermally shallow but optically insensitive traps, to become retrapped at the deeper optically sensitive traps. This may result in an experimentally measured rapid increase in the D_e value as function of the preheat temperature (see for example the experimental study of Kiyak and Canel [753], and references therein). In some experimental studies it has been shown that these thermal transfer effects can be avoided by using lower preheat temperatures. For young samples with D_e values of 5 Gy or less, Wintle and Murray [26] suggested using a lower preheat temperature of, for example, 200 °C for 10 s. Rhodes [754] studied TT-OSL signals in several glacigenic quartz samples. Laboratory-bleached samples were examined using a modified SAR protocol and clear evidence was found of TT-OSL signals which were related to the natural irradiation history of the samples. Several optically bleached glacigenic samples showed a significant rise in D_e as a function of preheat temperature, rising to several Gy for preheats of 10 s at 280 °C. In a separate experiment, Rhodes [754] established a strong linear correlation between the TL signal observed during preheating and the TT-OSL signal. This correlation was interpreted as indicating that the thermal transfer process is a simple conduction band process.

Kiyak and Canel [753] used a traditional SAR protocol to demonstrate thermal transfer effects in a variety of quartz samples. Prior to applying the SAR protocol, the samples were bleached twice for 40 s at RT with blue LED light, followed by a 4000 s pause to allow thermal decay of low temperature TL peaks. A test-dose of 0.2 Gy and a cutheat of 10 s at 220 °C was used, while the preheat temperature was varied between 150 °C and 300 °C. The experimental results of Kiyak and Canel [753] showed that the estimated value of D_e from the SAR protocol for these bleached samples was 0 Gy for preheat temperatures below 200 °C, while the estimated D_e value increased continuously up to a value of 0.025 Gy for preheats above 200 °C.

Several recent studies have documented a different type of TT-OSL signal, which has been used as the basis of a new OSL dating procedure [755–757]. By using this TT-OSL signal it was found possible to extend the dating range for fine-grained quartz extracted from Chinese loess by almost an order of magnitude. It is believed that the production of this type of TT-OSL signal involves a single charge-transfer mechanism involving a slowly bleachable OSL trap corresponding to a "source TL trap" at 315 °C. This single transfer mechanism was suggested by Adamiec *et al.* [758] and was modeled by Pagonis *et al.* [36].

The TT-OSL signal is usually obtained after a high temperature preheat (260 °C for 10 s), following an optical bleach at 125 °C for 270 s to deplete the fast and medium OSL components. The TT-OSL signal is typically measured for 90 s at 125 °C in order to avoid the effect of retrapping of electrons in the 110 °C TL trap. The luminescence sensitivity related to this TT-OSL measurement is determined by the OSL response to a subsequent test-dose. The TT-OSL signal is thought to consist of two components, a dose-dependent component termed the recuperated OSL (ReOSL), and a second component believed to be dose independent, which is termed the basic transferred OSL (BT-OSL). Wang *et al.* [755] proposed a method that used multiple aliquots to construct the dose-response curve and used it for D_e determination. Pagonis *et al.* [36] used a modified version of the model by Bailey [35], to simulate the dose response of the OSL, TT-OSL and BT-OSL signals, and compared directly the experimental results with the model output. Table 9.5 is an outline of the stages involved in the simulation of the new TT-OSL dating procedure by Pagonis *et al.* [36]. Specifically steps 1–4 are a simulation of the "natural" quartz sample as proposed by Bailey [35], and the rest of the steps are a simulation of the sequence of steps in the protocol of Wang *et al.* [755].

The TT-OSL model of Pagonis *et al.* [36] is based on the previous model by Bailey [35] and Figure 9.18 shows the corresponding energy level diagram. The original model by Bailey [35] consists of five electron traps and four hole centers, and has been used successfully to simulate a wide variety of TL and OSL phenomena in quartz. This model was expanded by Pagonis *et al.* [36] to include two additional levels. Levels 10 and 11 in Figure 9.18 are the two new levels added to the original Bailey model by Pagonis *et al.* [36], and were introduced in order to simulate the experimentally observed TT-OSL signals and

Table 9.5 *Steps in the simulation of the TT-OSL protocol of Wang et al. [755, 756], based on the simulations of Pagonis et al. [36]*

Step	Description
1	Geological dose - irradiation of 1000 Gy at 1 Gy s^{-1}
2	Geological time - heat to 350 °C
3	Illuminate for 100 s at 200 °C
4	Burial dose - 200 Gy at 220 °C at 0.01 Gy s^{-1}
5	Regenerative dose D_i at 20 °C and at 1 Gy s^{-1}
6	Preheat to 260 °C for 10 s
7	Blue stimulation at 125 °C for 270 s
8	Preheat to 260 °C for 10 s
9	Blue stimulation at 125 °C for 90 s (L_{TT-OSL})
10	Test dose = 7.8 Gy
11	Preheat to 220 °C for 20 s
12	Blue stimulation at 125 °C for 90 s (T_{TT-OSL})
13	Anneal to 300 °C for 10 s
14	Blue stimulation at 125 °C for 90 s
15	Preheating at 260 °C for 10 s
16	Blue stimulation at 125 °C for 90 s (L_{BT-OSL})
17	Test dose = 7.8 Gy
18	Preheat to 220 °C for 20 s
19	Blue stimulation at 125 °C for 90 s (T_{BT-OSL})
20	Repeat 1–19 for different regenerative doses $D_i = 0$–4000 Gy in step 5

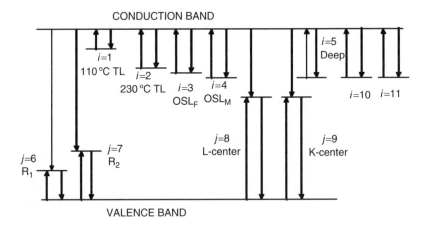

Figure 9.18 *The model of Pagonis et al. [36]. Levels 10 and 11 are the additional levels introduced to the original model by Bailey [35]. The model of Pagonis et al, reprinted from Pagonis, V., Wintle, A.G., Chen, R. and Wang, X.L., A theoretical model for a new dating protocol for quartz based on thermally transferred OSL (TT-OSL), Radiat. Meas. 43, 704–708. Copyright (2008) with permission from Elsevier. Levels 10 and 11 are the additional levels introduced to the original model by Bailey; from Bailey, R.M., Towards a general kinetic model for optically and thermally stimulated luminescence of quartz, Radiat. Meas. 33, 17–45. Copyright (2001) with permission from Elsevier*

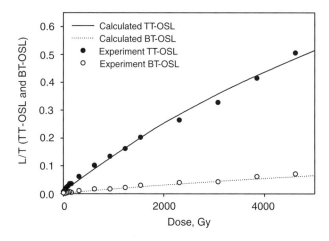

Figure 9.19 *The simulated dose-response curves for the TT-OSL signals are compared with the experimental data for sample IEE209. Reprinted from Pagonis, V., Wintle, A.G., Chen, R. and Wang, X.L., A theoretical model for a new dating protocol for quartz based on thermally transferred OSL (TT-OSL), Radiat. Meas. 43, 704–708. Copyright (2008) with permission from Elsevier*

BT-OSL signals. Level 10 in the model represents the source trap for the TT-OSL signal and is a slightly less thermally stable trap which saturates at very high doses. It is assumed that electrons are thermally transferred into the fast component trap (level 3) from level 10. This trap (level 10) is assumed to be emptied optically in nature by long sunlight exposure. Level 11 is believed to contribute most of the BT-OSL signal in quartz; these are believed to be light insensitive traps which are more thermally stable than either level 3 or level 10. Figure 9.19 shows a set of experimental data obtained for 4–11 μm quartz extracted from a young quartz sample (IEE209). This sample had a natural dose of 3 Gy as measured using OSL signals [759]. The sensitivity-corrected luminescence L/T signal corresponds to the first 5 s of the decay curve for the OSL, TT-OSL and BT-OSL signals measured using the new dating procedure. The difference between the two signals in Figure 9.19 is the ReOSL signal, which is used for calculating the equivalent dose. The solid lines represent the simulation output obtained using the sequence in Table 9.5. Good agreement between the simulated and the experimental data is obtained for all doses up to ∼4500 Gy.

Two possible mechanisms could give rise to the experimentally observed TT-OSL signals. These mechanisms are believed to involve either a double-step or a single-step charge transfer. The double-step mechanism for TT-OSL was suggested by Aitken [99], and uses a less optically sensitive "refuge trap". In this mechanism it is assumed that electrons are optically ejected from the fast OSL trap into the conduction band and then captured at an empty "refuge trap", one that has been emptied by the initial pre-heating at 260 °C for 10 s. From this trap, the electrons are then thermally ejected into the conduction band by the subsequent heat treatment (260 °C for 10 s) and captured in the now empty source traps of the fast OSL component. Aitken and Smith [556] suggested that this refuge trap was operative during stimulation at RT and gave rise to the 110 °C TL peak; this mechanism has been modeled by Pagonis *et al.* [760]. In the alternative single charge-transfer mechanism, the regenerated OSL signals are caused by the thermal transfer of electrons into the empty fast component trap from a less thermally stable trap with high dose saturation (level 10 in Figure 9.18). This trap is assumed to be emptied optically only by long sunlight exposure in nature. The results of the simulations by Pagonis *et al.* [36] are in agreement with a single charge-transfer mechanism being responsible for the TT-OSL signals measured by Wang *et al.* [755], rather than the more complex double-step mechanism hypothesized by Aitken [755]. More specifically the results of the simulation show that only a small percentage (2–10%) of the charge in this "source trap" transfers into the main fast OSL traps during the 10 s preheat at 260 °C.

Numerical modeling studies and associated experimental work have provided strong support for this single transfer TT-OSL mechanism [36, 758, 760]. In a recently published study, Adamiec *et al.* [761] carried out a detailed experimental study in order to identify the traps that are sources of the TT-OSL signal. During their study they determined the thermal stability of the signals by using TL signals associated with the TT-OSL signals. These authors used the information on thermal stability to develop a more appropriate ReSAR protocol that was tested using dose recovery tests.

In recent work, Wang *et al.* [762] developed a single aliquot measurement protocol which was applied to several samples of Chinese loess. In analogy to the well-established SAR protocol, the OSL responses to a test-dose are used to correct the TT-OSL signals for sensitivity changes occurring during the dating protocol. This single aliquot protocol based on TT-OSL signals has been termed the ReSAR protocol, and has been simulated by Pagonis *et al.* [763]. Pagonis *et al.* [763] carried out several simulations using the comprehensive quartz model developed previously by Pagonis *et al.* [36]. Furthermore, these authors were able to simulate the single aliquot ReSAR protocol of Wang *et al.* [762] and demonstrated that simulated doses of several thousand Gy could be recovered successfully within the model.

Perhaps the main drawback of the ReSAR protocol is its complexity, which makes it very time consuming from an experimental point of view. Several researchers have applied the ReSAR protocol to different sedimentary quartz samples, and suggested that it may be possible to improve its accuracy by making modifications, and that it may even be possible to simplify the protocol [757, 761, 764–766]. These simplified versions of the ReSAR protocol can result in very significant time savings during dating of samples using the TT-OSL signals. Tsukamoto *et al.* [757], in a comprehensive study of the TT-OSL ReSAR protocol, tested various versions of the original protocol by Wang *et al.* [762]. These authors suggested several modifications which may improve the accuracy of the ReSAR protocol. Furthermore, they studied the temperature dependence, dose response and bleaching characteristics of TT-OSL signals from several quartz loess samples from China and from coastal sands in South Africa. They suggested a modified protocol in which a step is added consisting of blue light stimulation at 280 °C for 100 s at the end of each ReSAR cycle. By adding this high temperature bleach step and by using a component separation of the ReOSL signal, it was found that the ReSAR protocol could recover given doses up to at least 1000 Gy.

Porat *et al.* [764] developed and tested successfully a simplified TT-OSL protocol which can also result in very significant time savings during the TT-OSL measurements. Stevens *et al.* [765] found difficulties in applying the ReSAR protocol to Chinese loess samples, including poor recycling, non linear L–T graphs and L–T graphs with negative intercepts. They suggested a modified protocol in which sensitivity changes are monitored by the response of the TT-OSL signal to a test-dose, and there is no correction for BT-OSL signals. Their protocol uses high temperature optical bleach in the middle and at the end of each cycle, in order to clear the traps and remove any remaining TT-OSL signal after each regenerative dose. Their simplified protocol was applied successfully to several quartz samples; however, they noted the presence of weak signals as one of the important limiting factors in applying the protocol to many of the studied samples.

In a recently published study, Kim *et al.* [766] studied the TT-OSL signals from seven loess-like fine grained samples from Korea, and examined the various procedures for separating the ReOSL and BT-OSL components. These authors also constructed and tested the use of a standardized growth curve for the ReOSL signal. In another notable study, Athanassas and Zacharias [767] studied raised marine sequences on the south-west coast of Greece during the Upper Quaternary. These authors tested the suitability of the ReSAR protocol by applying it to coarse-grained quartz aliquots from near shore outcrops. They studied the TT-OSL signal characteristics, dose response and sensitivity changes, recovery of known doses and the signals' bleachability by sunlight. The accuracy of the Re-OSL dates calculated for natural samples was compared with dates obtained from the SAR-OSL protocol.

10

Advanced Methods for Evaluating Trapping Parameters

10.1 Deconvolution

The subject of computerized curve-fitting analysis has become very popular during the last decade with the development of sophisticated glow curve deconvolution (GCD) techniques. The glow curves of TL materials are in most cases complex curves consisting of many overlapping glow-peaks. The deconvolution of complex glow curves into their individual components is widely applied for dosimetric purposes and for evaluating the trapping parameters E and s using curve-fitting methods.

Chen and McKeever [33] summarized the curve-fitting procedures commonly used to analyze multi-peak TL glow curves. They emphasized the primary importance of using a carefully measured TL glow curve, since any errors in measuring the glow curve can lead to the wrong results in the computerized procedures. Such procedures are more likely to yield accurate results in the case of linear superposition of first-order Randall–Wilkins type mathematical expressions. These authors concluded that curve fitting methods using a particular theoretical model should be applied with the utmost care, and extreme caution should be exercised when drawing conclusions from good curve-fitting results.

Horowitz and Yossian [291] provided an extensive and detailed review of the subject in a special issue of the journal Radiation Protection Dosimetry. This review covers most available information in the GCD literature of TL, and is a necessary tool for everyone wishing to use glow-curve analysis as a research and dosimetric analysis tool.

Bos *et al.* [386, 768, 769] in several important papers, presented the results of an evaluation of the capabilities of computer programs to analyze glow curves. This comparative study was carried out in the framework of the GLOw Curve ANalysis INtercomparison (GLO-CANIN) project. The papers contain the results of an analysis of thirteen different computer programs involving 11 participants from 10 countries, on both computer-generated and on experimentally measured glow curves. The intercomparison project concentrated on the

Thermally and Optically Stimulated Luminescence: A Simulation Approach, First Edition. Reuven Chen and Vasilis Pagonis. © 2011 John Wiley & Sons, Ltd. Published 2011 by John Wiley & Sons, Ltd.

goodness of fit, the determination of the peak area A, the temperature of the peak maximum T_M and the trapping parameters, i.e., activation energy E and frequency factor s. The following analytical expressions were evaluated within the GLOCANIN project

$$I(T) = As \exp\left(-\frac{E}{kT}\right) \exp\left[-\frac{skT^2}{\beta E} \exp\left(-\frac{E}{kT}\right)\left(1 - \frac{2kT}{E}\right)\right], \tag{10.1}$$

$$I(T) = I_M \exp\left[1 + \frac{E}{kT_M}\frac{T - T_M}{T_M} - \exp\left(\frac{E}{kT_M}\frac{T - T_M}{T_M}\right)\right], \tag{10.2}$$

$$I(T) = I_M \exp\left[1 + \frac{E}{kT}\frac{T - T_M}{T_M} - \exp\left(\frac{E}{kT}\frac{T - T_M}{T_M}\right)\right], \tag{10.3}$$

$$I(T) = \frac{A}{\sqrt{2\pi}\left[\sigma - \alpha\left(T - c\right)\right]} \exp\left[-\frac{(T - c)^2}{2\left[\sigma - \alpha\left(T - c\right)\right]^2}\right]. \tag{10.4}$$

In the above expressions A represents the area under the glow curve (in counts or arbitrary units), E is the activation energy (in eV), s is the frequency factor (in s^{-1}), T is the temperature (in K), k is the Boltzmann constant (eV K^{-1}), I_M is the maximum intensity of the glow peak (in counts K^{-1}) and T_M is the temperature at peak maximum (in K). Expression (10.1) can be easily derived from the Randall–Wilkins equation for first-order kinetics by using two terms in the series approximation of the integral in the first-order kinetics. Expressions (10.2) and (10.3) contain the experimental quantities I_M and T_M. Expression (10.3) is derived by approximating a linear heating rate function by a hyperbolic heating rate profile. It is sometimes referred to as the Hyperbolic Approximation. Expression (10.4) is termed a "modified Gaussian" curve and relies on a total of four adjustable parameters A, σ, α and c while the rest of the above expressions are based on three adjustable parameters. In this case of the modified Gaussian, the activation energy E is deduced from the width of the modified Gaussian.

Kitis *et al.* [770, 771] developed several new analytical expressions for use in GCD analysis, by using an approximation of the TL integral. The following expression was derived for the TL intensity of first-order TL glow curves

$$I(T) = I_M \exp\left[1 + \frac{E}{kT}\frac{T - T_M}{T_M} - \frac{T^2}{T_M^2}\left(1 - \frac{2kT_M}{E}\right)\exp\left(\frac{E}{kT}\frac{T - T_M}{T_M}\right) - \frac{2kT_M}{E}\right]. \tag{10.5}$$

In the case of general-order kinetics, these authors derived the following similar expression

$$I(T) = I_M b^{\frac{b}{b-1}} \exp\left(\frac{E}{kT}\frac{T - T_M}{T_M}\right)\left\{Z_M + (b - 1)(1 - \Delta)\left[\frac{T^2}{T_M^2}\exp\left(\frac{E}{kT}\frac{T - T_M}{T_M}\right)\right]\right\}^{-\frac{b}{b-1}}. \tag{10.6}$$

In this expression b is the kinetic order of the TL process, $\Delta = 2kT/E$ and $Z_M = 1 + (b - 1)2kT_M/E$. The accuracy of these expressions was tested by calculating the figure of merit (FOM) for synthetic glow curves, and it was found that they describe accurately glow peaks for first-, second- and general-order kinetics. Similar expressions were developed by the same authors for mixed-order kinetics and for continuous trap distributions. The advantage of using these equations to approximate the TL intensity is that they involve two quantities which are measured experimentally, namely the maximum TL intensity and the corresponding temperature (I_M and T_M). The activation energy E is treated in these expressions and during the computerized fitting procedure as an adjustable parameter.

Pagonis *et al.* [772, 773] presented analytical expressions for fitting first- and second-order kinetics glow-peaks based on the well-known Weibull and Logistic distribution functions. The proposed algorithms gave excellent fits to TL glow peaks, although they are not physically based. They used the expression

$$I(T) = 2.7131_M \left(\frac{T - T_M}{b} + 0.996 \right)^{15} \exp \left[- \left(\frac{T - T_M}{b} + 0.996 \right)^{16} \right], \quad (10.7)$$

for the Weibull approximation to the first-order glow peak. An analytical expression was also given which allows an accurate evaluation of the activation energy E, namely,

$$E = T_M \frac{k}{b} \left(-b + \sqrt{7b^2 + 242.036 T_M^2} \right) \quad (10.8)$$

where b is the width of the Weibull function.

Pagonis *et al.* [248] applied several expressions found in the GLOCANIN intercomparison program to synthetic first-order TL glow curves. These authors found that the expression which best fits the synthetic reference glow curves was Equation (10.5). In the same work Pagonis *et al.* [248] also compared several analytical expressions found in the literature for fitting general-order, as well as mixed-order TL peaks. Other notable efforts to improve deconvolution procedures for TL can be found in the literature [774–776].

Recently the GCD procedures for TL have also been applied to the peak-shaped LM-OSL composite curves. Kitis and Pagonis [313] studied the geometrical characteristics and symmetry factors of the peak-shaped LM-OSL curves, and pointed out the similarities and differences between TL and OSL deconvolution analysis. These authors studied both first-order and general-order peaks, and identified the presence of slowly varying pseudo constants which can be used to develop expressions for the optical cross-sections for OSL. In close analogy with the situation for TL, the following expressions were derived for first-order and general-order analysis

$$I(t) = 1.6487 I_m \frac{t}{t_m} \exp \left(- \frac{t^2}{2t_m^2} \right), \quad (10.9)$$

$$I(t) = I_m \frac{t}{t_m} \left(\frac{b-1}{2b} \frac{t^2}{2t_m^2} + \frac{b+1}{2b} \right)^{b/(1-b)}. \quad (10.10)$$

In these expressions, t is the time (s), t_m is the time at which the maximum LM-OSL intensity I_m occurs, and b is the kinetic order of the OSL process. Kitis and Pagonis [313] pointed out a major difference between TL and OSL deconvolution analysis. Once the time t_m of maximum LM-OSL intensity has been found in the experimental OSL curve, then the complete individual OSL curve can be identified within the composite LM-OSL curve. This simple property is very important for computerized curve decomposition analysis (CCDA), since for every given time t_m, there is one and only one corresponding LM-OSL peak. This is in contrast to the respective property of TL peaks, in which every peak maximum temperature T_M corresponds to an infinite number of possible TL peaks with different pairs of kinetic parameters (E, s). The usefulness of LM-OSL deconvolution techniques for analyzing LM-OSL curves from quartz for dating applications was examined by several researchers [74, 302, 303].

10.2   Monte-Carlo Methods

Theoretical models of TL are usually based on the assumption of a uniform spatial distribution of traps and recombination centers, resulting in models such as the single trap model (STM). However, several alternative models of TL are based on the existence of electron–hole pairs trapped close to each other, and are known as localized transition (LT) models. A third type of model has been proposed in the literature, which is based on spatial correlation of traps (T) and recombination centers (RC). In several experimental studies of correlated TSC/TL measurements, it was shown that the classic energy band models do not apply, because of the existence of such spatial correlations between traps and recombination centers. One may expect that spatially correlated systems would be found in materials with polycrystalline and low-dimensional structures. These models are of practical importance in dosimetric materials like the popular LiF:Mg,Ti, in which dosimetric peak 5a is believed to be related to localized transitions, while the main dosimetric peak 5 relates to carriers that escaped from the electron–hole pair system [231, 635]. Spatial correlation can also be expected in case of high energy and high dose irradiations, when large defects may be created in groups.

Initially calculations within these models were carried out by using Monte-Carlo techniques. These calculations showed the existence of unusual TL glow curves, in which the main TL peak is occasionally accompanied by a smaller peak called the displacement peak. Mandowski and Świątek [777] first introduced the use of Monte-Carlo techniques to simulate thermally stimulated relaxation processes in polycrystalline and two-dimensional solids. In such materials, each grain or plane can be assumed to act as a separate system. These authors suggested the use of Monte-Carlo simulations to test the range of applicability of the classical approach, in which a set of differential equations is solved instead. The basic physical assumption of the method is that the solid consists of a number of separate systems or "clusters", with the same trap depths and trap populations; initially all charge carriers are in traps, and the probabilities of possible transitions are given by

$$D_i = \nu_i \exp(-E_i/kT), \tag{10.11}$$

$$T_i = A_i(N_i - n_i), \tag{10.12}$$

$$R = mA_m, \tag{10.13}$$

where D_i represents the probability of detrapping of the trapped electron into the conduction band, T_i is the probability of a conduction electron becoming retrapped into the electron

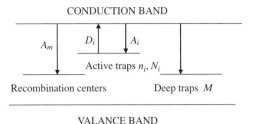

Figure 10.1 *The classical band model for thermally stimulated processes. Allowed transitions are retrapping (A_i), recombination (A_m) and detrapping (D_i). Reprinted from Mandowski, A. and Świątek, J., On the Influence of Spatial Correlation on the Kinetic Order of TL. Radiat. Prot. Dosim. 65, 25–28. Copyright (1996) with permission from Oxford University Press*

trap, and R is the recombination probability for an electron in the conduction band to recombine with a hole in the recombination center. N_i, n_i and m represent the number of traps, trapped electrons and holes, respectively. A_i and A_m are the retrapping and recombination probabilities. These transitions are shown in Figure 10.1. The Monte-Carlo calculations are performed with the total population of carriers simultaneously. The probability distribution of the transition times t_i is of the form

$$p(t_i) = \lambda(t_i) \exp\left(- \int_0^{t_i} \lambda(t')dt'\right), \tag{10.14}$$

and the transition time t_i is to be calculated from

$$\int_0^{t_i} \lambda(t')dt' = -\ln(\alpha_i), \tag{10.15}$$

where a_i is a homogeneous normalized random variable and $\lambda = D_i$, T_i, or R are the transition probabilities given above. In each step of the Monte-Carlo simulation, one finds the lowest transition time for all possible transitions, and this is the only transition which is executed. This time interval is calculated simultaneously for the whole group of allowed transitions of the same type. In order to decrease fluctuations in the numerical results, the calculations were repeated many times with the same initial parameters and the data was averaged over a time interval of 2 s.

Mandowski and Świątek [777] calculated the total concentration of carriers in the conduction band (n_c), as a function of the absolute concentration of initially trapped carriers in a single system (n_o). For the simulated heating of the sample, a constant heating rate of $1 \mathrm{K}\ \mathrm{s}^{-1}$ is used. Typical results for the simulated concentration of electrons in the conduction band are shown in Figure 10.2(a) for the following values of the parameters: activation energy $E = 0.9\,\mathrm{eV}$, frequency factor $\nu = 10^{10}\ \mathrm{s}^{-1}$, total concentration of traps $N = 10^6\ \mathrm{cm}^{-3}$, ratio of retrapping and recombination probabilities $r = A_t/A_m = 0.1$. The corresponding results for the negative rate of change of the number of recombination centers $-dm/dt$, which is proportional to the TL intensity, are given in Figure 10.2(b). Comparison of the results in

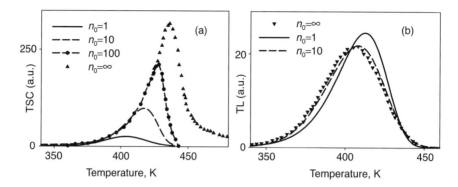

Figure 10.2   *Monte-Carlo simulations of the total number of electrons in the conduction band, for several values of the absolute number of initially trapped carriers in a single system (n_0). (a) The value of n_0 has a large effect on the number of electrons n_c in the conduction band during the heating process. (b) The value of n_0 does not affect significantly the TL peaks. Reprinted from Mandowski, A. and Świątek, J., Thermoluminescence and trap assemblies—results of Monte Carlo calculations. Radiat. Meas. 29, 415–419. Copyright (1998) with permission from Elsevier*

Figure 10.2(a) and (b) shows that while the value of n_0 does not affect significantly the TL peaks, it has a profound effect on the concentration of electrons n_c in the conduction band during the heating process.

Mandowski and Świątek [778] simulated TL spectra for different trap parameters and different correlations between traps and recombination centers, and compared the resulting TL glow curves with the empirical general-order model. The excitation stage was simulated by filling the traps using a Monte-Carlo approach; the results of this first stage in the simulations were then used as input for the linear heating stage. Their simulation was based on the parameters $E = 0.9\,\text{eV}$, $\nu = 10^{10}\,\text{s}^{-1}$, $n_0 = 1$, $A_m = 10^{-11}\,\text{cm}^3\,\text{s}^{-1}$ and a variable ratio of probabilities $r = A_t/A_m = 0, 1, 10$. These authors showed that in the case of $n_0 = 2$ and $r = 100$, the simulated glow curves consist of two closely situated peaks as shown in Figure 10.3. Although these TL glow curves were fitted very well with two first-order peaks, the resulting activation energies were found to be $E = 0.893\,\text{eV}$ and $E = 0.918\,\text{eV}$; these values have no physical meaning, and the authors suggested that extreme care be applied when analyzing TL glow curves which may be the result of a combination of localized and delocalized processes. For the case of one trap and one recombination center ($n_0 = 1$), the glow curves were found to be described always by first-order kinetics, in agreement with the experimental observation that most TL glow curves are described by first-order kinetics. For values of n_0 between 1 and 50 and for high values of the retrapping ratio r, these authors found very large differences between the modeled TL-TSC curves and the empirical general-order kinetics expressions.

Mandowski and Świątek [779] suggested a model in which the solid consists of groups or clusters with the same energy configuration, but separated by a distance and/or energy barriers. They discuss the case of an externally applied electric field, in which transitions of carriers between neighboring clusters cannot be neglected. In such cases, the whole system of clusters has to be considered simultaneously. They simulated a cubic network of cells

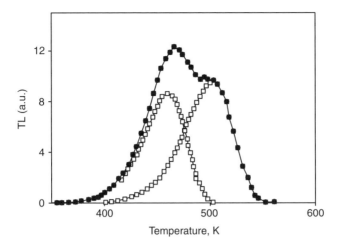

Figure 10.3   *Monte-Carlo simulations of TL glow curves for $n_0 = 2$ and $r = 100$. The simulated data can be fitted well by two first-order glow peaks, but the fitted activation energies have no physical meaning. Reprinted from Mandowski, A. and Świątek, J., On the Influence of Spatial Correlation on the Kinetic Order of TL. Radiat. Prot. Dosim. 65, 25–28. Copyright (1996) with permission from Oxford University Press*

with periodic boundary conditions. In addition to the three allowed transition probabilities in Equations (10.11–10.13), a carrier in the conduction band can also jump to an adjacent group, with a thermally activated transition probability. Simulations were carried out with various absolute numbers of carriers in a single separate group (n_0), and for both isothermal and non-isothermal conditions. Phosphorescence decay was simulated near the TL peak maximum for various values of n_0. In the case of no retrapping ($r = 0$) and for different values of n_0, significant changes were found only at high temperatures near 500 K, and for large isothermal times t. However, for high r values significant changes were observed, but these did not depend on the temperature of the measurement. For large clusters with $n_0 > 10$ all changes in simulated phosphorescence decay were insignificant. The simulations were also carried out for isothermal current decay (ICD), which is very sensitive to spatially correlated effects. Mandowski and Świątek showed results for n_c as a function of time for $n_0 = 1$–10^4. They also studied the effect of changing the barrier height within the cubic network. It was concluded that TSC/TL phenomena are more sensitive to cluster effects than isothermal experiments.

Mandowski and Świątek [779] considered the results of Monte-Carlo calculations within the context of commonly used localized and delocalized models for TL. For $n_0 = 1$ the Monte-Carlo results correspond to the localized model, while a homogeneous sample corresponds to $n_0 = \infty$. Their general conclusion was that for $n_0 > 1000$ the simulated TL coincided exactly with the delocalized model. They demonstrated results of the effect of deep traps for various concentrations of thermally disconnected traps. Their work showed conclusively that spatially correlated effects become prominent for low concentrations of thermally disconnected traps and for high recombination cases. They also showed contour plots of the deviations of the TL kinetics from standard theoretical models, as a function of n_0 and the retrapping ratio r. In the region where these large deviations occurred, the TL glow curves showed a rather complex structure. In the same work, Mandowski and Świątek also extended their simulations to a system with two possible energy levels. While the "classical" TL glow curves calculated from the standard delocalized TL models showed no structure, the spatially correlated TL peaks showed two well separated peaks. For intermediate cases of $n_0 = 2$ and $n_0 = 5$, the spatially correlated glow curves were very difficult to analyze, with the number of apparent peaks being larger than the number of trapping levels.

Mandowski and Świątek [239] studied the influence of the heating rate and initial filling ratio in spatially correlated systems. The authors applied the Hoogenstraaten heating-rate method and found a maximum error of 3% in the value of the calculated activation energy E in all cases studied. It was concluded that the heating-rate method is weakly sensitive to the presence of spatial correlation effects. A simulation of the fractional glow technique (FGT) showed that for high r and for full initial traps, the calculated E value is too low, in agreement with previous studies of this method for delocalized TL phenomena. The initial-rise method was found to have the same range of applicability as in the classical TL models, namely it will give accurate results only for the initial 10–15% of the peak maximum intensity, and for low initial occupancy of the traps. Fitting of the TL glow curves with general-order expressions gave satisfactory results only for the specific cases of first- and second-order kinetics. Mandowski suggested that if the presence of large clusters ($n_0 > 10$) is suspected, the TL should be measured with full initial trap filling ($n_0/N = 1$), to bring it closer to the simple model. However, if small clusters are suspected ($n_0 < 10$), TL should be measured at low initial filling to bring the system closer to first-order kinetics. Peak-shape methods should be used with care in other intermediate cases.

In a series of papers [239, 780, 781] the same authors studied the TL kinetics of one-dimensional (1D) structures which correspond to excitation of materials with high energy radiation. In some cases, it is thought that traps and recombination centers are populated along particle tracks. An important conclusion of these studies was that in some instances, an extra TL peak may appear at the high temperature region of the glow curve; this peak may be due to charge carriers that avoided nearest-neighbor recombination with adjacent recombination centers. Mandowski [239] presented several possible models of 1D systems. In the simplest model termed the fixed distance (FD) model, electrons can move to neighbor states with a transition probability given by

$$A_{tr} = \nu_{tr} \exp(-E_{tr}/kT), \tag{10.16}$$

where E_{tr} is the height of the potential barrier between two adjacent sites. In the second model, the transition probability can depend on the distance between trap levels, and is termed the variable distance (VD) model. In intermediate VD models, the transition probability may vary along the track, with two other models considered termed the variable distance models with intermediate states (VDIS). In these the chain of traps between active excited states is approximated by an additional level with an average lifetime. In order to include the effect of finite thickness in real TL detectors, an additional set of outer shallow traps was suggested. Examples were presented for the FD model, in which as the probability A_{tr} is increased, the two peaks merge into a simple second-order peak. For very low values of A_{tr} the second peak disappears, since low retrapping results in faster charge transportation along the chain, and it becomes easier for charges to reach a recombination center. Mandowski [780] showed also an example of the VD model, in which some of the localized pairs are joined together to form clusters of different sizes. In this case, the TL glow curve is the sum of individual peaks from single electron–hole pairs, as well as from groups with two pairs, three pairs, etc. The result is a broad featureless peak followed by a transport peak. Simulations of such 1D structures showed that as the frequency factor v_{tr} was increased, the transition probability increased as well, and the transport peaks shifted towards the main TL peak. For very low values of the transition probability, only a first-order TL peak was observed in the simulations, as expected from the classical TL theory. The FD model was studied also for a long chain with 250 localized electron–hole pairs, with boundary conditions by Mandowski [781], who gave examples of TL and thermally stimulated currents with and without an applied external electric field. Even with no applied electric field, extra peaks were apparent in the simulated TL glow curves. The effect of the external electric field was to increase and shift this extra peak. Low values of the electric field were found to change only the height and position of the displacement peak. High electric fields resulted in the two peaks merging into one, with the shape depending on the value of the electric field.

In a recent paper, Mandowski *et al.* [242] use the Monte-Carlo method for determining the thermal-quenching function by using variable heating-rate measurements (see e.g., Section 6.7). A way of calculating the quenching function and restoring the unquenched TL curve is presented. The reliability of the method is tested using computer-generated TL glow curves obeying the STM kinetics.

10.3 Genetic Algorithms

Adamiec *et al.* [729, 730] have developed a method for the application of genetic algorithms (GAs) to the problem of finding appropriate parameter values for quartz TL and OSL. The energy level model for TL and OSL has been developed by Bailey [532] and is discussed in some detail in Section 9.3. The model involving the quartz parameters is shown in Figure 9.6 and the set of simultaneous equations governing the process is (9.14–9.18). Finding all the relevant parameters consists of minimizing the difference between the experimental data in hand and the simulated data of the same sort, depending on the relevant model parameters. The experimental results in question may be the glow curve, namely TL intensity versus temperature, but it may also be any related phenomenon such as the thermal activation characteristic (TAC) as described by Adamiec *et al.* [729, 730] (see also Section 9.1). Applying traditional deterministic optimum search methods, such as the Levenberg–Marquardt method, usually leads to a local minimum rather than the desired global optimum, in particular in problems with such large number of parameters. Although GAs do not guarantee finding the global minimum, they have been proven to largely increase the probability of finding one or, at least come close to one, as compared with traditional search methods (see e.g. Goldberg [782]). Adamiec *et al.* [729, 730] use the experimental results of four samples of fired quartz displaying varying samples of TAC. They show that the form of the TAC could be reproduced and that it is possible to find model parameters that give a very good match of the TAC. The function minimized in the process was the sum of squared differences (SSD) between the experimental and simulated TACs.

The GA is a method of optimization based on the principle of biological evolution (see e.g. Holland [783]). It starts with a population of individuals, parameter sets randomly generated in a defined search space, and the optimum is searched by applying a numerical evolution mechanism. The fitter a given individual, the higher its chance of survival in an evolution step. Using an appropriate definition of the fitness function can have a significant influence on finding the desired solution. The advantage of the GAs is that they work on a whole population of possible solutions, thus improving the probability of finding the global optimum.

In the work by Adamiec *et al.* [729, 730], each individual consists of a set of parameter values for the model chosen to simulate the shape of the TAC. On one hand, one has an experimental TAC and on the other hand, a TAC obtained from the solution of the set of differential equations for a TAC protocol resembling the experimental one as closely as possible. The fitness is defined as a SSD between the experimental and simulated TAC and this is the function to be minimized. The evolution progresses according to a defined set of rules. As new generations of individuals are generated, only the fittest ones survive, leading to a gradually improved population of individuals. In the creation of new generations (points in the search space), the mechanisms of selection, crossover and mutation are involved.

Crossover combines the "genomes" (parameter sets) of the selected parents to produce an offspring. In each stage of the algorithm, the crossover is performed on a fixed number of individuals chosen randomly with a probability proportional to their fitness. Two crossover operators have been used. The first one obtains the parameters of the offspring genome as a weighted average of the corresponding parents' parameters using random weights (other weighting is also possible, e.g. with weights inversely proportional to the SSD). Here, the regular weighted average is calculated for the trap depth in TL, or for the optical detrapping probability in the case of OSL. For the other parameters, the weighted geometric average is used to allow many orders of magnitude of their values. If w_1 and w_2 are the weights and x_1 and x_2 are the values, then the new value is calculated as $\hat{x} = \exp[w_1 \ln(x_1) + w_2 \ln(x_2)]/(w_1 + w_2)$. The second crossover operator swaps random parameters or entire traps in randomly selected pairs of individuals.

The mutation operator modifies, with low probability, a random parameter from a randomly chosen individual. The GA is iterated and in each pass, a new generation of the same size is created. When the fitness function of the best individual in the population does not change significantly over several generations, the algorithm is stopped. The main advantage of GAs is that they are not limited by assumptions on the search space such as the differentiability and continuity of the relevant functions. GAs are much less prone to being caught in local minima than the traditional search algorithms. This feature is particularly valuable in complex search spaces with a large number of parameters to be adjusted, as is the case here with the Bailey model [35] for quartz.

Adamiec *et al.* [730] have used a software program consisting of three modules. The first one enables the creation of simulated measurement sequences, where for each step the initial temperature, final temperature, duration, light flux and dose rate are specified. The second module provides the solution of the charge transport equations (9.14–9.18), for a given sequence. The inputs are the model parameters, the simulation sequence and the initial charge concentrations. The differential equations are integrated using a stiff solver. The third module implements the GA, which is controlled by several key parameters, including the number of generations, crossovers and mutations per iteration. At the beginning of the procedure a set of parameter values is chosen which constitutes the first model generation. These values are generated from the ranges specified for each parameter, usually based on experience and on values found in the wider luminescence literature. Some of the parameters, namely, the frequency factor, the trapping and recombination probability coefficients as well as the concentrations of the traps and centers required special attention due to the broad possible span of parameter values, amounting to four to five orders of magnitude. Therefore, logarithms of the values were randomly generated from a uniform distribution over the given range. Concerning the trap thermal stability, Adamiec *et al.* [730] used the equivalent-peak temperature rather than the trap depth E, and from this the trap depth was calculated using a value of the frequency factor s and the chosen heating rate of $K s^{-1}$. This way, combinations of E and s leading to peaks outside the range of interest were avoided. These authors used somewhat different models consisting of several traps and recombination centers, depending on the sample in question. For the sample denoted SOT, a model with three trapping states and four kinds of recombination centers was used [34]. For the other samples, a model consisting of five electron and five hole traps was used.

For the TAC simulations, a dose of 1 Gy was chosen which produces 2.5×10^{10} electron–hole pairs. The dose rate was 0.4 Gy s^{-1}. The sensitivity was determined by an irradiation with a test-dose of 20 mGy, and heating to 180 °C. The irradiations were performed at 20 °C and were followed by a 5 s pause to allow system relaxation, namely the decrease of the charge populations in the conduction and valence bands. After each heating the cooling stages were also simulated to reflect the experimental situation as closely as possible. Adamiec *et al.* [730] report that sometimes, it was necessary to readjust the parameter ranges after an initial fit was obtained. In addition, the initial fits were performed with a smaller number of traps and after a fit was obtained, the parameters were used to define a new, smaller search space. Also, at this stage, an additional trap was added to reproduce some of the more detailed features of the TAC. For example, in the case of the Ark sample, the trap responsible for the slow component was not included. Only after a fit was reached, this trap was added and the procedure repeated. This way, the sensitivity decrease in the range of 400–500 °C could be accounted for.

The population size chosen consisted of 300 individuals. In each iteration, crossover was performed on 30–60 random pairs chosen from the whole population; 10–20 best pairs, 10–20 pairs consisting of the best individual and the following individuals and 10–20 pairs consisting of the best individual and individuals randomly chosen from the population. In 15 random pairs, the traps or single parameters were swapped. Up to 200 iterations were necessary to reach the presented fits.

The experimental and simulated TACs are given in Figure 10.4 and the parameter values obtained are shown in Table 2 in Adamiec *et al.* [730]. Very good agreement between experiment and simulations was achieved for all the investigated samples. The degree of sensitization varies largely between the samples. For the sample SOT, the maximum sensitization is around three, whereas for the Merck sample it is close to 300. It has been demonstrated that the relevant choice of trap populations and trapping and recombination probabilities can account for such a wide difference in sensitization properties. It has been seen that the finer features of the TAC could also be reproduced.

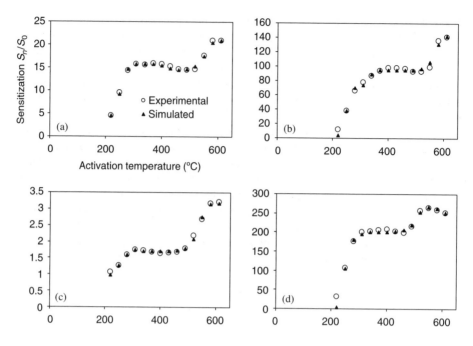

Figure 10.4 *The measured and simulated TACs of the investigated samples: (a) Ark; (b) BDH; (c) SOT; and (d) Merck. Reprinted from Adamiec, G., Bluszcz, A., Bailey, R. and Garcia-Talavera, M., Finding model parameters: Generic algorithms and the numerical modelling of quartz luminescence, Radiat. Meas. 41, 897–902. Copyright (2006) with permission from Elsevier*

Adamiec *et al.* [730] conclude that the numerical model employed is well suited to reproduce sensitivity changes in quartz. These results show that the sensitivity change appearing at high temperature is commensurate with the model, and that the TACs can be reproduced with high precision. The application of GAs opens new possibilities for numerical simulation of quartz properties. It provides a tool for semi-automated fitting of a model with a large number of parameters. It might also be possible to employ traditional minimization algorithms for further improvement of the fit, once the fit obtained using the GA is close (see Garcia-Talavera *et al.* [784]). The authors point out that for a given sample, a much larger data set needs to be considered. In addition to the shape of the TAC, they suggest to use in the future the shape of the OSL decay curves, isothermal sensitization curves, dose quenching, etc. In an ideal case, it might be possible to devise a protocol of measurements performed on single aliquots that would allow various aspects of luminescence data to be taken into account. In such cases, the definition of an appropriate fitness function is crucial, so that one numerical value would reflect the global quality of agreement between the experimental results and the model. The heterogeneity of quartz samples is another important issue that needs consideration. The method discussed has been applied under the assumption of homogeneity of a sample. In the case of a heterogeneous sample, the model needs to be much more complex.

10.4 Application of Differential Evolution to Fitting OSL Curves

Bluszcz and Adamiec [785] present the differential evolution (DE) algorithm and its application to the deconvolution of OSL decay curves into first-order components. Their underlying assumption is that under constant intensity stimulation, the intensity of quartz OSL is given by

$$I(t) = I_{01}e^{-\lambda_1 t} + I_{02}e^{-\lambda_2 t} + \cdots \qquad (10.17)$$

and the goal is to determine the number of components m and to estimate the value of the decay constants λ_k of each component, and the magnitude of each component, I_{0k}, related to the initial populations of electrons in traps. One can measure either the OSL decay curve or the LM-OSL (see Sections 7.6, 7.7 and 7.8). In the latter case, the LM-OSL curve is given as

$$I(t) = I'_{01}te^{-(1/2)\gamma_1 t^2} + I_{02}te^{-(1/2)\gamma_2 t^2} + \cdots . \qquad (10.18)$$

As the two results describe the same object under different stimulation conditions, there should be a correspondence between the parameters related to a certain component. As pointed out by Bluszcz and Adamiec [785], usually OSL intensities are not measured directly, but rather luminescence photons are counted over finite time intervals, i.e. the integrals over Equations (10.17) or (10.18) on those time intervals are recorded. Counting-time intervals may be of a constant length or may vary from shortest at the beginning of the OSL decay to longest at the end; this choice may optimize the resolution of the OSL curve when the total number of counting intervals is limited. In the case of an LM-OSL measurement, the stimulation light intensity is described by the stepwise function of the form

$$\varphi(t) = \frac{\varphi_{max}}{N-1} \sum_{j=1}^{N} H(t - t_j) \qquad (10.19)$$

consisting of N tiny steps of the same height and fixed or varying length, and occurring at t_j, where $H(x)$ is the Heaviside step function defined as

$$H(x) = \begin{matrix} 0, & x < 0 \\ 1, & x \geq 0 \end{matrix}. \qquad (10.20)$$

For the sake of mathematical simplicity, Bluszcz and Adamiec [785] took all counting periods to be of the same length of τ, and the measurement consisted of N numbers, the photon counts recorded in N equal periods over the interval $(0, N\tau)$. For the analysis, no background count in the experimental data is assumed; in practice, the background may be accounted for by subtracting it from the experimental data before the analysis in the CW-OSL case, or by subtracting a linearly increasing background from the LM-OSL. This requires a good knowledge of the machine background. The two cases mentioned have the following forms,

$$I_i = \sum_{k=1}^{m} a_k(e^{\lambda_k \tau} - 1)e^{-\lambda_k \tau i}, \quad i = 1, 2, 3, \ldots, N \qquad (10.21)$$

and

$$I_i = \sum_{k=1}^{m} b_k(e^{\gamma_k \tau^2 (i-1)} - 1)e^{-(1/2)\gamma_k \tau^2 (i-1)i}, \quad i = 1, 2, 3, \ldots, N, \tag{10.22}$$

for CW-OSL and LM-OSL, respectively.

It should be noted that although the two presentations have very different shapes, they depend on their parameters $\{\lambda_k, a_k\}_m$ and $\{\gamma_k, b_k\}_m$ in the same way, namely, they are linear combinations of exponential terms. More general treatments allowing for a varying length of counting intervals and excitation steps, have more complex forms but still depend on their parameters in a similar way. Therefore, Bluszcz and Adamiec [785] continue to deal with Equation (10.21) only, without any loss of generality. The expressions given depend linearly on a_k (or b_k) and non-linearly on λ_k (or γ_k). The problem is thus reduced to finding automatically, without human intervention, a statistically justified number of components m, and the best two sets of parameters, $\{\lambda_k\}$ and $\{a_k\}, k = 1, \ldots, m$. The weighted least squares method (LSM) is used to fit a series of N values $n_i; i = 1, 2, 3, \ldots, N$, which are the measured OSL photon counts. These authors presented the problem as finding the minimum of the χ^2 function of squared differences between measurements and model values, where it is further assumed that photon counts have Poisson variances, namely the weights are τ^2 / n_i for the ith interval in which the number of counts in n_i,

$$\chi^2 = \sum_{i=1}^{N} \frac{\tau^2}{n_i} \left[\frac{n_i}{\tau} - \sum_{k=1}^{m} a_k(e^{\lambda_k \tau} - i)e^{-\lambda_k \tau i} \right]^2 \tag{10.23}$$

with respect to the $2m$ parameters $\{\lambda_k\}$ and $\{a_k\}, k = 1, \ldots m$. A presence of a linear background component, either fixed or adjustable, does not change the reasoning discussed below, it merely increases the number of linear parameters and decreases the number of degrees of freedom of χ^2. Suppose that χ^2 has been minimized for $2m$ parameters. As described by Bevington and Robinson [786], the minimized χ^2 value has a chi-square distribution with $\nu = N - 2m$ degrees of freedom. Adding a component to the fitted model decreases the minimized χ^2 value by $\Delta\chi^2$ and its number of degrees of freedom by 2. Thus, the difference $\Delta\chi^2$ has 2 degrees of freedom and the new χ^2 value has $N - 2m - 2$ degrees of freedom. To test the justifiability of adding the $(m + 1)$th component, one may use the ratio

$$F = \frac{\Delta\chi^2}{\chi^2 / (N - 2m - 2)}, \tag{10.24}$$

which has an F distribution with 2 and $(N - 2m - 2)$ degrees of freedom. If the calculated F value is larger than a critical value F_α for a given significance level (e.g. 0.05 or 0.01), one may leave a newly added component in the model, otherwise one should revert to the previous one with m components. Thus, one can choose the number of components that can be justified by the experimental data.

Bluszcz and Adamiec [785] suggest that this procedure can be easily automated within a computer program. They propose an algorithm consisting of the following steps:

(1) Set the number of components $m = 1$ and minimize $\chi^2(m)$.
(2) Increase m by one and minimize $\chi^2(m + 1)$.

(3) Calculate $\Delta\chi^2 = \chi^2(m) - \chi^2(m+1)$ and the F statistics using Equation (10.24).

(4) If $F > F_\alpha$ proceed to step 2. Otherwise,

(5) Revert to m components and stop. Output the values of λ_k and a_k, and other statistics.

Due to the exponential dependence of the objective function (10.23) on the λ_k parameters, the minimization of the χ^2 function requires non-linear regression methods that have a serious common weakness. As pointed out in Chapter 5, the relevant minimization methods need a starting point in the parameter space, from which they proceed to the minimum. More often than not, these methods find the closest local minimum rather than the desired global one. Bluszcz and Adamiec [785] used a self-organizing optimization algorithm based on DE previously proposed by Storn and Price [787], to find the starting point for the Levenberg–Marquardt (LM) [286] χ^2 minimization method. DE is similar to GAs discussed in Section 10.3 in that it borrows basic features of natural evolution, but instead of a binary representation, uses a direct floating-point representation. The solution space searched by the GA is thus discrete, whereas DE searches for the best solution over a continuous space and therefore is better suited for this problem. The self organization of the DE algorithm includes automatic scaling of mutations adapted to the current topography of χ^2 hyper-surface over the parameter space. This feature makes DE superior to GAs or other continuous space searching algorithms, such as simulated annealing or Monte-Carlo methods, all of which need additional problem related information about scaling their random migration in different dimensions. The DE algorithm is also very concise; according to Price and Storn [788], its original C code is less than 20 lines long, yet it is fast and robust.

The DE algorithm stochastically searches a parameter space to find a quasi-optimum solution, in the present case being the global minimum of χ^2. Solutions are represented by vectors in n-dimensional parameter space, where n is a number of parameters of χ^2, and the coordinates of the vector consist of the parameter values. In the simplest case, the ith coordinate of the solution vector is the ith parameter. DE starts with a set of NP vectors $\mathbf{V}_i$ ($i = 1, 2, 3, \ldots, NP$) randomly chosen from the defined search space. The set of NP vectors is called "generation". The generation evolves into the descendant generation by means of three basic evolutionary operations: mutation, combination and selection. The mutation operation consists of adding to a given vector a weighted difference of two other vectors randomly selected from the current generation. If $\mathbf{A}$ is a vector undergoing mutation and $\mathbf{B}$ and $\mathbf{C}$ are two randomly selected vectors such that $A \neq B \neq C$, then the mutated vector $\mathbf{A}'$ is obtained as

$$\mathbf{A}' = \mathbf{A} + F \cdot (\mathbf{B} - \mathbf{C}), \tag{10.25}$$

where F is a weighting factor controlling the scale of mutation.

When a vector $\mathbf{A}$ combines with a vector $\mathbf{B}$, the child vector $\mathbf{C}$ inherits one randomly selected coordinate from $\mathbf{B}$ and each one of its remaining $n-1$ coordinates either from $\mathbf{A}$ with a probability $(1\text{-}CR)$, or from $\mathbf{B}$ with a probability CR. The combination operation is defined as

$$\mathbf{C} = R(\mathbf{A}, \mathbf{B}; CR). \tag{10.26}$$

The composition of the descendant generation is determined through NP competitions, such that in the ith competition, the ith vector competes against its child $\mathbf{V}'_i$ which is a product of mutation and a combination involving three selected vectors,

$$\mathbf{V}'_i = R[\mathbf{V_i}, \mathbf{V}_a + F \cdot (\mathbf{V}_b - \mathbf{V}_c); CR]. \tag{10.27}$$

Of the two competing vectors $\mathbf{V}_i$ and $\mathbf{V}'_i$, the one yielding a lower χ^2 value is selected to the next generation. For each vector $\mathbf{V}$, the DE algorithm calls an external function that returns the χ^2 value calculated for the fitted experimental data. When the generation is completed, the DE algorithm calls another external function that checks the "stop" condition. For example, it may check whether the number of generations reached a preset value or that the solutions do not improve significantly. If the stop condition is met, the best solution is printed out and otherwise, the algorithm proceeds to the next generation. In this way, the DE is controlled by three parameters: NP, F and CR, which are the number of vectors in the evolving population, the scale of mutation, and the crossing-over probability, respectively. Price and Storn [787, 788] recommended using NP to be 5–10 times the number of parameters n, $0 < F \leq 1.2$ and $0 \leq CR \leq 1.0$. Bluszcz and Adamiec [785] used the DE to optimize the values of NP, F, and CR for the discussed problem of OSL decay curves.

For the DE algorithm to be effective, it is necessary that it search the n-dimensional solution space exhaustively, which practically means that NP should be kept high especially as the number of dimensions (parameters) grows. On the other hand, this limits the efficiency of the algorithm, as it requires more computations of the χ^2 value. This means that reducing the number of parameters while using the DE algorithm may considerably reduce the number of computations and accelerate the procedure of finding the optimal solution. In the presently discussed case, it is possible to reduce the dimensionality of the problem by a factor of 2. For the given set of λ_k values, the best values of a_k may be found exactly by linear algebra methods. This is equivalent to the transformation of the independent variable, which does not change the χ^2 value, where current values of λ_k are used in the transformation [789]. Thus, half of the problem is linear and may be solved by linear regression methods whereas the other half is nonlinear and is first approximated by DE and then solved by a nonlinear LM regression method. The solution vector is divided into two parts. The first one is comprised of m decay constants λ_k and is denoted by $\mathbf{\Lambda}$, and the second one of all the remaining parameters a_k (including background parameters when appropriate), denoted by $\mathbf{A}$. The minimization of χ^2 is achieved by a hybrid algorithm that uses the evolutionary approach to optimize the $\mathbf{\Lambda}$ and linear regression to calculate the $\mathbf{A}$ vector. The linear regression calculations of the coordinates of the vector $\mathbf{A}$ are performed within the function returning the value of χ^2. Bluszcz and Adamiec [785] call this algorithm HELA, Hybrid Evolutionary Linear Algorithm. The HELA proposed for automated unassisted deconvolution of the OSL curve into m components can be presented as follows:

(1) Set the variability ranges for the m decay constants that are coordinates of the $\mathbf{\Lambda}$ vectors and randomly generate NP such vectors $\mathbf{\Lambda}_i$ ($i = 1, 2, 3, \ldots, NP$) to form the initial generation of the solution. For each $\mathbf{\Lambda}_i$, call an outside function that returns both the χ^2 value and the vector $\mathbf{\Lambda}_i$ whose coordinates are calculated by linear regression.

(2) Apply evolutionary operators to obtain the next generation of the solution. For each vector $\mathbf{\Lambda}_i$ produce an offspring $\mathbf{\Lambda}'_i$ by crossing-over and mutation [Equation (10.27)] and obtain the χ^2 value for it. From the competing pair $\mathbf{\Lambda}_i$ and $\mathbf{\Lambda}'_i$ select the one with the lower χ^2 value for the next generation.

(3) Check the "stop" condition and output the best solution. If the condition is not met, make the new generation incumbent and proceed to step 2.
(4) Use the best solution from step 3 as the starting point for the LM method to obtain the parameters of the m OSL components.

In order to check the method, Bluszcz and Adamiec have tested it on randomly generated data. The generated data consisted of components of known intensities and decay constants with an added background. For each time interval, the components were integrated, added, randomized and rounded to integer, so that the values simulate the counting mode of operation of the PM tube. As the stopping criterion, Bluszcz and Adamiec [785] used the magnitude of the variance of χ^2 within the whole population. The HELA was stopped when this variance reached a sufficiently small value, i.e., when all the solutions were very close to each other in the sense of the χ^2 value. In most of the numerical experiments, they chose this value to be 0.1. In most cases, 15–100 generations were necessary to get a satisfactory fit. In many cases, the LM algorithm did not provide better χ^2 values than the HELA algorithm. Bluszcz and Adamiec report that the HELA algorithm proved to be very effective in decomposing simulated OSL decay curves into constituent components. Thus, the known parameters could be retrieved very successfully. In real-life cases, the evaluated parameters can be used in estimating the absorbed doses for different components.

11

Simultaneous TL and Other Types of Measurements

11.1   Simultaneous TL and TSC Measurements; Experimental Results

TL and TSC have often been analyzed by the same methods (see e.g. Nicholas and Woods [790]), which is justified when one assumes a constant lifetime of the conduction carriers. This approach implies that the two curves have the same shape, and more specifically, that the maxima occur at the same temperature. Experimentally, however, simultaneous measurements of TL and TSC revealed in many cases a shift between the corresponding peaks; in most cases, the TL peak appeared at a lower temperature than its TSC counterpart [220, 264, 389, 790–796]. Chen [797] has shown that assuming that electrons are thermally raised into the conduction band (or holes into the valence band) prior to their thermoluminescent recombination with opposite sign carriers in centers, the TL peak should usually precede the accompanying TSC peak (see below). Furthermore, he gave a method for evaluating the recombination probability by the use of simultaneous measurements of these two phenomena.

The experimental material in the literature on simultaneous measurements of TL and TSC in various materials is quite abundant. In many cases, the appearance of the TL peak at a lower temperature than its TSC companion is reported, though the opposite effect sometimes takes place. Coterón *et al.* [798] reported on the results of TL-TSC simultaneous measurements in Al_2O_3 which was X-irradiated at RT. The results are shown in Figure 11.1. Apart from a minor TL peak at 75 °C with no TSC counterpart, five TL peaks occur at 120, 145, 172, 235 and 297 °C. Their TSC associated peaks are at 122, 148, 180, 238 and 300 °C, respectively. Thus, all the peaks appear in the expected order of TL first and TSC at a higher temperature. A later work by Lapraz *et al.* [799] reports on TL-TSC simultaneous measurements in α-Al_2O_3, X-irradiated at 77K. Here, two pairs of TL-TSC peaks are seen at ~300 and ~500 K the latter being probably the same as the 235–238 °C peaks in the mentioned work by Coterón *et al.* [798]; both pairs also appear in the "correct"

Thermally and Optically Stimulated Luminescence: A Simulation Approach, First Edition. Reuven Chen and Vasilis Pagonis.   © 2011 John Wiley & Sons, Ltd. Published 2011 by John Wiley & Sons, Ltd.

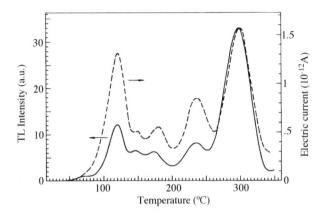

Figure 11.1 *TL (solid line) and TSC (dashed line) of Al₂O₃ X-irradiated at RT. Reprinted from Coterón, M.J., Ibarra, A. and Jiménez de Castro, M., Cryst. Latt. Def. Amorph. Mater. 17, 77–82. Copyright (1987) with permission*

order of TL first and TSC next. Bräunlich and Kelly [223, 800, 801] have studied the correlation between TL and TSC in the framework of the phenomenological theory of TL. They suggest that by simultaneous measurements of TL and TSC one cannot get sufficient information on all the parameters entering the relevant kinetic equations. Although this statement is strictly speaking correct, the vast amount of information that can be obtained by the thermally stimulated processes, including the simultaneous TL and TSC measurements, is quite impressive. In a later paper, Bräunlich *et al.* [802] sounded significantly more optimistic about the use of the TL-TSC technique, in view of the results of the work by Fillard *et al.* [803].

Another example of simultaneous measurements of the same kind was given by Barland *et al.* [804]; the results are shown at Figure 11.2. These authors studied KI:Tl, X-irradiated at 77 K. Here, the first TL and TSC peaks coincided, at 123 K. A second TSC peak occurred at 140 K with no TL companion. The next two peaks can be paired, with relatively large shifts between TL and TSC, the temperatures being 205 and 283 K in the former and 240 and 300 K in the latter.

Further experimental results confirming that the TL peak is usually shifted to lower temperature than the corresponding TSC peak were given by several researchers dealing with different materials. [227, 264, 389, 791, 793, 795, 805]. Böhm and Scharmann [806] reported on a pair of TL-TSC peaks at ∼140 K obtained from an X-rayed LiF crystal, where the TL peak preceded its TSC counterpart by ∼3 K. However, the same authors [807] reported two pairs of TL-TSC peaks in X-rayed Ca(NbO₃)₂ single crystals. The pair at ∼85 K appears in the "correct" order, whereas the positions of the pair at ∼140 K is inverted, namely the TSC maximum precedes the TL maximum.

Hillhouse and Woods [808] studied TL and TSC simultaneously measured in rutile (TiO₂) single crystals. They found six pairs of TL and TSC peaks in the temperature range of 100–330 K, and reported in each pair a shift between the maximum temperatures. This difference varied between 1.9 K and 7.1 K, and in all six pairs, the TL peak preceded its TSC companion.

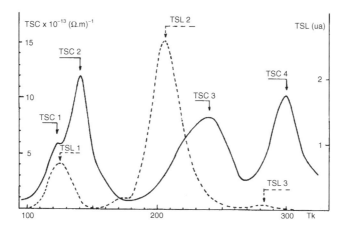

Figure 11.2   *TL (dashed line) and TSC (solid line) in a KI crystal with 0.01 mol% of Tl, after X-irradiation at 77K. Reprinted from Barland, M., Duval, E. and Nouailhat, A., J. Physique 41, 75–82 (1980). Copyright (1980) with permission from EDP Sciences*

Tuan *et al.* [809] studied TL and TSC peaks in X-irradiated KI crystals and found a TL peak at 138 K and a TSC one at 144 K. They discuss this result in the framework of the previously developed model by Käämbre and Bohun [810], in which the Auger effect played an important role in these thermally stimulated processes. They also simulated a series of 6 TL-TSC pairs according to their Auger effect model, and found shifts of 2–6 K between the TL and TSC peaks, and once again, TL occurred at lower temperatures.

A number of groups reported simultaneous TL-TSC measurements in different types of polyethylene. These include Nishitani *et al.* [811] who found four TSC and three TL peaks, in which three of the four TSC peaks correspond to the three TL maxima, and lagged behind them by several degrees. Ito and Nakakita [812] also found four peaks between 80 K and 420 K, and a similar shift of the TL maxima was reported. Markiewicz and Fleming [813] and Fleming [814] describe five TSC peaks, where only the first three are accompanied by TL peaks. The shifts reported are in the expected direction, namely TL first and TSC next.

Ikeya and Miki [815, 816] studied TL and TSC in LiH samples at low temperatures. A major TL peak was found at 57 K which was accompanied by the disappearance of an ESR signal, but no TSC peak was found in this temperature range. A large TSC peak was found at ∼187 K, accompanied by a TL peak at ∼178 K.

Abdurazakov *et al.* [817] reported on TL-TSC measurements in YAG:Nd^{3+} single crystals, and found 17 TL peaks and 14 TSC maxima, which corresponded to 14 of the TL peaks. In three peaks, TSC preceded the TL counterpart by 3–6 K; in one case, the two appeared at the same temperature, and in 10 of the cases, the TL peak occurred at lower temperature than the TSC peak, with shifts ranging between 1 K and 23 K. Possible reasons for the different kinds of temperature shift are discussed below.

It should be noted that important information can be revealed from simultaneous TL-TSC measurements even in cases where no correlation is found between the TL and TSC peaks. For example, Lapraz *et al.* [818] performed simultaneous measurements on calcium fluorapatite [$Ca_5(PO_4)_3F:Mn^{2+}$] single crystals, and found a TSC maximum at 70 K, accompanied by a TL maximum at the same temperature. In addition, three TL peaks were found at 173, 263 and 298 K, with no corresponding TSC maxima. The authors conclude that these peaks result from localized recombination processes.

Mariani *et al.* [819] reported on simultaneous TL-TSC measurements in Al_2O_3 which was X-irradiated at 80 K. Four peaks were found in the TL glow curve, and only two TSC peaks. At the low temperature range, TL occurred at 100 K, corresponding to a TSC peak at 105 K. The authors suggest that the following TL peaks at 220, 255 and 275 K, are associated with the same TSC maximum at 272 K. This would mean that for the last TL peak at 275 K, the TSC maximum preceded its TL counterpart. However, the situation is more complicated in the sense that this order changed to the expected one when the excitation dose was substantially decreased. Complementary work at lower temperatures was described by Böhm *et al.* [820] who reported simultaneous TL and TSC measurements in α-Al_2O_3 crystals X-irradiated at 10 K. A main pair of TL-TSC peaks was found at $\sim$60 K, where the TL maximum preceded its TSC companion by 1.5 K. Wong *et al.* [821] reported on TL-TSC measurements in Al_2O_3:Ti^{4+} (sapphire), UV irradiated at $\sim$100 K. Two TL peaks were found at $\sim$170 K and a major one at $\sim$267 K, which was explained as being associated with the 256 K TL maximum, with the shift being in the expected direction.

Martini *et al.* [822] studied the simultaneous TL and TSC results of $PbWO_4$ and found a close association of several TL and TSC peaks. The authors discussed the possible nature of the defects involved in the TL process, probably F- and F^+-centers.

Castiglioni *et al.* [823] reported on simultaneous measurements in synthetic crystalline quartz, X-irradiated at RT. TL peaks were found at 75 °C, 110 °C and 160 °C, accompanied by TSC peaks at 80 °C, 120 °C and 160 °C. For hydrogen-swept samples, additional TSC peaks were found at 180 °C and 275 °C, which were suggested to be of ionic nature.

Schrimpf *et al.* [824] reported the results of simultaneous TL-TSC measurements in a rather unusual material, namely, solid argon doped with gold or silver along with O_2 molecules in the temperature range of 5–30 K, following synchrotron or X-ray irradiation. A peak at $\sim$16 K was observed in the Au-doped material both in TL and TSC, with the former preceding the latter by $\sim$1 K. These authors used a first-order kinetics model to extract binding energies, and found values of the order of 50 meV.

Yukihara *et al.* [796] reported on simultaneous measurements of TL and TSC in Brazilian topaz irradiated by β-particles from a $^{90}Sr/^{90}Y$ source or with a 1.75 MeV van de Graaff electron beam. The TL and TSC peaks occur in the expected order, the former preceding the latter. The authors conclude that the recombination process occurs via the delocalized bands and that the recombination centers are characterized by thermal quenching of the luminescence in the temperature range of the lower two TL peaks. The dose dependencies of TL and TSC have also been studied, both showing some superlinearity. The TSC peak 1, however, started more superlinearly than the accompanying TL peak, but then around 1000 Gy it reached a maximum value and slightly decreased at higher doses (see discussion on non monotonic effects in Section 8.10). The authors do not have an explanation for this effect. Note that TL and TSC peaks may be sometimes uncorrelated in case one of the effects is due to the thermal release of electrons followed by recombination, and the other has to do with the motion of holes (see e.g. Sayer and Souder [825]).

11.2 Theoretical Considerations

Let us consider now the order of occurrence of TL-TSC companion peaks as expected from analytical considerations. From the three simultaneous differential equations (4.5–4.7) governing the transitions between a single trapping state, the conduction band and a recombination center, let us consider Equation (4.6) which is

$$I(t) = -\frac{dm}{dt} = Amn_c, \tag{11.1}$$

where A_m is the recombination probability coefficient which is now denoted by A for brevity. Equation (11.1) merely states that the rate of recombination is proportional to the concentrations of free electrons (n_c) and holes in centers (m), and that the TL intensity is equal, in suitable units, to this rate. Obviously, for the inverse case of hole-trapping states and electron recombination centers, an entirely analogous treatment can be given. Equation (11.1) has much more general bearing than just describing the situation when only one trapping state and one kind of recombination center are involved. As long as only transitions into one recombination center are *measured,* one does not care whether there are transitions into other centers. Such transitions may change the concentration of electrons in the conduction band, but Equation (11.1) holds true for the net instantaneous concentration n_c. Moreover, no information about the other levels contributing to n_c is needed as long as $n_c = n_c(t)$ can be measured simultaneously by a conductivity experiment.

Assuming that $n_c(t)$ is known independently, Equation (11.1) can be solved to yield

$$m = m_0 \exp\left[-A \int_0^t n_c(t')dt'\right], \tag{11.2}$$

where m_0 is the initial concentration of holes in centers. On inserting Equation (11.2) into (11.1), one obtains

$$I(t) = Am_0 n_c(t) \exp\left[-A \int_0^t n_c(t')dt'\right]. \tag{11.3}$$

It should be noted that the transformation of this equation to the conventional form of intensity versus temperature requires only the heating function $T = T(t)$. Among other possibilities, one can consider $T = $ constant, which in fact is the case of phosphorescence.

Chen [797] suggested the following method for evaluation of the parameter A from the mentioned simultaneous measurement of TL and TSC. Writing Equation (11.3) for two arbitrary points t_1 and t_2 and dividing one by the other, one has

$$\frac{I(t_2)}{I(t_1)} = \frac{n_c(t_2)}{n_c(t_1)} \exp\left[-A \int_{t_1}^{t_2} n_c(t)dt\right]. \tag{11.4}$$

All the quantities in Equation (11.4) except for A are measurable, and thus A can be evaluated from the equation. For the determination of $n_c(t)$, we have to measure the conductivity curve. This is given by $\sigma = e\mu n_c$, where e is the electron charge and μ the mobility. In many cases, the mobility and its dependence on temperature are known, and $n_c(t)$ follows immediately.

An alternative method for evaluating $n_c(t)$ is by thermally stimulated electron emission (TSEE) [795]. This method has the advantage that knowledge of the mobility is not needed,

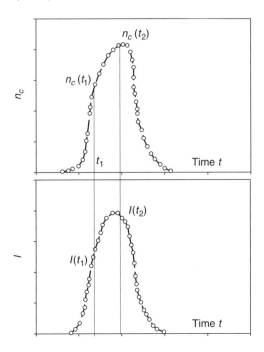

Figure 11.3 *Quantities measured to find the recombination probability coefficient. Chen, R., Simultaneous Measurement of Thermally Stimulated Conductivity and Thermoluminescence, J. Appl. Phys. 42, 5899. Copyright (1971) with permission from American Institute of Physics*

however experimental complications due to surface effects sometimes occur. Also, TSEE is usually limited to cases where the mobile carriers are electrons and not holes. An exception to this rule was mentioned by Tolpygo *et al.* [520] and also discussed by Huzimura and Matsumura [826] (see further discussion in Section 11.6).

Figure 11.3 shows schematically a peak of $n_c = n_c(t)$, the corresponding TL peak, the quantities $I(t_1)$, $I(t_2)$, $n_c(t_1)$, $n_c(t_2)$ and $\int_{t_1}^{t_2} n_c(t)dt$, which are necessary for evaluating the recombination probability A. It is to be noted that only relative values of $I(t)$ are needed, whereas absolute values of $n_c(t)$ are necessary for the integral. Some simplification can be achieved by choosing t_1 and t_2 such that $I(t_1) = I(t_2)$ or $n_c(t_1) = n_c(t_2)$. The main significance of this method is that the integration can be done in a limited range, as compared with the straight-forward method of finding A [264] by writing Equation (11.1) as $A = I(t)/[m(t)n_c(t)]$, and finding $m(t) = \int_t^\infty I(t')dt'$, namely from the area under the TL curve from t to infinity. The complication is that integration of the experimental curve to infinity is usually inaccurate. Moreover, the possible variations of A with temperature can be neglected only in limited ranges.

As an example, Chen [797] described the evaluation of the recombination probability coefficients in insulating and semiconducting diamonds. In the former, the value of A was found to be of the order of 10^{-5} cm^3 s^{-1}, and the cross-section for recombination was evaluated therefrom to be about 10^{-12} cm^2. In the semiconducting diamond the recombination probability coefficient was found to be $\sim 10^{-12}$ cm^3 s^{-1} and the cross-section for recombination $\sim 10^{-19}$ cm^2. The values were estimated only to an order of magnitude mainly because of experimental difficulties related to the geometrical shape of the samples and the uncertainty about the effective area of the electric contacts. To understand the large difference in the values for these two kinds of diamond, let us write Equation (11.1) as $A = I/(mn_c)$. One can see that if the values of I/m in the two samples differ by only a few orders of magnitude, as is the case here, the cross-sections are roughly inversely proportional to the conductivities. Thus, since the conductivity in semiconducting diamond is more than 10 orders of magnitude higher than in the insulator, a large difference between the cross-sections can be expected.

By equating the derivative of Equation (11.3) to zero, one has

$$\left(\frac{dn_c}{dt}\right)_m = A\,[n_c(t_m)]^2 , \qquad (11.5)$$

where the subscript m indicates that the values are taken at the maximum of the luminescence peak. The right-hand side of Equation (11.5) is always positive, and therefore, the left-hand side should also be positive. This means that when the TL peak reaches the maximum point, the curve of $n_c(t)$ is still increasing, so that $n_c(t)$ maximizes later (at a higher temperature) than $I(t)$. Thus, the TSEE peak would appear at a higher temperature than the corresponding TL peak. As for the accompanying TSC peak, since the dependence of the mobility on temperature is weak, it would also usually occur at a higher temperature than the TL peak, in agreement with the mentioned references. In fact, the order of occurrence of the TL and $n_c(T)$ peaks can be intuitively understood directly from Equation (11.1). The function $n_c(T)$ on the right-hand side is multiplied by $m(T)$ which, at least within the model of one trapping state and one recombination center (and actually in more complicated situations as well), must be a decreasing function of time and temperature. When a peak-shaped function is multiplied by a decreasing function, the result is usually another peak-shaped function which assumes its maximum at a lower value of the argument, hence the occurrence of the TL maximum before the $n_c(T)$ counterpart.

In some cases recombinations into more than one center contribute to the measured TL. Let the concentrations of holes in centers be $m_1, m_2, \ldots, m_k$ and the respective recombination probability coefficients $A_1, A_2, \ldots, A_k$. We now obtain

$$I_i(t) = -\frac{dm_i}{dt} = A_i m_i n_c. \qquad i = 1, 2, \ldots, k. \qquad (11.6)$$

Each equation can be solved independently using *the same* $n_c(t)$. The total measured emission would be the sum over i of the partial intensities $I_i(t)$. If the detector has different sensitivities for various i's (corresponding usually to various centers and therefore to various emission wavelengths), the I_i's are to be multiplied by constants α_i's, and

$$I(t) = \sum_{i=1}^{k} \alpha_i I_i(t) = \sum_{i=1}^{k} \alpha_i A_i m_{i0} n_c(t) \times \exp\left[-A_i \int_0^t n_c(t')dt'\right]. \qquad (11.7)$$

The condition for maximum would now be

$$\sum_{i=1}^{k} \alpha_i A_i m_{i0} \exp\left[-A_i \int_0^{t_m} n_c(t)dt\right] \times \left\{\left(\frac{dn_c}{dt}\right)_m - A_i\left[n_c(t_m)^2\right]\right\} = 0, \tag{11.8}$$

or

$$\alpha\left(\frac{dn_c}{dt}\right)_m = \beta\left[n_c(t_m)\right]^2, \tag{11.9}$$

where

$$\alpha = \sum_{i=1}^{k} \alpha_i A_i m_{i0} \exp\left[-A_i \int_0^{t_m} n_c(t)dt\right] \tag{11.10}$$

and

$$\beta = \sum_{i=1}^{k} \alpha_i A_i^2 m_{i0} \exp\left[-A_i \int_0^{t_m} n_c(t)dt\right]. \tag{11.11}$$

α and β are sums of positive terms and therefore are positive. The condition is thus similar to that obtained in Equation (11.5) and bears similar implications, namely that even under these general conditions, the TL peak should precede the corresponding TSC or TSEE ones. Opanowicz [827] has further studied the possibility of determining the recombination probability coefficient from simultaneous TL-TSC measurements, using specific models of thermal release of electrons from several kinds of active traps and recombination in a single center. In a later work, Opanowicz and Pietrucha [828] presented a method based on measurements at various heating rates, of determining the ratio between the recombination and trapping coefficients and the relative density of inactive traps by using simultaneous measurements of TL and TSC.

In a relatively small number of TL-TSC simultaneous measurements, the opposite order of appearance occurs, namely the TSC peak precedes the TL companion peak. As pointed out by Chen and McKeever [194], this may be just sheer coincidence, i.e., that the TL and TSC peaks originate from different sources. For example, it is possible that one is related to the release of trapped electrons and the other of trapped holes, and the two peaks just happen to occur in the same temperature range. Another possibility is that the TSC peak is distorted due to contact or space charge effects as discussed by Henisch [829] and others. A third possibility considered by Chen and Fleming [830], takes into account the possible influence of the temperature dependence of the recombination probability A and the mobility μ. As noted, the recombination probability coefficient A is the product of the cross-section for recombination and the thermal velocity. According to Bemski [55], Keating [258] and Lax [57], the former varies like $T^{-\kappa}$ with $0 \leq \kappa \leq 4$. The thermal velocity is proportional to $T^{1/2}$. One can therefore write $A = BT^a$ where $-7/2 \leq a \leq 1/2$. Equation (11.5) will be written in the general case of $A = A(T)$ [and through the heating function, $T = T(t)$, we also have $A = A(t)$], as

$$(dn_c/dt)_{max} = A_{max}n_{cmax}^2 - n_{cmax}(d\ln A/dt)_{max}, \tag{11.12}$$

where the subscript "max" is, in this context, the value of the recombination probability coefficient at the TL maximum. Equation (11.12) obviously reduces to Equation (11.5) for a temperature-independent recombination probability coefficient. For any monotonically increasing heating function $T = T(t)$, Equation (11.12) can be written as

$$(dn_c/dt)_{max} = A_{max}n_{cmax}^2/\beta_{max} - n_{cmax}(d \ln A/dT)_{max}, \tag{11.13}$$

where $\beta_{max} = (dT/dt)_{max}$ is the instantaneous heating rate at the TL maximum point. In order to determine whether the TL peak precedes the peak of $n_c(T)$ or not, it is necessary to examine the right-hand side of Equation (11.13). If this expression is negative, the $n_c(T)$ peak must occur at a lower temperature than that for TL. When $A = BT^a$ (and where B is constant), the right-hand side of Equation (11.13) is negative if

$$A_{max}T_{max}n_{cmax} < a\beta_{max}. \tag{11.14}$$

A necessary condition for this inequality to hold is $a > 0$. Chen and Fleming [830] gave a hypothetical, though possible, example in which the recombination cross-section is temperature independent, and therefore $a = 1/2$. Choosing the cross-section to be $S_m = 10^{-22} \, \text{m}^2$, $T_{max} = 300 \, \text{K}$ and $\beta_{max} = 0.5 \, \text{K s}^{-1}$, any $n_{cmax} < 7 \times 10^{13} \, \text{m}^{-3}$ would yield $(dn_c/dt)_{max} < 0$, implying the occurrence of the maximum of the $n_c(T)$ peak at a temperature lower than that of the corresponding TL peak.

In the present discussion, the TL and $n_c(T)$ peaks have been compared. The latter can be evaluated from conductivity measurements provided that the dependence of the mobility on temperature $\mu = \mu(T)$ is independently known, and for the measured conductivity $\sigma(T)$, the quantity $n_c(T) = \sigma(T)/[e\mu(T)]$ is calculated. If one wishes to compare peak temperatures of TL and TSC directly, the possible dependence of the mobility on temperature should be considered. According to Lax [57], the mobility of free carriers is given by $\mu = \mu_0 T^b$ where μ_0 is constant and b assumes value such as $-3/2$. Such a dependence of $\mu \propto T^{-3/2}$ was found, for example, by Benedict and Shockley [831] for electrons in germanium using contactless microwave measurements. Combining Equation (11.1) and $\sigma = e\mu n_c$, one gets

$$I(t) = -dm/dt = (A/\mu e)\sigma m = A'\sigma m \tag{11.15}$$

where $A' = A/\mu e$. Using the above mentioned dependence of A and μ on temperature one finds

$$A' = A/\mu e = (B/\mu_0 e)T^c \tag{11.16}$$

where $c = a - b$, and where $B/\mu_0 e$ is constant. One therefore has $-2 \leq c \leq 2$ for the previously mentioned temperature dependence of the recombination probability, and with the rather common case of $b = -3/2$. Because of the close similarity between Equations (11.1) and (11.16), the conclusions drawn from the former can be carried over to the latter [830]. Thus, the equation analogous to Equation (11.13) is

$$(d\sigma/dt)_{max} = A'_{max}\sigma_{max}^2/\beta_{max} - \sigma_{max}\left(d \ln A'/dT\right)_{max} \tag{11.17}$$

where A'_{max} is the value of A' at the TL maximum and σ_{max} the conductivity at the maximum. The order of occurrence of the TL and TSC peaks is governed by the sign of the right-hand side of Equation (11.17). If the expression is negative, and assuming that A' is given by

Equation (11.16), the condition for the TL peak to occur at a higher temperature than the $\sigma(T)$ counterpart is analogous to expression (11.14), namely

$$A'_{max} T_{max} \sigma_{max} < c\beta_{max}. \tag{11.18}$$

This is somewhat less stringent than inequality (11.14) since the range of possible positive values of c is broader because the necessary condition now is $c > 0$.

In summary, usually a TL peak and its counterpart $n_c(T)$ peak do not occur at exactly the same temperature. The normal case is that the TL peak appears at a lower temperature, which is substantiated by theoretical considerations and by experimental results. The situation is quite similar if one compares the $I(T)$ and $\sigma(T)$ curves, although the inclusion of the temperature-dependent mobility shifts $\sigma(T)$ to a slightly lower temperature than that of $n_c(T)$.

A different explanation for the inversion of the order of appearance of TL and TSC peaks has been given by Oczkowski [832] for the case of semiconductors. Having in mind mainly materials like Zn-annealed ZnSe crystals, Oczkowski used a model with a deep acceptor (A), deep donor (D), shallow donor (d) and a complex defect which is a metastable deep donor–acceptor (DA) pair. Using the DA model he has shown that for such semiconductors and under the mentioned circumstances, the inversion of the TL and TSC peaks temperatures comes out as a direct result of the trapping and compensation of access minority carriers. In a later work Opanowicz [833] challenged this model, and suggested that the TL and TSC in conductive ZnSe should be explained on the basis of the kinetic model with a deep electron trap, developed for a semi-insulating material. Work by Lee *et al.* [834] and McKeever *et al.* [227] on the simultaneous measurements of TL and TSC peaks in ZnSe showed that most of the peaks occur in the "correct" order, namely the TL peak precedes its TSC companion.

The relationship between TL and TSC has further been studied by Gasiot and Fillard [835, 836] and by Fillard *et al.* [837] and Gasiot *et al.* [838]. Observing Equation (11.1), one can immediately see that $I(t)/n_c(t)$ is proportional to the concentration m, and therefore its derivative with respect to time should be proportional to the TL intensity, $I(t)$. However, while comparing these two functions for the experimental results in SnO_2, a large discrepancy is observed. They conclude that something must be wrong with the assumption that $I(t) = -dm/dt$. This may indeed be the case in situations like the one mentioned in Section 6.11. In this case, the concentration of trapped holes in centers varies due to the capture of free holes and the recombination with free electrons, whereas only the latter is assumed to contribute to the measurable TL. Mandowski and co-workers [839–841] have further studied this subject. Mandowski and Świątek [839] presented a method for evaluation of the density of deep traps by the use of simultaneous measurements of TL and TSC. They studied a model of p active trapping states, k recombination centers and one or more kinds of deep disconnected traps. The relevant set of equations here is

$$-\frac{dn_i}{dt} = n_i D_i - n_c A_i (N_i - n_i), \qquad i = 1, ..., p \tag{11.19}$$

$$-\frac{dm_s}{dt} = B_s m_s n_c, \qquad , s = 1, ..., k \tag{11.20}$$

$$\sum_{s=1}^{k} m_s = \sum_{s=1}^{p} n_i + n_c + M, \tag{11.21}$$

where

$$D_i = s_i \exp(-E_{ti}/kT) \tag{11.22}$$

and where n_i and N_i are the level of filling and the capacity of the ith trap, respectively, A_i is the trapping probability coefficient of the ith trap, E_{ti} and s_i are its activation energy and frequency factor, m_s and B_s are the occupancy of holes in the sth center and the recombination probability into it respectively, n_c is the instantaneous concentration of free electrons in the conduction band, and M stands for the concentration of electrons in thermally disconnected deep traps. M is considered to be constant which means that the relevant trap or traps can neither release electrons thermally, nor can they trap free electrons, either due to a small trapping probability or since they are already full to capacity.

In the same way as done above with the solution of Equation (11.1), namely addressing $n_c(t)$ as a known function of time (or temperature) which yielded Equation (11.2), Mandowski and Świątek solved each of the equations in expression (11.19) for $i = 1, ..., p$ and found

$$n_i(t) = N_i \exp \left\{ -\int_0^t [D_i(\tau) + n_c(\tau)A_i]\, d\tau \right\}$$
$$\times \left(\eta_{i0} + \int_0^t n_c(\tau)A_i \exp \left\{ \int_0^\tau [D_i(\tau') + n_c(\tau')A_i]\, d\tau' \right\} d\tau \right), \tag{11.23}$$

where $\eta_{i0} = n_{i0}/N_i$ is the relative initial filling of the ith trap which can assume values between 0 and 1. Equation (11.2) is now extended to be k solutions of Equation (11.20), namely

$$m_s = m_{s0} \exp \left[-B_s \int_0^t n_c(\tau)d\tau \right], \quad s = 1, ..., k \tag{11.24}$$

where m_{s0} is the initial occupancy of the sth center. Mandowski and Świątek [839] defined the function

$$C(t) = \int_0^t n_c(\tau)d\tau \tag{11.25}$$

and its limit for very large values of t,

$$s = \lim_{t \to \infty} C(t). \tag{11.26}$$

Combining Equations (11.21), (11.23) and (11.25), they found the integral equation

$$\sum_{s=1}^{k} m_{s0} \exp(-B_s C) = \sum_{i=1}^{p} N_i \exp(-A_i C) \exp\left(-\int_0^t D_i d\tau\right)$$

$$\times \left[\eta_{i0} + \int_0^t A_i \dot{C} \cdot \exp(A_i C) \exp\left(\int_0^\tau D_i d\tau'\right) d\tau\right] + \dot{C} + M, \qquad (11.27)$$

where, according to Equation (11.25), $\dot{C} = n_c$. The solutions of these equations are expected to be the same as those of the set of equations (11.19–11.22). However, one should note that merely being able to write one equation instead of $k + p + 1$ simultaneous equations does not really help in solving them or even to draw some useful conclusions from them. Mandowski and Świątek [839] first turned to the relatively simple case of $p = 1$ and $k = 1$, and showed that for $t \to \infty$, one should take

$$\lim_{t\to\infty} n_c(t) = \lim_{t\to\infty} \dot{C}(t) = 0, \qquad (11.28)$$

and therefore, Equation (11.27) reduces to

$$m_0 \exp(-BS) = M. \qquad (11.29)$$

The initial condition in this case is $m_0 = n_0 + M$, and with this, Equation (11.29) can be written as

$$S = \frac{1}{B} \ln\left(1 + \frac{n_0}{M}\right), \qquad (11.30)$$

which yields the dependence of the area under the TSC curve on the initial concentration of trapped carriers. A slightly different way of writing this expression is

$$S(\eta_0) = \frac{1}{B} \ln\left(1 + \frac{\eta_0}{\omega}\right) \qquad (11.31)$$

where $\eta_0 = n_0/N$ and $\omega = M/N$.

Mandowski and Świątek [839] extended the discussion to the more interesting case of one kind of recombination center and a series of trapping states. Instead of Equation (11.30) they now obtained

$$S = \frac{1}{B} \ln\left(1 + \sum_{i=1}^{p} n_{i0}/M\right). \qquad (11.32)$$

Denoting now $\sum_{i=1}^{p} n_{i0}/\sum_{i=1}^{p} N_i$ by η and $M/\sum_{i=1}^{p} N_i$ by ω, they found the same relation as Equation (11.31) with η replacing η_0, namely

$$S(\eta) = \frac{1}{B} \ln\left(1 + \frac{\eta}{\omega}\right). \qquad (11.33)$$

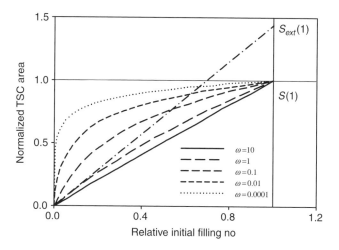

Figure 11.4 *Plot of the area S against the relative initial filling of traps η_0 for different relative populations of carriers in thermally disconnected traps ω. Values of $S(1)$ and $S_{ext}(1)$ denote, respectively, the TSC area for $\eta_0 = 1$ directly measured in the experiment and that extrapolated from the linear initial part of the curve. Reprinted from Mandowski, A. and Świątek, J., A method for determining the density of deep traps by using the simultaneous TL/TSC measurement technique, J. Phys. D: Appl. Phys. 25, 1829–1833. Copyright (1992) with permission from Institute of Physics*

From the simultaneous measurements of TL and TSC, one can determine S from the TSC curve and η from the TL curve [839]. The plot of $S(\eta)$ can therefore be evaluated experimentally and compared with Figure 11.4. By curve fitting, the best of the curves in the figure can be chosen, and ω can be determined. Alternatively, one can extrapolate the linear range occurring at low values of η_0/ω so as to get $S_{ext}(1)$ as seen in Figure 11.4. One gets now

$$\frac{S(1)}{S_{ext}(1)} = \omega \ln\left(1 + \frac{1}{\omega}\right), \tag{11.34}$$

and once S_1 and $S_{ext}(1)$ are known experimentally, Equation (11.34) can be solved numerically to yield ω. In addition, one can evaluate η_0/ω by measuring two areas S_1 and S_2 for the relative initial populations of carriers η_1 and η_2, respectively. In cases where the parameters $a = \eta_2/\eta_1$ and $\chi = S_1/S_2$ are known, the ratio $\eta_0/\omega (= n_0/M)$ can be found numerically from

$$\left(1 + a\frac{\eta_0}{\omega}\right)^\chi = \left(1 + \frac{\eta_0}{\omega}\right). \tag{11.35}$$

Thus, the solution of Equations (11.34) and (11.35) together yields both $\eta_0 = n_0/N$ and $\omega = M/N$.

11.3 Numerical Analysis of Simultaneous TL-TSC Measurements

In a later work, Mandowski *et al.* [840] describe the numerical analysis of simultaneous TL-TSC measurements. Following Gasiot and Fillard [836], Mandowski defined the ratio

$$R(t) = J_{TL}/J_{TSC}, \tag{11.36}$$

where J_{TL} and J_{TSC} are the TL and TSC intensities, respectively. As pointed out above, the TL intensity which is denoted here by J_{TL} is associated with $-dm/dt$ [see e.g. Equation (4.6)]. However, the intensity can be slightly dependent on temperature through the radiative efficiency; Mandowski assumed that an appropriate correction can be made so that

$$J_{TL} = -\frac{dm}{dt}/\chi_L, \tag{11.37}$$

where χ_L is a constant. He also assumed that a correction can be made to the temperature dependence of the mobility so that the TSC intensity is directly related to the concentration of free carriers n_c,

$$J_{TSC} = n_c/\chi_C, \tag{11.38}$$

where χ_C is another constant. Using Equation (4.6) (and denoting the recombination-probability coefficient by B), Equation (11.36) can be written as

$$R(t) = \frac{\chi_C B}{\chi_L} m, \tag{11.39}$$

and therefore, $dR/dt \propto -J_{TL}$ as previously shown by Gasiot and Fillard [836]. Thus, in the framework of the STM, $R(t)$ must always be a decreasing function. A numerically generated example is shown in Figure 11.5.

These authors discussed the validity of the quasi-steady condition by numerical simulation, and for a variety of trap parameters [842, 843]. In analogy with Equation (11.30), they derived the approximation

$$S(t) = \int_0^t n_c(t')dt' \approx \frac{1}{B} \ln \left[1 + \frac{n_0 - n(t)}{M + n(t)} \right], \tag{11.40}$$

which could be solved for $n(t)$,

$$n(t) = (M + n_0) \exp\left[-BS(t)\right] - M, \tag{11.41}$$

and differentiating with respect to t yields

$$\dot{n} = Bn_c(t)\left[n(t) + M\right]. \tag{11.42}$$

With the solution of Equation (4.6),

$$m(t) = m_0 \exp\left[-BS(t)\right], \tag{11.43}$$

and using Equation (4.7), Mandowski *et al.* [840] show that Equations (11.42) and (11.43) are equivalent to the classical quasi-steady (QS) conditions, i.e., $n_c \ll n$ and $|\dot{n}_c| \ll |\dot{n}|$. Mandowski *et al.* [833, 840, 842, 843] report that the validity of the QS conditions was

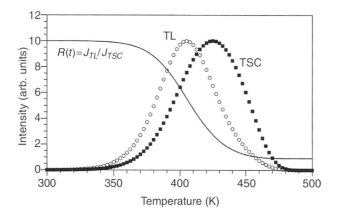

Figure 11.5 *Typical TL-TSC curves and the ratio $R(t) = J_{TL}/J_{TSC}$ simulated in the framework of the STM defined by Equations (4.5–4.7). The curves were simulated with $E = 0.9\,eV$; $s = 10^{10}\,s^{-1}$ and the heating rate $\beta = 1K\ s^{-1}$. Reprinted from Mandowski, A., Świątek, J., Iacconi, P. and Bindi, R., Numerical Analysis of Simultaneous TL/TSC Measurements, Radiat. Prot. Dosim. 100, 187–190. Copyright (2002) with permission from Oxford University Press*

tested numerically for a variety of trap parameters and suggest that QE probably prevails in all common phosphors.

Using Equations (11.40) and (11.41), the authors were able to write equations for TL and TSC as generalized initial-rise expressions. Furthermore, it was numerically verified that for most cases the validity of these approximations covered the whole measurable range, much beyond the initial-rise region. The generalized initial-rise TL equation is given as

$$-\frac{E}{kT(t)} = \ln\left[L(t)\right] + \ln\left\{\frac{1}{M + n_0 - U(t)}\left[\frac{1}{n_0 - U(t)} + \frac{B - A}{AN + BM}\right]\right\} + \ln\left(\frac{AN + BM}{vB}\right).$$

(11.44)

where the function $L(t)$ is defined as

$$L(t) = -\dot{m}(t) = \chi_L J_{TL},$$

(11.45)

and $U(t)$ is

$$U(t) = \int_0^t L(t')dt' = m_0 - m(t) = n_0 + M - \frac{\chi_L}{\chi_C B}R(t),$$

(11.46)

where m_0 and n_0 indicate the initial values of m and n, respectively. Inserting Equations (11.45) and (11.46) into Equation (11.44) yields

$$\ln\left[J_{TSC}(t)\right] + \ln\left[\delta + \frac{1}{R(t)/R_f - 1}\right] = -\frac{E}{kT} - \ln K,$$

(11.47)

where δ, R_f and K are constants,

$$\delta = \frac{1-r}{1+r/\omega},\tag{11.48}$$

$$R_f = \frac{BM\chi_C}{\chi_L},\tag{11.49}$$

$$K = \frac{B\chi_C}{\nu}(1+r/\omega).\tag{11.50}$$

In these equations, $\omega = M/N$ denotes the relative concentration of deep traps and $r = A/B$ is the ratio of retrapping and recombination coefficients. For given TL-TSC measurement data, the plot of the left-hand side of Equation (11.50) as a function of $1/kT$ should yield a straight line. Mandowski *et al.* [840] suggest that as a measure of linearity, one may use the correlation coefficient, which in the present case depends only on the two parameters δ and R_f. Thus, it is easy to fit the parameters so as to minimize the correlation coefficient r_c, which should be as close as possible to -1. The authors point out that Equation (11.47) coincides with a similar one previously derived by Bindi *et al.* [516] for the second-order kinetics. This coincidence occurs in the case of $R(t) \gg R_f$, which is evident for a low number of thermally disconnected traps, $\omega \ll 1$ and for $\delta \ll 1$ which corresponds to $r \geq 1$.

Mandowski *et al.* [840] present an example of fitting Equation (11.47) using experimental TL-TSC curves in α-alumina. They use a Monte-Carlo method previously mentioned by Mandowski and Świątek [844] in order to find a global minimum of the difference between experimental and theoretical values, within the physically allowed range of parameters. The basic idea here is to generate pairs of the parameters (δ, R_f) for which the correlation coefficient r_c is calculated. In all the considered cases, the number of 10^6 trials was found to be high enough to yield accurate results. Each point (δ_i, R_{fi}) determines the value of the activation energy E_i which is calculated using a standard linear regression method. The final values of E, δ and R_f are estimated by averaging several best-fit results, namely points for which r_c is closest to -1.

Mandowski *et al.* [840] applied the proposed method to the analysis of simultaneous TL-TSC measurements in an α-alumina undoped single crystal. The measured curves are shown in Figure 11.6 along with the $R(t)$ function. The authors state that in order to analyze the data, one has to estimate the applicability of the STM, namely that only one trapping state and one kind of recombination center are involved in the process within a certain temperature range [see Equations 4.5–4.7]. The necessary condition for this is that $R(t)$ be a decreasing function of time. In Figure 11.6, the ranges where $R(t)$ is indeed decreasing are denoted by thick lines. The two ranges of temperature where this takes place, i.e. 280–350 K and 500–590 K correspond to the well known peaks B' and D' of α-alumina. The activation energy obtained for B' is $E_{B'} = 0.53 \pm 0.04$ eV which is in agreement with previous works [799, 845], and $\delta_{B'} = 0.07 \pm 0.10$. The second peak D' is more problematic because the determined values are more scattered. Mandowski *et al.* [840] report that the analysis in a broader range of 500–580 K yields $E_{D'} = 0.95 \pm 0.04$ eV, slightly smaller than previous results. The analysis performed inside the region, from 520 to 560 K gives much higher

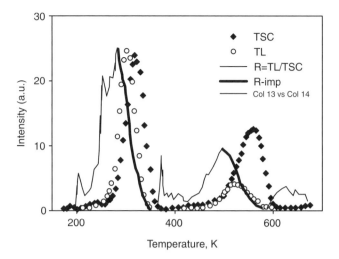

Figure 11.6 *TL-TSC curves and the ratio R(t) measured from an α-alumina single crystal. The thick line shows the regions where the function R(t) decreases. Reprinted from Mandowski, A., Świątek, J., Iacconi, P. and Bindi, R., Numerical Analysis of Simultaneous TL/TSC Measurements, Radiat. Prot. Dosim. 100, 187–190. Copyright (2002) with permission from Oxford University Press*

values of $E_{D'} = 1.3$–2.4 eV. The authors conclude that using all the available data, the values of $E_{D'} = 1.3 \pm 0.5$ eV and $\delta_{D'} = 0.25 \pm 0.23$ are obtained.

Opanowicz [846] has performed numerical simulations of TL and TSC by solving the relevant simultaneous equations associated with complex systems including a number of trapping states. He found that due to the possible interaction between traps, electrons thermally released from traps of one kind may be trapped by traps of another kind, and thus may influence peaks occurring at higher temperatures. This process of transfer was shown to alter the features of the TL and TSC peaks and therefore, the accuracy of some of the methods for evaluating the activation energy, in particular the initial-rise and peak-shape methods, is reduced. One should also mention the work by Kulkarni [847] who used a Monte-Carlo simulation (see also Section 10.2) for the simultaneous calculation of TL and TSC.

11.4 Thermoluminescence and Optical Absorption

In many cases, the concentration of trapped carriers may be associated with the optical absorption (OA) of a crystal. Figure 11.7 shows the results of simultaneous measurements of a glow curve and an absorbance curve of an X-ray excited calcite sample as measured in the range of 260–350 nm by Medlin [233]. The main maximum of TL at 360 K is clearly seen to be in the same temperature range as the fast change of absorbance. This indicates that for the derivative of the concentration with respect to temperature, one has $I \propto -dn/dT$ where n is the concentration. For a constant heating rate, this is equivalent to $I \propto -dn/dt$, which is expected [see e.g. Equation (4.10)]. Medlin reported that the thermal bleaching of

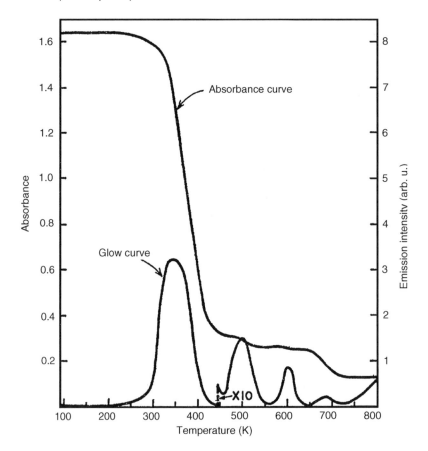

Figure 11.7 *Bleaching curve and simultaneously measured glow peak in calcite. Reprinted from Medlin, W.L., Phys. Rev. 135, A1770–1779. http://prola.aps.org/abstract/PR/v135/i6A/pA1770_1. Copyright (1964) with permission from American Physical Society*

these color centers occur in four stages, which is seen in the main variation of the absorbance curve at ~380 K, and by the three wiggles at higher temperatures and, in parallel by the presence of the main TL peak and three smaller ones in the glow curve. More detailed work by Medlin on the calcite samples, based on polarized absorption spectra, indicated that the absorption bands are due to four kinds of trapped electron centers and one kind of trapped hole center. TL is produced when the recombination energy excites Mn^{2+} impurity ions.

Braner and Halperin [848] reported the results of a similar investigation of X-ray colored KCl crystals. Figure 11.8 shows some of their results where curve (a) is TL, and curve (b) is the negative first-order derivative of the absorption curve with respect to temperature. The agreement between the temperatures of occurrence of the peaks is very good. Scaramelli [849] described the correlation between TL and photo-transferred TL (PTTL) on one hand, and OA on the other hand in F' and M' traps in alkali halides.

Thomas and Houston [172] reported on simultaneous measurements of TL and OA in X-irradiated MgO. Following excitation at RT, TL peaks were found at 410 and 520 K.

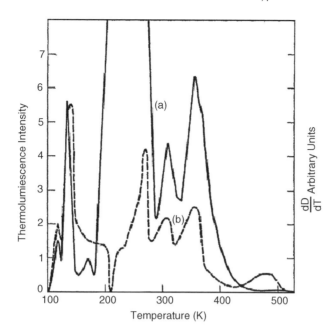

Figure 11.8   *(a) Glow curve and (b) temperature derivative of the thermal bleaching curve of the same crystal of X-irradiated KCl. Reprinted from Braner, A.A. and Halperin, A., Thermal Bleaching of F Centers and Its Correlation to Thermoluminescence in X-Ray Colored KCl Crystals, Phys. Rev. 108, 932–936. http://prola.aps.org/abstract/PR/v108/i4/p932_1. Copyright (1957) with permission from American Physical Society*

Absorption peaks at 2.3 and 3.0 eV were shown to be linked with the 410 and 520 K peaks, respectively. Diéguez and Cabrera [850] studied the TL and OA of X-irradiated KH_2OP_4 (KDP) crystals. A strong dichroism is known to occur at low temperatures in the absorption spectra of the $\parallel c$ plates, and therefore, Diéguez and Cabrera [850] measured the thermal bleaching of the σ and π polarizations separately. This was done by recording the decay of the optical density maxima at 550 nm and 510 nm, respectively, on warming up the crystals at the same rate of $\sim 5\,K\,min^{-1}$ as used in the TL experiment, following an excitation at 13 K. The results are shown in Figure 11.9 where quite different behaviors can be observed for the σ and π polarizations. In the π case, the absorption remains constant up to 75 K, then it decreases rather quickly until $\sim 125\,K$ and bleaches out completely at about 230 K. The σ absorption has two steps of rapid decrease at ~ 40 and $\sim 100\,K$; from 220 K up, no OA is detected. Two glow peaks centered at ~ 73 and $\sim 123\,K$ are seen. The 73 K peak seems to correlate with the σ polarization curve and the 123 K peak with both the σ and π polarization curves. It should be noted that the latter temperature coincides with the Curie temperature of KDP. The dependence of the TL peaks on the irradiation dose is similar to that found for the optical density. The authors [850] explain their results as due to the existence of two centers, which they denote σ and π centers. They suggest that the σ center does not contribute to the π absorption, whereas the π center does make some contribution to the σ absorption. Apparently, the σ center is associated with the B

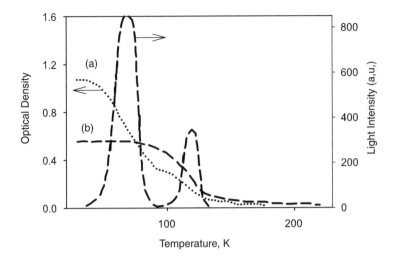

Figure 11.9 *Thermal bleaching curves of a ∥ c plate in X-irradiated KDP. (a) σ polarization, λ = 550 nm. (a) π polarization , λ = 510 nm. The TL glow peaks are shown as dashed lines. The irradiation time for both experiments was 360 min. Diéguez, E. and Cabrera, J.M., Optical absorption and thermoluminescence of X-irradiated KDP, J. Phys. D: Appl. Phys. 14, 91–97. Copyright (1981) with permission from Institute of Physics*

radical, $\dot{P}O_4H_2$ in KDP. They also suggest that since the light emission of the π center occurs at T_c, a close relationship seems to exist between the annihilation and the phase transition mechanism.

Hageseth [851] has studied the TL and OA of X-irradiated KBr crystals. A peak at 152 °C is proposed to be due to the liberation and recombination of electrons from F-centers. The OA measurements indicate that F- and V_2-centers are formed in the crystal during the X-irradiation. TL and OA in X-irradiated KBr were also discussed by Bangaru *et al.* [852]. Delgado and Alvarez-Rivas [853] have studied TL and the annealing of OA in NaCl:Cu. They found a one-to-one correspondence between the TL peaks and the annealing steps of the F and Cu$^-$ absorption band, and with anion vacancies. In the latter case short lived V_k-centers are formed, which in turn recombine with Cu$^-$ ions; thus, the crystal lattice is brought back to its pre-irradiation state. Bandyopadhyay and Summers [854] studied the OA and TL of the same materials at higher temperatures. Here the OA decreases rather abruptly between 350K and 430 K, and is entirely annealed above 433 K. Some difference was observed between unbleached samples and those bleached to remove F-centers before heating. In both samples a TL peak occurring at ∼450 K seemed to be uncorrelated with the OA. In the bleached sample, another peak was observed at ∼415 K which is in the range of a strong decrease in the absorption. López *et al.* [855, 856] investigated the TL and OA of NaCl:Mn^{2+}. They found close correlation between the TL peaks at ∼170 K and ∼220 K, and a strong decrease in the absorption with temperature in the same temperature range. Delgado and Alvarez-Rivas [857] performed more measurements on this material. They explained their results as being due to the recombination of thermally released interstitials with anion vacancies. The resulting short-lived V_k-centers recombine with electrons trapped at impurities, thus yielding TL. Simultaneously, F-centers are annealed by luminescent recombination with interstitials. Similar measurements were reported by Prasad and Rao [858] in laser excited KCl crystals previously irradiated by X-rays. They found two TL peaks at 143 °C and 255 °C, both associated with the destruction of the F-centers, previously formed by the X-ray irradiation. Jiménez de Castro and Alvarez-Rivas [859] also reported on measurements of TL, TSC and OA of X-irradiated KCl. They concluded that eight out of nine glow peaks found between 80K and 300 K are due to recombination of mobile interstitial-halogen atoms which are thermally released from traps with F centers. Self-trapped excitons are formed in the light-emitting stage of this recombination. The ninth glow peak is due to holes liberated from traps which might be V_k centers. Murti and Murthy [860] also reported on TL and OA measurements in some alkali halides. In KCl:Cu, TL peaks at 207 and 243 K correlated with OA associated with V_k- and Cu^0-centers, respectively. Further discussion on TL of colored alkali halides has been given by Murti and Sucheta [861].

Jackson and Harris [862] reported on the correlation between TL and OA in LiF. Correlations have been made of the 3.23 eV absorption band to the glow peaks at 115 °C and 160 °C, and the 3.99 eV absorption band to the peaks at 195 °C and 210 °C. Sagastibelza and Alvarez-Rivas [863] reported on the results of TL and optical absorption measurements in X-irradiated LiF:Mg, Ti (TLD-100) samples. They studied the correlation between the TL curve and the negative derivative of the absorption curve, and found very close resemblance between them. These authors [863] could identify this correlation between the corresponding curves, both in the F and M absorption bands. Further extensive work on the relationship between TL and OA in TLD-100 was performed by Issa *et al.* [864, 865]. They identify the 5.45 eV OA band as a likely candidate for the recombination stage competitor for composite peak 5 supralinearity. An additional OA band at 5.7 eV is also identified as a possible recombination stage competitor. Nail *et al.* [635] also discussed the correlation between the absorption bands and the glow peaks in LiF:Mg, Ti. Weiss *et al.* [866, 867] gave a recombination kinetic model of the LiF:Mg, Ti glow peak 5 TL system. Using numerical simulation, they demonstrate the connection between the TL peaks and the change in OA. They find supralinear dose dependence of the maximum TL on the dose and find variations in the supralinearity with varying heating rate. Further work on LiF:Mg, Ti (TLD-100) TL and OA has been given by Lakshmanan and Madhusoodanan [868] who studied the two effects following irradiations at different temperatures between 77 K and 523 K. Petö and Kelemen [869] and Petö [870] also described the relation between TL and OA in LiF:Mg, Ti. They show that the 310, 350 and 380 nm absorption bands strongly correlate with peaks 2 and 3, and show no significant bleaching at temperatures above that of peak 3. Nakajima [871] compared the results of TL and thermal bleaching of color centers in LiF. The author did not find a close correlation between the TL peaks and the color centers in LiF. Finally, Yazici [872] presented TL-like curves of $-d\alpha/dT$ in LiF:Mg, Ti as a function of temperature, where α is the absorption coefficient.

Jain [873] has studied charge-carrier trapping and TL in CaF_2-based phosphors. For CaF_2:Gd he shows very close similarity between the TL peaks and the negative temperature derivative of the OA at 315 and 540 nm. Sastry and Moses-Kennedy [874] reported on the simultaneous measurements of TL and OA of SrF_2:Pb; they found that a TL peak at ~653 K occurs in the same temperature range as the strong decline of the ESR signal.

Martín [875] *et al.* reported on the results of a study of X-irradiation effect on tetrachlorides. Both in Rb_2ZnCl_4 and $[(CH_3)_4N]_2 ZnCl_4$, also termed $(TMA)_2ZnCl_4$, a major TL peak at 200 K in the former and 150 K in the latter, is associated with the very fast decrease of the OA in the respective temperature-dependent curves.

Lecoq *et al.* [876] studied the effects of different irradiations on $Bi_4Ge_3O_{12}$ (BGO) monocrystals. They used UV light, ^{60}Co γ-rays as well as 18 MeV electrons for excitation, and studied simultaneously the variation of OA with temperature, TL and TSC. In some samples they found a TL peak at 100 °C associated with a strong decrease of the optical absorption. Another decrease of the absorption was accompanied by a smaller TL peak at $\sim$140 °C, as well as a major TSC peak at $\sim$150 °C. The authors suggest that the two TL peaks are related, one to a localized transition and the other to holes released into the valence band. The latter is associated with the mentioned TSC companion, observed when holes are in the valence band.

Baba *et al.* [877] investigated the simultaneous results of TL and OA of $NaBr:S^{2-}$ crystals. They found that the glow peak with an IR emission of 815 nm appears at 230 K and is accompanied by a significant change in the absorption. The authors ascribe the effect to the formation and decomposition of F_H-centers, namely, F-centers associated with a neighboring sulfur ion. Pokorný and Ibarra [878] studied TL and OA in X-ray excited $Al_2O_3:Cr,Ni$. The dominant TL peak at 330 °C was shown to be associated with a strong decrease of the absorption bands at 6.1, 4.4 and 2.7 eV as a function of annealing temperature. It was proposed that the high-intensity peak is due to oxygen vacancies, induced by the presence of Ni, and that the main recombination center is caused by Cr. Cruz-Zaragoza *et al.* [879] reported on optical absorption and TL in single NaCl:Cu crystals following exposure to ^{60}Co and UV light. Optical bleaching with F light showed that the F-centers are the main defects responsible for the TL emission. They suggest that the electron released from F-centers by optical or thermal excitations are eventually captured by different copper complex centers. Dolivo and Gäumann [880] described the measurements of TL, TSC and OA of methylcyclohexane (MCH) γ-irradiated at 77 K. The TL and TSC peaks were found to occur at nearly the same temperatures. As for the OA, during the heating of the irradiated sample, the 1650 nm absorption disappeared around 93 K and the absorption at 255 nm at about 98 K, temperatures at which the main TL peaks occurred. A third TSC peak which is not accompanied by TL does not correspond to any change in the OA. An interesting study of TL and OA in CsI has been performed by Okada *et al.* [881]. The sample has been self-excited by ^{134}Cs which had been produced from ^{133}Cs by thermal neutrons.

Some other aspects of the relationship between TL/OSL and OA were reported by Hoogenstraaten [264], Chen *et al.* [461] and Wallinga and Duller [882]. Also of importance is the work by Wieser *et al.* [604] dealing with TL, OA and ESR (see Section 11.5) in fused silica, all depending in a similar superlinear manner on the excitation dose.

Yang *et al.* [614] reported on simultaneous measurements of TL and OA in carbon-doped $Y_3Al_5O_{12}$ (YAG) crystals. The strong absorption bands of YAG:C are correlated with the five measured TL glow peaks.

11.5   Simultaneous Measurements of TL and ESR (EPR)

Another absorption technique is ESR (EPR). ESR is used for the study of materials which exhibit paramagnetism because of the magnetic moment of unpaired electrons. ESR spectra are usually presented as plots of the absorption or dispersion of the energy of an oscillating magnetic field of fixed radio frequency versus the intensity of an applied static magnetic field. Among the wide variety of paramagnetic substances to which ESR spectroscopy has been applied, the one that is of interest in the present context is that of impurity centers in solids, mainly single crystals. The same impurity centers may be associated with TL, either acting as the charge-carrier traps, or as recombination centers. They may also be responsible for OA. Since ESR is normally capable of identifying the impurities in the crystal, in cases where the simultaneous TL-ESR measurements show a direct relationship between the two phenomena, the identification by ESR serves as a direct proof for the identity of the impurity involved in the TL process. Similarly to the OA case, it is normal that the ESR signal shows a drop over a certain temperature range where the trapping state becomes unstable. The instability may be associated with either the thermal release of charge carriers from the paramagnetic impurity, or, alternatively, with the filling of paramagnetic impurities which serve as TL recombination centers. In both cases it is expected that the TL peak will resemble the negative derivative of the ESR signal with a close resemblance to OA.

Figure 11.10, taken from Kortov *et al.* shows results of thermal annealing of paramagnetic centers, TL and OA at 340 nm of BeO ceramics, following β-excitation. The signal of paramagnetic centers (curve 1) and the induced OA at 340 nm (curve 2) are seen to decrease between 350 K and 450 K. The TL peak (curve 3) appears at $\sim$380 K, where the two other curves have the largest (negative) derivative. The results are explained in terms of lithium impurity which is abundant in this material, the paramagnetic properties of which are commensurate with the reported results. The role of Li as well as Al and B, which

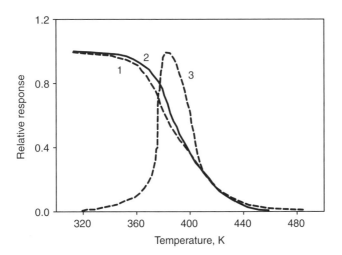

Figure 11.10 *Thermal anneal of paramagnetic centers (1), induced OA at 340 nm (2) and TL (3) of the transparent BeO ceramics. TL was recorded at a heating rate of 3K min^{-1}. Reprinted from Kortov, V.S., Milman, I.I., Slesarev, A.I. and Kijko, V.S., New BeO Ceramics for TL ESR Dosimetry, Radiat. Prot. Dosim. 47, 267–270. Copyright (1993) with permission from Oxford University Press*

are present in smaller amounts in BeO on the ESR as well as the OA and TL results is also discussed.

Born *et al.* [883] show in a similar figure the variation of ESR with temperature, its negative derivative and the TL intensity in X-ray excited $CaWO_4$ crystals. The peak of the negative first derivative is shifted by ~ 17 K to a lower temperature than its companion TL maximum. The reason seems to be the use of different heating rates in this case, where the measurement was not genuinely simultaneous. Jaszczyn-Kopec and Lambert performed simultaneous measurements of TL and ESR in mixed polytype ZnS:Cu, Cr, Cl. The glow peaks were identified as being produced by three Cr^+-centers also observed by ESR; one cubic and two axial (H_1 and H_2). The authors compared the negative derivative of the ESR signal with respect to temperature and the glow curve, and found similar curves slightly shifted from one another. They explain this shift as being the result of the emptying of some chromium traps without giving rise to light emission.

Other cases of a TL peak occurring simultaneously with the strong decrease of the ESR signal have been reported. Tzalmona and Pershan [884] performed such simultaneous measurements in CaF_2:Eu. The range of strong decrease of ESR occurs at ~ 165 K, where the TL peak takes place in samples with different Eu doping. Jahan *et al.* [885] reported on TL and ESR measurements of X-irradiated CaF_2:Mn single crystals. The emission peaks at 345 and 485 nm are attributed to V_k-center holes recombining with both electrons and Mn^+ ions. Cooke and Alexander [886] studied the TL curve as well as the temperature dependence of ESR in L-alanine:Cr^{3+}. The TL peak occurs at ~ 160 K and the ESR represents the concentration of the cation in this material, X-irradiated at 77 K. The TL peak looks very much the same as the negative derivative of the ESR signal, but the former is shifted by ~ 20 K above the latter. Here too, the shift is attributed to the different heating rates used. Another indication to the common origin of TL and ESR is that if UV light is applied following the X-irradiation, both signals are bleached quite substantially. Alybakov *et al.* [887] investigated ESR and TL glow curves of X-irradiated KCl:CrO_4^-, Ca^{2+}. As a result of the X-irradiation, the Cr^{6+} impurity centers turned into a paramagnetic pentavalent state, and showed ESR spectra of two types, termed A_1 and A_2. The A_1 spectrum appeared to be due to the Cr^{5+} ions generated in complexes with anion vacancies, whereas the A_2 spectrum was associated with Ca^{2+}. A comparison with TL results shows that heating the sample results in the destruction of the A_1-center, whereas a peak at 486 K is correlated with the destruction of the A_2-center.

Boas [888] and Boas and Filbrow [889] investigated TL and ESR in CaO and CaO:Mg crystals. The resonances were produced by illumination at < 750 nm and < 220 K, and were associated with the TL peak at 180 K, with the emission spectrum centered at $\lambda_{max} = 690$ nm. Orera *et al.* [890] studied ESR and TL in X-irradiated CaO at higher temperatures, and found two glow peaks at 360 and 400 K. These were associated with the thermal destruction of two new donor-type paramagnetic defects of extrinsic character. The ESR spectrum of the defect which was associated with the first glow peak consisted of an intense central line and six weak transitions, corresponding to the hyperfine interaction with a ^{25}Mg nucleus. The second defect showed an interaction with an ^{27}Al nucleus. The thermal annealing of the ESR signals occurred between 350 K and 400 K, the same range in which the two TL peaks were seen.

Tatumi *et al.* [891] investigated TL and ESR in aragonite, a trimorph of $CaCO_3$, along with calcite and vaterite. Natural aragonite shows a prominent TL peak at 613 K. Additional

γ-irradiation enhanced this glow peak but shifted it down to 593 K and induced two new peaks at 393 and 493 K. Pre-annealing at 653 K and subsequent irradiation enhanced Pb^{3+} and CO_3^- hole centers observed in ESR. An isochronal thermal anneal sequence indicated that these ESR centers were related to the two glow peaks at 393 and 493 K. The experimental results indicated that the TL glow peaks in aragonite can be explained by second-order kinetics, and the concentrations of the Pb^{3+} and CO_3^- centers decreased with increasing anneal temperature up to 603 K, at which temperature the two TL glow peaks at 393 and 493 K are exhausted. These glow peaks appeared to be related to a recombination between holes in the Pb^{3+} ions and electrons thermally released from the CO_3^- and CO_2^- electron traps. The difference in the dose response of the two peaks indicated a different dependence of the filling process of these two traps on the dose.

Sullasi *et al.* [892] reported on the γ-radiation effects on TL and ESR in natural zircon. They show a very strong correlation between the 140 °C TL peak and the ESR signal, which indicates that the same defect centers are responsible for both effects. Cano *et al.* [893] studied TL, ESR and reflectance from natural diopside ($CaMgSi_2O_6$) crystals. In this case, no correlation was found between ESR and reflectance on one hand and TL on the other, indicating that different centers are involved in these phenomena. Cano *et al.* [894] reported on ESR and TL in sodalite. Out of five TL peaks, they found correlation between a peak at 365 °C and an ESR line.

Seshagiri *et al.* [895] discussed the results in γ- irradiated crystals of uranium-doped $K_2Ca_2(SO_4)_3$. ESR measurements showed the formation of the radical ions SO_4^-, SO_3^-, SO_2^- and O_3^-. The thermal destruction of the SO_2^-, and SO_4^- ions was found to be associated with the TL peaks at $\sim$400 and $\sim$435 K, respectively. It was found that the uranate (UO_6^{6-}) ion serves as the luminescence center for the observed TL glow. It should be noted that similarly to uranium compounds, self-excitation of TL and ESR may occur by uranium impurity in this material. Rao *et al.* [896] reported on simultaneous measurements of $BaSO_4$:Eu,P and associated two TL peaks at 170 and 215 °C with two radicals, assigned to SO_2^-.

López *et al.* reported on simultaneous measurements of TL and ESR in X-irradiated NaCl:Mn^{2+}. Peak 2 is clearly related to manganese-vacancy dipoles and peak 1 can be roughly associated with free cation vacancies. Peak 4 appears to relate to large manganese aggregates, whereas peak 5 is intrinsic and unrelated to impurities.

Some authors performed triple simultaneous measurements such as of TL, TSC and ESR. Katzir *et al.* [898] carried out such experiments in boron nitride. They observed both TL and TSC peaks and found, as expected, that the two TL peaks preceded their TSC counterparts, and appeared in ranges of fast decrease of the ESR amplitude. Böhm *et al.* [899] performed TL, TSC and ESR measurements in $BaWO_4$ crystals. They used ESR to identify electrons and holes trapped at WO_4 complexes in $BaWO_4$, X-irradiated at 80 K. The thermal decay of the intrinsic hole centers at $\sim$100 K is accompanied by a simultaneous decrease of $(WO_4)^{3-}$ electron traps and maxima for both TL and TSC. They explained this connection by a thermally activated hopping of the $(WO_4)_2^{3-}$ hole centers, followed by a radiative recombination with electrons trapped at $(WO_4)^{3-}$.

Huzimura and Atarashi [900] studied ESR, TL and TSEE in X-irradiated $CaSO_4$:Tm. They identified four types of electron centers related to SO_3^- and three types of hole centers associated with SO_4^-, and one type of axial center. By comparing the step-by-step annealing of these centers with the TL and TSEE curves, they reached the following conclusions. Electrons are thermally freed from SO_3^- at about 100 °C, partly to be emitted as exoelectrons and mostly to be trapped at Tm^{3+} sites, with which holes from axial centers recombine to produce a minor TL peak. The major TL peak was found to be composed of two parts, one related to the thermal decay of SO_4^-, and the other to a relaxation process of the SO_3^- centers around 200 °C.

Anderson *et al.* [901] reported on TL, ESR and OA in X-irradiated Ge-doped quartz. Strong bleaching of the 280 nm OA and the A- and C-center ESR signals coincided with maxima of the TL glow curves. Jani *et al.* [902] investigated point defects in crystalline SiO_2 using TL, ESR and IR OA. Variations of the results have been observed following sweeping in a hydrogen atmosphere. This was done at 400 °C with an electric field of $1000V\,cm^{-1}$ applied along the *z*-axis for 18 h. To sweep lithium or sodium into the quartz, the appropriate salt was deposited on one face of the crystal and an electric field was applied for 1–2h. Jani and Halliburton [903] have studied TL, ESR and OA in $Bi_{12}SiO_{20}$ (BSO) and $Bi_{12}GeO_{20}$ (BGO). Excitation at 77 K with 350 nm light converts Fe^{3+} to Fe^{2+} ions. The source of electrons in this process is unknown; the possibilities mentioned include transition-metal ion impurities, neutral oxygen vacancies, germanium or silicon vacancy complexes and antisite bismuth ions.

Halperin *et al.* [904] reported on simultaneous measurements of TL and ESR in quartz. The main result here is the full correlation between the 190 K peak and the four-line ESR spectrum due to a defect which they labeled $[SiO_4/Li]^0$ center. Both were formed under exactly the same conditions, both were absent after subjecting the quartz crystal to heavy doses of irradiation near RT, and both were eliminated by hydrogen sweeping. A similar relationship between a TL peak at a higher temperature of $\sim$180 °C and a paramagnetic center has been reported by Pogorelov *et al.* [905].

Andreev [906] investigated the annealing of radiation-generated paramagnetic centers in γ-irradiated $KTiOPO_4$ (KTP) single crystals by measuring TL, TSC, ESR and OA. A complete annealing of the radiation-induced centers and color centers occurred at 430 K. A TL peak observed at $\sim$330 K was found to be due to the annealing of radiation-produced hole centers of the PO_4^- type, and a TSC peak was found to be due to annealing of radiation-induced paramagnetic electron in Ti^{3+} centers.

Vedda *et al.* [907] studied the recombination processes ascribed to rare earth activators in $YAlO_3$ single crystal host. Three out of seven glow peaks due to intrinsic hole traps are related to different variants of O^--centers, as demonstrated by the correspondence between the thermal annihilation of these centers detected by ESR and the glow-peak temperatures. To interpret their results, these authors suggest the existence of defect complexes involving intrinsic defects coupled to Ce^{3+}, Pr^{3+} and Tb^{3+} ions, in which carriers can be transferred from intrinsic levels to rare-earth ion levels where the radiative recombination occurs.

Daling *et al.* [908] report on the investigation of TL, ESR and positron annihilation in γ-ray irradiated magnesium sulfate. The TL and ESR dose responses in the range between 1 Gy and 20 kGy at RT are found to be nonlinear. The f peak of ESR, attributed to SO_3^- radicals is sublinear, whereas the g peak attributed to SO_4^- radicals is linear. The TL dose response of the four glow peaks is superlinear. The authors conclude that the generation of the TL signal includes a complex, multistage process in which not only anion radicals but also Mg^{2+} vacancies and impurities might be involved. This is supported by the measured positron lifetime parameters of $MgSO_4$. Danby *et al.* [909] studied the TL and ESR in X-irradiated $CaSO_4$ single crystals, and found two related radiation damage centers. Step annealing measurements suggest a connection between these centers and TL glow peaks at 395 and 465 K.

Vaĭner *et al.* [910] discuss the simultaneous measurements of TL and ESR in Y_2O_3. The simultaneous results help them to attribute a specific TL peak to each known center.

Kortov *et al.* [911] described the results of simultaneous TL and ESR measurements in MgO, excited by ionizing and non ionizing UV radiations. Whereas the TL grew linearly with the excitation dose, the ESR grew superlinearly with the dose, behaving approximately like $\sim D^3$ (see Section 8.2). Singh *et al.* [912] reported on simultaneous measurements of TL and ESR in $ZnAl_2O_4$ phosphors doped with *Er* and *Yb*. Two centers seen in the ESR spectrum appear to be associated with O^- ions and F^+-centers, and correlated with TL peaks at 174 °C and 483 °C, respectively.

Finally, the measurements of TL and ESR in ThO_2 by Rodine and Land [610] should be mentioned. They report that some glow peaks in the undoped ThO_2 correlate with the annealing of ESR spectra associated with trapped electrons. Note that in materials like this, both TL and ESR (as well as OA), may be caused by internal irradiation by the radioactive ingredient, in this case thorium.

11.6   Simultaneous Measurements of TL and TSEE

One may think that since both in TL and TSEE (see Section 6.11), the measured effect is related to the thermal release of electrons, the peaks showing up in the two effects should be very similar. There are, however, some reasons why this is not always the case:

(1) In TL, when the thermal-quenching effect (see Section 9.3) takes place, the expression for TL is to be multiplied by a decreasing function of temperature, thus causing the TL peak to shift to a higher temperature and changing its shape to some extent.

(2) In TSEE electrons and not holes are released from the surface of the material, whereas TL may just as well result from the thermal release of holes from traps into the valence band, and their subsequent recombination with electrons in recombination centers. However, as mentioned above, in the Auger effect the thermal release of holes may also result in TSEE. As opposed to the Maxwell-tail theory, which simply describes the effect as the thermionic emission of conduction electrons liberated thermally from traps, the Auger-type model considers the recombination of thermally freed holes, which releases energy sufficient to emit electrons in the form of TSEE. The Auger effect of having TSEE as a result of thermal freeing of trapped holes in phosphate glasses has been reported by Zatsepin *et al.* [235].

(3) Almost always TL is an effect occurring in the bulk of the sample whereas TSEE is a surface effect. It is therefore very interesting to follow those cases in which there is an obvious correlation between the two phenomena. Once this happens, one can conclude that the thermally released agents are electrons. Even in this case, one cannot expect an exact correlation between the TL and TSEE peaks. The shift expected between the TL peak and the $n_c(T)$ peak mentioned above plays a role here, since the TSEE curve depends on $n_c(T)$. In addition, the function $\exp(-\phi/kT)$ associated with the work function ϕ, also causes a shift (see e.g. Chen and McKeever [33]).

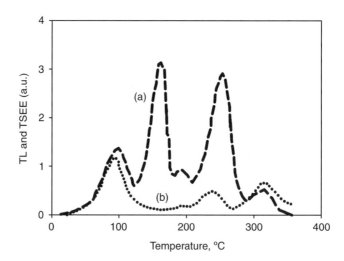

Figure 11.11   *TL (a) and TSEE (b) in X-irradiated $Li_2B_4O_7$:Cu crystallized glass. Reprinted from Kutomi, Y., Tomita, A. and Takeuchi, N., TL and TSEE in Crystallised $Li_2B_4O_7$:Cu Glass, Radiat. Prot. Dosim. 17, 499–502. Copyright (1986) with permission from Oxford University Press*

(4) One cannot always conclude with certainty that the initially released carriers are electrons. Auger processes (see Section 6.11) may lead to the emission of electrons from the surface, following the thermal release of holes from traps.

Figure 11.11 shows an example, taken from Kutomi *et al.* [913]. These authors studied TL and TSEE in different kinds of $Li_2B_4O_7$:Cu glasses. Here, four out of five TL peaks are accompanied by TSEE companions, two of the latter appearing at slightly lower temperatures than the former, and two at slightly higher temperatures. It seems that in this case $\exp(-\phi/kT)$ has a rather small influence on the TSEE maximum, and the order of appearance seems to be related to effects similar to those discussed in regard to the TL-TSC results. It also appears that the peak at $\sim160\,°C$ is related to the release of holes from a trapping state into the valence band, and their subsequent recombination with electrons in a center. When the holes are in the valence band they do not contribute to TSEE, as do the electrons released into the conduction band in the process leading to the appearance of the other peaks. The thermally stimulated processes in phosphate glasses have been discussed by Zatsepin *et al.* [235] who reported on the results of TL, TSEE, ESR and OA and associated it with the participation of localized electronic states in both discrete and continuous centers.

Pipinis and Grigas [795] reported on such simultaneous measurements in cathode-ray-excited NaCl−AgCl. The TL peaks in their measurements corresponded to the TSEE peaks, but were shifted by a few degrees to lower temperatures. Thus, in this case it seems that the TSEE resembles $n_c(T)$, and the additional factor $\exp(-\phi/kT)$ mentioned above does not shift the TSEE by much. This indicates that the value of ϕ for this material is small. An example in which the TSEE peaks precede their TL counterparts is given by Kawamoto *et al.* [914] and Kutomi *et al.* [913]. The same group of researchers reported on the results of such TL-TSEE measurements in different materials. Tomita *et al.* [915, 916] performed such measurements in NaCl:Ca, Cu, which yielded a large number of peaks. In these results too, one can see TL peaks unaccompanied by TSEE, which therefore seem to be related to the thermal release of holes, as well as TL-TSEE pairs which definitely are associated with the thermal release of electrons. Further results of the same nature were reported by Fukuda *et al.* [917] in MgB_4O_7 samples, by Fukuda [918] in $Ca_3(PO_4)_2$:Ce samples, by Tomita *et al.* [919] in KCl:Ag single crystals and by Tomita *et al.* [920] in $LiNbO_3$.

Aoki *et al.* [921–923] studied simultaneous TL-TSEE in LiF and CaF_2 evaporated thin films. They reported a rather unusual effect of TL and TSEE observable on the first heating of the sample with no pre-excitation. The TSEE peaks in these samples were shifted by ~50 K above their TL counterparts. This seems excessive even when the two reasons for such a shift mentioned above are taken into consideration. The authors attribute these first-heating peaks to the crystallization of the film evaporated in vacuum, which means that they have close correlation with the growth of many microscopic bordered domains which appear on the surface of the film. Iacconi *et al.* [924] investigated TL and TSEE in ZrO_2 samples. A number of peaks were observed in each of the two phenomena but, as reported by the authors, no correspondence was seen between the TL and TSEE peaks in this case.

Sorkin and Käämbre [925] reported on the results of simultaneous TL-TSEE measurements in CsBr:Tl, X-irradiated at 80 K. Three pairs of peaks were reported; in each pair the TSEE member was shifted to a higher temperature than that of the TL peak. Taking into consideration all the available information on this sample, the authors conclude that migrating V_k holes recombine with Tl^+ trapped electrons emitting the recombination energy, partly as TL light and partly by transferring it to other Tl^0-centers, ionizing them and thus initiating TSEE in the 120–170 K range. Lesz *et al.* [275] reported the TL-TSEE results of BeO thin-film samples. A pair of matching TL-TSEE peaks was reported at ~290 °C with the TSEE peak occurring at a slightly higher temperature. In addition, another TL peak appeared at ~200 °C and a TSEE peak at ~400 °C. In a later work, however, Lesz *et al.* [276] did not observe the ~300 °C TL peak, and concluded that there is no correlation between TSEE and TL in BeO thin films. Bichevin *et al.* [926] studied excimer-laser-induced TL-TSEE of alkali halides, as well as X-ray excited processes in these materials. In KCl:Ag, a main TL peak was seen at ~400 K and TSEE at ~430 K in the laser-excited case. The authors associate the observed peaks with Ag^o centers. Peoglos and Christodoulides [927] studied the applicability of the thermionic model for exoelectron emission to a thin film of KCl by measuring the activation energy for TSEE and TL, as well as electron affinity of this solid. The latter was measured through the study of the energy distribution of the secondary electrons. The exoelectron emission was found to be strongly correlated with TL. The samples were irradiated with 600 eV electrons and the TL-TSEE peaks occurred between −60 °C and 120 °C.

Kortov *et al.* [928, 929] as well as Iacconi *et al.* [930] studied simultaneously TL and TSEE in Al_2O_3 samples. Kortov *et al.* [929] reported on results from X-irradiated Al_2O_3:Mg. Here the main peaks are ascribed to the Mg impurity since they are not observed in the nominally pure corundum samples. It is proposed that that the TSEE peak is related to thermally free holes in the Auger process, similar to the explanation given above. Triple simultaneous measurements of TL, TSC and TSEE of anion-defective corundum were reported by Nikiforov *et al.* [37]. They provide an energy-level model which validates the effect of deep traps on specific features of TL, TSC and TSEE. Akselrod and Kortov [928] performed similar measurements on α-Al_2O_3:C. The main TL-TSEE system here is

~175 °C where the TL peak precedes its TSEE counterpart. Iacconi *et al.* [930] studied the TL and TSEE properties of α-Al_2O_3 samples following UV or X-ray excitations. They report that there is a reasonably good match between the TL and TSEE peaks, and that the TSEE maxima are at slightly higher temperatures. A TL peak at ~87 K is also reported, which is not related to a TSEE maximum. Their conclusion is that this TL peak is associated with the thermal release of holes. Vincellér *et al.* [691] reported on the measurements of TL, TSC and TSEE in Al_2O_3. They describe the thermal quenching of TL in this material, and the dependence of all three thermally stimulated effects on the heating rate. An energy model is proposed to describe simultaneously these effects, explaining the results as due to competition for the localization of electrons. Also, Berkane-Krachai *et al.* [693] gave a model accompanied by numerical analysis of the simultaneous TL and TSC phenomena of α-Al_2O_3 single crystals. Here the shift of the TSEE peak to higher temperature than its TL counterpart is clearly shown to be associated with the electron affinity χ (eV). Further work along the same lines has been presented by Bindi *et al.* [516] and by Guissi *et al.* [931]. These authors consider the simultaneous measurement of TL, TSC and TSEE in a model of a single type of electron trap, a single type of recombination center and thermally disconnected traps. TSEE is described as resulting from the thermionic effect, and therefore the model applies only to thin solid films ~10 nm thick. The influence of various parameters in the model on the peak shape and position was investigated.

Holzapfel and Krystek [932] have reported on the simultaneous measurements of TL and TSEE in $BaSO_4$ and $SrSO_4$ doped with europium. In both materials the shapes of the TL and TSEE curves are very similar with no shift, and so is the dose dependence. The peaks are found to be of first order, pointing to negligible retrapping during the heating of the samples. Finally, Sakurai *et al.* [933] described a method for the analysis of simultaneous results of TL and TSEE and distinguished between cases where TL and TSEE peaks were correlated with each other, and cases where no such correlation took place.

12

Applications in Medical Physics

12.1 Introduction

Radiation is widely used in medicine, both for diagnosis such as X-ray imaging, and for therapy such as cancer treatment. Consequently, dosimetry on live subjects (*in vivo*) is important. Furthermore, these methods are supported by medical research and development which may be performed either *in vivo* or *ex vivo*, i.e. on materials called phantoms, which are designed to simulate the radiological properties of live subjects. This chapter will provide examples of medical dosimetry applications using thermoluminescence (TL), optically stimulated luminescence (OSL), and radiophotoluminescence (RPL). Medicine also has need for real-time measurements and this chapter will discuss the theory and practice of such measurements using radioluminescence (RL).

Ensuring that the correct dose is delivered is particularly important for the treatment of cancer. Medical treatment of cancer begins, when possible, with surgical removal of cancerous tumors. This is followed by treatment to kill the remaining cancer cells. This can be done by chemotherapy which has strong undesirable side-effects. If the area to be treated is confined to a small volume and that volume is in a suitable location, radiation treatment is used instead. After a physician decides on the radiation dose to be delivered, a medical dosimetrist selects radiation sources and appropriate geometries using simulation software.

The next section provides an overview of various applications of luminescence detectors in medical dosimetry, including a discussion of the suitability of various dosimeter materials for different medical applications, and specifically their energy response. This is followed by a section on several examples of *in-vivo* applications which have become popular during the past decade. The last section in this chapter presents a model for a real-time dosimetic system which uses the RL signal from Al_2O_3:C.

Thermally and Optically Stimulated Luminescence: A Simulation Approach, First Edition. Reuven Chen and Vasilis Pagonis.   © 2011 John Wiley & Sons, Ltd. Published 2011 by John Wiley & Sons, Ltd.

12.2 Applications of Luminescence Detectors in Medical Physics

A key issue for biological and medical applications is how well the dosimeter material mimics the response of biological materials to radiation. At the simplest level, this is evaluated by comparing the effective atomic number (for high energy photons) of the dosimetric material to the biological materials of interest. An example of the effective atomic number for a variety of materials is shown in Table 12.1. It is seen that LiF is a fairly good match to soft tissue such as muscle, while several other TL and RPL materials are closer to the Z_{eff} value of bone. The response of dosimeter materials at lower photon energies is also important. Some materials with the correct Z_{eff} at high photon energies over-respond at lower (<0.3 MeV) photon energies. For example Figure 12.1 shows the spectral response of four dosimeter materials, two traditional TL materials and two fiber optic materials. Observe that the fiber optic materials over-respond at low energies by as much as a factor of 15, while LiF provides the flattest response. The response of two LiF TLD materials is shown in Figure 12.2. It is seen that LiF:Mg, Cu, P provides a flatter response than LiF:Mg, Ti. These responses will be modified by any material used to cover the dosimeter [934–936].

The wide variety of medical applications of radiation require a wide variety of radiation sources. For cancer treatment these sources can be divided into two broad categories based on size. Electron accelerators, for example, are large. They must be placed outside the patient and must produce radiation with sufficient penetrating power to enter deeply enough into the patient. Therapy with radiation sources outside the body is called *teletherapy*. If a radiation source is small in size, such as certain nuclear isotopes, it may be placed in the body near the targeted tumor. This technique is called *brachytherapy*. Because the source is near the target, the radiation may have a short penetration depth and still reach the target. The advantage of

Table 12.1 *Effective atomic numbers (Z_{eff}) for various materials*

Material	Z_{eff}
Radiochromic film	6.0–6.5
Soft tissue of composition $C_5H_{40}O_{18}N$	7.22
Excised muscle specimens	7.42
Striated muscle (calculated)	7.46
Lithium tetraborate ($Li_2B_4O_7$)	7.23
$Li_2B_4O_7$:Mn	7.4
Tissue	7.4
Air	7.6
LiF	8.2
Al_2O_3:C	10.2
RPL glass (Dose Ace)	12.0
Bone	13.3–13.8
Bone (experimental)	13.8
$CaSO_4$:Mn	15.3
RPL glasses	14–15
$CaSO_4$:Dy	close to bone
Calcium phosphate TLD	close to bone

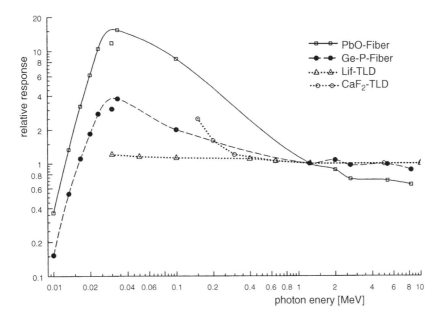

Figure 12.1 *The relative sensitivity, normalized to ^{60}Co, is shown for LiF and CaF$_2$ thermoluminescence dosimeter (TLD) materials, along with two optical fibers selected for radioluminescence dosimetry. Reprinted from Gripp, S., Haesing, F-Wg, Bueker, H., and Schmitt, G., Clinical in vivo dosimetry using optical fibers, Radiat. Oncol. Invest., 6, 3, 142–149. Copyright (1998) with permission from Wiley-Liss, Inc.*

this method is that, with a short penetration depth, the radiation will not reach, and therefore will not damage, organs further away [937].

In a recent review of luminescence dosimetry materials Olko [15] summarized recent developments in the applications of luminescence detectors in the field of medical dosimetry. The author discussed the requirements for luminescence dosimeters in radiotherapy, and found that typical clinical recommendations for the accuracy of dose delivery are around 5%. Modern radiotherapy protocols use active dosimeters, such as ionization chambers, silicon diodes and MOSFET detectors since they allow real-time read-out of results, easy operation and minimum maintenance. The most common applications of TLDs in radiotherapy are mailed dosimetric services of radiotherapy units, and this area historically has used TLD-100 detectors. Yukihara *et al.* [17] demonstrated that an OSL system based on Al$_2$O$_3$:C detectors is also a potentially useful system for depth dose distribution measurements. Along the same lines, RPL glasses were also tested recently for possible mailed dosimetric services, and their performance was compared with LiF:Mg, Ti (Rah *et al.* [938]). Measurements in dosimetric phantoms typically use TLD materials like LiF:Mg, Ti, LiF:Mg, Cu, P and Li$_2$B$_4$O$_7$ [15].

During brachytherapy treatment, as opposed to teletherapy, a source of short–range radiation is placed next to the site to be treated with the intention of reducing radiation damage to other organs. Because the source must be small, radioisotopes are used. We will present an example of this type of treatment, in which RPL glasses are used for dosimetry purposes. Nose *et al.* [939] used RPL glass dosimeters to study the treatment of 61 head and neck

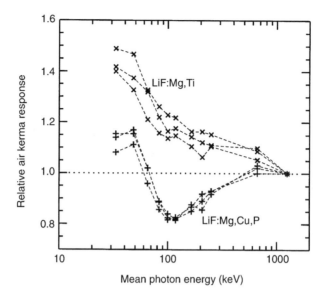

Figure 12.2   *The absorbed dose in air is compared for two LiF TLD materials. Air is close to tissue equivalent. Reprinted from Davis, S. D., Ross, C. K., Mobit, P. N., Van der Zwan, L., Chase, W. J. and Shortt, K. R., The response of LiF thermoluminescence dosemeters to photon beams in the energy range from 30 kV X rays to ^{60}Co gamma rays. Radiat. Prot. Dosim., 106, 33–43. Copyright (2003) with permission from Oxford University Press*

cancer patients at Osaka Medical Center. The irradiation source was a Varisource (Varian TEM) with a 10 Ci source of ^{129}Ir isotope. Some patients were treated to brachytherapy alone, in which case typical doses are 48–54 Gy. Other patients were treated to a combination of brachytherapy and teletherapy, in which case their brachytherapy doses were 18–36 Gy. This study chose RPL glass dosimeters. Advantages of RPL for *in-vivo* dosimetry include the ruggedness and non-toxicity of RPL glass. Their response is nearly independent of photon energy for photons with energies above 0.3 MeV, but they over-respond at lower energies relative to biological tissue [939]. A comparison of measured and calculated doses is shown in Figure 12.3. The results in Figure 12.3(a) show that measured doses for targeted organs are on average only 87% of the planned (calculated) doses. In some cases, the measured doses were as little as half the calculated dose. The authors attribute this to the failure of the software to account for the proximity of the head and neck organs to air, both inside and outside the patient. The software assumed the organs to be surrounded by unbounded water–equivalent media which would have had higher radiation backscatter. Another factor was that the software assumed that the radiation source was cylindrically symmetric, while the actual source strength varied with angle by a factor of two [939]. For the nontarget but neighboring organs, the ratio between measured and calculated dose ranged from 0.39 to 1.84. This wide variation was attributed to unplanned movement of the dosimeters relative to the radiation sources. Because the measured doses can be nearly double the planned dose, these authors recommended the use of radioprotection such as lead shields, even when planned doses appear small. More recently Nose *et al.* [940] have

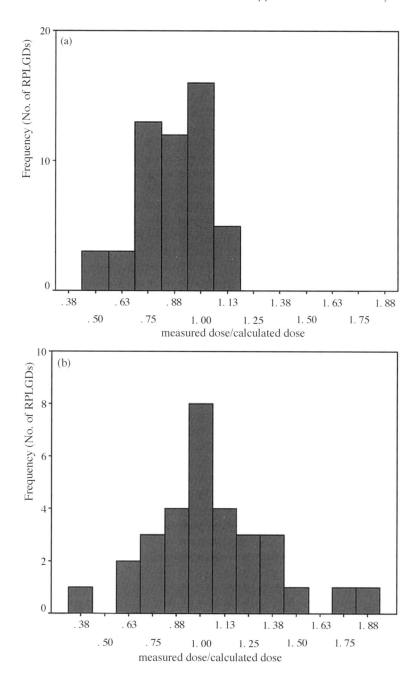

Figure 12.3 Histograms are shown for the ratio of measured to calculated dose for RPL dosimeters used on 61 head and neck cancer patients. (a) shows data for the dose delivered to the targeted organs, while (b) shows dose for neighboring but non-targeted organs. After Nose et al. [939]. Reprinted from Nose, T., et al., In vivo dosimetry of high-dose-rate brachytherapy: Study on 61 head-and-neck cancer patients using radiophotoluminescence glass dosimeter, Int. J. Radiat. Oncology, 61, 945–953. Copyright (2005) with permission from Elsevier

used RPL glass for dosimetry during brachytherapy on malignancies of the pelvic region; 1004 points were measured in 66 patients. For this study the measured doses for non-target organs differed from the planned doses by $\pm 20\%$. This is much less than the 32% deviation observed for head and neck patients. This change is attributed to patients having less voluntary control over pelvic region muscles than over head and neck muscles, and hence there was less unplanned motion of the dosimeters in the pelvic non target organs. Consequently a lesser deviation is observed between measured and planned doses.

We also mention a few selected applications of OSL dosimetry. Aznar [941] report having made *in-vivo* measurements on a patient during radiation treatment for head and neck cancer. They used a Al_2O_3:C fiber and performed RL and OSL measurements. Optical fibers were inserted through the patient's nose, and positioned down the oesophagus. Positioning of the fibers was assisted with the use of lead markers attached to the fibers. The measured dose was 1.76 ± 0.05 Gy while the software predicted a dose which is two standard deviations higher than the measured dose average of 1.85 Gy. Marckmann *et al.* [942] have demonstrated that these doses can be monitored in a clinical setting using Al_2O_3:C fiber optic probes, while real-time response can be monitored with RL. Afterwards, the integrated dose can be assessed with OSL. In the case of a radiological or nuclear event, there will be a need to assess quickly individuals for their level of exposure as part of a medical triage process. In such cases, one would need to use materials at hand. Yukihara *et al.* [943] have investigated the possibility of using dental enamel as a dosimeter. They found that dental enamel exhibits an OSL signal, and obtained the strongest signal using blue/green stimulation and by monitoring luminescence in the UV. The sensitivity was limited by signal strength and was found to be 4–6 Gy.

As a final example, we mention a phantom study in space programs. A key issue with space travel, particularly interplanetary space travel, is the radiation exposure that the astronauts will receive over long voyages. To better understand radiation exposure in low-Earth orbit, a life-size Rando® human phantom was flown on a Shuttle-Mir mission (STS-91). The phantom was instrumented with both RPL glass (FD-7/SC-1 from Asahi Technoglass) and a thermoluminescent material (Mg_2SiO_4:Tb from Kasei Optonics Inc.) at 59 locations. The phantom was fixed via bungee cords onto a rack in the Spacehab module of Space Shuttle Discovery. The mission lasted 9.8 days including 4 days during which the shuttle was docked at the Mir space station. Upon return to Earth, doses were measured using both the RPL and TL materials at each location. After having calibrated both materials relative to ^{137}Cs γ-rays, the readings from RPL were systematically lower than those from the TL material. This was attributed to the dose being produced more from charged particles than from γ-rays [944].

Other novel developments in the use of luminescence detectors in medical dosimetry include CCD-TLD readers for 2D TL dosimetry [945] and a 2D TL dosimetry system consisting of TL foils and two large-area TLD-CCD readers (Olko *et al.* [946]).

12.3   Examples of *in-vivo* Dosimetric Applications

In this section we provide a brief overview of *in-vivo* dosimetric applications. Recent extensive reviews of *in-vivo* OSL systems have been published by Akselrod *et al.* [695] and by Yukihara and McKeever [947]. Several *in-vivo* dosimetry systems are available, with various advantages and disadvantages. Some of the desirable characteristics of such dosimetry systems are small size, safe use, low cost, easy calibration and use, accuracy, precision, and real-time estimation of imparted doses. Aznar *et al.* [948] reviewed some of the most popular *in-vivo* dosimetry devices and summarized their main characteristics (see their Table 1.1). These authors provided also an overview of the main application areas of *in-vivo* systems, namely radiation therapy and diagnostics. They discussed specific examples of dosimetry applications in external radiation therapy and mammography.

Typically, radiation therapy treatments involve high-energy X-rays (4–25 MeV), but high energy electron beams (6–22 MeV) are also used. Two types of treatments are: conventional external therapy; and intensity modulated radiation therapy (IMRT). In conventional external therapy the process involves using computerized tomography (CT) images of the patient before the beginning of the treatment, and a computerized treatment planning system (TPS) is used to calculate a 3D dose distribution [948]. During IMRT, several modulated beams are used to produce the desired dose delivery pattern to the target. During such treatments it is often desirable to perform a dosimetry check using either appropriate phantoms or by using *in-vivo* dosimetry. The second major application area of *in-vivo* dosimetry is in diagnostic radiology and mammography. Mammography is the imaging of the human breast, and the absorbed doses measured are of the order of 1 mGy. TLDs on the breast may be used as a simple method of determining the dose received by the patient.

Commonly used dosimetric materials for *in-vivo* dosimetry are: TLDs, semiconductor diodes and films. Aznar *et al.* [948] provided a brief overview of their characteristics, and highlighted their relative advantages and disadvantages. Ionization chambers are considered to be the "state of the art" of radiation dosimeters providing both high accuracy and high precision. However, ionization chambers require several correction factors, and their current use is mostly limited to phantom measurements. For an example of a comparative study of TLDs and ionization chambers, the reader is referred to the work by Soares *et al.* [949].

TLDs are used both in radiation therapy and in diagnostic radiology [950]. TLDs offer the best option for *in-vivo* measurements in mammography because of their high sensitivity and small size: they can be placed anywhere on the skin or inside the body. However, TLDs suffer from sensitivity variations which require annealing treatments and calibrations after each measurement. With great care in manipulation from an experienced technician, a reproducibility of about 2% can be reached [951]. The overall TLD handling process is a time consuming procedure, which can take up to several hours. Another major problem with TLDs is their relatively low precision in daily clinical uses. As a result, the total uncertainty on dose determination can be high: one study reported deviations up to 7% for *in-vivo* TLD measurements of patients over a 7-year period [952].

Semiconductor diodes have been widely accepted in medical dosimetry because they are robust, relatively simple to operate, have high sensitivity and can be also read-out quickly [948]. However, these detectors also require several correction factors to be applied to their signal. Additional novel techniques have been reported in the literature, and the most significant ones are the use of radiographic film, metal oxide semiconductor field-effect transistors (MOSFETs), diamond detectors, and optical fiber dosimetry systems. Aznar *et al.* [941, 948] describe the development of a point dosimeter which is highly accurate and offers real-time dose information about the dose delivered. This technique uses optical fibers and monitors both the OSL and RL signals emitted by Al_2O_3:C during irradiation; a model for this type of measurement is described in another section of this chapter.

The luminescent crystal is attached to an optical fiber cable, which is used for transmitting both the stimulation light and the emitted OSL signal. The crystal can be attached to the body or, because of the very low diameter of the cable (2 mm), introduced into body cavities. Their small size and negligible temperature dependence give some advantage to active OSL dosimeters over the presently used semiconductor diodes. The OSL system can operate in active mode, measuring the prompt RL signal, or in integration mode. A number of systems are available as working prototypes and were tested in clinical conditions. In another interesting application of the system, it has been used to verify dose delivery in remotely afterloaded brachytherapy, allowing for automatic online *in-vivo* dosimetry directly in the tumor region using thin OSL probes placed in catheters [953].

12.4 Radioluminescence

The phenomenon of RL has been reported in a number of materials including Al_2O_3:C, which is one of the main dosimetric materials. Over the last decade, attention has been given to applications of the dosimetric material Al_2O_3:C in medical dosimetry. One such application is *in-vivo* dose verification in radiotherapy of cancer patients. Andersen *et al.* [954] developed a system for *in-vivo* dose measurements during radiotherapy. Their system uses the RL and OSL signals from small Al_2O_3:C crystals attached to long optical fiber cables. During radiation therapy, the RL signal provides a real-time measurement of the dose rate at the position of the crystal, and immediately after the treatment, the CW-OSL signal is used to determine the integrated dose (see, for example, Aznar *et al.* [941]). These experimental studies have established several empirical results for RL measurements carried out in Al_2O_3:C crystals. Several such empirical results are shown in Figure 12.4 which shows experimental data from Aznar [948], indicating that the RL intensity during a short irradiation pulse increases linearly with time. The experimental data in Figure 12.4(b) indicate that the corresponding initial RL intensity I_0 varies linearly with the dose rate X, while the inset demonstrates that the slope of the linear part of the RL intensity varies quadratically with the dose rate X.

Pagonis *et al.* [488] explained these experimental findings by using the same model previously used to describe the TL and OSL dose response of Al_2O_3:C crystals (see, for example, Chen *et al.* [686] and Pagonis *et al.* [38]). The model includes two trapping states, N_1 and N_2 and two kinds of hole recombination centers, M_1 and M_2 as shown in Figure 12.5. N_1 is the main dosimetric trap from which stimulating light can release electrons, and N_2 represents a competitor in which electrons can be trapped, but the stimulating light cannot release electrons from it. Two recombination centers are assumed to exist, M_1 which is radiative and M_2, a nonradiative competitor. The various transitions in the model taking place during an RL measurement are indicated by arrows in Figure 12.5.

The set of coupled differential equations governing the process during the irradiation process is

$$-\frac{dm_1}{dt} = A_{m1}n_cm_1 - B_1n_v(M_1 - m_1),$$ (12.1)

$$-\frac{dm_2}{dt} = A_{m2}n_cm_2 - B_2n_v(M_2 - m_2),$$ (12.2)

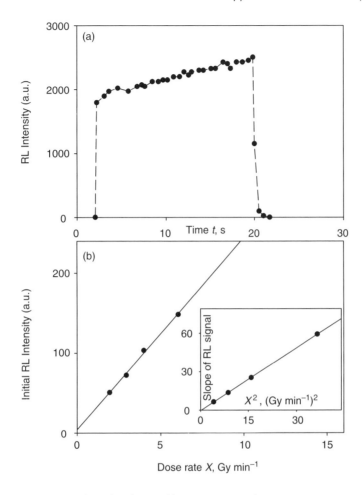

Figure 12.4 *Experimental results obtained by Aznar [948] during RL experiments. (a) The RL pulses as a function of time are characterized by a linear region. (b) The initial RL intensity I_0 of the RL pulses in (a) is found empirically to depend linearly on the dose rate X. The inset shows the slope of the RL pulses depends linearly on the square of the dose rate X^2 used during the RL experiments. Reprinted from Aznar, M.C., PhD Thesis, pp134, Risø-PhD-12(EN).http://www.risoe.dtu.dk/rispubl/NUK/nukpdf/ris-phd-12.pdf (2005). Copyright Risø*

$$\frac{dn_1}{dt} = A_{n1}n_c(N_1 - n_1), \tag{12.3}$$

$$\frac{dn_2}{dt} = A_{n2}n_c(N_2 - n_2), \tag{12.4}$$

$$\frac{dn_v}{dt} = X - B_1 n_v(M_1 - m_1) - B_2 n_v(M_2 - m_2), \tag{12.5}$$

$$\frac{dm_1}{dt} + \frac{dm_2}{dt} + \frac{dn_v}{dt} = \frac{dn_1}{dt} + \frac{dn_2}{dt} + \frac{dn_c}{dt}. \tag{12.6}$$

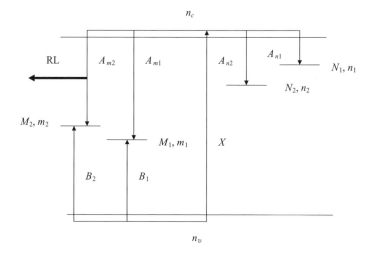

Figure 12.5 *The energy level diagram of two electron-trapping states and two kinds of hole recombination centers used in Pagonis et al. [488]. Transitions occurring during irradiation of the sample are shown by arrows. Reprinted from Pagonis, V., et al., Radioluminescence in Al₂O₃: C–analytical and numerical simulation results, J. Phys. D: Appl. Phys. 42, 175107. Copyright (2009) with permission from Institute of Physics*

The RL intensity is associated with the recombination into m_1; therefore, the intensity $I(t)$ is given by

$$I(t) = A_{m1}m_1 n_c. \tag{12.7}$$

Pagonis *et al.* [488] were able to arrive at an analytical solution to the system of equations (12.1–12.6) by assuming only that the quasi-steady conditions hold, and that the initial concentrations of electrons and holes in the model are $m_1(0) = m_{10}$ and $n_{10} = n_{20} = m_{20} = 0$. Experimentally it is possible to achieve these initial conditions by annealing the material at a high temperature, as is commonly done before using a dosimetric material. After a very short transitional period, a dynamic balance is established between the irradiation process creating pairs of electrons and holes on the one hand, and the relaxation process of electrons and holes into the various energy levels on the other. According to the quasi-static assumption commonly made in kinetic models, one can assume that during this dynamic balance the concentrations of electrons in the conduction band and of holes in the valence band change very slowly, so that $dn_v/dt = 0$ and $dn_c/dt = 0$. By substituting $dn_c/dt \simeq 0$ into Equation (12.5), it can be solved for n_v to yield

$$n_v = \frac{X}{B_2(M_2 - m_2) + B_1(M_1 - m_1)}. \tag{12.8}$$

The initial concentration n_{v0} of holes in the valence band at time $t = 0$ can be obtained by replacing the concentrations m_1 and m_2 in Equation (12.8) with their initial values at time $t = 0$, i.e. $m_2 = m_{20} = 0$ and $m_1 = m_{10}$. Equation (12.8) then yields

$$n_{v0} = \frac{X}{B_2 M_2 + B_1(M_1 - m_{10})}. \tag{12.9}$$

In a similar manner by assuming that during dynamic balance $dn_c/dt = 0$, Equation (12.6) can be solved for n_c to yield

$$n_c = \frac{X}{A_{n1}(N_1 - n_1) + A_{n2}(N_2 - n_2) + A_{m1}m_1 + A_{m2}m_2}. \tag{12.10}$$

The initial concentration n_{c0} of electrons in the conduction band at time $t = 0$ is obtained by replacing the concentrations n_1, n_2, m_1 and m_2 with their initial concentrations in Equation (12.10) to obtain

$$n_{c0} = \frac{X}{A_{n1}N_1 + A_{n2}N_2 + A_{m1}m_{10}}. \tag{12.11}$$

For short irradiation pulses, the RL intensity $I(t)$ can be expanded as a Taylor series about $t = 0$

$$I(t) = C_1 + C_2 t + O(t^2), \tag{12.12}$$

with the coefficients C_1 and C_2 given by

$$C_1 = I(0), \tag{12.13}$$

$$C_2 = \frac{dI}{dt}|_{t=0}. \tag{12.14}$$

The calculation of C_1 is straightforward by using Equation (12.7) for $I(t)$ and Equation (12.11) for the initial quasi-steady equilibrium value of n_{c0}

$$C_1 = I(0) = A_{m1}m_{10}n_{co} = A_{m1}m_{10}\frac{X}{A_{n1}N_1 + A_{n2}N_2 + A_{m1}m_{10}}. \tag{12.15}$$

Equation (12.15) indicates that the initial RL intensity I_0 will be proportional to the dose rate X used during the RL experiment, in agreement with the empirical data shown in Figure 12.4(b). The calculation of the constant coefficient C_2 in Equation (12.14) is much more complex algebraically, and is derived in detail in Pagonis *et al.* [488]. The final analytical result expresses the coefficient C_2 in the general form

$$C_2 = EX^2 + F, \tag{12.16}$$

where the coefficients E and F are functions of the transition probabilities A_{n1}, A_{n2}, A_{m1}, A_{m2} and of the concentrations $N_1, N_2, M_1, M_2, m_{10}$. Equations (12.12) and (12.15) show that for a constant dose rate X the RL intensity $I(t)$ will be increasing linearly with time t, in agreement with the empirical data shown in Figure 12.4(a). The slope of this linear dependence of $I(t)$ is given by the constant C_2, and is proportional to the square of the dose rate X^2. This prediction from the model is in agreement with the empirical data shown in

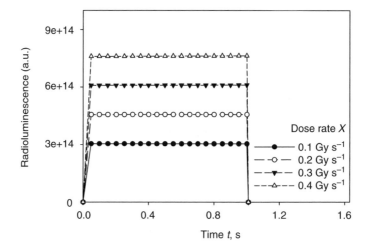

Figure 12.6   *Simulated RL signal as a function of irradiation time t for several different values of the dose rate X. The solid lines show the calculated RL signal using the analytical expressions in Pagonis et al. [488]. Reprinted from Pagonis, V., et al., Radioluminescence in Al_2O_3: C – analytical and numerical simulation results, J. Phys. D: Appl. Phys. 42, 175107. Copyright (2009) with permission from Institute of Physics*

the inset of Figure 12.4(b). One can also obtain analytical expressions for the electron and hole concentrations $m_1(t)$, $n_1(t)$, $n_2(t)$, $m_2(t)$ as shown in detail in Pagonis *et al.* [488].

The numerical values used in the simulations by Pagonis *et al.* [488] are those obtained previously by Pagonis *et al.* [38] for sample Chip101, and were based on the experimental data of Yukihara *et al.* [456]. The simulation consists of two stages; first, in the irradiation stage, the differential equations (12.1–12.7) are solved for an irradiation time t_D. In the next stage of the simulation, a relaxation period was simulated by setting the excitation dose rate X to zero, and solving the same set of equations for a short period of time so as to have the concentrations of the electrons, n_c, and free holes, n_v, go to negligible values. The initial values of the concentration functions for the relaxation stage are the final values at the excitation stage. Figure 12.6 depicts the simulated RL intensity for a total irradiation time of 1s, using the parameters of sample Chip101 and for several different values of the dose rate X. The simulated RL signals in Figure 12.6 are similar in shape to the experimentally measured RL signals shown in Figure 12.4(a). The lines through the simulated data points represent the analytical solutions obtained in Pagonis *et al.* [488]. The agreement between the simulation results and the analytical expressions is very good for all times during the short irradiation pulses.

Figure 12.7 shows the simulated initial RL signal I_0 obtained from Figure 12.6 plotted as a function of the dose rate X, yielding a linear dependence. This result is in agreement with the empirical dose-rate dependence of the initial RL intensity. Similarly, the simulated dependence of the slope of the linear part of the RL signal in Figure 12.6, as a function of the square of the dose rate X^2 was found to agree with the analytical equations. This is shown in the simulated results of Figure 12.8. An important experimental consideration is whether the $Al_2O_3 : C$ samples can be reused during a medical radiotherapy session. For example,

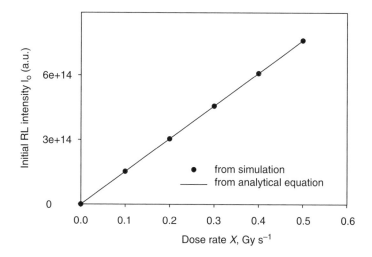

Figure 12.7 *The simulated initial RL signal I_0 plotted as a function of the dose rate X, yielding a linear dependence. This result is in agreement with the empirical experimental data shown in Figure 12.4(b), and with the exact analytical expression in Pagonis et al. [488], shown as a solid line. Reprinted from Pagonis, V., et al., Radioluminescence in Al_2O_3: C – analytical and numerical simulation results, J. Phys. D: Appl. Phys. 42, 175107. Copyright (2009) with permission from Institute of Physics*

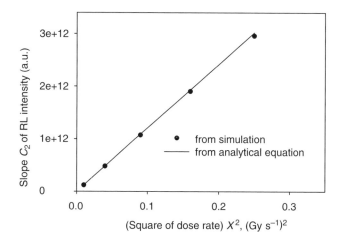

Figure 12.8 *The slope of the linear part of the simulated RL signals in Figure 12.7 is plotted as a function of the square of the dose rate X^2. This dependence is linear, as observed empirically previously in the experimental data of Figure 12.4(b). The solid line represents the exact analytical expression given in Equation (12.16). From Pagonis et al. [488]*

the system developed by Andersen *et al.* [954] provides *in-vivo* measurements of the dose rate X during radiotherapy. Immediately after the treatment the CW-OSL signal is used to determine the integrated dose, by optically bleaching the $Al_2O_3 : C$ samples (see, for example, Aznar *et al.* [941]). Pagonis *et al.* [488] simulated the delivery of 100 successive RL pulses, with each pulse followed by an optical bleach. These authors found that there was a very small change in the shape of the RL pulse.

Polf *et al.* [694] explained the linear dependence of the initial RL intensity I_0 on the dose rate X as due to sensitivity changes taking place in the material during irradiation. Specifically, they described a similar increase in the RL intensity and stated that 'in a general model, an increase in the RL is expected with the filling of all electron traps, including shallow and deep electron traps, as the trapping probability decreases and more electrons become available for recombination'. The results presented by Pagonis *et al.* [488] show that at least within the range of parameters used in their model, this linear dependence of I_0 on the dose rate X is actually due to the increase in the concentration of holes (m_1) in the radiative recombination center during irradiation. This increase leads to an apparent change in the RL sensitivity of the sample. However, it is quite possible that in a different sample of this material, in which the traps and centers are closer to saturation, the experimentally observed sensitivity change is caused by the filling of the traps and centers as saturation is approached.

There have been additional efforts to simulate RL phenomena in quartz, feldspars and other materials. Trautmann *et al.* [955] investigated the RL signals from different types of feldspars by using spectral measurements, and suggested the possibility of using these signals for a new dating technique. These authors interpreted the production of the RL signal as a competition between charge trapping and recombination, and carried out simulations using a simple OTOR model. They simulated both the trap filling and the luminescent recombination for different parameters in their simple model. In another work, Trautmann *et al.* [956] presented a more complex model for feldspar RL, consisting of one trap and two recombination centers. Using this model they were able to simulate RL signals in both the IR (IR-RL) and visible (VIS-RL) regions of the spectrum. By employing the principle of detailed balance, an analytical expression was derived for the intensity of the IR-RL signal. In later work [957], these authors discussed the IR-RL signal from K-feldspars, and suggested this as the basis of a more precise and accurate method for sediment dating. They simulated both the RL and IR-OSL signals on the basis of an updated band model containing also a localized transition, and were able to simulate the process of charge accumulation using a very low dose rate corresponding to the slow irradiation of the sediments in nature.

Edmund and Andersen [958] used a generalized OSL model proposed by McKeever and co-workers [74, 560, 561], to simulate the effect of temperature on OSL response of Al_2O_3:C. Their model includes a shallow trap, the main dosimetric trap, a deep electron trap, a radiative center and a non-radiative center. The simulations showed that the OSL response changes with both irradiation and stimulation temperature, as well as with the chosen OSL integration time. However, it was found that the RL response only depends on the irradiation temperature.

13

Radiophotoluminescence

When subjected to short-wavelength radiation such as UV light, many materials emit light of a longer wavelength. For some materials such luminescence increases strongly after the material has been exposed to radiation. When this change in luminescence is relatively permanent, the luminescence is called radiophotoluminescence (RPL). It is assumed that the defects involved in the RPL process are created by the radiation in a non-ionizing process. The luminescence signal from the RPL process is due to a transition between the excited and the ground state of the defect. Like TL and OSL, this effect can be used for dosimetry. Unlike TL or OSL, the read-out process does not remove or reduce the RPL signal. A distinguishing feature of RPL is that read-out of the dosimeter can be done repeatedly with almost no loss of signal [74].

In the next section we first present the development and use of radiophotoluminescent materials as a tool in dosimetric applications. Next, a simple model is presented which demonstrates the basic principle behind RPL dosimetry.

13.1 Development and Use of RPL Materials

Although the RPL phenomenon has been known for almost 100 years, interest in RPL for dosimetry was limited because the early materials had less favorable properties than TL and OSL materials. For one, the energy response to ionizing radiation did not simulate biological response as well as TL materials. In addition, the materials suffered from poor sensitivity and there were problems with interpreting low-dose signals. Recently, there have been important improvements in both areas. LiF is one the dosimetric materials which has been used as a high-dose RPL dosimeter [959–961]. Piesch and co-workers [962–965] developed phosphate glass dosimeters for use in personal dosimetry.

From the early studies of RPL glasses it was observed experimentally that there is some RPL signal even at zero dose. The existence of a nonzero signal, called the *RPL predose*, makes accurate measurement of small dose levels problematic. It should be noted that the

Thermally and Optically Stimulated Luminescence: A Simulation Approach, First Edition. Reuven Chen and Vasilis Pagonis. © 2011 John Wiley & Sons, Ltd. Published 2011 by John Wiley & Sons, Ltd.

predose effect in RPL and the predose effect in *quartz* discussed elsewhere in this book, represent two physically different and unrelated phenomena. One possible explanation of the predose in RPL glasses is incomplete annealing of the traps. Although annealing is intended to empty the trap prior to irradiation, this might not be always the case here. For example, it is known that annealing does not empty the radiative center in Al_2O_3:C. Similarly, it might be that the initial concentration of traps in RPL materials is not zero and this could cause an RPL predose effect. However, it is observed in time-resolved experiments that much of the RPL predose fluorescence decays at different time scales than the RPL fluorescence. This indicates that the RPL predose fluorescence may have a different physical origin. For one, there are other fluorescence centers whose excitation spectra and emission spectra may overlap those of the RPL centers. This effect can be reduced but not eliminated, by the use of optical filters to select excitation wavelengths, by using a broadband source, and by selecting emission wavelengths [966, 967]. It has also been found that contamination on the surface of the glass can increase RPL predose emissions. This is attributed to fluorescence centers being formed on the surface. As this is a surface phenomenon, while RPL is a volume phenomenon, the importance of this signal is geometry dependent. To counter this effect, various cleaning procedures have been developed [968].

In an early experimental study of RPL dosimetry systems Piesch *et al.* [963] described different UV pulsed laser excitation modes for RPL glasses, as opposed to using mercury lamps. These authors presented data on the dosimetric properties, photon energy response, lowest detectable doses and random uncertainties in RPL automated systems. Piesch *et al.* [964] pointed out that the pulsed UV laser excitation allows the predose signal to be subtracted internally and directly during the read-out process. This reduces the predose value by two orders of magnitude and makes cleaning of the glass unnecessary. The UV pulses have a duration of 4 ns, and the intensity of the RPL signal is integrated in two time intervals. The signal between 40 μs and 45 μs is the radiation-independent long-lived component, and a short-lived component is measured between 2 μs and 7 μs. The fully automated RPL system manufactured by Toshiba also offers automatic calibration using the same dosimeter glass, automatic subtraction of PM dark current, simultaneous measurements of different radiation quantities, indication of radiation quality and estimation of the angle of incidence. The system is equipped with two main diaphragms which allow the dosimeter to be used in three different read-out modes. By using horizontal and vertical scanning of the RPL glass volume, the system obtains information about radiation quality and about the angle of radiation incidence. The authors concluded that the performance of the RPL system was comparable with the performance of highly sensitive TLD systems.

In an extensive review of RPL glass dosimetry Piesch *et al.* [965] discussed the originally limited use of these materials in dosimetry as due to the high predose effect and also due to their strongly energy-dependent response. They pointed out the advantages of RPL glasses as: simplicity of read-out, very good batch uniformity, good dosimetric properties, and capability for repeated read-out and long-term dose accumulation and stability of the signal over time. These authors discussed the properties of the automated Toshiba reader which is based on intrinsic suppression of the predose effect, and as a result allows accurate dose measurements in the range 10–100 mSv, and long-term stability of the read-out signal. Piesch *et al.* [965] discussed the results of the IAEA intercomparison program for personnel dosimeters, which compared the performance of several dosimetric systems (TLD, TLD plus film, film, TLD and RPL). It was concluded that RPL glasses are ideal for long-term monitoring periods of 6–12 months or longer, and for detection of low level exposures in individual and environmental radiation monitoring.

One dosimetry area in which RPL dosimeters seem to have a clear advantage is as environmental dosimeters, which are used to monitor the external radiation dose to the general public. In recent years, RPL glass dosimeters (RPLGDs) have also attracted attention due to their good dosimetric characteristics. When a RPLGD is irradiated by ionizing radiation, stable centers (Ag^0, Ag^{2+}) are formed; under illumination from the 337 nm pulsed UV laser, these centers will be excited. Upon subsequent return to a stable energy level, they emit an orange RPL signal. The RPL intensity is proportional to the dose received by the glass dosimeter.

Lee *et al.* [969] compared the dosimetric characteristics of environmental RPLGD and TLD systems used at the Institute of Nuclear Energy Research in Taiwan. As expected, the TLD signal was destroyed by the reading process, while RPLGDs can be repeatedly read over time. Fading for the RPL dosimeters was found to be about 1% within 30 days after exposure to ionizing radiation. The cumulative doses from the environment were measured by both types of dosimeters and were compared at the same monitoring points over a period of 2 years. The main difference found between results with the two types of dosimeters was the stronger dependence on the ambient temperature for TLDs. These authors concluded that both types of dosimeters could meet their test criteria. During this study a pressurized ionization chamber was used to monitor the dose rate, and the cumulative doses determined by RPLGD and TLD were compared with those of the ionization chamber. In general, this study found good correlation between the results from the two types of dosimeters, except the TLD-values tend to become lower within a year due to fading effects. Seasonal variations in the average values of the measured doses were correlated to ambient-temperature changes.

It is rather surprising that there are very few published studies of the RPL mechanisms in glasses, even though this type of dosimeter has been in use for more than 50 years. As a result, the RPL mechanism is not yet fully understood. Some of the early mechanism studies were reported in the book by Perry [967]. Recently Miyamoto *et al.* [970] investigated the RPL emission mechanism in silver-doped phosphate glasses. These authors presented experimental data on absorption spectra of the X-ray irradiated glasses, as well as data on their RPL emission and excitation spectrum. They also studied the changes taking place in the RPL spectrum after X-ray irradiation and the associated ''build-up effect''. Finally these authors presented fluorescent spectra of the Ag-doped phosphate glass before and after femtosecond pulsed laser irradiation. The authors proposed that the emission mechanism for the two RPL peaks measured at 460 nm and 560 nm involves electrons and holes produced by X-ray irradiation. These carriers are trapped at Ag^+ ions to produce Ag^0 and Ag^{2+} ions, respectively.

In a recent study Rah *et al.* [971] studied the dosimetric properties of commercially available glass dosimeters, in order to evaluate their use for *in-vivo* dosimetry. This type of dosimetric study is usually performed using a variety of radiation detectors, such as silicon diodes, TLDs, OSL dosimeters, MOSFET dosimeters, and radiochromic films. The authors reported on the precision of dose measurements carried out in patients who received a total body irradiation (TBI) treatment. In addition to the glass dosimeters, this study used also TLDs and MOSFETs. The results from this study demonstrated that the glass dosimeter measurements are accurate and reproducible during TBI treatment of patients.

Hsu *et al.* [972] studied four series of glass dosimeters which were prepared from reagent powders of appropriate chemical compounds. They reported on some of their important physical characteristics including reproducibility of read-out values, dose linearity, energy dependence and angular dependence. Rah *et al.* [938] studied the dosimetric properties of a commercially available RPL glass rod dosimeter, and compared them with those of LiF-TLD powder. They found that the RPL glass rod dosimeter had better reproducibility than the TLD powder for a ^{60}Co beam, and also for the clinically used photon beam. This comparative study of several dosimeters concluded that the RPL glass rod dosimeter is suitable for a mailed dosimetry program. In another experimental study, Ugumori [973] reported on intensity-dependent effects. Their study used an N_2 laser ($\lambda = 337$ nm) with a pulse width of 5 nm and a peak intensity of about 1 MW cm^{-2}. Lommler *et al.* [974] reported that an argon ion laser ($\lambda = 334$ nm) can create RPL centers via two-photon absorption.

In the next section we present a simple model for RPL phenomena, and demonstrate that the RPL signal is indeed proportional to the radiation dose received by the glass dosimeter.

13.2   The Simplest RPL Model

To understand RPL, one can consider the energy level diagram shown in Figure 13.1. The material is subjected to a source of short wavelength light of intensity I_{in}. Electrons trapped in trap N_1 are excited into the excited state represented by trap N_2 during the irradiation process which takes place with a dose rate X. Those electrons excited into N_2 may decay radiatively back into N_1 by emitting a longer-wavelength light of intensity I_{out}. The governing equations are

$$\frac{dm}{dt} = B(M - m)n_v, \tag{13.1}$$

$$\frac{dn_1}{dt} = A(N_1 - n_1)n_c, \tag{13.2}$$

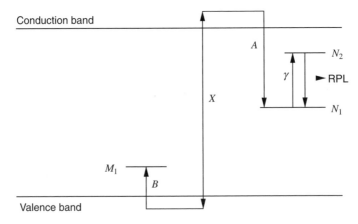

Figure 13.1 *The energy level diagram is shown for a one-electron-trap and one-hole-trap model of RPL*

$$\frac{dn_c}{dt} = X - A(N_1 - n_1)n_c, \tag{13.3}$$

$$\frac{dn_v}{dt} = X - B(M - m)n_v. \tag{13.4}$$

In these equations A and B represent the transition probabilities for the processes shown in Figure 13.1. N_1, N_2, M and n_1, n_2, m represent the total and instantaneous concentrations of carriers in the ground state, excited state and recombination center, respectively. These equations can be re-arranged and integrated. To start, let us substitute Equation (13.3) in Equation (13.2) to find

$$\frac{dn_1}{dt} = X - \frac{dn_c}{dt}. \tag{13.5}$$

We can immediately integrate this from $t = 0$, which is before irradiation starts, up to some time t_r after the irradiation and relaxation periods, to find

$$n_1(t_r) = D + n_1(0) + n_c(0) - n_c(t_r), \tag{13.6}$$

where D is the total dose

$$D = \int_0^\infty X(t)dt . \tag{13.7}$$

To simplify this, note that there are no free electrons before irradiation, $n_c(0) = 0$. Similarly the free electron concentration has again dropped to zero after relaxation, so that $n_c(t_r) = 0$. Also, we assume that the electron trap population was emptied before irradiation by annealing, so that $n_1(0) = 0$. Consequently

$$n_1(t_r) = D. \tag{13.8}$$

A similar analysis shows that the hole trap concentration after relaxation is

$$m(t_r) = D. \tag{13.9}$$

That concludes the analysis of the irradiation stage.

Now let us consider the read-out stage. The material is subjected to a source of short wavelength light of intensity I_{in}. As a result, some electrons in trap N_1 are excited into trap N_2. Those electrons excited into N_2 may decay radiatively back into N_1 by emitting a longer-wavelength light of intensity I_{out}. The governing equations during the read-out process are

$$\frac{dn_1}{dt} = \beta n_2 - \alpha I_{in} n_1, \tag{13.10}$$

$$I_{out} = \beta n_2. \tag{13.11}$$

In these equations α and β represent the transition probabilities for the transition processes I_{in} and I_{out} shown in Figure 13.1. When the intensity of the exciting beam I_{in} is constant, we can look for the steady–state solution which is defined by $dn_1/dt = dn_2/dt = 0$. It follows from these equations that

$$n_2 = \alpha I_{in} n_1/\beta, \tag{13.12}$$

and

$$I_{out} = \alpha I_{in} n_1. \tag{13.13}$$

From conservation of charge we know that $n_1 + n_2$ during read-out must equal $n_1(t_r)$ at the end of the irradiation. Combining this with Equation (13.8), we have

$$D = n_1 + n_2 = n_1 + \alpha I_{in} n_1/\beta, \tag{13.14}$$

or

$$n_1 = \frac{D}{1 + \alpha I_{in}/\beta}. \tag{13.15}$$

This can be combined with Equation (13.13) to find the RPL intensity as a function of the dose D,

$$I_{out} = \frac{\alpha I_{in}}{1 + \alpha I_{in}/\beta} D. \tag{13.16}$$

If the light source used for I_{in} were an intense pulsed laser, then the denominator above will be important. However, if the light sources used during the RPL experiment are lamps or LEDs, these sources are of low enough intensity that the RPL signal will likely be well approximated by

$$I_{out} \approx \alpha I_{in} D. \tag{13.17}$$

Equation (13.17) is the result that we were looking for. It shows that the RPL signal I_{out} is proportional to the excitation intensity I_{in}, and also proportional to the accumulated dose D.

Experimentally it is observed that unlike the ideal RPL intensity of Equation (13.17), there is some RPL signal even at zero dose. Specifically it is found empirically that the RPL intensity obeys the equation

$$I_{out} = \alpha I_{in}(D + D_p), \tag{13.18}$$

where D_p is a constant, often in the range of $1\,\text{mGy}$–$100\,\text{mGy}$. The existence of a non-zero D_p, called the RPL predose effect, made accurate measurement of small dose levels problematic in early RPL dosimetry.

14

Effects of Ionization Density on TL response

14.1 Modeling TL Supralinearity due to Heavy Charged Particles

It has been suggested by Horowitz et al. [975, 976] that microdosimetric phenomena, not addressed by conventional kinetic calculations in the framework of conduction-band/valence-band models, play an important role in the TL dose response. These authors explain how the track-interaction model (TIM), which incorporates first, second and third nearest neighbor interactions, describes the α-particle induced linear–supralinear behavior in some peaks in LiF:Mg,Ti. The increasing supralinearity with glow-peak temperature is shown to arise from increasing charge carrier migration distances in the luminescent recombination stage, 150, 300, 3000 and 5000 Å for peaks 5, 7, 8 and 9, respectively. Along the same lines, McKeever [320] has pointed out that in the kinetic model previously suggested by Chen and Bowman [617] and by Chen et al. [977], it is impossible to predict the linear-supralinear behavior in LiF:Mg, Ti. The reason is that if the supralinearity is ascribed to the excitation stage, one would expect a supralinear dose dependence of the optical absorption (OA) in LiF:Mg, Ti, which has not been observed. McKeever proposes, in the framework of the 'defect interaction' model, the existence of spatially associated trapping center-luminescence center (TC-LC) configurations which trap electron-hole pairs, and which lead to the linear-supralinear behavior. Horowitz [975] points out that kinetic models deal with average concentrations of trapped and released carriers in the conduction band and valence band, as if these concentrations were independent of position within the crystal. In other words, they deal with homogeneous, uniform systems and can therefore directly apply only to homogeneous, uniformly irradiated systems. This requirement is achieved to a good approximation by γ- and electron-irradiated systems at an initial radiation energy of a few hundred keV or more. Lower energy radiations below a few hundred keV, neutrons and heavy charged particles create non-uniform microscopic patterns of ionization density. This is especially true in the case of heavy charged particles (HCPs); for example, for low

Thermally and Optically Stimulated Luminescence: A Simulation Approach, First Edition. Reuven Chen and Vasilis Pagonis. © 2011 John Wiley & Sons, Ltd. Published 2011 by John Wiley & Sons, Ltd.

energy α-particles the dose varies by as much as seven orders of magnitude over a distance of 200 Å measured from the track axis.

In the heating stage, the charge carriers are released into the conduction band only in the vicinity of the track. The charge-carrier concentrations will be in saturation near the track axis and very high in or near the track. Between the tracks, the charge-carrier concentrations will be non zero if there is significant charge-carrier diffusion. The strong dependence of TL superlinearity in LiF:Mg, Ti on ionization density implies that the wave packet of the thermally freed charge carriers is localized around the HCP track. Therefore, any discussion of TL supralinearity in non-uniformly irradiated systems must take into account the microscopic distribution of ionization density in the radiation absorption stage. The track-interaction model is a competition-during-heating model with the defect distribution modified in accordance with the ionization density of the radiation used. Horowitz [975] explains, however, that this microscopic dose distribution determines the dose-dependent increase in luminescent recombination efficiency and helps understand the complex behavior of TL supralinearity in LiF:Mg, Ti. This supralinearity depends on the ionization density on one hand, and on the glow-peak temperature on the other. The dependence on the ionization density manifests itself in several ways. One of them is the dependence of supralinearity on γ-ray energy. Recalling the supralinearity index (factor) defined in Section 8.1 which measures the degree of supralinearity, Horowitz [975] reports that for ^{60}Co and ^{137}Cs energies the supralinearity factor of peak 5 is $\sim$3.5, does not increase at higher energies but begins to decrease below $\sim$300 keV. For α-particles (1 MeV per amu) and neutrons of very high ionization density, the TL dose response is strictly linear–sublinear. As the α-particle energy, is increased to 16.8 MeV, however, the TL dose response again becomes linear–superlinear and the supralinearity increases with increasing α-particle energy, as pointed out by Mische and McKeever [601]. Almost the same behavior has been observed for peak 7. The ^{60}Co dose response is strongly supralinear and no linear dose response region is observed. At 95 keV X-ray energy, however, the supralinearity up to a dose level of 1 Gy has almost completely disappeared. The α-particle induced TL dose response shows only mild supralinearity. At very high temperatures (peaks 8 and 9), both γ and α-particles induce strong supralinearity and the γ-induced supralinearity is much stronger.

The supralinearity in LiF:Mg, Ti increases with increasing glow-peak temperature. This is true for both γ-rays and low energy α-particles. For 4 MeV α-particles the supralinearity factor at a fluence of 2×10^{10} cm^{-2} is 1.0, 1.15, 1.9, and 2.7 for peaks 5, 7, 8, and 9, respectively. The increase of supralinearity with heating rate reported by McKeever [978] is additional evidence of this temperature dependence, since the glow-peak temperature increases with heating rate.

Horowitz and his group [975] developed the TIM first proposed by Attix *et al.* [495, 496]. They first concentrated on HCPs, the HCP near straight-line track and the nearly cylindrical radial dose distribution creates an amenable geometry for the calculation of track interaction effects [976, 979–981]. In the TIM, illustrated in Figure 14.1, the dose dependence of the luminescence recombination efficiency arises naturally from inter-track recombination, i.e. the possibility of a charge carrier liberated within one track, escaping from the track, avoiding capture in a nonluminescent center between the tracks and recombining in the oppositely charged carrier trapped in a neighboring track. The distances between the tracks at low doses are large enough for the recombination during glow-curve heating to occur between charge carriers and the activated luminescence centers in the same track only.

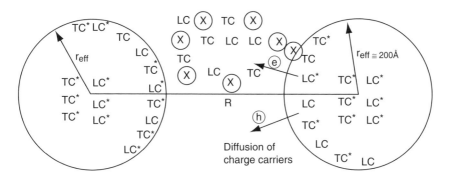

Figure 14.1 *A schematic presentation of the TIM including long-range charge-carrier migration between nearest neighbor tracks. Second and third nearest neighbors also contribute to the supralinearity. TC and LC are the trapping centers and the luminescence centers, respectively, and TC* and LC* are the same centers following activation by charge-carrier capture. ⊗ are unidentified competing centers in the unirradiated region between the tracks. Reprinted from Horowitz, Y.S., Mathematical Modelling of TL Supralinearity for Heavy Charged Particles Radiat. Prot. Dosim. 33, 75–81. Copyright (1990) with permission from Oxford University Press*

The TL dose response is linear in this region since the TL is strictly proportional to the number of tracks. In order for inter-track migration not to occur at low dose levels, the unirradiated regions between the tracks must be inhabited by non radiative competing centers. At higher doses, when the distances between the tracks become comparable with the average separation of the activated centers within the tracks, the probability of inter-track recombination increases and the TL dose response rises more rapidly than linearly. As the incident γ-ray energy decreases, or if the irradiation is via HCP's or neutrons, the tracks become more localized requiring even greater dose levels to initiate the track interaction. Thus, the TIM naturally accommodates the ionization density dependence of TL supralinearity.

As explained by Horowitz [975], since the capture cross-section for Coulombic attractive traps and recombination centers is believed to follow a $T^{-\kappa}$ dependence (see Lax [57] and Section 11.2), the effectiveness of the competing centers may decrease with increasing glow-peak temperature. Alternatively, if the mobile carriers are actually interstitial atoms, as is a possibility for some of the peaks in LiF:Mg, Ti [863], then one would expect that the mobility of the charge carriers would increase with temperature anyway. For both possibilities, the TIM naturally accommodates the generally observed increasing supralinearity with increasing glow-peak temperature. One should note that in order to explain TL supralinearity, the TIM does not require sublinear growth with the dose of the available competitors, in agreement with similar conclusions reached by Mische and McKeever [601] concerning the defect interaction model (DIM). The dose dependence of the competitor could be sublinear, linear or even supralinear and still allow for TL supralinearity. The point is that TIM demonstrates that increased luminescence recombination efficiency can result from dose-dependent changes in the probability distribution function of the distances between the trapping and recombination centers, rather than a dose-dependent decrease in the number of available competitors.

Figure 14.1 shows two neighboring tracks separated by a distance R considerably larger than $2r_0$, where r_0 defines the track radius in which most of the dose is deposited. For 4 MeV α-particles stopping in LiF, 98% of the dose is deposited within 200 Å of the track axis. The probability of a released charge carrier migration from one track to its nearest neighbors can be approximated by the product of a geometrical configuration factor $g(r_0, R)$, and an exponential term $\exp(-\alpha R)$ representing the probability of the migrating charge carrier not being captured by the mentioned competing centers TC and unidentified competing centers denoted by $\otimes$ in Figure 14.1. In his calculation, Horowitz [975] approximates $g(r_0, R)$ by a one-dimensional solid angle factor $2r_0/R$. A parameter α defined as the inverse of the mean-free path λ, is expected to be dependent on both temperature and the identity of the charge carrier. The increase in the TL yield due to the track interactions is calculated by multiplication of $g(r_0, R) \exp(-\alpha R)$ by the nearest neighbor probability functions and integrating over all possible values of R from r_0 to ∞,

$$f(n) = 1 + \int_{r_0}^{\infty} g(r_0, R) e^{-\alpha R} \left[\sum_i P_i(n, R) \right] dR. \tag{14.1}$$

The unity on the right-hand side represents the TL arising from an isolated track. $P_i(n, R)$ is the ith nearest neighbor probability distribution function representing the probability that an α-particle track has its ith nearest neighbor at distance R. n is the α-particle fluence (cm^{-2}). Horowitz and Rosenkrantz [976] have evaluated expression (14.1) for first, second and third nearest neighbor interactions, and the resulting TL fluence response is given by

$$f(n) = 1 + \left[\frac{\pi^2 n^2 r_0^4}{6} - \frac{\alpha \pi n r_0^3}{12} + \left(\frac{\alpha^2}{24} - \frac{5\pi n}{4} \right) r_0^2 - \left(\frac{\alpha^3}{48\pi n} + \frac{17\alpha}{24} \right) r_0 \right]$$
$$\times \exp\left[-\left(\pi n r_0^2 + \alpha r_0 \right) \right] \tag{14.2}$$

$$+ \left[\frac{21\pi\sqrt{n}}{8} + \frac{3\alpha^2}{8\sqrt{n}} + \frac{\sqrt{\pi}\alpha^4}{96(\pi n)^{3/2}} \right] \times r_0 \exp(\alpha^2/4\pi n) \times \mathrm{erfc}\left(\sqrt{\pi n}r_0 + \frac{\alpha}{2\sqrt{\pi n}} \right).$$

This expression is usually not as complicated as it looks since the terms in r_0^4, r_0^3, $n^{-1/2}$ and $n^{-3/2}$ in the brackets are generally negligible. At a fluence of 10^{10} particles cm^{-2}, the relative contribution of the first, second and third nearest-neighbor interactions is 70, 24 and 6%, respectively, for typical values of r_0 and λ. Higher order track interactions contribute negligibly to the supralinearity.

The lower limit of integration in Equation (14.1) is taken as r_0 in order to distinguish between long-range track interactions, adequately described by the nearest neighbor probability distribution functions and the solid-angle factor, and short range track interactions where the two tracks strongly overlap. Horowitz *et al.* [979] have shown that dose overlap in the radiation absorption stage cannot lead to significantly increased TL, due to the counter mechanism of enhanced saturation. Horowitz [975] states that Equations (14.1) and (14.2) are valid up to approximately 2×10^{10} particles cm^{-2} after which multiple track overlap leading to rapid saturation begins to dominate the TL dose response. Horowitz has shown graphs of the TL dose response $f(n)$ evaluated for representative values of r_0 and λ. The threshold for supralinearity appears at approximately 10^8 particles cm^{-2} and, as expected, strongly depends on λ. An increase in r_0 leads to an increase in the maximum value of $f(n)$ due to an increase in the solid-angle factor, as well as a shift of the maximum supralinearity to lower particle fluence levels. For example, at a fluence of 10^{10} α-particles cm^{-2} and a mean-free path of 1000 Å, the supralinearity factor is 1.3, 1.55 and 1.7 for r_0 equal to 100, 200 and 300 Å, respectively. This is consistent with the experimental results of Mische and McKeever [601] who observed an increase in the supralinearity factor as the α-particle energy increased. This arises due to the increasing energy of the secondary electrons liberated by the α-particles leading to an increase in r_0.

Further work on the TIM as related to the TLD-100 supralinear dose dependence has been given by Rodríguez-Villafuerte [982] who used Monte-Carlo simulation. Excellent agreement was found between the experimental data and the simulations which suggests that the contribution of electrons migrating beyond the third neighbor is important, and that saturation effects must be taken into account.

14.2   Defect Interaction Model

Mische and McKeever [601] developed the DIM. As described by Mahajna and Horowitz [983], for isotropically ionizing γ- or electron-induced dose response, recombination within the TC–LC complex (geminate recombination) (see Section 4.6), dominates at low dose levels and leads to a linear response, whereas charge-carrier migration leads to inter-complex recombination, i.e., increased luminescence efficiency at higher dose levels. The same basic physical mechanism leads to supralinearity, i.e., a localized trapping entity (such as the track for HCPs and the TC–LC complex for γ-rays). Due to competitive centers, at low dose levels recombination occurs exclusively within the trapping entity, leading to a linear dose response. At higher dose levels, the efficiency of the competitive centers decreases due to the decreasing distance between adjacent TC–LC complexes, thereby leading to increased luminescent efficiency, i.e. supralinearity. For LiF:Mg, Ti, abundant experimental evidence exists to indicate that the traps (Mg trimers) and the recombination centers (TiOH$_n$) complexes, are spatially correlated, namely, bounded together by some form of long-range interaction into a large complex. The evidence comes from measurements of TL emission spectra [168], X-ray induced luminescence spectra [984] and ionic thermocurrents [985]. It has also been demonstrated that the activation energy of peak 5, as measured by the total peak-shape analysis using the mixed-order kinetic model and computerized glow-curve deconvolution [986], appears to be a function of the Ti concentration [987]. This indicates an influence of the availability of Ti ions on the Mg-related trapping structure.

In the DIM, as developed by Mische and McKeever [601], a certain fraction K of the electron- and hole-trapping states exist in pairs. Within the pairs, the electron and hole traps are separated by a distance s. The distance s is assumed to be small enough so that no other center of any other type can exist between the pair, thus, recombination within the pair cannot be affected by the competitive centers. K, however, is assumed to be independent of dose or γ-ray energy, and is assumed to be set by the equilibrium constant for the reaction leading to the complex during annealing and read–out. In other words, every trapping pair is assumed to be occupied by an electron-hole pair following irradiation. The distance s is also dose independent and γ-ray energy independent, since it is fixed by the material properties and defect structure. The TL intensity as a function of the dose is given by

$$F(D) = \mu n \left[KS/4\pi s^2 + (1 - KS/4\pi s^2)Smd \right], \tag{14.3}$$

where μ is a proportionality constant, S is the recombination trapping cross-section and m and n are, respectively, the dose-dependent occupation probabilities of the hole and electron traps. The mean-free path d is given by

$$d = 1/(Sm + S_c n_c), \tag{14.4}$$

where S_c is the trapping cross-section of the competitive center and n_c is the concentration of the available competitive traps. In the formulation, where N_c is the total concentration of competitive centers, the neutrality condition $(N_c - n_c) + n = m$ is used to determine n_c. Use of the neutrality condition reduces the number of free parameters by one (the filling of the competitor), but limits the generality of the formulation to a single trapping center and recombination center. As stated by Mahajna and Horowitz [983], this is a highly unphysical assumption for LiF:Mg, Ti with its complex glow curve in the temperature range of peak 5. These authors avoid this assumption in their development of the unified interaction model (UNIM) described below. In Equation (14.3), the first term is linear with the dose and the second is nonlinear. Depending on the value of K, the linear term may dominate initially. As the dose increases, the nonlinear term becomes more important and eventually dominates, leading to supralinearity. The exact shape of $F(D)$ depends, of course, on K and the functions $n(D)$, $m(D)$ and $n_c(D)$.

Mische and McKeever [601] showed that Equation (14.3) was capable of predicting the linear–supralinear behavior, but no attempt was made to fit the experimental TL dose response of peak 5 in TLD-100, to demonstrate a capability to predict the abrupt departure from linearity with reasonable values of the model parameters, nor was there an attempt to explain the strong γ-ray energy dependence of $F(D)$.

14.3   The Unified Interaction Model

Mahajna and Horowitz [983] describe the development of a unified theory, the UNIM, which incorporates both TIM and DIM in an identical and consistent conceptual and mathematical formalism. This unified theory is capable of describing all the previously mentioned important features of TL supralinearity and sensitization. The theoretically derived expression for $F(D)$ is applied to the experimentally measured ^{60}Co (1.25 MeV) supralinearity of peak 5 in LiF:Mg, Ti, as well as to the supralinearity of peak 5 as a function of γ-ray energy below 0.5 MeV. When the UNIM is applied to γ-induced supralinearity, the TC–LC complex takes the role of the trapping recombination entity instead of the HCP track, as shown in Figure 14.2. The same formulation, employing the nearest neighbor probability

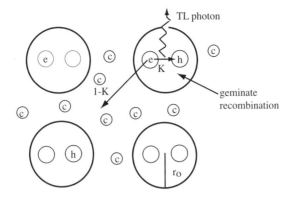

Figure 14.2 *A schematic representation of the UNIM model applied to spatially correlated TC–LC pairs following γ-irradiation. k is the fraction of electrons undergoing geminate recombination, and not subject to competitive processes. The circles labeled c represent competitive unoccupied electron-trapping states in the regions between the trapping states. The large circles represent the TC–LC pairs in the four possible configurations following irradiation. Reprinted from Mahajna, S. and Horowitz, Y.S., The unified interaction model applied to the gamma ray induced supralinearity and sensitization of peak 5 in LiF:Mg, Ti (TLD-100) J. Phys. D: Appl. Phys. 30, 2603–2619. Copyright (1997) with permission from Institute of Physics*

distribution functions used to describe the increase in TL efficiency for track interaction, but modified for defect interaction is given by

$$F(D) = ksn_e + (1 - ks)n_e \sum_{i=1}^{n_h} \int_{r_0}^{\infty} g(r_h, R_i) e^{-\alpha R_i} P_i(n_h, R_i) dR_i, \qquad (14.5)$$

where k is the probability of geminate recombination within the TC–LC complex and s is the fraction of these complexes which have simultaneously captured an electron–hole pair. Note that k and s here are not the same as in the DIM model discussed above. n_e is the concentration of trapped holes.

Following irradiation, the TC–LC complex can exist in four different trapping configurations:

(i) an electron and a hole have been trapped with relative probability s;
(ii) an electron only has been trapped with relative probability $1 - s$;
(iii) a hole only has been trapped with relative probability s;
(iv) no charge has been trapped.

It should be noted that the given probabilities are conditional, where the absolute probability at a certain dose level may not be known. The occupation probability of the first configuration is represented by s and the second configuration by $(1 - s)$. The occupation probability of the third configuration may also be $(1 - s)$, but this is not necessarily required since the model can accommodate more than one type of recombination center. In their formulation of the UNIM, Mahajna and Horowitz [983] allow s to be γ-ray energy dependent, due to the different and more localized patterns of microscopic ionization density as the γ-ray energy decreases. The distance s is not required to be dose dependent due to previous evidence that in the linear–supralinear dose region, multiple capture of charge carriers due to macroscopic increase in the ionization density does not seem to be significant. In other words, the value of s is dependent of the particulars of the microscopic pattern of the ionization density of a single track created by the γ-ray, and does not represent the possible effects of overlapping-intersecting γ-tracks at high levels of dose. Thus, ks is the fraction of trapped electrons which contribute to the linear part of the dose response (not subject to competitive processes), and 1 − ks is the fraction which contributes to supralinearity. This arises because the TL photons can originate from three different processes:

(i) geminate recombination occurring between a trapped electron–hole pair in the same complex and not subject to competitive processes, occurring with probability ks;

(ii) recombination between an electron released by heating from the first configuration occurring with probability s(1 − k), which then migrates in the conduction band and recombines with a hole trapped at a luminescent center and is subject to competitive processes;

(iii) recombination between an electron released by the second configuration upon heating, which migrates in the conduction band and recombines with a hole trapped at a recombination center and is also subject to competition. This process occurs with probability (1 − s).

The fraction of released electrons subject to competition is given by the sum of the second and third processes, i.e., s(1 − k) + 1 − s = 1 − ks, as given in Equation (14.5). Mahajna and Horowitz [983] have defined the normalized dose-response function, f(D) as

$$f(D) = \left[F(D)/D\right] / \left[F(D_L)/D_L\right], \tag{14.6}$$

where $F(D)$ is the TL signal intensity as a function of the dose D and $F(D_L)$ is measured at the initial linear region of $F(D)$. Substituting Equation (14.5) into Equation (14.6) one gets

$$f(D) = \frac{D_l}{D_h} \cdot \frac{n_e(D_h)}{n_e(D_l)} \times \left[1 + \frac{1 - ks}{ks} \int_{r_0}^{\infty} \sum_{i=1}^{n_h} g(n_h, R_i) e^{-\alpha R_i} P_i(n_h, R_i) dR_i\right]. \tag{14.7}$$

The expression inside the square brackets describes the defect interaction and its similarity to Equation (14.1) (amended by Mahajna and Horowitz by multiplying the integral by N_e/N_w, the ratio between the number of electrons which escape the parent track and those undergoing radiative recombination within the parent track during heating) for the HCP track interaction (TIM) is obvious. The only major difference is that (1−ks)/ks replaces N_e/N_w as the fraction of charge carriers contributing to the track–defect interaction. When the UNIM is applied to HCPs, r_0 represents the radius beyond which electrons have a significant probability of escaping from the parent track. In the UNIM applied to γ-rays [Equation (14.7)], r_0 represents a critical radius, beneath which carriers have a significant probability of geminate recombination without being thermally ejected into a transport band. The assumption of a critical radius r_0 seems reasonable while applied to the process of geminate recombination because of the well known strong dependence of tunneling probabilities on distance.

Comparing Equations (14.5) and (14.7), one can see that the evaluation of the solid-angle factor and the distance between nearest neighbor probability density functions undergo some modifications. The solid-angle factor between neighboring spheres, represented by the TC–LC trapping entity and separated by a distance R is now given in a 3D approximation by

$$g(r_h, R) = \pi r_h^2 / 4\pi R^2, \tag{14.8}$$

where πr_h^2 represents the recombination cross-section S_h. The first nearest neighbor probability density function is given in 3D calculation by

$$P_i(n_h, R)dR = n_H C_l 4\pi \cdot R^2 dR \left(1 - \frac{4}{3}\pi R\right)^{n_h - 1}, \tag{14.9}$$

compared with the 2D probabilities used by Horowitz *et al.* [988] for the TIM.

The definition of the mean-free path λ remains physically unaltered in the two applications of the UNIM. For HCPs (TIM), the mean free path is given by

$$\lambda_{HCP} = 1/S_c N_c, \tag{14.10}$$

and is independent of the dose since the region between the well defined HCP track is unirradiated. For γ-rays, the created pattern of ionization density is nearly isotropic so that

$$\lambda_g = 1/S_c(N_c - n_c) = 1/\left[S_c N_c \exp(-\gamma D)\right] = (1/S_c N_c)\exp(\gamma D). \tag{14.11}$$

In accordance with the dose-response behavior of all the absorption bands in irradiated TLD-100, Equation (14.11) assumes that the competitors are filled exponentially, i.e., start linearly and continue toward saturation, where γ is a dose proportionality constant. Comparing Equations (14.10) and (14.11), one obtains

$$\lambda_g = \lambda_{HCP} \exp(\gamma D). \tag{14.12}$$

Equation (14.12) shows how the UNIM can be used to reduce the number of unconstrained parameters by a comparison of TL dose-response data using both HCPs and γ-rays. This comparison is not possible in the formulation of DIM. In Equation (14.11), the mean-free path is a function of three unconstrained parameters, S_c, N_c and γ. The use of Equation (14.12) leaves a single unconstrained parameter γ. The use of expression (14.12) leaves a single unconstrained parameter γ, if for example, $\lambda_{HCP} = 300$–400 Å for peak 5 of TLD-100 is employed, as deduced from near-parallel beam, α-particle induced TL supralinearity as shown by Horowitz *et al.* [988]. The Equations (14.7), (14.9) and (14.12) illustrate the complexity of γ-induced supralinearity which requires knowledge of the filling rate of the electron traps (α'), the hole recombination centers (β) as well as the competitive centers (γ). Based on the filling behavior of the optical absorption bands in LiF:Mg, Ti, these are assumed to be given, respectively, by

$$n_e = N_e \left[1 - \exp\left(-\alpha' D\right)\right], \tag{14.13}$$

$$n_h = N_h \left[1 - \exp\left(-\beta D\right)\right], \tag{14.14}$$

and

$$n_c = N_c \left[1 - \exp\left(-\gamma D\right)\right]. \tag{14.15}$$

Mahajna and Horowitz [983] mention that $n_c(D)$, as well as $n_h(D)$, could be sublinear, linear or supralinear, or various combinations of these, and still allow TL supralinearity in the framework of UNIM. UNIM demonstrates that increased luminescence recombination efficiency can arise from dose-dependent changes in the probability distribution functions of the distances between the trapping and recombination centers, rather than a dose-dependent decrease in the number of available competitors.

Substitution of Equations (14.8), (14.9) and (14.13–14.15) as well as the nearest neighbor probability distribution functions into (14.7) yields an expression with the parameters [983]: $\alpha', kS, N_h, \beta, r_0, r_h, S_c, N_c$, and γ. Unlike the expressions for HCPs, the integrals in Equation (14.7) and in the combined expression must be evaluated numerically.

Mahajna and Horowitz [983] conclude that the model combines the features of the DIM for γ-rays (uniformly ionizing radiation) and the TIM for densely ionizing HCPs, into a unified framework which allows the same mathematical formalism and physical concepts to explain the unique features of peak 5 supralinearity. In particular, it explains the strictly linear, then supralinear behavior and the dependence of the supralinearity on ionization density, i.e., γ-ray energy and radiation type. Both these features arise from a localized trapping entity, the track for HCPs and the spatially correlated TC–LC pairs for γ-rays and electrons, which dominates the dose response at low dose. At low dose, competitive processes during the glow-curve heating stage suppress luminescent recombination between nearest neighbors leading to the linear response. At higher doses, the efficiency of radiative recombination increases, due to the decreasing distances between the nearest neighbors and possibly to early filling of the competitors. The UNIM is based on both absorption stage and recombination stage mechanisms. A significant achievement of the model is its ability to explain the γ-energy dependence of the supralinearity, in terms of the increasing population of the spatially correlated TC–LC complex by an electron–hole pair as the γ-energy decreases.

In later publications, Horowitz and co-workers [635, 989, 990] further discuss the models dealing with the effects of ionization density, considering successes, failures and conflicts within these models. Superlinearity as well as sensitization have been discussed. More work on the UNIM and its application in the investigation of the X-ray dose response of TLD-100 has been reported by Livingstone *et al.* [991].

15

The Exponential Integral

15.1 The Integral in TL Theory

In TL, when the conventional linear heating function is utilized, the exponential integral or its variations should be evaluated. To begin with, for any heating scheme, the relevant expressions involve the integral

$$\int_0^t s \exp(-E/kT)dt. \tag{15.1}$$

This expression occurs in situations of first order, second order, general order and mixed order as mentioned earlier, and it takes place in TL, TSC and TSEE discussed in this book as well as in other thermally stimulated phenomena such as differential thermal analysis (DTA), thermally stimulated depolarization currents (TSDC), thermogravimetry (TG) and thermal desorption. In many cases, the frequency factor is considered to be constant, in which case the expression to be evaluated is

$$s \int_0^t \exp(-E/kT)dt. \tag{15.2}$$

In the usual case where the linear heating function, $T = T_0 + \beta t$ is utilized, Equations (15.1) and (15.2) change into

$$(s/\beta) \int_{T_0}^T \exp(-E/k\theta)d\theta, \tag{15.3}$$

which appeared in the previous chapters several times. As mentioned in the literature [460, 545–547, 992], a hyperbolic heating scheme,

---

Thermally and Optically Stimulated Luminescence: A Simulation Approach, First Edition. Reuven Chen and Vasilis Pagonis. © 2011 John Wiley & Sons, Ltd. Published 2011 by John Wiley & Sons, Ltd.

$$T = T_0/(1 - \beta' t) \tag{15.4}$$

where β' (s^{-1}) is a constant, has a significant theoretical advantage, since it changes the integral in Equations (15.2) into

$$\alpha \int_{T_0}^{T} (1/\theta)^2 \exp(-E/k\theta)d\theta \tag{15.5}$$

where $\alpha = \beta'/T_0$. This is an elementary integral since

$$\int_{T_0}^{T} (1/\theta)^2 \exp(-E/k\theta)d\theta = (k/E)[\exp(-E/kT) - \exp(-E/kT_0)]. \tag{15.6}$$

However, most investigators of TL and other thermally stimulated phenomena employ a linear heating function, and the integral (15.3) must be dealt with. As shown below, the additional computational effort needed for the evaluation of the integral with the linear heating scheme is a moderate price to pay for avoiding the experimental difficulty in producing a hyperbolic heating function.

The integral in Equation (15.3) can obviously be written as

$$\int_{T_0}^{T} \exp(-E/k\theta)d\theta = F(T, E) - F(T_0, E) \tag{15.7}$$

where

$$F(T, E) = \int_{0}^{T} \exp(-E/k\theta)d\theta. \tag{15.8}$$

The problem thus reduces to the evaluation of the integral in (15.8) since the integral in (15.7) has been shown here to be the difference between two expressions of the same form, one with T and the other with T_0. In fact, it will be shown below that $F(T, E)$ is a very fast increasing function of T, and unless $T \cong T_0$, the approximation made by taking

$$\int_{T_0}^{T} \exp(-E/k\theta)d\theta \cong F(T, E) \tag{15.9}$$

is usually very good.

15.2   Asymptotic Series

As shown by Grossweiner [254] and Haake [993], repeated integration of the integral in Equation (15.8) yields

$$F(T, E) = T \exp(-E/kT) \sum_{n=1}^{} (kT/E)^n (-1)^{n-1} n! \tag{15.10}$$

The number of terms to be taken in this summation for best accuracy will be discussed below. It can be seen that, as suggested above, due to the exponent, $F(T, E)$ is a very fast growing function of T for a given value of E. We are going to continue discussing Equation (15.10), however, a slightly different notation has been given by Biegen and Czanderna [994] who wrote

$$F(T, E) = T \exp(-E/kT) + (E/k)E_i(-E/kT) \tag{15.11}$$

where $E_i(-x)$ is the exponential integral defined by

$$-E_i(-x) = \int_0^\infty (e^{-u}/u)du \tag{15.12}$$

which can be written as the asymptotic series

$$-E_i(-x) = e^{-x}(1/x - 1/x^2 + 2!/x^3 - 3!/x^4 + \ldots). \tag{15.13}$$

In fact, Equations (15.10) and (15.13) are practically the same. The main feature of these expansions is that the terms go to infinity in absolute value as $n \to \infty$; thus, an asymptotic series is divergent, and obviously, taking a large number of terms may not yield the desired result. However, these series have properties that permit a very good approximation for the integral in question. In fact, under certain circumstances explained below, even the first few terms may give a reasonably good approximation. This approximation has been used (and sometimes misused) in some approximate semi-analytical expressions for the thermally stimulated formulae. Gartia *et al.* [995] showed, however, that although taking only the first term in the approximation yields a relatively small change in the appearance of the peak, the consequences in terms of the evaluated activation energy as found by the curve-fitting method may be substantial. The reason is that the integral may appear in an exponential, such as in Equation (4.11), and therefore small errors in its value may result in large variations in the full expression. This problem may apply to other analytical methods for evaluating the trapping parameters. Therefore, the use of more terms in the series is advisable.

Before discussing the best approximation that one can achieve using the asymptotic series, let us present the formal definition of an asymptotic series due to Poincaré (see Dingle [996]). Given a function $f(z)$, the series $\sum_{r=1}^n a_r/z^{r-1}$ is its asymptotic expansion if

$$\lim_{|z|\to\infty} \left[f(z) - \sum_{r=1}^n a_r/z^{r-1} \right] = 0, \tag{15.14}$$

for every value of n. We will be more interested here in the more restricted, and easier to handle case of alternating sign asymptotic series as seen in Equations (15.10) and (15.13). As shown by Airey [997], when truncating such a series at a certain term n, the possible deviation from the desired function does not exceed the $(n+1)th$ term. The absolute values of the terms in Equations (15.10) and (15.13) decrease with n from $n = 1$ to a certain $n = N$, from which the terms start increasing in an unbounded manner. This can be seen by looking at the absolute value of the ratio between two subsequent terms,

$$|a_{n+1}/a_n| = \left[(kT/E)^{n+1}(n+1)! \right] / \left[(kT/E)^n n! \right], \tag{15.15}$$

which yields

$$|a_{n+1}/a_n| = (kT/E)(n+1). \tag{15.16}$$

The range of interest in the study of TL and other thermally stimulated processes is normally $E/kT \geq 10$. In most cases the value of this quantity is in fact 20 or more, in which case taking just two terms in Equation (15.10) results in an error which does not exceed 1.5%. The utilization of two terms has been reported by some researchers in the development of approximate expressions for thermally stimulated processes. However, one would like to study the asymptotic series more deeply and find the best accuracy that one can achieve by using it [33]. Inspecting Equation (15.16), one can immediately see that the absolute values of the terms decrease from $n = 1$ to the term for which $n(kT/E) \cong 1$, namely, $n \cong E/kT$, which as stated above has a value of 10 or more. Thus, it would be a good strategy to continue the series down to the smallest term in absolute value, namely the one term with $N = [E/kT]$. The possible error in truncating the series in this way would be

$$|R_N| = |a_{N+1}| = (kT/E)^N(N+1)! \tag{15.17}$$

The next step, as explained by Chen [998] is to add $\frac{1}{2}a_{N+1}$ to the series up to the term a_N. For $N \cong E/kT$, a_N and a_{N+1} differ only slightly from one another, but have opposite signs. Thus, the partial sums S_{N-1} and S_{N+1} have about the same value, differing from S_N by $\sim a_N$. Since the possible errors in S_{N-1}, S_N and S_{N+1} do not exceed a_{N+1} by much, adding $\frac{1}{2}a_{N+1}$ to S_N (or adding $\frac{1}{2}a_N$ to S_{N-1}) would reduce the possible error by at least a factor of 2.

A more elaborate discussion on such truncated asymptotic series was given by Dingle [996] who showed that improved results may be achieved by multiplying the last term taken, namely, the smallest in absolute value, by a "converging factor". This was given as a power series, but as shown by Chen [999], the sums of these series attain values of 0.49–0.51 for the present case of alternating sign asymptotic series for $E/kT \geq 10$. Thus, adding $\frac{1}{2}a_{N+1}$ to S_N proves to be much better than estimated before, yielding a possible error of

$$|R_N| \cong \frac{1}{100}(kT/E)^{N+1}(N+1). \tag{15.18}$$

By the use of Stirling formula for $(N+1)!$, Chen [998] has shown that when N is the index of the smallest term, the absolute value of this term is approximately

$$|a_{N+1}| \cong \sqrt{2\pi E/kT} \exp(-E/kT), \tag{15.19}$$

which gives the possible expected accuracy without actually calculating the series. It is more interesting, however, to evaluate the *relative* possible error, namely the ratio of the possible error to the sum of the series. This is not very different than the ratio of the possible error to the first term, namely,

$$|R_N/a_1| \cong \frac{1}{100}\sqrt{2\pi(E/kT)^3} \exp(-E/kT). \tag{15.20}$$

For $E/kT = 10$, this gives about 3.6×10^{-5}, and it gets smaller very fast as E/kT grows larger. Thus, for $E/kT = 15$, $|R_N/a_1| \cong 4.5 \times 10^{-7}$ [1000].

Looking again at Equation (15.1), different authors [203], showed that s may be slightly temperature dependent and a common temperature behavior is $s = s''T^a$ where s'' is a constant and $-2 \leq a \leq 2$. This makes Equation (15.1)

$$s'' \int_{t_0}^{t} T^a \exp(-E/kT)dt. \tag{15.21}$$

The expression for the conventional constant heating rate is

$$(s'')/\beta \int_{T_0}^{T} \theta^a \exp(-E/k\theta)d\theta, \tag{15.22}$$

which in the specific case of $a = -2$ yields the very simple situation in which the integral is elementary, a similar situation to that with $a = 0$ and hyperbolic heating function.

A similar treatment to that described above for the case of $a = 0$, was given by Chen [1001] for the more general integral in Equation (15.22). The integration in parts yields now

$$\int_{0}^{T} \theta^a \exp(-E/k\theta)d\theta = \left[T^{a+1}/(a+1)\right] \exp(-E/kT)$$

$$- \left[E/k(a+1)\right] \int_{0}^{T} \theta^{a-1} \exp(-E/k\theta)d\theta, \tag{15.23}$$

which can be utilized as a recurrence equation for $-1 < a \leq 1$. This may help us to reduce the problem with any given a to a problem with $0 \leq a \leq 1$, unless a is an integer smaller than -2. The expression for the integral, which is now denoted by $F(T, E, a)$, is given by

$$F(T, E, a) = kT^{a+2} \exp(-E/kT)$$

$$\times \left[1 + \frac{1}{\Gamma(a+2)} \sum_{n=2}(kT/E)^{n-1}(-1)^{n-1}\Gamma(a+n-1)\right], \tag{15.24}$$

where $\Gamma(y)$ is the gamma function of y, namely, the extension of the factorial to non-integer values of the argument, which reduces to the factorial for integer numbers k, $\Gamma(k+1) = k!$. Equation (15.24) looks easier for application if we remember that

$$\Gamma(a+n+1)/\Gamma(a+2) = (a+2)(a+3)\cdots(a+n). \tag{15.25}$$

Using the same strategy of going down to the smallest term in absolute value, and adding one half of the next term, one gets the relative possible error as [999]

$$|R_N/a_1| \cong \frac{1}{100} \sqrt{2\pi}(E/kT)^{a+3/2} \exp(-E/kT)/\Gamma(a+2). \tag{15.26}$$

15.3   Other Methods

In cases where E/kT is large enough, the first two terms in Equation (15.10) give a reasonably good approximation to $F(T, E)$. This is (see e.g., Balarin [1002])

$$F(T, E) \cong (kT^2/E)(1 - 2kT/E)\exp(-E/kT). \tag{15.27}$$

It has been shown by Gorbachev [1003] that a somewhat better two-term approximation can be given by

$$F(T, E) \cong \left[kT^2/(E + 2kT)\right]\exp(-E/kT). \tag{15.28}$$

A further improvement has been given by Balarin [1004] who suggested the use of

$$F(T, E) \cong (kT^2/E)\exp(-E/kT)/\sqrt{1 + 4kT/E}. \tag{15.29}$$

Perhaps the best two-term approximation is that used in Bos *et al.* [386], namely

$$F(T, E) \cong (kT^2/E)\exp(-E/kT)\left(0.992 - 1.620\frac{kT}{E}\right). \tag{15.30}$$

As for the case of the temperature-dependent frequency factor, Gorbachev [1005] has extended the approximation in Equation (15.28) and gave the expression

$$F(T, E, a) \cong \left\{kT^{a+2}/[E + (a + 2)kT]\right\}\exp(-E/kT). \tag{15.31}$$

Another approach was used by Krystek [1006] who used the Taylor expansion for the exponential integral, Equation (15.12) (see also Abramowitz and Stegun [1007]), namely

$$E_i(-x) = c + \sum_{i=0}^{\infty}(-x)^i/(i \times i!) \tag{15.32}$$

with $c = 0.5772\ldots$ being the Euler constant. Although this is a convergent series rather than a divergent asymptotic series, it is useful mainly for small values of x and therefore is not very efficient in our case. For $x = 10$, for example, the first term in the series is $(-10)^1/(1 \times 1!) = -10$ and the 10th term is $(-10)^{10}/(10 \times 10!) = 276$. One should, therefore go to much higher values of i before convergence is reached, and thus, the computational effort needed for a given accuracy is much larger than with the asymptotic series. The situation gets even worse in the present case for larger values of x, as opposed to to the asymptotic series for which the accuracy improves significantly for higher values of $x = E/kT$, when the same number of terms is utilized.

Previous Books and Review Papers

Books

1 J.R. Cameron, N. Suntharalingam and G.N. Kenney, Thermoluminescent Dosimetry, The University of Wisconsin Press (1968).
2 M.J. Aitken, Physics and Archaeology, Clarendon Press (1974).
3 S.J. Fleming, Thermoluminescence Techniques in Archaeology, Clarendon Press, Oxford University Press (1979).
4 P. Bräunlich (editor) Thermally Stimulated Relaxation in Solids, Springer (1979).
5 R. Chen and Y. Kirsh, Analysis of Thermally Stimulated Processes, Pergamon Press (1981).
6 M. Oberhofer and A. Scharmann (editors) Applied Thermoluminescence Dosimetry, Adam Hilger (1981).
8 Y.S. Horowitz (editor) TL and TL Dosimetry, Vols I, II and III, CRC Press (1984).
9 M.J. Aitken, TL Dating, Academic Press (1985).
10 K. Mahesh and D.R. Vij (editors) Techniques in Radiation Dosimetry, John Wiley & Sons Ltd (1985).
11 S.W.S. McKeever, TL of Solids, Cambridge University Press (1985).
12 K. Mahesh, P.S. Weng and C. Furetta, Thermoluminescence in Solids and its Applications, Nuclear Technology Publishing (1989).
13 D.R. Vij, Thermoluminescent Materials, PTR Prentice Hall (1993).
14 M. Oberhofer and A. Scharmann, Techniques and Management of Personnel Thermoluminescence Dosimetry Services, Eurocourses: Health Physics and Radiation Protection (1993).
15 S.W.S. McKeever, M. Moscovitch and P.D. Townsend, TL Dosimetry: Properties and Uses, Nuclear Technology Publishing (1995).
16 R. Chen and S.W.S. McKeever, Theory of TL and Related Phenomena, World Scientific (1997).
17 M.J. Aitken, An Introduction to Optical Dating, Oxford University Press (1998).
18 C. Furetta and P.S. Weng, Operational Thermoluminescence Dosimetry, World Scientific (1998).
19 C. Furetta, Handbook of Thermoluminescence, World Scientific (2003).
20 L. Bøtter-Jensen, S.W.S. McKeever and A.G. Wintle, Optically Stimulated Dosimetry, Elsevier (2003).

Thermally and Optically Stimulated Luminescence: A Simulation Approach, First Edition. Reuven Chen and Vasilis Pagonis.   © 2011 John Wiley & Sons, Ltd. Published 2011 by John Wiley & Sons, Ltd.

21 V. Pagonis, G. Kitis and C. Furetta, Numerical and Practical Exercises in Thermoluminescence, Springer (2006).
22 Y.S. Horowitz (editor) Microdosimetric Response of Physical and Biological Systems to Low- and High-LET Radiations, Elsevier (2006).
23 S.W.S. McKeever and E.G. Yukihara, Optically Stimulated Luminecence: Fundamentals and Applications, Wiley (2011).

Review Papers

1 R. Chen, Methods for Kinetics Analysis of Thermally Stimulated Luminescence, J. Mater. Sci. 11, 1521–1641 (1976).
2 K.S.V. Nambi, Thermoluminescence: Its Understanding and Applications, Instituto de Energia Atomica, São Paulo (1977).
3 Y.S. Horowitz, The Theoretical and Microdosimetric Measurement during Thermoluminescence and Applications to Dosimetry, Phys. Med. Biol. 26, 765–824 (1981).
4 P.D. Townsend and Y. Kirsh, Spectral Measurement during Thermoluminescence-an Essential Requirement, Contemp. Phys. 30, 337–354 (1989).
5 Y. Horowitz and S.W.S. McKeever, Charge Carrier Trapping in Lithium Fluoride, Radiat. Phys. Chem. 36, 35–46 (1990).
6 Y. Kirsh, Kinetic Analysis of Thermoluminescence–Theoretical and Practical Aspects, Phys. Status Solidi A 129, 15–48 (1992).
7 Y.S. Horowitz and D. Yossian, Computerized Glow-Curve Deconvolution: Application to Thermoluminescence Dosimetry, Radiat. Prot. Dosim. 60, 3–114 (1995).
8 S.W.S. McKeever and R. Chen, Luminescence Models, Radiat. Meas. 27, 625–661 (1997).
9 M. Martini and F. Meinardi, Thermally Stimulated Luminescence: New Perspectives in the Study of Defects in Solids, Riv. Nuovo Cim. 20, 1–71 (1997).
10 Y.S. Horowitz, S. Mahajna, Y. Weizman, D. Satinger and D. Yossian, The Unified Interaction Model Applied to the Gamma Induced Supralinearity and Sensitisation of Peaks 4 and 5 in LiF:Mg, Ti (TLD-100), Radiat. Prot. Dosim. 78, 169–194 (1998).
11 S.W.S. McKeever, Optically Stimulated Luminescence Dosimetry, Nucl. Instrum Methods 184, 29–54 (2001).
12 A.G. Wintle and A.S. Murray, A Review of Quartz OSL Characteristics and their Relevance in Single-Aliquot Protocols, Radiat. Meas. 41, 369–391 (2006).
13 A.J.J. Bos, Theory of Thermoluminescence, Radiat. Meas. 41, S45–S56 (2007).
14 A.G. Wintle, Fifty Years of Luminescence Dating, Archaeometry 50, 276–312 (2008).

Appendix A
Examples

In this Appendix we provide examples of several models commonly used to simulate the luminescence behavior of complex physical systems. First we present the computer code for the OSL simulation within the OTOR model frequently discussed in this book. The code can be easily modified to demonstrate the behavior of the model during TL measurements. We will also provide an example of a more complex system, the interactive multiple trap system (IMTS) commonly used to model both OSL and TL measurements. Finally the computer code is presented for the more complex quartz model by Bailey which was discussed in some detail in this book.

A.1 Simulation of OSL Experiments Using the OTOR Model

We will use Mathematica to integrate numerically the differential equations describing the OSL process for the OTOR model. The equations for the model during the OSL read-out stage are

$$\frac{dn}{dt} = -\lambda n + (N - n)A_h n_c, \tag{A.1}$$

$$\frac{dn_c}{dt} = -\frac{dn}{dt} - A_h n_c(n + n_c), \tag{A.2}$$

$$\frac{dn_h}{dt} = \frac{dn}{dt} + \frac{dn_c}{dt}, \tag{A.3}$$

$$I_{OSL} = -\frac{dn_h}{dt}, \tag{A.4}$$

Thermally and Optically Stimulated Luminescence: A Simulation Approach, First Edition. Reuven Chen and Vasilis Pagonis. © 2011 John Wiley & Sons, Ltd. Published 2011 by John Wiley & Sons, Ltd.

where λ is the probability of optical excitation of the electrons from the trap, E is the thermal activation energy of the trap (in eV), s is the frequency factor (in s^{-1}), T is the temperature of the sample (in K), k is the Boltzmann constant (in eV K^{-1}), is the total concentration of the traps in the crystal (in cm^{-3}), n is the concentration of filled traps in the crystal (in cm^{-3}) n_0 is the initial concentration of filled traps at time $t = 0$ (in cm^{-3}), A_n is the probability of electron retrapping in the traps (in $cm^3 s^{-1}$), and A_h is the probability of electron recombining with holes in the recombination centers (in $cm^3 s^{-1}$). The initial conditions at time $t = 0$ are assumed to be $n_0 = N$ (full electron traps), $n_{c0} = 0$ and $n_{h0} = N$ (charge balance requires that equal numbers of electrons and holes exist in the system at any time t).

The following program solves the system of differential equations (A.1–A.4) using a set of numerical values for the constants λ, N, A_h, A_n and n_{10}:

```
Remove["Global`*"];
E1 = 1.`; s1 = 10^12; k1 = 8.617`/10^5; λ = 0.1;

N1 = 10^10; An = 1/10^7; Ah = 100 An; no = N1; Npoints = 400;

sol = NDSolve[{n1'[x] == -n1[x] λ + An (N1 - n1[x]) nc[x],
    nc'[x] == -n1'[x] - Ah nc[x] (n1[x] + nc[x]),
    nh'[x] == n1'[x] + nc'[x], n1[0] == no, nc[0] == 0, nh[0] == n1[0] + nc[0]},
    {n1, nc, nh}, {x, 0, Npoints}];
Plot[Evaluate[{n1[x]} /. sol], {x, 0, Npoints}, Frame → True,
    FrameLabel → {"Temperature T, C", "n(T)"}, RotateLabel → True,
    BaseStyle → {FontFamily → "Arial", FontSize → 16}, ImageSize → 75 5, PlotRange → All]
Plot[Evaluate[{nc[x]} /. sol], {x, 0, Npoints}, Frame → True,
    FrameLabel → {"Temperature T, C", "nc(T)"}, RotateLabel → True,
    BaseStyle → {FontFamily → "Arial", FontSize → 16}, ImageSize → 75 5, PlotRange → All]
Plot[Evaluate[{nh[x]} /. sol], {x, 0, Npoints}, Frame → True,
    FrameLabel → {"Temperature T, C", "nh(T)"}, RotateLabel → True,
    BaseStyle → {FontFamily → "Arial", FontSize → 16}, ImageSize → 75 5, PlotRange → All]
Plot[Evaluate[{-nh'[x]} /. sol], {x, 0, Npoints}, Frame → True,
    FrameLabel → {"Temperature T, C", "OSL(T)"}, RotateLabel → True,
    BaseStyle → {FontFamily → "Arial", FontSize → 16}, ImageSize → 75 5, PlotRange → All]
```

The output of this simple program is shown in Figure A.1.

As Figure A.1 shows, the concentrations of holes $n_h(t)$ and of electrons $n(t)$ in the trap decrease with the illumination time t, and are equal to each other at all times. The concentration of electrons in the conduction band $n_c(t)$ decreases at a much slower rate, while the OSL signal $OSL(t)$ follows a typical decay curve with time.

A.2   Simulation of OSL Experiments Using the IMTS Model

In this section we present a more complex situation in which we integrate numerically the differential equations describing the OSL process for the IMTS. The IMTS model consists

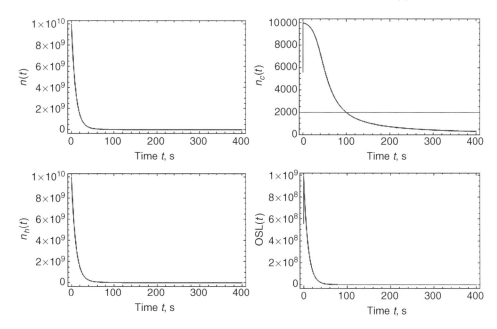

Figure A.1

of two competing trapping states and a recombination center. The first trap characterized by n, N is the active trap, while the second trap is characterized by m, M and is considered to be thermally disconnected. The equations for the model during the OSL read-out stage are

$$\frac{dn}{dt} = -\lambda n + (N - n)A_h n_c, \tag{A.5}$$

$$\frac{dm}{dt} = (M - m)A_m n_c, \tag{A.6}$$

$$\frac{dn_c}{dt} + \frac{dn}{dt} + \frac{dm}{dt} = -(n + m + n_c)A_h n_c, \tag{A.7}$$

$$I_{OSL} = (n + m + n_c)A_h n_c. \tag{A.8}$$

The following program solves the system of differential equations (A.5–A.8) using a set of numerical values for the constants λ, N, M, A_h, A_n, m_{10} and n_{10}:

```
Remove["Global`*"];
k1 = 8.617` / 10^5 ; Npoints = 250; Ah = 1 / 10^5 ; λ = 0.1;
N1 = 10^10; M = 10^10; no = 10^9; mo = 10^9; An = 1 / 10^7 ; Am = 1 / 10^5 ;
sol = NDSolve[{n1'[x] == -n1[x] *λ + An (N1 - n1[x]) nc[x],
    nc'[x] == -n1'[x] - Ah nc[x] (m[x] + n1[x] + nc[x]) - nc[x] (M - m[x]) Am,
    m'[x] == nc[x] (M - m[x]) Am, n1[0] == no, nc[0] == 0, m[0] == mo},
  {n1, nc, m}, {x, 0, Npoints}, MaxSteps → 50000];
Plot[Evaluate[n1[x] /. sol], {x, 0, Npoints}, PlotRange → All,
  Frame → True, FrameLabel → {"Time, s", "n(t)"}, RotateLabel → True,
  BaseStyle → {FontFamily → "Arial", FontSize → 16}, ImageSize → 75 5]
Plot[Evaluate[nc[x] /. sol], {x, 0, Npoints}, PlotRange → All,
  Frame → True, FrameLabel → {"Time, s", "nc(t)"}, RotateLabel → True,
  BaseStyle → {FontFamily → "Arial", FontSize → 16}, ImageSize → 75 5]
Plot[Evaluate[m[x] /. sol], {x, 0, Npoints}, PlotRange → All,
  Frame → True, FrameLabel → {"Time, s", "m(t)"}, RotateLabel → True,
  BaseStyle → {FontFamily → "Arial", FontSize → 16}, ImageSize → 75 5]
Plot[Evaluate[nc[x] (m[x] + n1[x] + nc[x]) Ah /. sol], {x, 0, Npoints},
  PlotRange → All, Frame → True, FrameLabel → {"Time, s", "OSL(t)"},
  RotateLabel → True, BaseStyle → {FontFamily → "Arial", FontSize → 16}, ImageSize → 75 5]
```

The output of this simple program is shown in Figure A.2.

As Figure A.2 shows, the concentrations of electrons $n(t)$ in the trap and the OSL signal $OSL(t)$ decrease with the illumination time t at very similar decay rates. The concentration

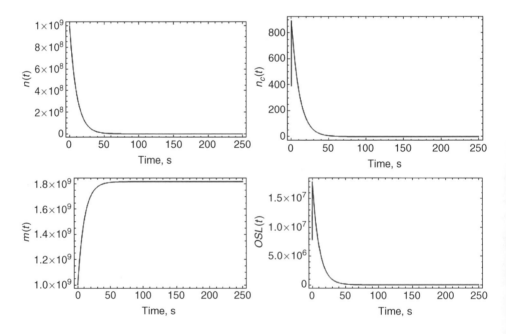

Figure A.2

of electrons in the competing trap $m(t)$ increases slowly with time, while the concentration of electrons $n_c(t)$ in the conduction band is negligible during the sample illumination.

A.3   Simulation of TL Experiment Using the Bailey Model

In this section we present the computer code and typical simulation results for the TL glow curve in quartz samples, using the comprehensive model by Bailey which is discussed in the book. The simulation shown here assumes certain values of the initial concentrations in the nine energy levels in the model denoted by $n_{10}, n_{20}, \ldots n_{90}$ as well as initial concentrations $n_{c0} = 0$ and $n_{v0} = 0$ in the conduction and valence bands, respectively. The equations in the model are:

$$\frac{dn_i}{dt} = n_c(N_i - n_i)A_i - n_i P\theta_{0i}e^{\left(-\frac{E_i^{th}}{k_B T}\right)} - n_i s_i e^{\left(-\frac{E_i}{k_B T}\right)} \quad (i = 1, \ldots, 5), \tag{A.9}$$

$$\frac{dn_j}{dt} = n_v(N_j - n_j)A_j - n_j s_j e^{\left(-\frac{E_j}{k_B T}\right)} - n_c n_j B_j \quad (j = 6, \ldots, 9), \tag{A.10}$$

$$\frac{dn_c}{dt} = R - \sum_{i=1}^{5}\left(\frac{dn_i}{dt}\right) - \sum_{j=6}^{9} n_c n_j B_j, \tag{A.11}$$

$$\frac{dn_v}{dt} = \frac{dn_c}{dt} + \sum_{i=1}^{5}\left(\frac{dn_i}{dt}\right) - \sum_{j=6}^{9}\left(\frac{dn_j}{dt}\right), \tag{A.12}$$

and the luminescence is defined as

$$L = n_c n_8 B_8 \eta(T). \tag{A.13}$$

The following program solves the system of differential equations (A.9–A.13):

```
Remove["Global`*"];
  k = 8.617` * 10^-5; s1 = 5 10^12; E1 = 0.97`; N1 = 1.5` 10^7; A1 = 10^-8; th1 = 0.75`;
  E1th = 0.1`; s2 = 5 10^14; E2 = 1.55`; N2 = 1 10^7; A2 = 10^-8; th2 = 0; E2th = 0;
  s3 = 5 10^13; E3 = 1.7`; N3 = 1 10^9; A3 = 10^-9; th3 = 6; E3th = 0.1`; s4 = 5 10^14;
  E4 = 1.72`; N4 = 2.5` 10^8; A4 = 5 * 10^-10; th4 = 4.5`; E4th = 0.13`;
  s5 = 1 10^10; E5 = 2; N5 = 5 10^10; A5 = 10^-10; th5 = 0; E5th = 0; s6 = 5 10^13; E6 = 1.43`;
  N6 = 3 10^8; A6 = 5 * 10^-7; B6 = 5 * 10^-9; s7 = 5 10^14; E7 = 1.75`; N7 = 1 10^10; A7 = 10^-9;
  B7 = 5 * 10^-10; s8 = 1 10^13; E8 = 5; N8 = 1 10^11; A8 = 10^-9; B8 = 10^-10;
  s9 = 1 10^13; E9 = 5; N9 = 5 10^9; A9 = 10^-10; B9 = 10^-10;
  n10 = 1.1251 x 10^6; n20 = 7.6 * 10^5; n30 = 1.34 * 10^8; n40 = 7.92 * 10^6; nc0 = 0; nv0 = 0;
  n50 = 2.69 * 10^10; n60 = 3.17 * 10^7; n70 = 6.82 * 10^7; n80 = 2.68 * 10^10; n90 = 1.59 * 10^8;
  R = 0; P = 0; βheat = 5; irrTemp = 20; tfinal18 = (450 - irrTemp) / βheat;
  sol = NDSolve[{n1'[t] == -n1[t] s1 Exp[-E1 / (k (273 + irrTemp + βheat t))] + nc[t] (N1 - n1[t]) A1
        - n1[t] (P th1 Exp[-E1th / (k (273 + irrTemp + βheat t))]),
      n2'[t] == -n2[t] s2 Exp[-E2 / (k (273 + irrTemp + βheat t))]
        + nc[t] (N2 - n2[t]) A2 - n2[t] (P th2 Exp[-E2th / (k (273 + irrTemp + βheat t))]),
```

```
n3'[t] == -n3[t] s3 Exp[-E3 / (k (273 + irrTemp + βheat t))] + nc[t] (N3 - n3[t]) A3
   - n3[t] (P th3 Exp[-E3th / (k (273 + irrTemp + βheat t))]),
n4'[t] == -n4[t] s4 Exp[-E4 / (k (273 + irrTemp + βheat t))] + nc[t] (N4 - n4[t]) A4
   - n4[t] (P th4 Exp[-E4th / (k (273 + irrTemp + βheat t))]),
n5'[t] == -n5[t] s5 Exp[-E5 / (k (273 + irrTemp + βheat t))] + nc[t] (N5 - n5[t]) A5
   - n5[t] (P th5 Exp[-E5th / (k (273 + irrTemp + βheat t))]),
n6'[t] == -n6[t] s6 Exp[-E6 / (k (273 + irrTemp + βheat t))] + nv[t] (N6 - n6[t]) A6
   - nc[t] n6[t] B6,
n7'[t] == -n7[t] s7 Exp[-E7 / (k (273 + irrTemp + βheat t))] + nv[t] (N7 - n7[t]) A7
   - nc[t] n7[t] B7,
n8'[t] == -n8[t] s8 Exp[-E8 / (k (273 + irrTemp + βheat t))] + nv[t] (N8 - n8[t]) A8
   - nc[t] n8[t] B8,
n9'[t] == -n9[t] s9 Exp[-E9 / (k (273 + irrTemp + βheat t))] + nv[t] (N9 - n9[t]) A9
   - nc[t] n9[t] B9, nc'[t] == R - n1'[t] - n2'[t] - n3'[t] - n4'[t] - n5'[t]
   - nc[t] (n6[t] B6 + n7[t] B7 + n8[t] B8 + n9[t] B9),
nv'[t] == R - n6'[t] - n7'[t] - n8'[t] - n9'[t] - nc[t] (n6[t] B6 + n7[t] B7 + n8[t] B8
      + n9[t] B9),
n1[0] == n10, n2[0] == n20, n3[0] == n30, n4[0] == n40, n5[0] == n50, n6[0] == n60,
n7[0] == n70,
n8[0] == n80, n9[0] == n90, nc[0] == nc0, nv[0] == nv0}, {n1, n2, n3, n4, n5, n6, n7,
n8, n9, nc, nv},
{t, 0, tfinal18}, MaxSteps → 50000];
tlList3 = Table[{βheat t + irrTemp,
   First[Evaluate[ (n8[t] nc[t] B8) / (1 + 2.8` 10^7 Exp[-0.64` / (k (273 + irrTemp + βheat t))]) /. sol]]},
{t, 0, tfinal18, 0.2`}]; ListPlot[tlList3, Joined → True,
AxesLabel → {"Temp,deg C)", "TL(T)"}, PlotStyle → {RGBColor[1, 0, 0]},
PlotRange → {{0, Automatic}, {0, Automatic}}, ImageSize → 724,
PlotLabel → "TL vs Temperature"]
```

The TL glow curve calculated using this computer program is shown in Figure A.3.

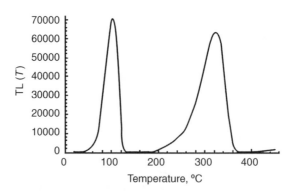

Figure A.3

References

[1] Arnold, W.A., Photosynth. Res. 27, 73–82 (1991).
[2] Heckelsberg, L.F., Health Phys. 39, 391–393 (1980).
[3] Wick, F.G., JOSA 14, 33–44 (1927).
[4] Wick, F.G., JOSA 21, 223–231 (1931).
[5] Johnson, R.P., J. Opt. Soc. Am. 29, 387–391 (1939).
[6] Randall, J.T. and Wilkins, M.H.F., Proc. R. Soc. London, 184, 347–365, 366–389, 390–407 (1945).
[7] Daniels, F. and Saunders, D.F., Science 111, 461–469 (1950).
[8] Daniels, F., Boyd, C.A. and Saunders, D.F., Science 117, 343–349 (1953).
[9] Kennedy, G.C. and Knopf, L., Archaeology 13, 147–148 (1960).
[10] Aitken, M.J., Thompson, J. and Fleming, S.J., Proc. 2nd Int. Conf. Lumin. Dosim., 364 (1968).
[11] Antonov-Romanovskiĭ, V.V., Keirum-Markus, M.S., Porkina, M.S. and Trapeznikova, Z.A., UASEC Report AEC-tr-2435, 239 (1956).
[12] Huntley, D.J., Ancient TL 3, 20–21 (1985).
[13] McKeever, S.W.S., Moscovitch, M. and Townsend, P.D., Thermoluminescence Dosimetry Materials: Properties and Uses, Nuclear Technology Publishing, (1995).
[14] Kortov, V., Radiat. Meas. 42, 576–581 (2007).
[15] Olko, P., Radiat. Meas. 45, 506–511 (2010).
[16] McKeever, S.W.S. and Moscovitch, M., Radiat. Prot. Dosim. 104, 263–270 (2003).
[17] Yukihara, E.G., Mittani, J.C., Vanhavere, F. and Akselrod, M.S., Radiat. Meas. 43, 309–314 (2008).
[18] Lommler, B., Pitt, E., Scharmann, A. and Guldbakke, S., Glasses 47, 285–288 (1993).
[19] Sykora, G.J., Salasky, M. and Akselrod, M.S., Radiat. Meas. 43, 1017–1023 (2008).
[20] Sykora, G.J. and Akselrod, M.S., Radiat. Meas. 45, 594–598 (2010).
[21] Oster, L., Horowitz, Y.S. and Podpalov, L., Radiat. Prot. Dosim. 128, 261–265 (2008).
[22] Akselrod, M.S., Agersnap Larsen, N. and McKeever, S.W.S., Radiat. Meas. 32, 215–225 (2000).
[23] Budzanowski, M., Olko, P. and Golnik, N., Radiat. Prot. Dosim. 119, 259–262 (2006).
[24] Marczewska, B., Bilski, P., Czopyk, L., Olko, P., Waligórski M.P.R. and Zapotoczny, S., Radiat. Prot. Dosim. 120, 129–132 (2006).
[25] Wintle, A.G., Archaeometry 50, 276–312 (2008).
[26] Wintle, A.G. and Murray, A.S., Radiat. Meas. 41, 369–391 (2006).
[27] Garlick, G.F.J. and Gibson, A.F., Proc. Phys. Soc. A 60, 574–590 (1948).
[28] Hill, J.J. and Schwed, P., J. Chem. Phys. 13, 652–658 (1955).

[29] May, C.E. and Partridge, J.A., J. Chem. Phys. 40, 1401–1409 (1964).

[30] Halperin A. and Braner, A.A., Phys. Rev. 117, 408–415 (1960).

[31] Levy, P.W., Nucl. Tracks Radiat. Meas. 10, 547–556 (1985).

[32] McKeever, S.W.S. and Chen, R., Radiat. Meas., 27, 625–661, (1997).

[33] Chen, R. and McKeever, S.W.S., Theory of Thermoluminescence and Related Phenomena, World Scientific (1997).

[34] Adamiec, G., Radiat. Meas. 19, 175–189 (2005).

[35] Bailey, R.M., Radiat. Meas. 33, 17–45 (2001).

[36] Pagonis, V., Wintle, A.G., Chen, R. and Wang, X.L., Radiat. Meas. 43, 704–708 (2008).

[37] Nikiforov, S.V., Milman, I.I. and Kortov, V.S., Radiat. Meas. 33, 547–551 (2001).

[38] Pagonis, V., Chen, R. and Lawless, J.L., Radiat. Meas. 42, 198–204 (2007).

[39] Pagonis, V., Chen, R. and Lawless, J.L., Radiat. Meas. 43, 175–179 (2008).

[40] Kittel, C., Introduction to Solid State Physics, 8th Edn, John Wiley & Sons, Ltd (2004).

[41] Ibach, H., and Lüth, H., Solid-State Physics: An Introduction to Principles of Materials Science, Springer-Verlag (2009).

[42] Williams, F.E., Ann. Rev. Phys. Chem. 3, 339–358 (1952).

[43] Williams, F.E., J. Chem. Phys. 57, 780–784 (1953).

[44] Norgett, M.J., Stoneham, A.M. and Pathak, A.P., J. Phys. C: Solid State Phys. 10, 555–565 (1977).

[45] Lagos, M., Solid State Commun. 53, 51–54 (1985).

[46] Testa, A., Stoneham, A.M., Catlow, C.R.A., Song, K.S., Harker, A.H. and Harding, J.H., Radiat. Eff. Defects Solids 119, 27–32 (1991).

[47] Harrison, W.A., Solid State Theory, McGraw-Hill (1970).

[48] Stoneham, A.M., Contemp. Phys. 20, 535–545 (1979).

[49] Stoneham, A.M., Theory of Defects in Solids: Electronic Structure of Defects and Semiconductors, Oxford University Press (2001).

[50] Dorenbos, P., J. Phys.: Condens. Matter 15, 8417–8434 (2003).

[51] Dorenbos, P. and Bos, A.J.J., Radiat. Meas. 43, 139–145 (2008).

[52] Bos, A.J.J., Dorenbos, P., Bessière, A. and Viana, B., Radiat. Meas. 43, 222–226 (2008).

[53] Bos, A.J.J., Poolton, N.R.J., Wallinga, J., Bessière, A. and Dorenbos, P., Radiat. Meas. 45, 343–346 (2010).

[54] Rose, A., Concepts in Photoconductivity and Allied Problems, Interscience (1963).

[55] Bemski, G., Phys. Rev. 111, 1515–1518 (1958).

[56] Alekseeva, V.G., Kalashnikov, S.G., Kalnach, L.P., Karpova, I.V. and Morozov, A.I., Sov. Phys. Tech. Phys. 2, 1794–1801 (1957).

[57] Lax, M., Phys. Rev. 119, 1502–1523 (1960).

[58] Mitonneau, A., Mircea, A., Martin, G.M. and Pons, D., Rev. Phys. Appl. 14, 853–861 (1979).

[59] Arrhenius, S.A., Z. Phys. Chem. 4, 226–248 (1889).

[60] Gibbs, J.H., J. Chem. Phys. 57, 4473–4478 (1972).

[61] Curie, D., Luminescence in Crystals, John Wiley & Sons, Ltd (1963).

[62] Peacock-López, E. and Suhl, H., Phys. Rev. B 26, 3774–3782 (1982).

[63] Böhm, M., Erb, O. and Scharmann, A., Appl. Phys. A 37, 165–170 (1985).

[64] Böhm, M. and Scharmann, A., Appl. Phys. A 43, 29–35 (1987).

[65] Tavgin, V.L. and Stepanov, A.V., Phys. Status Solidi B 161, 123–130 (1990).

[66] Stepanov, A.V., J. Mol. Struct. THEOCHEM 538, 179–188 (2001).

[67] Stepanov, A.V., J. Mol. Strruct. THEOCHEM 805, 87–90 (2007).

[68] Grant, W.J.C., Phys. Rev. B 4, 648–663 (1971).

[69] Maxia, V., Phys. Rev. B 17, 3262–3268 (1978).

[70] Maxia, V., Phys. Rev. B 21, 749–759 (1980).

[71] Maxia, V., Lett. Nuov. Cim. 31, 27–32 (1981).

[72] Swandic, J.R., Phys. Rev. B 45, 622–634 (1992-II).

[73] Swandic, J.R., Phys. Rev. B 53, 2352–2366 (1996-I).

[74] Bøtter-Jensen, L., McKeever, S.W.S. and Wintle, A.G., Optically Stimulated Luminescence Dosimetry, Elsevier, (2003).

[75] Wintle, A.G., Geophys. J.R. Astr. Soc. 41, 107–113 (1975).

[76] Smith, B.W., Rhodes, E.J., Stokes, S. and Spooner, N.A., Rad. Prot. Dosim. 34, 75–78 (1980).

[77] Spooner, N.A., Radiat. Meas. 23, 593–600 (1994).

[78] Akselrod, M.S., Agersap Larsen N., Whitley, V. and McKeever, S.W.S., J. Appl. Phys. 84, 3364–3373 (1998).

[79] Sanderson, D.C.W. and Clark, R.J., Radiat. Meas. 23, 633–639 (1994).

[80] Bailiff, I.K., Radiat. Meas. 32, 401–405 (2000).

[81] Tsukamoto, S., Denby, P.M., Murray, A.S. and Bøtter-Jensen, L., Radiat. Meas. 41, 790–795 (2006).

[82] Denby, P.M., Bøtter-Jensen, L., Murray, A.S., Thomsen, K.J. and Moska, P., Radiat. Meas. 41, 774–779 (2006).

[83] Chithambo, M.L., Ogundare, F.O. and Feathers, J., Radiat. Meas. 43, 1–4 (2008).

[84] Ankjærgaard, C., Jain, M., Kalchgruber, R., Lapp, T., Klein, D., McKeever, S.W.S., Murray, A.S. and Morthekai, P., Radiat. Meas. 44, 571–581 (2009).

[85] Pagonis, V., Mian, S.M., Chithambo, M.L., Christensen, E. and Barnold, C., J. Phys. D: Appl. Phys. 42, 055407 (2009).

[86] Galloway, R.B., Radiat. Meas. 35, 67–77 (2002).

[87] Chithambo M.L., Radiat. Meas. 37, 167–175 (2003).

[88] Chithambo M.L., Radiat. Meas. 41, 862–865 (2006).

[89] Chithambo, M.L., J. Phys. D: Appl. Phys. 40, 1874–1879 (2007).

[90] Chithambo, M.L., J. Phys. D: Appl. Phys. 40, 1880–1889 (2007).

[91] Pagonis, V., Ankjærgaard, C., Murray, A.S., Jain, M., Chen, R., Lawless, J.L., and Greilich, S., J. Lumin. 130, 902–909 (2010).

[92] Mott, N.F. and Gurney, R.W., Electronic Processes in Ionic Solids, 2nd Ed., Oxford University Press, (1948).

[93] Evans, B.D., J. Nucl. Mater., 219, 202–223 (1996).

[94] Milman, I.I., Kortov, V.S. and Nikiforov, S.V., Radiat. Meas. 29, 401–410 (1998).

[95] Boyd, C.A., J. Chem. Phys. 17, 1221–1226 (1949).

[96] Watanabe, K., Phys. Rev. 83, 785–791 (1951).

[97] Aitken, M.J., Physics and Archaeology, 2nd Ed., Oxford University Press (1974).

[98] Huntley, D.J., Godfrey-Smith, D.I. and Thewalt, M.L.W., Nature 313, 105–107 (1985).

[99] Aitken, M.J., An Introduction to Optical Dating, Oxford University Press (1998).

[100] Bräunlich, P., Gasiot, J. and Fillard, J.P., Appl. Phys. Lett. 39, 769–771 (1981).

[101] Gasiot, J., Bräunlich, P. and Fillard, J.P., J. Appl. Phys. 53, 5200–5209 (1982).

[102] Mathur, V.K., Brown, M.D. and Bräunlich, P., Radiat. Prot. Dosim. 6, 163–165 (1985).

[103] Grupen-Shemansky, M.E., Kearfortt, K.J. and Hirleman, E.D., J. Appl. Phys. 66, 3407–3409 (1989).

[104] Kearfortt, K.J. and Grupen-Shemansky, M.E., Med. Phys. 17, 429–435 (1990).

[105] Kearfortt, K.J., Han, S., Wagner, E.C., Samei, E. and Wang, C.K.C., Appl. Radiat. Isot. 52, 1419–1429 (2000).

[106] Jones, S.C., Sweet, J.A., Fehl, D.L., Sujka, B.R., Vehar, D.W. and Westfall, R.L., Rev. Sci. Instrum. 63, 4898–4900 (1992).

[107] Gayer, O. and Katzir, A., J. Lumin. 113, 151–155 (2005).

[108] Lawless, J.L., Lam, S.K. and Lo, D., Opt. Express 10, 291–296 (2002).

[109] Ditcovski, R., Gayer, O. and Katzir, A., J. Lumin. 130, 141–144 (2010).

[110] Ueta, M., Sumigoto, H. and Nagasawa, I., J. Phys. Soc. Jpn. 17, 1465–1473 (1962).

[111] Gabrysh, A.F., Kennedy, J.M., Eyring, H. and Johnson, V.R., Phys. Rev. 131, 1543–1548 (1963).

[112] Hook III, J.W. and Drickamer, H.G., J. Appl. Phys. 49, 2503–2507 (1978).

[113] Joshi, R.V., Dhake, K.P. and Joshi, T.R., J. Lumin. 31–32, 136–138 (1984).

[114] Kos, H.-J. and Mieke, S., Phys. Status Solidi A 50, K165-K168 (1978).

[115] Petralia, S. and Gnani, G., Lett. Nuovo Cim. 4, 483–486 (1972).

[116] Bradbury, M.H. and Lilley, E., J. Mater. Sci. 11, 1849–1856 (1976).

[117] Petit, J.-R, and Duval, P., Solid State Commun. 19, 475–477 (1976).

[118] Reitz, R.A. and Thomas, B.R., Phys. Rev. B 13, 1802–1804 (1976).

[119] Nyswander, R.E. and Cohn, B.E., Phys. Rev. 36, 1257–1260 (1930).

[120] Nambi, K.S.V., Nucl. Instrum. Methods 197, 453–457 (1982).
[121] Templer, R.H., Nucl. Tracks Radiat. Meas. 10, 531–537 (1985).
[122] Templer, R.H., Radiat. Prot. Dosim. 17, 235–239 (1986).
[123] Pashchenko, L.P., Pérez Sales, R., Aceves, R. and Barboza-Flores, M., Appl. Phys. Lett. 66, 3126–3127 (1995).
[124] Pashchenko, L.P., Pérez Sales, R., Aceves, R. and Barboza Flores, M., Appl. Phys. Lett. 67, 3266–3268 (1995).
[125] Araki, K., Endo, M. and Yahagi, K., Jpn. J. Appl. Phys. 13, 1787–1791 (1974).
[126] Levinson, J., Halperin, A. and Bar, V., J. Lumin. 6, 1–10 (1973).
[127] Miyashita, K. and Henisch, H.K., Solid-State Electron. 9, 29–33 (1966).
[128] Arnold, W., Research in Photosynthesis, Eds H. Gafron et al., John Wiley & Sons, Ltd, pp 128–133 (1957).
[129] Arnold, W. and Azzi, J.R., Proc. Natl Acad. Sci. USA 61, 29–35 (1968).
[130] Arnold, W., J. Phys. Chem. 69, 788–791 (1965).
[131] Arnold, W., Science 154, 1046–1049 (1966).
[132] Vass, I., Photosynth. Res. 76, 303–318 (2003).
[133] Rappaport, F. and Lavergne, J., Photosynth. Res. 101, 205–216 (2009).
[134] Keller, S.P., Mapes, J.E. and Cheroff, G., Phys. Rev. 108, 663–676 (1957).
[135] Riehl, N., J. Lumin. 1, 1–16 (1970).
[136] Baur, G., Freydorf, E.V. and Koschel, W.H., Phys. Status Solidi A 21, 247–251 (1974).
[137] Sanborn, E.N. and Beard, E.L., Proc. Int. Conf. Lumin. Dosim. 183-191 (1966).
[138] Cameron, J.R., Suntharalingam, N. and Kenney, G.N., Thermoluminescent Dosimetry, The University of Wisconsin Press (1968).
[139] Bräunlich, P., Schäfer, D. and Scharmann, A., Proc. Int. Conf. Lumin. Dosim. 57–73 (1966).
[140] Pradhan, A.S., Chandra, Bhuwan and Bhatt, R.C., Radiat. Prot. Dosim. 5, 159–162 (1983).
[141] Allen, P. and McKeever, S.W.S., Radiat. Prot. Dosim. 33, 19–22 (1990).
[142] Nanto, H., Murayama, K., Usada, S., Taniguchi, S. and Takeuchi, N., Radiat. Prot. Dosim. 47, 281–284 (1993).
[143] Nanto, H., Usada, T., Murayama, S., Nakamura, S., Inabe, K. and Takeuchi, N., Radiat. Prot. Dosim. 47, 293–296 (1990).
[144] Manfred, M.E., Gabriel, N.T., Yukihara, E.G. and Talghader, J., Radiat. Prot. Dosim. 139, 560–564 (2010).
[145] Nahum, J. and Halperin, A., J. Phys. Chem. Solids 23, 345–358 (1962).
[146] Halperin, A. and Nahum, J., J. Phys. Chem. Solids 18, 297–306 (1961).
[147] Israeli, M. and Kristianpoller, N., Phys. Status Solidi B 45, K29-K33 (1971).
[148] Israeli, M. and Kristianpoller, N., Solid State Commun. 9, 1749–1753 (1971).
[149] Kuhn, R., Trautmann, T., Singhvi, A.K., Krbetschek, M.R., Wagner, G.A. and Stolz, W., Radiat. Meas. 12, 653–657 (2000).
[150] L. Bøtter-Jensen, L., Agersnap Larsen, N., Markey, B.G. and McKeever, S.W.S., Radiat. Meas. 27, 295–298 (1997).
[151] Whitley, V.H. and McKeever, S.W.S., J. Appl. Phys. 87, 249–256 (2000).
[152] Markey, B.G., Colyott, L.E. and McKeever, S.W.S., Radiat. Meas. 24, 457–464 (1995).
[153] Walker, F.D., Colyott, L.E., Agersap Larsen, N. and McKeever, S.W.S., Radiat. Meas. 26, 711–718 (1996).
[154] Duller, G.A.T. and Bøtter-Jensen, L., Radiat. Meas. 26, 603–609 (1996).
[155] Bøtter-Jensen, L., Duller, G.A.T. and Poolton, N.R.J., Radiat. Meas. 23, 613–616 (1994).
[156] Clark, R.J. and Sanderson, D.C.W., Radiat. Meas. 23, 641–646 (1994).
[157] Krbetschek, M.R., Gotze, J., Dietrich, A. and Trautmann, T., Radiat. Meas. 27, 695–748 (1997).
[158] Prescott, J.R., Fox, P.J., Akber, R.A. and Jensen, H.E., Appl. Opt. 27, 3496–3502 (1988).
[159] Haschberger, P., Appl. Radiat. Isot. 42, 797–800 (1991).
[160] Luff, B.J. and Townsend, P.D., Meas. Sci. Technol. 4, 65–71 (1993).
[161] Townsend, P.D., Radiat. Meas. 23, 341–348 (1994).
[162] Piters, T.M., Meulemans, W.H. and Bos, A.J.J., Rev. Sci. Instrum. 64, 109–117 (1993).
[163] Townsend, P.D. and Kirsh, Y., Contemp. Phys. 30, 337–354 (1989).

[164] Karali, T., Rowlands, A.P., Townsend, P.D., Prokić, M. and Olivares, J., J. Phys. D: Appl. Phys. 31, 754–765 (1998).

[165] Spooner, N.A. and Franklin, A.D., Radiat. Meas. 35, 59–66 (2002).

[166] Schlesinger, M. and Whippey, P.W., Phys. Rev. 162, 286–290 (1967).

[167] Sridaran, P., Bhunia, R.C. and Ratnam, V.V., Phys. Status Solidi B 104, 459–468 (1981).

[168] Townsend, P.D., Ahmed, K., Chandler, P.J., McKeever, S.W.S. and Whitlow, H.J., Radiat. Eff. 72, 245–257 (1983).

[169] Huntley, D.J., Godfrey-Smith, D.I. and Haskell, E.H., Radiat. Meas. 18, 127–131 (1991).

[170] Martini, M., Fasoli, M. and Galli, A., Radiat. Meas. 44, 458–461 (2009).

[171] Huntley, D.J., McMullan, W.G., Godfrey-Smith, D.I. and Thewalt, M.L.W., J. Lumin. 44, 41–46 (1989).

[172] Thomas, B. and Houston, E., Br. J. Appl. Phys. 15, 953–958 (1964).

[173] Halperin, A. and Chen, R., Phys. Rev. 148, 839–845 (1966). http://prola.aps.org/abstract/PR/v148/i2/p839_1.

[174] Morris, M.F. and McKeever, S.W.S., Radiat. Meas. 23, 323–327 (1994).

[175] Chen, R., Hornyak, W.F. and Mathur, V.K., J. Phys. D: Appl. Phys. 23, 724–728 (1990).

[176] Morris, M.F. and McKeever, S.W.S., Radiat. Prot. Dosim. 47, 637–641 (1993).

[177] Subramanian, U., Mukherjee, M.L. and Acharya, B.S., Cryst. Latt. Def. Amorph. Mat. 10, 221–228 (1984).

[178] Spooner, N.A., Prescott, J.R. and Hutton, J.T., Quat. Sci. Rev. 7, 325–329 (1988).

[179] Benabdesselam, M., Iacconi, P., Trinkler, L., Berzina, B. and Butler, J.E., Radiat. Prot. Dosim. 119, 390–393 (2006).

[180] Chruścińska, A., Oczkowski, H.L. and Przegiętka, K.R., J. Phys. D: Appl. Phys. 34, 2939–2944 (2001).

[181] Secu, M., Jipa, S., Secu, C.E., Zaharescu, T., Georgescu, R. and Cutrubinis, M., Phys. Status Solidi B 245, 159–162 (2008).

[182] Ambrico, P.F., Ambrico, M., Schiavulli, L., Ligonzo, T. and Augelli, A., Appl. Phys. Lett. 94, 051501 (2009).

[183] Dotzler, C., Williams, G.V.M., Rieser, U. and Robinson, J., J. Appl. Phys. 105, 023107 (2009).

[184] Elle, D.R., Fehringer, D.J., Vetter, R.J. and Ziemer, P.L., IEEE Trans. Microwave Theory Tech. MTT-21, 836–837 (1973).

[185] Nagpal, J.S., Varadharajan, G. and Gangadharan, P., Phys. Med. Biol. 27, 145–148 (1982).

[186] Antonov-Romanovskiĭ, V.V., Izv. Akad. Nauk. SSSR Fiz. Ser. 15, 637–641 (1951).

[187] Lushchik, C.B., Sov. Phys. JETP 3, 390–395 (1956).

[188] Denks, V.P., Leiman, V.I., Lukantsever, N.L. and Savikhin, F.A., Sov. Phys. Solid State 12, 1142–1146 (1970).

[189] Kantorovich, L.N., Fogel, G.M. and Gotlib, V.I., J. Phys. D: Appl. Phys. 21, 1008–1014 (1988).

[190] Kantorovich, L.N. and Fogel, G.M., J. Phys. D: Appl. Phys. 22, 817–824 (1989).

[191] Kantorovich, L.N., Fogel, G.M. and Gotlib, V.I., J. Phys. D: Appl. Phys. 23, 1219–1226 (1990).

[192] Kantorovich, L.N., Livshicz, A.I. and Fogel, G.M., J. Phys.: Condens. Matter 5, 7503–7514 (1993).

[193] McKeever, S.W.S., Thermoluminescence in Solids, Cambridge University Press, (1985).

[194] Chen, R. and McKeever, S.W.S., Radiat. Meas. 23, 667–673 (1994).

[195] Adirovitch, E.I., J. Phys. Radium 17, 705–707 (1956).

[196] Sunta, C.M., Feria Ayta, W.E., Kulkarni, R.N., Piters, T.M. and Watanabe, S., J. Phys. D: Appl. Phys. 30, 1234–1242 (1997).

[197] Sunta, C.M., Kulkarni, R.N., Piters, T.M., Feria Ayta, W.E., and Watanabe, S., J. Phys. D: Appl. Phys. 31, 2074–2081 (1998).

[198] Sunta, C.M., Feria Ayta W.E., Piters, T.M. and Watanabe, S., Radiat. Meas. 30, 197–201 (1999).

[199] Levy, P.W., Eur. PACT J., 6, 224–242 (1982).

[200] Levy, P.W., Eur. PACT J., 9, 109–122 (1983).

[201] Levy, P.W., J. Lumin. 31–32, 133–135 (1984).

[202] Hornyak, W.F., Levy, P.W. and Kierstead, J.A., Nucl. Tracks Radiat. Meas., 10, 557–563 (1985).

[203] Chen, R., J. Appl. Phys. 40, 570–585 (1969).

[204] Chen, R., J. Electrochem. Soc. 116, 1254–1257 (1969).

[205] Opanowicz, A., Phys. Status Solidi A 116, 343–348 (1989).

[206] Rasheedy, M.S., J. Phys.: Condens. Matter 5, 633–636 (1993).

[207] Kanunnikov, L.A., J. Appl. Spectrosc. 28, 597–599 (1978).

[208] Takeuchi, N., Inabe, K. and Nanto, H., Solid State Commun. 17, 1267–1269 (1975).

[209] Xu, Z., Zhang, F., Wang, L., Georgobiani, A.N. and Xu, X., J. Appl. Phys. 101, 033518 (1–3) (2007).

[210] Christodoulides, C., Phys. Status Solidi A 118, 333–342 (1990).

[211] Rasheedy, M.S., J. Phys.: Condens. Matter 8, 1291–1300 (1996).

[212] Chen, R., Kristianpoller, N., Davidson, Z. and Visocekas, R., J. Lumin. 23, 293–303 (1981).

[213] Schlesinger, M. and Menon, A.K., Can. J. Phys. 47, 1637–1638 (1969).

[214] Visocekas, R., Thèse, Université Pierre et Marie Curie (1979).

[215] Singh, W.S., Singh, S.D. and Mazumdar, P.S., J. Phys.: Condens. Matter 10, 4937–4946 (1998).

[216] Sunta, C.M., Feria Ayta, W.E., Chubaci, J.F.D. and Watanabe, S., Radiat. Meas. 35, 47–57 (2002).

[217] Fuchs, W. and Taylor, A., Phys. Rev. B 2, 3393–3403 (1970).

[218] Fuchs, W. and Taylor, A., Phys. Status Solidi 38, 771–775 (1970).

[219] Davidson, Z. and Kristianpoller, N., Solid State Commun. 33, 79–85 (1980).

[220] Saunders, I.J., Brit. J. Appl. Phys. 18, 1219–1220 (1967).

[221] Chen, R., Br. J. Appl. Phys. (J. Phys. D) 2, 371–375 (1969).

[222] Shenker, D. and Chen, R., J. Phys. D: Appl. Phys. 4, 287–291 (1971).

[223] Kelly, P., Laubitz, M.J. and Bräunlich, P., Phys. Rev. B 4, 1960–1968 (1971).

[224] Opanowicz, A. and Przybyszewski, K., Proc. SPIE 2373, 236–240 (1995).

[225] Lewandowski, A.C. and McKeever, S.W.S., Phys. Rev. B 43, 8163–8178 (1991).

[226] Lewandowski, A.C., Markey, B.G. and McKeever, S.W.S., Phys. Rev. B 49, 8029–8047 (1994).

[227] McKeever, S.W.S., Markey, B.G. and Lewandowski, A.C., Nucl. Tracks Radiat. Meas. 21, 57–64 (1993).

[228] McKeever, S.W.S., Lewandowski, A.C. and Markey, B.G., Radiat. Prot. Dosim. 47, 9–16 (1993).

[229] Chen, R. and Kirsh, Y., Analysis of Thermally Stimulated Processes, Pergamon Press (1981).

[230] Bull, R.K., J. Phys. D: Appl. Phys. 22, 1375–1379 (1989).

[231] Horowitz, Y.S., Oster, L., Biderman, S. and Einav, Y., J. Phys. D: Appl. Phys. 36, 446–459 (2003).

[232] Kirsh, Y., Townsend, P.D. and Shoval, S., Nucl. Tracks Radiat. Meas. 13, 115–119 (1987).

[233] Medlin, W.L., Phys. Rev. 135, A1770–1779 (1964). http://prola.aps.org/abstract/PR/v135/i6A/pA1770_1.

[234] Kirsh, Y. and Townsend, P.D., J. Phys. C: Solid State Phys. 20, 967–980 (1987).

[235] Zatsepin, A.F., Kortov, V.S. and Shchapova, J.V., J. Lumin. 65, 355–362 (1996).

[236] Hagston, W.E., J. Phys. C: Solid State Phys. 6, 784–796 (1973).

[237] Ronda, C.R., van der Meer, J.H., van Heuzen, A.A. and Haas, C., J. Solid State Chem. 70, 3–11 (1987).

[238] Mandowski, A., J. Phys. D: Appl. Phys. 38, 17–21 (2005).

[239] Mandowski, A., Radiat. Meas. 33, 745–749 (2001).

[240] Pagonis, V., J. Phys. D: Appl. Phys. 38, 2179–2186 (2005).

[241] Pagonis, V. and Kulp, C., J. Phys. D: Appl. Phys. 43, 175403 (2010).

[242] Mandowski, A., Bos, A.J.J., Mandowska, E. and Orzechowski, J., Radiat. Meas. 45, 284–287 (2010).

[243] Mandowski, A., Orzechowski, J. and Mandowska, E., Opt. Mater. 32, 1568–1572 (2010).

[244] Chen, R. and Haber, G.A., Chem. Phys. Lett. 2, 483–485 (1968).

[245] Halperin, A., Braner, A.A., Ben-Zvi, A. and Kristianpoller, N., Phys. Rev. 117, 416–422 (1960).

[246] Muntoni, C., Rucci, A. and Serpi, A., La Ric. Scient. 38, 762–764 (1968).

[247] Ilich, B.M., Sov. Phys. Solid State 21, 1880–1882 (1979).
[248] Pagonis, V., Kitis, G. and Furetta, C., Numerical and Practical Exercises in Thermoluminescence, Springer, (2006).
[249] Nahum, J. and Halperin, A., J. Phys. Chem. Solids 24, 823–834 (1963).
[250] Gobrecht, H. and Hoffman, D., J. Phys. Chem. Solids 27, 509–522 (1966).
[251] Tale, I.A., Phys. Status Solidi A 66, 65–75 (1981).
[252] Firszt, F. and Oczkowski, H.L., Phys. Status Solidi A 111, K 113-K 117 (1989).
[253] Chruścińska, A., J. Lumin. 62, 115–121 (1994).
[254] Grossweiner, L., J. Appl. Phys., 24, 1306–1309 (1953).
[255] Dussel G.A. and Bube, R.H., Phys. Rev. 155, 764–771 (1967).
[256] Balarin, M., Solid State Commun. 34, 419–422 (1980).
[257] Fleming, R.J., J. Phys. D: Appl. Phys. 23, 950–954 (1990).
[258] Keating, P.N., Proc. Phys. Soc. 78, 1408–1415 (1961).
[259] Land, P.L., J. Phys. Chem. Solids 30, 1681–1692; 1693–1708 (1969).
[260] Singh, S.J., Gartia, R.K. and Mazumdar, P.S., J. Phys. D: Appl. Phys. 22, 467–469 (1989).
[261] Booth, A.H., Can. J. Chem. 32, 214–215 (1954).
[262] Bohun, A., Czech. J. Phys. 4, 91–93 (1954).
[263] Parfianovitch, I.A., J. Exp. Theor. Phys. SSSR 26, 696 (1954).
[264] Hoogenstraaten, W., Philips Res. Repts. 13, 515–593 (1958).
[265] Chang, I.F. and Thioulouse, P., J. Appl. Phys. 53, 5873–5875 (1982).
[266] Osada, K., J. Phys. Soc. Jpn, 15, 145–149 (1960).
[267] Kitis, G., Chen, R., Pagonis, V., Carinou, E. and Kamenopoulou, V., J. Phys. D: Appl. Phys. 39, 1500–1507 (2006).
[268] Kitis, G., Chen, R., Pagonis, V., Carinou, E., Ascounis, P. and Kamenopoulou, V., J. Phys. D: Appl. Phys. 39, 1508–1514 (2006).
[269] Chen, R. and Winer, S.A.A., J. Appl. Phys. 41, 5227–5232 (1970).
[270] Singh, S.J., Mazumdar, P.S. and Gartia, R.K., J. Phys. D: Appl. Phys. 23, 562–566 (1990).
[271] Bell, F. and Sizmann, R., Phys. Status Solidi 15, 369–376 (1966).
[272] Chvoj, Z., Czech. J. Phys. B 27, 957–958 (1977).
[273] Chvoj, Z. and Pokorný, P., Czech. J. Phys. B 28, 446–455 (1978).
[274] Chvoj, Z. and Plichta, J., Czech. J. Phys. B 31, 572–577 (1981).
[275] Lesz, J., Scharmann, A. and Holzapfel, G., Jpn. J. Appl. Phys. Suppl. 24, 254–256 (1985).
[276] Lesz, J., Kriegseis, W. and Scharmann, A., Radiat. Prot. Dosim. 17, 513–517 (1986).
[277] Betts, D.S., Couturier, L., Khayrat, A.H., Luff, B.J. and Townsend, P.D., J. Phys. D: Appl. Phys. 26, 843–849 (1993).
[278] Betts, D.S. and Townsend, P.D., J. Phys. D: Appl. Phys. 26, 849–857 (1993).
[279] Kitis, G., Spiropulu, M., Papadopoulos, J. and Charalambous, S., Nucl. Instrum. Methods 73, 367–372 (1993).
[280] Piters, T.M. and Bos, A.J.J., J. Phys. D: Appl. Phys. 27, 1747–1756 (1994).
[281] Furetta, C., Kitis, G., Kuo, J.H., Vismara, L. and Weng, P.S., J. Lumin. 75, 341–351 (1997).
[282] Kitis, G. and Tuyn, J.W.N., J. Phys. D: Appl. Phys. 31, 2065–2073 (1998).
[283] Kitis, G. and Tuyn, J.W.N., Radiat. Prot. Dosim. 84, 371–374 (1999).
[284] Mohan, N.S. and Chen, R., J. Phys. D: Appl. Phys. 3, 243–247 (1970).
[285] Bull, R.K., McKeever, S.W.S., Chen, R., Mathur, V.K. and Brown, M.D., J. Phys. D: Appl. Phys. 19, 1321–1334 (1986).
[286] Marquardt, D.W., J. Soc. Indust. Math. 11, 431–441 (1963).
[287] Avriel, M., Nonlinear Programming: Analysis and Methods, Prentice-Hall Inc. (1976).
[288] van Laarhoven, P.J.M. and Aarts, E.A., Simulated Annealing. Theory and Applications, Reidel (1987).
[289] Battiti, R. and Tecchiolli, G., ORSA J. Comput. 6, 126–140 (1994).
[290] Glover, F., Discr. Appl. Math. 49, 231–255 (1994).
[291] Horowitz, Y.S. and Yossian, D., Radiat. Prot. Dosim. 60, 1–102 (1995).
[292] Hoogenboom, J.E., de Vries, W., Dielhof, J.B. and Bos, A.J.J., J. Appl. Phys. 64, 3193–3200 (1988).
[293] Sakurai, T., J. Phys. D: Appl. Phys. 28, 2139–2143 (1995).

[294] Sakurai, T., J. Phys. D: Appl. Phys. 34, L105-L107 (2001).

[295] Chen, R., J. Mater. Sci. 11, 1521–1541 (1976).

[296] Lucovsky, G., Solid State Commun. 3, 299–302 (1965).

[297] Lucovsky, G., Solid State Commun. 88, 879–882 (1993).

[298] Huntley, D.J., Short, M.A. and Dunphy, K., Can. J. Phys. 74, 81–91 (1996).

[299] Bulur, E., Bøtter-Jensen, L. and Murray, A.S., Radiat. Meas. 32, 407–411 (2000).

[300] Bulur, E., Bøtter-Jensen, L. and Murray, A.S., Radiat. Meas. 33, 715–719 (2001).

[301] Singarayer, J.S. and Bailey, R.M., Radiat, Meas. 38, 111–118 (2004).

[302] Singarayer, J.S. and Bailey, R.M., Radiat. Meas. 37, 451–458 (2003).

[303] Jain, M., Murray, A.S. and Bøtter-Jensen, L., Radiat. Meas. 37, 441–449 (2003).

[304] Bulur, E., Duller, G.A.T., Solongo, S., Bøtter-Jensen, L. and Murray, A.S., Radiat. Meas. 35, 79–85 (2002).

[305] Yoshida, H., Roberts, R.G. and Olley, J.M., Quat. Sci. Rev. 22, 1273–1278 (2003).

[306] Kiyak, N.G., Polymeris, G.S. and Kitis, G., Radiat. Meas. 42, 144–155 (2007).

[307] Kitis, G., Polymeris, G.S. and Kiyak, N.G., Radiat. Meas. 42, 1273–1279 (2007).

[308] Kiyak, N.G., Polymeris, G.S. and Kitis, G., Radiat. Meas. 43, 263–268 (2008).

[309] Kuhns, C.K., Agersnap larsen, N. and McKeever S.W.S., Radiat. Meas. 32, 413–418 (2000).

[310] Choi, J.H., Duller, G.A.T. and Wintle, A.G., Ancient TL 24, 9–20 (2006).

[311] Chruścińska, A., Radiat. Meas. 45, 991–999 (2010).

[312] Noras, J.M., J. Phys. C: Solid State Phys. 13, 4779–4789 (1980).

[313] Kitis, G. and Pagonis, V., Radiat. Meas. 43, 737–741 (2008).

[314] Garlick, G.F.J., Encyclopedia of Physics, Vol. 26, Springer Verlag, pp. 1–28 (1958).

[315] Chen, R. and Kristianpoller, N., Radiat. Prot. Dosim. 17, 443–446 (1986).

[316] Cook, J.R., Proc. Phys. Soc. 71, 422–429 (1958).

[317] Rzepka, E., Lefrant, S., Taurel, L. and Chapelle, J.P., J. Phys. C: Solid State Phys. 10, 2285–2293 (1977).

[318] Rácz, Z., J. Lumin. 23, 255–260 (1981).

[319] Delgado, A., Gómez-Ros, J.M., Bacci, C., Furetta C. and Rispoli, B., J. Phys. D: Appl. Phys. 23, 1132–1134 (1990).

[320] McKeever, S.W.S. and Horowitz, Y.S., Radiat. Phys. Chem. 36, 35–46 (1990).

[321] Mahajna, S., Yossian, D., Horowitz, Y.S. and Horowitz, A., Radiat. Prot. Dosim. 47, 73–77 (1993).

[322] Yossian, D., Mahajna, S., Ben-Shachar, B. and Horowitz, Y.S., Radiat. Prot. Dosim. 47, 129–133 (1993).

[323] Horowitz, Y.S., Ben-Shachar, B. and Yossian, D., J. Phys. D: Appl. Phys. 26, 1475–1481 (1993).

[324] Kitis, G., Tzima, A., Cai, G.G. and Furetta, C., J. Phys. D: Appl. Phys. 29, 1601–1612 (1996).

[325] Furetta, C., Nucl. Tracks Radiat. Meas. 11, 141–145 (1986).

[326] Jain, M., Bøtter-Jensen, L., Murray, A.S., Denby, P.M., Tsukamoto, S. and Gibling, M.R., Ancient TL 23, 9–24 (2005).

[327] Jain, M., Duller, G.A.T. and Wintle, A.G., Radiat. Meas. 42, 1285–1293 (2007).

[328] Chen, R. and Hag-Yahya, A., Radiat. Prot. Dosim. 65, 17–20 (1996).

[329] Chen, R. and Hag-Yahya, A., Radiat. Meas. 27, 205–210 (1997).

[330] Chen, R., Leung, P.L. and Stokes, M.J., Radiat. Meas. 32, 505–511 (2000).

[331] Bull, C. and Garlick, G.F.J., Proc. Phys. Soc. London, 63, 1283–1291 (1950).

[332] Schulman, J.H., Ginther, R.J., Gorbics, S.G., Nash, A.E., West, E.J. and Attix, F.H., Appl. Radiat. Isot. 20, 523–527 (1969).

[333] Kieffer, F., Meyer, C. and Rigaut, J., Chem. Phys. Lett. 11, 359–361 (1971).

[334] Wintle, A.G., Nature 245, 143–144 (1973).

[335] Visocekas, R. and Zink, A., Quat. Geochron. 18, 271–278 (1999).

[336] Visocekas, R., Radiat. Meas. 32, 499–504 (2000).

[337] Jaek, I., Molodkov, A. and Vasilchenko, V., Estonian J. Earth Sci., 56, 167–178 (2007).

[338] Molodkov, A., Jaek, I. and Vasilchenko, V., Geochronometria 26, 11–17 (2007).

[339] Gong, G.L., Liu, S.S., Xia, B., Sun, W.D. and Huang, B.L., Sci. China, Ser. G 51, 225–231 (2008).

[340] Li, B. and Li, S.H., J. Phys. D: Appl. Phys. 41, 225502 (15pp) (2008).

[341] Bailiff, I.K., Nature, 264, 531–533 (1976).

[342] Lamothe, M. and Auclair, M., Earth Planet. Sci. Lett. 171, 319–323 (1999).

[343] Lamothe, M., Auclair, M., Hamzaoui, C. and Huot, S., Radiat. Meas. 37, 493–498 (2003).

[344] Wallinga, J., Bos, A.J.J., Dorenbos, P., Murray, A.S. and Schokker, J., Quat. Geochron. 2, 216–221 (2007).

[345] Polymeris, G.S., Tsirliganis, N., Loukou, Z. and Kitis, G., Phys. Status Solidi A 203, 578–590 (2006).

[346] Spooner, N.A., Quat. Sci. Rev. 11, 139–145 (1992).

[347] Visocekas, R., Nucl. Tracks Radiat. Meas. 21, 175–178 (1993).

[348] Visocekas, R., Ouchene, M. and Gallois, B., Nucl. Instrum. Methods 214, 553–555 (1983).

[349] Visocekas, R., Nucl. Tracks Radiat. Meas. 10, 521–529 (1985).

[350] Clark, P.A. and Templer, R.H., Archaeometry 30, 19–36 (1988).

[351] Vedda, A., Martini, M., Meinardi, F., Chval, J., Dusek, M., Mares, J.A., Mihokova, E. and Nikl, M., Phys. Rev. B 61, 8081–8086 (2000).

[352] de Lima, J.F., Valerio, M.E.G. and Okuno, E., Phys. Rev. B 64, 014105 (6pp) (2001).

[353] Templer, R.H., Radiat. Prot. Dosim. 17, 493–497 (1986).

[354] Kitis, G., Bousbouras, P., Antypas, C. and Charalambous, S., Nucl. Tracks Radiat. Meas. 18, 61–65 (1991).

[355] Trautmann, T., Krbetschek, M.R., Dietrich, A. and Stolz, W., Radiat. Meas. 32, 487–492 (2000).

[356] Hasan, F.A., Keck, B.D., Hartmetz, C. and Sears, D.W.G., J. Lumin. 34, 327–335 (1986).

[357] Meisl, N.K., and Huntley, D.J., Ancient TL 23, 1–7 (2005).

[358] Tyler, S. and McKeever, S.W.S., Nucl. Tracks Radiat. Meas. 14, 149–156 (1988).

[359] Sears, D.W.G., Myers, B.M., Hartmetz, C. and Hasan, F.A., Nucl. Tracks Radiat. Meas. 17, 583–586 (1990).

[360] Tsirliganis, N., Polymeris, G.S., Loukou, Z. and Kitis, G., Radiat. Meas. 41, 954–960 (2006).

[361] Fragoulis, D. and Stoebe, T.G., Radiat. Prot. Dosim. 34, 65–68 (1990).

[362] Visocekas, R., Tale, V., Zink, A., Spooner, N.A. and Tale, I., Radiat. Prot. Dosim. 66, 391–394 (1996).

[363] Moharil, S.V., Patey, S.M. and Deshmukh, B.T., J. Lumin. 42, 325–329 (1989).

[364] Visocekas, R. and Geoffroy, A., Phys. Status Solidi A 41, 499–503 (1977).

[365] Visocekas, R., Eur. PACT J. 3, 258–265 (1977).

[366] Mikhailov, A.J., Dokl. Phys. Chem. 197, 223–225 (1971).

[367] Delbeq, C.J., Toyozawa, Y. and Yuster, P.H., Phys. Rev. B 10, 4497–4505 (1974).

[368] Templer, R.H., D. Phil. Thesis, University of Oxford (1986).

[369] Huntley, D.J., J. Phys.: Condens. Matter 18, 1359–1365 (2006).

[370] Cordier, P., Delouis, J.F., Kieffer, F., Lapersonne, C. and Rigaut, J., C. R. Acad. Sci. C 279, 589–591 (1974).

[371] Debye, P. and Edwards, J.O., J. Chem. Phys. 20, 236–239 (1952).

[372] Huntley, D.J. and Lamothe, M., Can. J. Earth Sci. 38, 1093–1106 (2001).

[373] Visocekas, R., Eur. PACT J. 3, 258–265 (1979).

[374] Visocekas, R., Ceva, T., Marti, C., Lefaucheux, F. and Robert, M.C., Phys. Status Solidi A 35, 315–327 (1976).

[375] Visocekas, R., Radiat. Prot. Dosim. 100, 45–54 (2002).

[376] Haynes, J.R. and Hornbeck, J.A., Phys. Rev. 90, 152–153 (1953).

[377] Zimmerman, D.W., Rhyner, C.R. and Cameron, J.R., Health Phys. 12, 525–531 (1966).

[378] Blak, A.R. and Watanabe, S., Proc. 4th Int. Conf. on Lumin. Dosim. 169–189 (1974).

[379] Johnson, T.L., Proc. 4th Int. Conf. Lumin. Dosim. 197–217 (1974).

[380] Taylor, G.C. and Lilley, E., J. Phys. D: Appl. Phys. 11, 567–581 (1978).

[381] Gorbics, S.G., Attix, F.H. and Pfaff, J.A., Int. J. Appl. Radiat. 18, 625–630 (1967).

[382] Pohlit, W., Biophysik 5, 341–356 (1969).

[383] Fairchild, R.G., Mattern, P.L., Lengweiler, K. and Levy, P.W., IEEE Trans. Nucl. Sci. NS-21, 366–372 (1974).

[384] Townsend, P.D., Taylor, G.C. and Wintersgill, M.C., Radiat. Eff. 41, 11–16 (1979).

[385] McKeever, S.W.S., Radiat. Prot. Dosim., 6, 49–51 (1984).

[386] Bos, A.J.J., Piters, T.M., Gómez-Ros, J.M. and Delgado, A., GLOCANIN, An Intercomparison of Glow Curve Analysis Computer Programs, IRI-93-005 (1993).

[387] Tatake, V.G., Desai, T.S., Govindjee and Sane, P.V., Photochem. Photobiol. 33, 243–250 (1981).

[388] DeVault, D., Govindjee and Arnold, W., Proc. Natl Acad. Sci. USA 80, 983–987 (1983).

[389] Sayer, M. and Souder, A.D., Can. J. Phys. 47, 463–471 (1969).

[390] Winer, S.A.A., Kristianpoller, N. and Chen, R., Luminescence of Crystals, Molecules and Solutions, Ed. F. Williams, Plenum, pp. 473–477 (1973).

[391] Pilkuhn, M.H., Proc. 13th Int. Conf. Phys. Semicond. 61–70 (1976).

[392] Wintle, A.G. and Huntley, D.J., Nature 279, 710–712 (1979).

[393] Huntley, D.J., Phys. Chem. Miner. 12, 122–127 (1985).

[394] Singhvi, A.K., Sharma, Y.P. and Agarwal, D.P., Nature 295, 313–315 (1982).

[395] Shlukov, A.I. and Shakhovetz, S.A., Ancient TL 5, 11–15 (1987).

[396] Dharamsi, A.N. and Joshi, R.P., J. Phys. D: Appl. Phys. 24, 982–987 (1991).

[397] Mejdahl, V., Radiat. Prot. Dosim. 17, 219–227 (1988).

[398] Mejdahl, V., Quat. Sci. Rev. 7, 357–360 (1988).

[399] Balescu, S., Packman, S.C. and Wintle, A.G., Quat. Res. 35, 91–102 (1991).

[400] Balescu, S. and Lamothe, M., Quat. Sci. Rev. 11, 45–51 (1992).

[401] Xie, J. and Aitken, M.J. Ancient TL 9, 21–25 (1991).

[402] Hornyak, W., Franklin, A. and Chen, R., Ancient TL 11, 21–26 (1993).

[403] Kikuchi, T., J. Phys. Soc. Jpn. 13, 526–532 (1958).

[404] Medlin, W.L., Phys. Rev. 123, 502–509 (1961).

[405] Bosacchi, A., Bosacchi, B., Franchi, S. and Hernandez, L., Solid State Commun. 13, 1805–1809 (1973).

[406] Bosacchi, A., Franchi, S. and Bosacchi, B., Phys. Rev. B 10, 5235–5238 (1974).

[407] Simmons, J.G., Taylor, G.W. and Tam, M.C., Phys. Rev. B 7, 3714–3719 (1973).

[408] Guzzi, M. and Baldini, G., J. Lumin. 9, 514–522 (1975).

[409] Fleming, R.J. and Pender, L.F., Phys. Rev. B 18, 5900–5902 (1978).

[410] Shousha, A.H.M., Thin Solid Films 20, 33–41 (1974).

[411] Srivastava, J.K. and Supe, S.J., J. Phys. D: Appl. Phys. 16, 1813–1818 (1983).

[412] Srivastava, J.K. and Supe, S.J., J. Phys. D: Appl. Phys. 22, 1537–1543 (1989).

[413] Zahedifar, M., Karimi, L. and Kavianinia, M.J., Nucl. Instrum. Methods Phys. Res. A 564, 515–520 (2006).

[414] Srivastava, J.K. and Supe, S.J., Radiat. Prot. Dosim. 6, 41–44 (1983).

[415] Sakurai, T. and Gartia, R.K., J. Appl. Phys. 82, 5722–5727 (1997).

[416] Sakurai, T., Shoji, K., Itoh, K. and Gartia, R.K., J. Appl. Phys. 89, 2208–2212 (2001).

[417] Hagekyriakou, J. and Fleming, R.J., J. Phys. D: Appl. Phys. 15, 163–176 (1982).

[418] Hagekyriakou, J. and Fleming, R.J., J. Phys. D: Appl. Phys. 16, 1343–1352 (1983).

[419] Hornyak, W.F. and Franklin, A., Nucl. Tracks. Radiat. Meas. 14, 81–89 (1988).

[420] Hornyak, W.F., Chen, R. and Franklin, A., Phys. Rev. B 1, 46, 8036–8049 (1992). http://prb.aps.org/abstract/PRB/v46/i13/p8036_1.

[421] Bonfiglioli, G., Brovetto, P. and Cortese, C., Phys. Rev. 114, 951–955 (1959).

[422] Hornyak, W.F. and Chen, R., J. Lumin. 44, 73–81 (1989).

[423] Stoddard, A.E., Phys. Rev. 120, 114–117 (1960).

[424] Braner, A.A. and Israeli, M., Phys. Rev. 132, 2501–2505 (1963).

[425] Kos, H.-J. and Nink, R., Phys. Status Solidi A 44, 505–510 (1977).

[426] Wang, H.E., Lin, S.W., Weng, P.S. and Hsu, P.C., Appl. Radiat. Isot. 46, 869–873 (1995).

[427] Schlesinger, M., J. Phys. Chem. Solids 26, 1761–1766 (1965).

[428] Crippa, P.R., Paracchini, C. and Félszerfalvi, J., J. Phys. Soc. Jpn. 24, 92–95 (1968).

[429] Caldas, L.V.E. and Mayhugh, M.R., Health Phys. 31, 451–452 (1976).

[430] Pradhan, A.S. and Bhatt, R.C., J. Phys. D: Appl. Phys. 18, 317–320 (1985).

[431] Goyet, D., Lapraz, D., and Iacconi, P., Radiat. Prot. Dosim. 65, 317–320 (1996).

[432] Puite, K.J., Int. J. Appl. Radiat. Isot. 19, 397–402 (1968).

[433] Lakshmanan, A.R. and Vohra, K.G., Nucl. Instrum. Methods 159, 585–592 (1979).

[434] Wheeler, G.C.W.S., Quat. Sci. Rev. 7, 407–410 (1988).

[435] Milanovich-Reichhalter, I., and Vana, N., Radiat. Prot. Dosim. 33, 211–213 (1990).

[436] Alexander, C.S. and McKeever, S.W.S., J. Phys. D: Appl. Phys. 31, 2908–2920 (1998).

[437] Wintle, A.G. and Murray, A.S., Radiat. Meas. 27, 611–624 (1997).

[438] Sunta, C.M. and Watanabe, S., J. Phys. D: Appl. Phys. 9, 1271–1278 (1976).

[439] Moharil, S.V. and Kathuria, S.P., J. Phys. D: Appl. Phys. 18, 691–701 (1985).

[440] Bhasin, B.D., Kathuria, S.P. and Moharil, S.V., Phys. Status Solidi A 106, 271–276 (1988).

[441] Osvay, M., Ranogajec-Komor, M. and Golder, F., Radiat. Prot. Dosim. 33, 135–138 (1990).

[442] Yazici, A.N., Karali, T., Townsend, P.D. and Murat, A., J. Phys. D: Appl. Phys. 37, 3165–3173 (2004).

[443] Lima, J.F., Trzesniak, P., Yoshimura, E.M. and Okuno, E., Radiat. Prot. Dosim. 33, 143–146 (1990).

[444] Liu, B., Shi, C., Wei, Y., Wu, C, Li, Y. and Hu, G., J. Phys. Condens. Matter 14, 7065–7069 (2002).

[445] Iacconi, P., Lapraz, D., Keller, P., Portal, G. and Barthe, J., Radiat. Prot. Dosim. 17, 475–478 (1986).

[446] Lapraz, D., Iacconi, P., Sayadi, Y., Keller, P., Barthe, J. and Portal, G., Phys. Status Solidi A 108, 783–794 (1988).

[447] Kharita, M.H., Stokes, R. and Durrani, S.A., Radiat. Meas. 23, 493–496 (1994).

[448] Panizza, E., Phys. Lett. 10, 37–38 (1964).

[449] Alexander, C.S., Morris, M.F. and McKeever, S.W.S., Radiat. Meas. 27, 153–159 (1997).

[450] Kristianpoller, N., Chen, R. and Israeli, M., J. Phys. D: Appl. Phys., 7, 1063–1072 (1974).

[451] Bøtter-Jensen, L., Agersap Larsen, N., Mejdahl, V., Poolton, N.R.J., Morris, M.F. and McKeever, S.W.S., Radiat. Meas. 24, 535–541 (1995).

[452] McKeever, S.W.S., Nucl. Tracks Radiat. Meas. 18, 5–12 (1991).

[453] McKeever, S.W.S., Radiat. Meas. 23, 267–276 (1994).

[454] Böer, K.W., Survey of the Physics of Semiconductors, Van Nostrand Reinhold (1990).

[455] Umisedo, K.U., Yoshimura, E.M., Gasparian, P.B.R, and Yukihara, E.G., Radiat. Meas. 45, 151–156 (2010).

[456] Yukihara, E.G., Whitley, V.H., Polf, J.C., Klein, D.M., McKeever, S.W.S., Akselrod, A.E. and Akselrod, M.S., Radiat. Meas. 37, 627–638 (2003).

[457] Nikiforov, S.V. and Kortov, V.S., Radiat. Meas. 45, 527–529 (2010).

[458] Kitis, G., Phys. Status Solidi A 191, 621–627 (2002).

[459] Subedi, B., Kitis, G. and Pagonis, V., Phys. Status Solidi A 207, 1216–1226 (2010).

[460] Halperin, A., Leibovitz, M. and Schlesinger, M., Rev. Sci. Instrum. 33, 1168–1170 (1962).

[461] Chen, R., Israeli, M. and Kristianpoller, N., Chem. Phys. Lett. 7, 171–172 (1970).

[462] Johnson, E.J., Semiconductors and Semimetals, Vol. 3, Eds R.K. Willardson and A.C. Beer, Academic Press, Ch. 6 (1967).

[463] Johnson, P.D. and Williams, F.E., J. Chem. Phys. 18, 323–326 (1950).

[464] Johnson, P.D. and Williams, F.E., J. Chem. Phys. 18, 1477–1483 (1950).

[465] Johnson, P.D. and Williams, F.E., J. Chem. Phys. 21, 125–130 (1953).

[466] Ewles, J. and Lee, N., J. Electrochem. Soc. 100, 392–398 (1953).

[467] Schulman, J.H., Claffy, E.W. and Potter, R.J., Phys. Rev. 108, 1398–1401 (1957).

[468] Van Uitert, L.G., J. Electrochem. Soc. 107, 803–806 (1960).

[469] Van Uitert, L.G., Linares, R.C., Soden, R.R. and Ballman, A.A., J. Chem. Phys. 36, 702–705 (1962).

[470] Van Uitert, L.G. and Johnson, L.F., J. Chem. Phys. 44, 3514–3522 (1966).

[471] Van Uitert, L.G., Dearborn, E.F. and Rubin, J.J., J. Chem. Phys. 46, 420–425 (1967).

[472] Van Uitert, L.G., J. Electrochem. Soc. 114, 1048–1053 (1967).

[473] Van Uitert, L.G., Proc. Int. Conf. Lumin. 1588–1603 (1966).

[474] Medlin, W.L., J. Chem. Phys. 30, 451–458 (1959).

[475] Medlin, W.L., J. Chem. Phys. 34, 672–677 (1961).

[476] Rossiter, M.J., Rees-Evans, D.B., Ellis, S.C. and Griffiths, J.M., J. Phys. D: Appl. Phys. 4, 1245–1251 (1971).

[477] Nambi, K.S.V., Bapat, V.N. and Ganguly, A.K., J. Phys. C: Solid State Phys. 7, 4403–4415 (1974).

[478] Nambi, K.S.V., Eur. PACT J., 3, 293–310 (1979).

[479] Wachter, W., J. Appl. Phys. 53, 5210–5215 (1982).

[480] Lai, L.J., Sheu, H.S., Lin, Y.K., Hsu, Y.C. and Chu, T.C., J. Appl. Phys. 100, 103508 (2006).

[481] Tajika, Y. and Hashimoto, T., Radiat. Meas. 41, 809–812 (2006).

[482] Vij, A., Lochab, S.P., Singh, S., Kumar, R. and Singh, N., J. Alloys Compd. 486, 554–558 (2009).

[483] Vij, A., Lochab, S.P., Kumar, R. and Singh, N., J. Alloys Compd. 490, L33-L36 (2010).

[484] Sharma, G., Chawla, P., Lochab, S.P. and Singh, N., Chalcogenide Lett. 6, 705–712 (2009).

[485] Lawless, J.L., Pagonis, V. and Chen, R., Mater. Sci. Eng. Eurodim 2010, Pécs, Hungary, Book of Abstracts, 8.5 (2010).

[486] Chen, R. and Leung, P.L., Radiat. Meas. 33, 475–481 (2001).

[487] Yukihara, E.G., Whitley, V.H., McKeever, S.W.S., Akselrod, A.E. and Akselrod, M.S., Radiat. Meas. 38, 317–330 (2004).

[488] Pagonis, V., Chen, R., Lawless, J.L. and Andersen, C., J. Phys. D: Appl. Phys. 42, 175107 (2009).

[489] Carter Jr, J.R., J. Phys. Chem. Solids 31, 2405–2416 (1970).

[490] Inabe, K., Miyanaga, T. and Takeuchi, N., J. Mater. Sci. Lett. 4, 925–927 (1985).

[491] Hackenschmied, P., Li, H., Epelbaum, E., Fasbender, R., Batentschuk, M. and Winnacker, A., Radiat. Meas. 33, 669–674 (2001).

[492] Yawata, T., Takeuchi, T. and Hashimoto, T., Radiat. Meas. 41, 841–846 (2006).

[493] Medlin, W.L., Thermoluminescence of Geological Materials, Ed. D.J. McDougall, Academic Press, pp. 193–224 (1968).

[494] Schmidt, K., Linemann, H. and Giessing, R., Proc. 4th Int. Conf. Lumin. Dosim. 237–253 (1974).

[495] Claffy, E.W., Klick, C.C. and Attix, F.H., Proc. 2nd Int. Conf. on Lumin. Dosim. 302–309 (1968).

[496] Attix, F.H., J. Appl. Phys. 46, 81–88 (1975).

[497] Bull, R.K., Radiat. Prot. Dosim. 17, 459–463 (1986).

[498] Kristianpoller, N. and Israeli, M., Phys. Rev. B 2, 2175–2182 (1970).

[499] Israeli, M., Kristianpoller, N. and Chen, R., Phys. Rev. B 6, 4861–4867 (1972).

[500] Kristianpoller, N., Schriever, R. and Schwentner, N., J. Phys.: Condens. Matter 2, 6939–6944 (1990).

[501] Pooley, D., Proc. Phys. Soc. 87, 245–256 (1966).

[502] Hersh, H.N., Phys. Rev. 148, 928–932 (1966).

[503] Chen, R., Pagonis, V. and Lawless, J.L., Mater. Sci. Eng. 15, 012071 (2010).

[504] Kristianpoller, N. and Katz, I., Solid State Commun. 11, 1057–1061 (1972).

[505] Israeli, M. and Kristianpoller, N., Solid State Commun. 7, 1131–1133 (1969).

[506] Nishidai, T., Onoyama, Y., Abe, M., Takahashi, M. and Suyama, S., Nippon Acta Radiol. 33, 889–902 (1973).

[507] Popov, A.I., Nucl. Instrum. Methods Phys. Res., Sect. B 65, 521–524 (1992).

[508] Panova, A.N., Kosinov, N.N., Shplinskaya, L.N. and Kovaleva, L.V., Opt. Spectrosk. 68, 37–39 (1990).

[509] McKeever, S.W.S., Rhodes, J.F., Mathur, V.K., Chen, R., Brown, M.D. and Bull, R.K., Phys. Rev. B 32, 3835–3843 (1985).

[510] Zimmerman, J., Phys. C: Solid State Phys. 4, 3265–3276, 3277–3291 (1971).

[511] Chen, R., Pagonis, V. and Lawless, J.L., Radiat. Meas. 43, 162–166 (2008).

[512] Lawless, J.L., Chen, R. and Pagonis, V., J. Phys. D: Appl. Phys. 42, 155409 (2009).

[513] Lawless, J.L., Chen, R. and Pagonis, V., Radiat. Meas. 44, 606–610 (2009).

[514] Tomita, A., Hirai, N. and Tsutsumi, K., Jpn. J. Appl. Phys. 15, 1899–1908 (1976).

[515] Kuzakov, S.M., Radiat. Prot. Dosim. 33, 115–117 (1990).

[516] Bindi, R., Iacconi, P., Lapraz, D. and Petel, F., J. Phys. D: Appl. Phys. 30, 137–143 (1997).

[517] Rosenman, G., Naich, M., Molotskii, M., Dechtiar, Yu. and Noskov, V., Appl. Phys. Lett. 80, 2743–2745 (2002).

[518] Oster, L. and Haddad, J., Mater. Sci. 9, 297–302 (2003).
[519] Kappers, L.A., Bartram, R.H., Hamilton, D.S., Brecher, C. and Lempicki, A., Nucl. Instrum. Methods Phys. Res., Sect. A 537, 443–445 (2005).
[520] Tolpygo, E.I., Tolpygo, K.B. and Sheinkman, M.K., Bull. Acad. Sci. USSR, Phys. Ser. (English translation) 30, 1980–1984 (1966).
[521] McKeever, S.W.S., Akselrod, M.S. and Markey, B.G., Radiat. Prot. Dosim. 65, 267–272 (1996).
[522] McKeever, S.W.S. and Akselrod, M.S., Radiat. Prot. Dosim. 84, 317–320 (1999).
[523] Bøtter-Jensen, L. and Murray, A.S., Radiat. Prot. Dosim. 84, 307–316 (1999).
[524] Godfrey-Smith, D.I., J. Phys. D: Appl. Phys. 27, 1737–1746 (1994).
[525] Roberts, R.G., Spooner, N.A. and Questiaux, D., Radiat. Meas. 23, 647–653 (1994).
[526] Banerjee, D., Radiat. Meas. 33, 47–57 (2001).
[527] Jursinic, P.A., Med. Phys. 37, 132–140 (2010).
[528] Lowick, S.E., Preusser, F. and Wintle, A.G., Radiat. Meas. 45, 975–984 (2010).
[529] Chen, R. and Leung, P.L., J. Appl. Phys. 89, 259–263 (2001).
[530] McKeever, S.W.S., Bøtter-Jensen, L., Agersap Larsen, N. and Duller, G.A.T., Radiat. Meas. 23, 647–653 (1994).
[531] Chen, R., Huntley, D.J. and Berger, G.W., Phys. Status Solidi A 79, 251–261 (1983).
[532] Bailey, R.M., Radiat. Meas. 38, 299–310 (2004).
[533] Bailey, R.M., Armitage, S.J. and Stokes, S., Radiat. Meas. 39, 347–359 (2005).
[534] Chen, R., Radiat. Prot. Dosim. 100, 71–74 (2002).
[535] Akselrod, M.S. and McKeever, S.W.S., Radiat. Prot. Dosim. 81, 167–176 (1999).
[536] Bulur, E., Radiat. Meas. 26, 701–709 (1996).
[537] Wrzesinska, A., Acta Phys. Polon. 15, 151–162 (1956).
[538] Bulur, E. and Göksu, H.Y., Radiat. Meas. 30, 203–206 (1999).
[539] Mishra, D.R., Kulkarni, M.S., Rawat, N.S., Muthe, K.P., Gupta, S.K., Bhatt, B.C. and Sharma, D.N., Radiat. Meas. 43, 1177–1186 (2008).
[540] Bos, A.J.J. and Wallinga, J., Phys. Rev. B 79, 195118 (2009).
[541] Whitley, V.H. and McKeever, S.W.S., J. Appl. Phys. 90, 6073–6083 (2001).
[542] Kitis, G., Liritzis, I. and Vafeiadou, A., J. Radioanal. Nucl. Chem. 254, 143–149 (2002).
[543] Kitis, G., Kiyak, N., Polymeris, G.S. and Tsirliganis, N.C., J. Lumin. 130, 298–303 (2010).
[544] Bulur, E. and Yeltik, A., Radiat. Meas. 45, 29–34 (2010).
[545] Arnold, W. and Sherwood, H., J. Phys. Chem. 63, 2–4 (1959).
[546] Stammers, K., J. Phys. E, 12, 637–639 (1979).
[547] Kelly, P.J. and Laubitz, M.J., Can. J. Phys. 45, 311–321 (1967).
[548] Chen, R. and Pagonis, V., J. Phys. D: Appl. Phys. 41, 035102 (2008).
[549] Chen, R., Pagonis, V. and Lawless, J.L., Radiat. Meas. 44, 344–350 (2009).
[550] Chen, R., Pagonis, V. and Lawless, J.L., Radiat. Meas. 45, 147–150 (2010).
[551] Chen, R., Pagonis, V. and Lawless, J.L., Radiat. Meas. 45, 277–280 (2010).
[552] Bulur, E., Radiat. Meas. 32, 141–145 (2000).
[553] Bulur, E., Bøtter-Jensen, L. and Murray, A.S., Nucl. Instrum. Methods B179, 151–159 (2001).
[554] Poolton, N.R.J., Bøtter-Jensen, L., Andersen, C.E., Jain, M., Murray, A.S., Malins, A.E.R. and Quinn, F.M., Radiat. Meas. 37, 639–645 (2003).
[555] Smith, B.W. and Rhodes, E.J., Radiat. Meas. 23, 329–333 (1994).
[556] Aitken, M.J. and Smith, B.W., Quat. Sci. Rev. 7, 387–393 (1988).
[557] Ankjærgaard, C. and Jain, M., J. Phys. D: Appl. Phys. 43, 255502 (2010).
[558] Chen, R., J. Lumin. 102–103, 510–518 (2003).
[559] Poolton, N.R.J., Bøtter-Jensen, L. and Jungner, H., Radiat. Meas. 24, 543–549 (1995).
[560] McKeever, S.W.S., Agersap Larsen, N., Bøtter-Jensen, L.L. and Mejdahl, V.L., Radiat. Meas. 27, 75–82 (1997).
[561] McKeever, S.W.S., Bøtter-Jensen, L., Agersap Larsen, N. and Duller, G.A.T., Radiat. Meas. 27, 161–170 (1997).
[562] Bailey, R.M., Radiat. Meas. 32, 233–246 (2000).
[563] Akselrod, M.S., Lucas, A.C., Polf, J.C. and McKeever, S.W.S., Radiat. Meas. 29, 391–399 (1998).

[564] Ankjærgaard, C., Jain, M., Thomsen, K.J. and Murray, A.S., Radiat. Meas. 45, 778–785 (2010).

[565] Chen, R. and Leung, P.L., Radiat. Meas. 37, 519–526 (2003).

[566] Kohlrausch, R., Pogg. Ann. Phys. Chem. 91, 179–213 (1854).

[567] Williams, G. and Watts, D.C., Trans. Faraday Soc. 66, 80–85 (1970).

[568] Brinkman, W.F., Physics through the 1990s, National Academy of Science (1986).

[569] Klafter, J. and Schlesinger, M.J., Proc. Natl Acad. Sci. USA 83, 848–851 (1986).

[570] Cardona, M., Chamberlin, R.V. and Marx, W., Ann. Phys. 16, 842–845 (2007).

[571] Chen, X., Henderson, B. and O'Donnell, K.P., Appl. Phys. Lett. 60, 2672–2674 (1992).

[572] Pavesi, L. and Ceschini, M., Phys. Rev. B 48, 17625–17628 (1993).

[573] Pavesi, L., J. Appl. Phys. 80, 216–225 (1996).

[574] Huber, D.L., Personal communication (2002).

[575] Ankjærgaard, C., Denby, P.M., Murray, A.S. and Jain, M., Radiat. Meas. 43, 273–277 (2008).

[576] Pagonis, V., Ankjærgaard, C., Murray, A.S. and Chen, R., J. Lumin. 129, 1003–1009 (2009).

[577] Oster, L. and Haddad, J., Phys. Status Solidi A 196, 471–476 (2003).

[578] Wintle, A.G. and Murray, A.S., Radiat. Meas. 29, 81–84 (1998).

[579] Rhodes E.J., Quat. Sci. Rev. 7, 395–400, (1988).

[580] Duller G.A.T. and Wintle A.G., Nucl. Tracks Radiat. Meas. 18, 379–384 (1991).

[581] Bailiff, I.K. and Poolton, N.R.J., Nucl. Tracks Radiat. Meas. 18, 111–119 (1991).

[582] Duller G.A.T. and Bøtter-Jensen L., Radiat. Prot. Dosim. 47, 683–688 (1993).

[583] Duller, G.A.T., Radiat. Meas. 23, 281–285 (1994).

[584] Short, M.A. and Tso, M.Y.W., Radiat. Meas. 23, 335–338 (1994).

[585] Li, S.H., Tso, M.Y.W. and Wong, N.W., Radiat. Meas. 27, 43–47 (1997).

[586] Li, S.H. and Chen, G., J. Phys. D: Appl. Phys. 34, 493–498 (2001).

[587] Li, B. and Li, S.H., J. Phys. D: Appl. Phys. 39, 2941–2949 (2006).

[588] Pagonis, V. and Chen, R., Geochronometria 30, 1–7 (2008).

[589] Pagonis, V., Wintle, A.G. and Chen, R., Radiat. Meas. 42, 1587–1599 (2007).

[590] Neely, W.C., Mandal, K. and Peters, C.R., J. Lumin. 29, 341–347 (1984).

[591] Vakulenko, O.V. and Shutov, B.M., Sov. Phys. Semicond. 12, 603–604 (1978).

[592] Otaki, H., Kido, H., Hiratsuka, A., Fukuda, Y. and Takeuchi, N., J. Mater. Sci. Lett. 13, 1267–1269 (1994).

[593] Mikado, T., Tomimasu, T., Yamazaki, T. and Chiwaki, M., Nucl. Instrum. Methods 157, 109–116 (1978).

[594] Yamazaki, T., Tomimasu, T., Mikado, T. and Chiwaki, M., J. Appl. Phys. 49, 4929–4931 (1978).

[595] Matrosov, I.I. and Pogorelov, Yu.L., J. Appl. Spectrosc. 29, 1327–1331 (1978).

[596] Felix, C. and Singhvi, A.K., Radiat. Meas. 27, 599–609 (1997).

[597] Piesch, E., Burgkhardt, B. and Sayed, A.M., Proc. Int. Conf. Lumin. Dosim., 1201–1212 (1974).

[598] Wintersgill, M.C. and Townsend, P.D., Radiat. Eff. 38, 113–118 (1978).

[599] Waligórski M.P.R. and Katz, R., Nucl. Instrum. Methods 172, 463–470 (1980).

[600] Horowitz, Y.S., Phys. Med. Biol. 26, 765–824 (1981).

[601] Mische, E.F. and McKeever, S.W.S., Radiat. Prot. Dosim. 29, 159–175 (1989).

[602] Nail, I., Horowitz, Y.S., Oster, L., Brandan, M.E., Rodríguez-Villafuerte, M., Buenfil, A.E., Ruiz-Trejo, C., Gamboa-deBuen, I., Avila, O., Tovar, V.M., Olko, P. and Ipe, N., Radiat. Prot. Dosim. 119, 180–183 (2006).

[603] Mitchell, P.V., Wiegand, D.A. and Smoluchowski, R., Phys. Rev. 121, 484–498 (1961).

[604] Wieser, A., Göksu, H.Y., Regulla, D.F. and Waibel, A., Nucl. Tracks Radiat. Meas. 18, 175–178 (1991).

[605] Gulamova, R.R., Gasanov, E.M. and Sazonova, E.V., Phys. Status Solidi A 135, 109–117 (1993).

[606] Watanabe, S., Ayta, W.E.F., Paião, J.R.B., Ferraz, G.M., Farias, T.M.B. and Cano, N.F., J. Phys. D: Appl. Phys. 41, 105401 (1–6) (2008).

[607] McKeever, S.W.S., J. Appl. Phys. 68, 724–731 (1990).

[608] Stoebe, T.G. and Watanabe, S., Phys. Status Solidi A 29, 11–29 (1975).

[609] Lakshmanan, A.R., Chandra, B. and Bhatt, R.C., J. Phys. D: Appl. Phys. 15, 1501–1517 (1982).

[610] Rodine, E.T. and Land, P.L., Phys. Rev. B 4, 2701–2727 (1971).

[611] Nakajima, T., Jpn. J. Appl. Phys. 15, 1179–1180 (1976).

[612] Nakajima, T., J. Appl. Phys. 48, 4880–4885 (1977).

[613] Nakajima, T., J. Appl. Phys. 49, 6189–6191 (1978).

[614] Yang, X.B., Xu, J., Li, H.J., Bi, Q.Y., Cheng, Y. and Tang, Q., J. Appl. Phys. 106, 033105 (5pp) (2009).

[615] Cameron, J.R. and Zimmerman, D.W., Ret. COO-1105–102, USAEC (1965).

[616] Suntharalingam, N. and Cameron, J.R., Phys. Med. Biol. 14, 397–410 (1969).

[617] Chen, R. and Bowman, S.G.E., Eur. PACT J. 2, 216–230 (1978).

[618] Duboc, C.A., Br. J. Appl. Phys. 6, S107-S111 (1955).

[619] Bowman, S.G.E. and Chen, R., J. Lumin. 18–19, 345–348 (1979).

[620] Chen, R., Yang, X.H. and McKeever, S.W.S., J. Phys. D: Appl. Phys. 21, 1452–1457 (1988).

[621] Savikhin, F.A., J. Appl. Spectrosc 17, 889–893 (1972).

[622] Zavt, G.S. and Savikhin, F.A., Izv. Akad. Nauk SSSR, Ser. Fiz. 38, 1325–1329 (pp. 190–194 in the English translation) (1974).

[623] Sunta, C.M., Kulkarni, R.N., Yoshimura, E.M., Mol, A.W., Piters, T.M. and Okuno, E., Phys. Status Solidi B 186, 199–207 (1994).

[624] Chen, R., Fogel, G. and Lee, C.K., Radiat. Prot. Dosim. 65, 63–68 (1996).

[625] Mady, F., Lapraz, D. and Iacconi, P., Radiat. Meas. 43, 180–184 (2008).

[626] Faïn, J. and Monnin, M., J. Electrostat 3, 289–296 (1977).

[627] Faïn, J., Sanzelle, S., Miallier, D., Monret, M. and Pillere, Th., Radiat. Meas. 23, 287–291 (1994).

[628] Fleming, S.J., Archaeometry 15, 13–30 (1973).

[629] Fleming, S.J. and Thompson, J., Health Phys. 18, 567–568 (1970).

[630] Martini, M., Sibilia, E., Spinolo, G. and Vedda, A., Nucl. Tracks 10, 497–501 (1985).

[631] Lee, C.K., Wong, H.K. and Leung, Y.L., Radiat. Meas. 44, 215–222 (2009).

[632] Petralia, S., Lett. Nuovo Cim. 2, 413–417 (1971).

[633] Lakshmanan, A.R., Bhatt, R.C. and Vohra, K.G., Phys. Status Solidi A 53, 617–625 (1979).

[634] Moharil, S.V., Phys. Status Solidi A 79, 599–605 (1983).

[635] Nail, I., Horowitz, Y.S., Oster, L. and Biderman, S., Radiat. Prot. Dosim. 102, 295–304 (2002).

[636] Srivastava, J.K. and Supe, S.J., Nucl. Instrum. Methods Phys. Res., Sect. A243, 567–571 (1986).

[637] Tournon, J., Phys. Status Solidi 10, K7-K9 (1965).

[638] Kirsh, Y., Townsend, J.E. and Townsend, P.D., Phys. Status Solidi A 739–747 (1989).

[639] Murthy, K.B.S., Sunta, C.M., Khatri, D.T. and Soman, S.D., J. Phys. D: Appl. Phys. 11, 561–565 (1978).

[640] Hübner, S. and Göksu, H.Y., Appl. Radiat. Isot. 48. 1231–1235 (1997).

[641] Koul, D.K. and Chougaonkar, M.P., Radiat. Meas. 42, 1265–1272 (2007).

[642] Koul, D.K., Chougaonkar, M.P. and Polymeris, G.S., Radiat. Meas. 45, 15–21 (2010).

[643] Chen, R., and Halperin, A., Proc. Int. Conf. Lumin. Budapest, 1414–1417 (1966).

[644] Chen, R., Abu-Rayya, M. and Kristianpoller, N., J. Lumin. 48, 833–837 (1991).

[645] Chen, R., Fogel, G. and Kristianpoller, N., Radiat. Meas. 23, 277–280 (1994).

[646] Marrone, M.J. and Attix, F.H., Health Phys. 10, 431–436 (1964).

[647] Karzmark, C.J., White, J. and Fowler, J.F., Phys. Med. Biol. 9, 273–286 (1964).

[648] Tochilin, E. and Goldstein, N., Health Phys. 12, 1705–1713 (1966).

[649] Facey, R.A., Health Phys. 12, 715–717 (1966).

[650] Groom, P.J., Durrani, S.A., Khazal, K.A.R. and McKeever, S.W.S., Eur. PACT J. 2, 200–210 (1978).

[651] Hsu, P.C. and Weng, P.S., Nucl. Instrum. Methods 147, 453–453 (1977).

[652] Urbina, M., Millán, A., Beneitez, P. and Calderón, T., J. Lumin. 79, 21–28 (1998).

[653] Shlukov, A.I., Shakhovets, L.T. and Lyashenko, M.G., Nucl. Instrum. Methods Phys. Res., Sect. B 73, 373–381 (1993).

[654] Shlukov, A.I., Usova, M.G., Voskovskaya, L.T. and Shakhovets, S.A., Quat. Sci. Rev. 20, 875–878 (2001).

[655] Kvasnička, J., Int. Appl. Radiat. Isot. 34, 713–715 (1983).

[656] Valladas, G. and Ferreira, J., Nucl. Instrum. Methods 175, 216–218 (1980).

[657] Valladas, G. and Valladas, H., Eur. PACT J. 6, 281–286 (1982).

[658] Gastélum, S., Cruz-Zaragoza, E., Meléndrez, R., Chernov, V. and Barboza-Flores, M., Radiat. Eff. Defects Solids 162, 587–595 (2007).

[659] Boustead, I. and Charlesby, A., Proc. R. Soc. London, Ser. A 315, 271–286 (1970).

[660] Goldstein, F.T., Solid State Commun. 4, 621–625 (1966).

[661] Goldstein, F.T., Phys. Status Solidi 20, 379–383 (1967).

[662] Durand, P., Farge, Y. and Lambert, M., J. Phys. Chem. Solids 30, 1353–1374 (1969).

[663] Farge, Y., J. Phys. Chem. Solids, 30, 1375–1382 (1969).

[664] Lakshmanan, A.R., Bhatt, R.C. and Supe, S.J., J. Phys. D: Appl. Phys. 14, 1683–1706 (1981).

[665] Rasheedy, M.S. and Amry, A.M.A., Nucl. Instrum. Methods Phys. Res., Sect. A350, 561–565 (1994).

[666] Chen, R. and Fogel, G.M., Radiat. Prot. Dosim. 47, 23–26 (1993).

[667] Lee, C.K. and Chen, R., J. Phys. D: Appl. Phys. 28, 408–412 (1995).

[668] Chen, R. and Leung, P.L., J. Phys. D: Appl. Phys. 33, 846–850 (2000).

[669] Avila, O., Gamboa-deBuen, I. and Brandan, M.E., J. Phys. D: Appl. Phys. 32, 1175–1181 (1999).

[670] Charlesby, A. and Partridge, R.H., Proc. R. Soc. London, Ser. A 271, 170–206 (1963).

[671] Jain, V.K., Kathuria, S.P. and Ganguly, A.K., J. Phys. C: Solid State Phys. 8, 2191–2197 (1975).

[672] Horowitz, Y.S., Moscovitch, M. and Wilt, M., Nucl. Instrum. Methods 556–564 (1986).

[673] Ichikawa, Y., Jpn. J. Appl. Phys. 7, 220–226 (1968).

[674] Damm, J.Z. and Opyrchal, H., Cryst. Latt. Def. 6, 21–26 (1975).

[675] David, M., Sunta, C.M. and Ganguly, A.K., Ind. J. Pure Appl. Phys. 15, 277–280 (1977).

[676] Schulman, J.H., Etzel, H.W. and Allard, J.G., J. Appl. Phys. 28, 792–795 (1957).

[677] Freytag, E., Health Phys. 20, 93–94 (1971).

[678] Tesch, K., Radiat. Prot. Dosim. 6, 347–349 (1983).

[679] Böhm, M., Pitt, E. and Scharmann, A., Radiat. Prot. Dosim. 8, 139–145 (1984).

[680] Zeneli, D., Tavlet, M. and Conincks, F., Radiat. Prot. Dosim. 66, 205–207 (1996).

[681] Horowitz, Y.S., Fraier, I. and Kalefezra, J., Phys. Med. Biol. 24, 832–834 (1979).

[682] Mukherjee, B. and Vana, N., Nucl. Instrum. Methods 226, 572–573 (1984).

[683] Bloom, D., Evans, D.R., Holmstrom, S.A., Polf, J.C., McKeever, S.W.S. and Whitley, V.H., Radiat. Meas. 37, 141–150 (2003).

[684] Yukihara, E.G., Gaza, R., McKeever, S.W.S. and Soares, C.G., Radiat. Meas. 38, 59–70 (2004).

[685] Hughes, A.E. and Pooley, D., J. Phys. C: Solid State Phys. 4, 1963–1976 (1971).

[686] Chen, R., Lo, D. and Lawless, J.L., Radiat. Prot. Dosim. 119, 33–36 (2006).

[687] Petrov, S.A. and Bailiff, I.K., J. Lumin. 65, 289–291 (1996).

[688] Kortov, V.S., Milman, I.I. and Nikiforov, S.V., Radiat. Prot. Dosim. 100, 75–78 (2002).

[689] Kortov, V.S., Milman, I.I., Nikiforov, S.V. and Moiseĭkin, E.V., Phys. Solid State 48, 447–452 (2006).

[690] Agersap Larsen, N., Bøtter-Jensen, L. and McKeever, S.W.S., Radiat. Prot. Dosim. 84, 87–90 (1999).

[691] Vincellér, S., Molnár, G., Berkane-Krachai, A. and Iacconi, P., Radiat. Prot. Dosim. 100, 79–82 (2002).

[692] Molnár, G., Benabdesselam, M., Borossay, J., Iacconi, P., Lapraz, D. and Akselrod, M., Radiat. Prot. Dosim. 100, 139–142 (2002).

[693] Berkane-Krachai, A., Iacconi, P. and Bindi, R., Radiat. Prot. Dosim. 100, 143–146 (2002).

[694] Polf, J.C., Yukihara, E.G., Akselrod, M.S. and McKeever, S.M.S., Radiat. Meas. 38, 227–240, (2004).

[695] Akselrod, M.S., McKeever, S.W.S. and Bøtter-Jensen, L., Radiat. Meas. 41, S78–S99 (2007).

[696] Pagonis, V., Chen, R. and Lawless, J.L., Radiat. Meas. 41, 903–909 (2006).

[697] Chen, R., Pagonis, V. and Lawless, J.L., J. Appl. Phys. 99, 033511 (2006).

[698] Chruścińska, A., J. Phys. D: Appl. Phys. 39, 2321–2329 (2006).
[699] Chruścińska, A., Radiat. Meas. 42, 727–730 (2007).
[700] Chruścińska, A., Radiat. Meas. 43, 213–217 (2008).
[701] Chruścińska, A., Radiat. Meas. 44, 329–337 (2009).
[702] Bailiff, I.K. and Haskell, E.H., Radiat. Prot. Dosim. 6, 245–248 (1984).
[703] Haskell, E.H., Kaipa, P.L., and Wrenn, M.E., Nucl. Tracks Radiat. Meas. 14, 113–120 (1988).
[704] Haskell, E.H., Radiat. Prot. Dosim. 47, 297–303 (1993).
[705] Hütt, G., Brodksi, L., Bailiff, I.K., Göksu, Y., Haskell, E., Jungner, H. and Stoneham, D., Radiat. Prot. Dosim. 47, 307–311 (1993).
[706] Haskell, E.H., Bailiff, I.K., Kenner, G.H., Kaipa, P.L. and Wrenn, M.E., Health Phys. 66, 380–391 (1994).
[707] Aitken, M.J., Thermoluminescence Dating, Academic Press, p. 158 (1985).
[708] Bailiff, I.K., Radiat. Meas. 23, 471–479 (1994).
[709] Bailiff, I.K., Radiat. Meas. 27, 923–941 (1997).
[710] Stoneham, D., Nucl. Tracks Radiat. Meas., 10, 509–512 (1985).
[711] Göksu, H.Y., Stoneham, D., Bailiff, I.K. and Adamiec, G., Appl. Radiat. Isot. 49, 99–104 (1998).
[712] Bailiff, I.K. and Holland, N., Radiat. Meas. 32, 615–619 (2000).
[713] Bailiff, I.K., Bøtter-Jensen, L., Correcher, V., Delgado, A., Göksu, H.Y., Jungner, H. and Petrov, S.A., Radiat. Meas. 32, 609–613 (2000).
[714] Göksu, H.Y., Schwenk, P. and Semiochkina, N., Radiat. Meas. 33, 785–792 (2001).
[715] Galli, A., Martini, M., Montanari, C., Panzeri, L. and Sibilia, E., Radiat. Meas. 41, 1009–1014 (2006).
[716] Chen, R., Eur. PACT J. 3, 325–335 (1979).
[717] Chen, R. and Leung, P.L., J. Phys. D: Appl. Phys. 31, 2628–2635 (1998).
[718] Chen, R. and Leung, P.L., Radiat. Prot. Dosim. 84, 43–46 (1999).
[719] Pagonis, V. and Carty, H., Radiat. Prot. Dosim. 109, 225–234 (2004).
[720] Pagonis, V., Balsamo, E., Barnold, C., Duling, K. and McCole, S., Radiat. Meas. 43, 1343–1353 (2008).
[721] Itoh, N., Stoneham, D. and Stoneham, A.M., J. Phys.: Condens. Matter 13, 2201–2210 (2001).
[722] Martini, M., Spinolo, G. and Vedda, A., J. Appl. Phys. 61, 2486–2488 (1987).
[723] Yang, X.H. and McKeever, S.W.S., Nucl. Tracks Radiat. Meas. 14, 75–79 (1988).
[724] Itoh, N., Stoneham, D. and Stoneham, A.M., J. Appl. Phys. 92, 5036–5044 (2002).
[725] Chen, R. and Pagonis, V., J. Phys. D: Appl. Phys. 37, 159–164 (2004).
[726] Pagonis, V., Kitis, G. and Chen, R., Radiat. Meas. 37, 267–274 (2003).
[727] Polymeris, G.S., Kitis, G. and Pagonis, V., Radiat. Meas. 41, 554–564 (2006).
[728] Pagonis, V., Chen, R., and Kitis, G., Radiat. Prot. Dosim. 119, 111–114 (2006).
[729] Adamiec, G., Garcia-Talavera, M., Bailey, R.M. and Iniguez de la Torre, P., Geochronometria 23, 9–14 (2004).
[730] Adamiec, G., Bluszcz, A., Bailey, R. and Garcia-Talavera, M., Radiat. Meas. 41, 897–902 (2006).
[731] Yang, X.H. and McKeever, S.W.S., J. Phys. D: Appl. Phys. 23, 237–244 (1990).
[732] David, M. and Sunta, C.M., Ind. J. Pure Appl. Phys. 21, 659–660 (1983).
[733] Haskell, E.H. and Bailiff, I.K., Radiat. Prot. Dosim. 34, 195–197 (1990).
[734] Chen, G., Li, S.H. and Murray, A.S., Radiat. Meas. 32, 641–645 (2000).
[735] Figel, M. and Goedicke, C., Radiat. Prot. Dosim. 84, 433–438 (1999).
[736] Fleming, S.J., Thermoluminescence Techniques in Archaeology, Clarendon Press, p. 84 (1979).
[737] Stoneham, D. and Stokes, S., Nucl. Tracks Radiat. Meas. 18, 119–123 (1991).
[738] Stokes, S., Radiat. Meas. 23, 601–605 (1994).
[739] Murray, A.S. and Roberts, R.G., Radiat. Meas. 29, 503–515 (1998).
[740] Bailey, R.M., Smith, B.W. and Rhodes, E.J., Radiat. Meas. 27, 123–136 (1997).
[741] Nanjundaswamy, R., Lepper, K. and McKeever, S.W.S., Radiat. Prot. Dosim. 100, 305–308 (2002).
[742] Wintle, A.G., J. Lumin. 33, 333–334 (1985).

[743] Petrov, S.A. and Bailiff, I.K., Radiat. Meas. 27, 185–191 (1997).

[744] Murray, A.S. and Wintle, A.G., Radiat. Meas. 32, 57–73 (2000).

[745] Murray, A.S., Olley, J.M. and Caitcheon, G.C., Quat. Sci. Rev. 14, 365–371 (1995).

[746] Olley, J.M., Caitcheon, G.C. and Murray, A.S., Quat. Sci. Rev. 17, 1033–1040 (1998).

[747] Bailey, S.D., Wintle, A.G., Duller, G.A.T. and Bristow, C.S., Quat. Sci. Rev. 20, 701–704 (2001).

[748] Murray, A.S. and Clemmensen, L.B., Quat. Sci. Rev. 20, 751–754 (2001).

[749] Hilgers, A., Murray, A.S., Schlaak, N., Radtke, U., Quat. Sci. Rev. 20, 731–736 (2001).

[750] Rhodes, E.J. Geochronometria 26, 19–29 (2007).

[751] Banerjee, D., Murray, A.S. and Foster, I.D.L., Quat. Sci. Rev. 20, 715–718 (2001).

[752] Wallinga, J., Murray, A.S., Duller, G.A.T. and Törnqvist, T.E., Earth Planet. Sci. Lett. 193, 617–630 (2001).

[753] Kiyak, N.G. and Canel, T., Radiat. Meas. 41, 917–922 (2006).

[754] Rhodes, E.J. Radiat. Meas. 32, 595–602 (2000).

[755] Wang, X.L., Lu, Y.C. and Wintle, A.G., Quat. Geochron. 1, 89–100 (2006a).

[756] Wang, X.L., Wintle, A.G. and Lu, Y.C., Radiat. Meas. 41, 649–658 (2006b).

[757] Tsukamoto, S., Duller, G.A.T. and Wintle, A.G., Radiat. Meas. 43, 1204–1218 (2008).

[758] Adamiec, G., Bailey, R.M., Wang, X.L., Wintle, A.G., J. Phys. D: Appl. Phys. 41, 135503 (2008).

[759] Lu, Y.C., Wang, X.L., Wintle, A.G., Quat. Res. 67, 152–160 (2007).

[760] Pagonis, V., Wintle, A.G. and Chen, R., J. Phys. D: Appl. Phys. 40, 998–1006 (2007).

[761] Adamiec, G., Duller, G.A.T., Roberts, H.M. and Wintle, A.G., Radiat. Meas. 45, 768–777 (2010).

[762] Wang, X.L., Wintle, A.G. and Lu, Y.C., Radiat. Meas. 42, 380–391 (2007).

[763] Pagonis, V., Wintle, A.G., Chen, R., Wang, X.L., Radiat. Meas. 44, 634–638 (2009).

[764] Porat N., Duller, G.A.T., Roberts, H.M. and Wintle, A.G., Radiat. Meas. 44, 538–542 (2009).

[765] Stevens, T., Buylaert, J.P., Murray, A.S., Radiat. Meas. 44, 639–645 (2009).

[766] Kim, J.C., Duller, G.A.T., Roberts, H.M., Wintle, A.G., Lee, Y.I. and Yi, S.B., Radiat. Meas. 44, 132–143 (2009).

[767] Athanassas, C. and Zacharias, N., Quat. Geochron. 5, 65–75 (2010).

[768] Bos, A.J.J., Piters, T.M., Gómez-Ros, J.M. and Delgado, A., Radiat. Prot. Dosim. 47 473–477 (1993).

[769] Bos, A.J.J., Piters, T.M., Gómez-Ros, J.M. and Delgado, A., Radiat. Prot. Dosim. 51, 257–264 (1994).

[770] Kitis, G., Gómez-Ros, J.M. and Tuyn, J.W.N., J. Phys. D: Appl. Phys. 31, 2636–2641 (1998).

[771] Kitis, G., Gómez-Ros, J.M. , Nucl. Instrum. Methods Phys. Res., Sect. A 440, 224–231 (2000).

[772] Pagonis, V., Mian, S.M. and Kitis, G., Radiat. Prot. Dosim. 93, 11–17 (2001).

[773] Pagonis, V. and Kitis, G., Radiat. Prot. Dosim. 95, 225–229 (2001).

[774] Chung, K.S., Choe, H.S., Lee, J.I., Kim, J.L. and Chang, S.Y., Radiat. Prot. Dosim. 115, 343–349 (2005).

[775] Chung, K.S., Choe, H.S., Lee, J. I. and Kim, J.L., Radiat. Meas., 42, 731–734 (2007).

[776] Puchalska, M. and Bilski, P., Radiat. Meas. 41, 659–664 (2006).

[777] Mandowski, A. and Świątek, J. Phil. Mag. B65, 729–732 (1992).

[778] Mandowski, A. and Świątek, J., Radiat. Prot. Dosim. 65, 25–28 (1996).

[779] Mandowski, A. and Świątek, J., Radiat. Meas. 29, 415–419 (1998).

[780] Mandowski, A. and Świątek, J., Synth. Met. 109, 203–206 (2000).

[781] Mandowski, A., J. Electrostat. 51–52, 585–589 (2001).

[782] Goldberg, D.E., Genetic Algorithms in Search, Optimization and Machine Learning, Addison-Wesley (1989).

[783] Holland, J.H., Adaptation in Natural and Artificial Systems, The University of Michigan Press (1975).

[784] Garcia-Talavera, M. and Ulicny, B., Nucl. Instrum. Methods Phys. Res., Sect A 512, 585–594 (2003).

[785] Bluszcz, A. and Adamiec, G., Radiat. Meas. 41, 886–891 (2006).

[786] Bevington, P.R. and Robinson, D.K., Data Reduction and Error Analysis for the Physical Sciences, McGraw-Hill (1992).

[787] Storn, R. and Price, K., J. Global Optim. 11, 341–359 (1997).

[788] Price, K. and Storn, R., Dr. Dobb's J. 22, 18–24 (1997).

[789] Bluszcz, A., Geochronometria 13, 135–141 (1996).

[790] Nicholas, K.H. and Woods, J., Br. J. Appl. Phys. 15, 783–795 (1964).

[791] Bose, D.N., Phys. Status Solidi 30, K57-K60 (1968).

[792] Bräunlich, P. and Scharmann, A., Z. Physik 177, 320–336 (1964).

[793] Broser, I. and Warminski, R., Br. J. Appl. Phys. Suppl. 4, 90–94 (1955).

[794] Neumark, G.F., Phys. Rev. 103, 41–46 (1956).

[795] Pipinis, P.A. and Grigas, B.P. Opt. Spectrosk. (English translation) 18, 43–46 (1965).

[796] Yukihara, E.G., McKeever, S.W.S., Okuno, E. and Yoshimura, E.M., Radiat. Prot. Dosim. 100, 361–364 (2002).

[797] Chen, R., J. Appl. Phys. 42, 5899–5901 (1971).

[798] Coterón, M.J., Ibarra, A. and Jiménez de Castro , M., Cryst. Latt. Def. Amorph. Mater. 17, 77–82 (1987).

[799] Lapraz, D., Boutayeb, S., Iacconi, P., Bindi, R. and Rostaing, P., Phys. Status Solidi A 136, 497–507 (1993).

[800] Bräunlich, P. and Kelly, P., Phys. Rev. B 1, 1596–1603 (1970).

[801] Kelly, P. and Bräunlich, P., Phys. Rev. B 1, 1587–1595 (1970).

[802] Bräunlich, P., Kelly, P. and Fillard, J.P., Top. Appl. Phys. 37, 35–92 (1979).

[803] Fillard, J.P., Gasiot, J. and Manifacier, J.C., Phys. Rev. B 18, 4497–4507 (1978).

[804] Barland, M., Duval, E. and Nouailhat, A., J. Physique 41, 75–82 (1980). http://jphys. journaldephysique.org/index.php?option=article&access=standard&Itemid= 129&url= /articles/jphys/abs/1980/01/jphys_1980__41_1_75_0/jphys_1980__41_1_ 75_0.html.

[805] Bräunlich, P. and Scharmann, A., Phys. Status Solidi 18, 307–316 (1966).

[806] Böhm, M. and Scharmann, A., Phys. Status Solidi A 4, 99–104 (1971).

[807] Böhm, M. and Scharmann, A., Phys. Status Solidi A 22, 143–147 (1974).

[808] Hillhouse, R.W.A. and Woods, J., Phys. Status Solidi A 6, 119–128 (1981).

[809] Tuan, B.T., Velický, B. and Bohun, A., Z. Phys. 251, 289–299 (1972).

[810] Käämbre, H. and Bohun, A., Czech. J. Phys. B 14, 54–62 (1964).

[811] Nishitani, T., Yoshino, K., and Inuishi, Y., Jpn. J. Appl. Phys. 14, 721–722 (1975).

[812] Ito, D. and Nakakita, T., J. Appl. Phys. 51, 3273–3277 (1980).

[813] Markiewicz, A. and Fleming, R.J., J. Phys. D: Appl. Phys. 21, 349–355 (1988).

[814] Fleming, R.J., IEEE Trans. Elect. Insul. 24, 523–531 (1989).

[815] Ikeya, M. and Miki, T., J. Physique C 6, 312–314 (1980).

[816] Miki, T. and Ikeya, M., J. Phys. Soc. Jpn, 51, 2862–2868 (1982).

[817] Abdurazakov, A., Antonov, V.A. and Arsen'ev, P.A., Zhurnal Prik. Spekt. 36, 26–30 (1981).

[818] Lapraz, D., Gaume, F. and Barland, M., Phys. Status Solidi A 89, 249–253 (1985).

[819] Mariani, D.F. and Jiménez de Castro, M., J. Phys. Chem. Solids, 50, 125–131 (1989).

[820] Böhm, M., Iacconi, P., Kromm, K.D. and Scharmann, A., Phys. Status Solidi A 146, 757–764 (1994).

[821] Wong, W.C., McClure, S.A., Basun, S.A. and Kokta, M.R., Phys. Rev. B 51, 5682–5692; 5693–5698 (1995).

[822] Martini, M., Rosetta, E., Spinolo, G., Vedda, A., Nikl, M., Nitsch, K., Dafinei, I. and Lecoq, P., J. Lumin. 72–74, 689–690 (1997).

[823] Castiglioni, M., Martini, M., Spinolo, G. and Vedda, A., Radiat. Meas. 23, 361–365 (1994).

[824] Schrimpf, A., Boekstiegel, C., Stökmann, H.J., Borenman, T., Ibbeken, K., Kraft, J. and Herkert, B., J. Phys.: Condens. Matter 8, 3677–3689 (1996).

[825] Sayer, M. and Souder, A., Phys. Lett. 24A, 246–247 (1965).

[826] Huzimura, R. and Matsumura, K., Jpn. J. Appl. Phys. 13, 1079–1084 (1974).

[827] Opanowicz, A., J. Phys. D: Appl. Phys. 33, 1635–1642 (2000).

[828] Opanowicz, A. and Pietrucha, P., J. Appl. Phys. 93, 957–967 (2003).

[829] Henisch, H.K., J. Electrostat. 3, 233–239 (1977).

[830] Chen, R. and Fleming, R.J., J. Appl. Phys. 44, 1393–1394 (1973).

[831] Benedict, T.S. and Shockley, W., Phys. Rev. 89, 1152–1153 (1957).

[832] Oczkowski, H.L., Phys. Status Solidi A 103, 499–509 (1987).

[833] Opanowicz, A., Phys. Status Solidi A 108, K47-K51 (1988).

[834] Lee, C.H., Jeon, G.N., Yu, S.C. and Ko, S.Y., J. Phys. D: Appl. Phys. 28, 1951–1957 (1995).

[835] Fillard, J.P. and Gasiot, J., Phys. Status Solidi 32, K85-K88 (1975).

[836] Gasiot, J. and Fillard, J.P., J. Appl. Phys. 48, 3171–3172 (1977).

[837] Fillard, J.P., Gasiot, J. and de Murcia, M., J. Electrostat. 3, 99–104 (1977).

[838] Gasiot, J., de Murcia, M., Fillard, J.P. and Chen, R., J. Appl. Phys. 50, 4345–4349 (1979).

[839] Mandowski, A. and Świątek, J., J. Phys. D: Appl. Phys. 25, 1829–1833 (1992).

[840] Mandowski, A., Świątek, J., Iacconi, P. and Bindi, R., Radiat. Prot. Dosim. 100, 187–190 (2002).

[841] Mandowski, A., Physica B 358, 166–173 (2005).

[842] Mandowski, A. and Świątek, J., Proc. SPIE 2373, 225–230 (1995).

[843] Mandowski, A. and Świątek, J., Radiat. Prot. Dosim. 65, 55–58 (1996)

[844] Mandowski, A. and Świątek, J., J. Phys. III, 7, 2275–2280 (1997).

[845] Bindi, R., Lapraz, D., Iacconi, P. and Boutayeb, S., J. Phys. D: Appl. Phys. 27, 2395–2400 (1994).

[846] Opanowicz, A., J. Appl. Phys. 84, 5218–5228 (1998).

[847] Kulkarni, R.N., Radiat. Prot. Dosim. 51, 95–105 (1994).

[848] Braner, A.A. and Halperin, A., Phys. Rev. 108, 932–936 (1957). http://prola.aps.org/abstract/PR/v108/i4/p932_1.

[849] Scaramelli, P., Nuov. Cim. X45, 119–131 (1966).

[850] Diéguez, E. and Cabrera, J.M., J. Phys. D: Appl. Phys. 14, 91–97 (1981).

[851] Hageseth, G.T., Phys. Rev. B 5, 4060–4064 (1972).

[852] Bangaru, S., Muralidharan, G. and Brahmanandhan, G.M., J. Lumin. 130, 618–622 (2010).

[853] Delgado, L. and Alvarez-Rivas, J.L., Phys. Rev. B 23, 6699–6710 (1981).

[854] Bandyopadhyay, P.K. and Summers, G.P., Solid State Commun. 83, 227–230 (1992).

[855] López, F.J., Cabrera, J.M. and Agulló-López, F., J. Phys. C: Solid State Phys. 12, 1221–1238 (1979).

[856] López, F.J., Aguillar, M. and Agulló-López, F., Phys. Rev. B 23, 3041–3049 (1981).

[857] Delgado, L. and Alvarez-Rivas, J.L., J. Phys. C: Solid State Phys. 15, 1591–1600 (1982).

[858] Prasad, G. and Rao, K.V., Phys. Status Solidi A 94, 659–663 (1986).

[859] Jiménez de Castro, M. and Alvarez-Rivas, J.L., Phys. Rev. B 19, 6484–6492 (1979).

[860] Murti, Y.V.G.S. and Murthy, K.R.N., J. Phys. C: Solid State Phys. 74, 1918–1928 (1974).

[861] Murti, Y.V.G.S. and Suchet, N., Phys. Status Solidi B 109, 325–334 (1982).

[862] Jackson, J.H. and Harris, A.M., J. Phys. C: Solid State Phys., 3, 1967–1977 (1970).

[863] Sagastibelza, F. and Alvarez-Rivas, J.L., J. Phys. C: Solid State Phys. 14, 1873–1889 (1981).

[864] Issa, N., Horowitz, Y.S. and Oster, L., Radiat. Meas. 33, 491–496 (2001).

[865] Issa, N., Oster, L. and Horowitz, Y.S., Radiat. Prot. Dosim. 100, 107–110 (2002).

[866] Weiss, D., Horowitz, Y.S. and Oster, L., Radiat. Meas. 43, 190–193 (2008).

[867] Weiss, D., Horowitz, Y.S. and Oster, L., J. Phys. D: Appl. Phys. 41, 185411 (2008).

[868] Lakshmanan, A.R. and Madhusoodanan, U., Phys. Status Solidi A 139, 229–240 (1993).

[869] Petö, A. and Kelemen, A., Radiat. Eff. Defects Solids 119–121, 81–86 (1991).

[870] Petö, A., Radiat. Prot. Dosim. 47, 79–82 (1993).

[871] Nakajima, T., J. Phys. C: Solid State Phys. 4, 1060–1068 (1971).

[872] Yazici, A.N., J. Phys. D: Appl. Phys. 36, 1418–1422 (2003).

[873] Jain, V.K., Radiat. Phys. Chem. 36, 47–57 (1990).

[874] Sastry, S.B.S. and Moses-Kennedy, S.M., Phys. Status Solidi A 144, 479–484 (1994).

[875] Martín, A., López, F.J., Diéguez, E. and Agulló-López, F., Cryst. Latt. Def. Amorph. Mater. 17, 69–75 (1987).

[876] Lecoq, P., Li, P.J. and Rostaing, P., Nucl. Instrum. Methods Phys. Res., Sect. A 300, 240–258 (1991).

[877] Baba, M., Shibata, K., Kamei, Y. and Ikeda, T., Jpn. J. Appl. Phys. 31, 1085–1091 (1992).

[878] Pokorný, P. and Ibarra, A., J. Phys.: Condens. Matter 5, 7387–7396 (1993).

[879] Cruz-Zaragoza, E., Barboza-Flores, M., Chernov, V., Meléndrez, R., Ramos, B., Negrón-Mendoza, A. Hernández, J.M. and Murrieta, H., Radiat. Prot. Dosim. 119, 102–105 (2006).

[880] Dolivo, G. and Gäumann, T., Radiat. Phys. Chem. 10, 207–218 (1977).

[881] Okada, M., Nakagawa, M., Atobe, K., Itatani, N. and Ozawa, K., Phys. Status Solidi A 167, 253–259 (1998).

[882] Wallinga, J. and Duller, G.A.T., Quat. Sci. Rev. 19, 1035–1042 (2000).

[883] Born, G.K., Grasser, R.J. and Scharmann, A.O., Phys. Status Solidi 28, 583–588 (1968).

[884] Tzalmona, A. and Pershan, P.S., J. Chem. Phys. 55, 4804–4811 (1971).

[885] Jahan, M.S., Cooke, D.W. and Alexander Jr, C., J. Lumin. 40–41, 153–154 (1988).

[886] Cooke, D.W. and Alexander Jr, C., J. Chem. Phys. 65, 3515–3519 (1976).

[887] Alybakov, A.A., Toichiev, N. and Akchalov, Sh., Phys. Status Solidi B 109, 295–301 (1982).

[888] Boas, J.F., Radiat. Prot. Dosim. 6, 58–60 (1984).

[889] Boas, J.F. and Pilbrow, J.R., Phys. Rev. B 32, 8258–8263 (1985).

[890] Orera, V.M., Sanjuá, M.L. and Alonso, P.J., J. Phys. C: Solid State Phys. 19, 4763–4769 (1986).

[891] Tatumi, S.H., Nagatamo, T., Matsuoka, M. and Watanabe, S., J. Phys. D: Appl. Phys. 26, 1482–1486 (1993).

[892] Sullasi, H.S.L., Watanabe, S. and Sanchez, M.A.E., Phys. Status Solidi C 2, 596–599 (2005).

[893] Cano, N.F., Watanabe, S., Mittani, J.C.R., Ayta, W.E.F. and Blak, A.R., Phys. Status Solidi C 4, 1305–1308 (2007).

[894] Cano, N.F., Blak, A.R. and Watanabe, S., Phys. Chem. Miner. 37, 57–64 (2010).

[895] Seshagiri, T.K., Natarajan, V. and Sastry, M.D., Pramana, 44, 211–217 (1995).

[896] Rao, T.K.G., Shinde, S.S., Bhatt, B.C., Srivastava, J.K. and Nambi, K.S.V., J. Phys.: Condens. Matter 7, 6569–6581 (1995).

[897] López, F.J., Jaque, F., Fort, A.J. and Agulló-López, F., J. Phys. Chem. Solids, 38, 1101–1109 (1977).

[898] Katzir, A., Süss, J.T., Zunger, A. and Halperin, A., Phys. Rev. B 11, 2370–2377 (1975).

[899] Böhm, M., Cord, B., Hofstaetter, A., Scharmann, A. and Parot, P., J. Lumin. 17, 291–300 (1978).

[900] Huzimura, R. and Atarashi, K., Phys. Status Solidi A 70, 649–657 (1982).

[901] Anderson, J.H., Feigel, F.J. and Schlesinger, M., J. Phys. Chem. Solids 35, 1425–1428 (1974).

[902] Jani, M.G., Halliburton, L.E. and Kohnke, E.E., J. Appl. Phys. 54, 6321–6328 (1983).

[903] Jani, M.G. and Halliburton, L.E., J. Appl. Phys. 64, 2022–2025 (1988).

[904] Halperin, A., Jani, M.G. and Halliburton, L.E., Phys. Rev. B 15, 34, 5702–5707 (1986).

[905] Pogorelov, Yu.L., Mashkovtsev, R.I. and Tarashchan, A.N., Zh. Prik. Spektrosk. 34, 1084–1087 (1980).

[906] Andreev, B.V., Sov. Phys. Solid State 34, 1304–1307 (1992).

[907] Vedda, A., Fasoli, M., Nikl, M., Laguta, V.V., Mihokova, E., Pejchal, J., Yoshikawa, A. and Zhuravleva, M., Phys. Rev. B 80, 045113 (2009).

[908] Daling, L., Chungxiang, Z., Leung, P.L., Zoupling, D. and Stokes, M.J., J. Phys. D: Appl. Phys. 31, 906–912 (1998).

[909] Danby, R.J., Boas, J.F., Calvert, R.L. and Pilbrow, J.R., J. Phys. C: Solid State Phys. 15, 2483–2493 (1982).

[910] Vaĭner, V.S., Veĭnger, A.I. and Polonskiĭ, Yu. A., Sov. Phys. Solid State 18, 237–238 (1976).

[911] Kortov, V.S., Milman, I.I., Monakhov, A.V. and Slesarov, A.I., Radiat. Prot. Dosim. 47, 273–276 (1993).

[912] Singh, V., Watanabe, S., Gunda Rao, T.K., Chubaci, J.F.D., Ledoux-Rak, I. and Kwak, H.-Y., Appl. Phys. B 98, 165–172 (2010).

[913] Kutomi, Y., Tomita, A. and Takeuchi, N., Radiat. Prot. Dosim. 17, 499–502 (1986).

[914] Kawamoto, T., Tomita, A. and Kawanishi, M., Jpn. J. Appl. Phys. Suppl. 24, 242–245 (1985).

[915] Tomita, A., Inabe, K. and Takeuchi, N., Radiat. Prot. Dosim. 6, 359–361 (1983).

[916] Tomita, A. and Takeuchi, N., Phys. Status Solidi A 89, 609–615 (1985).

[917] Fukuda, Y., Tomita, A. and Takeuchi, N., Phys. Status Solidi A 114, K245-K247 (1989).

[918] Fukuda, Y., Radiat. Prot. Dosim. 33, 151–154 (1990).

[919] Tomita, A., Fukuda, Y., Kutomi, Y. and Takeuchi, N., Phys. Status Solidi B 158, 383–398 (1990).

[920] Tomita, A., Fukuda, Y. and Takeuchi, N., Phys. Status Solidi A 134, 279–292 (1992).

[921] Aoki, M., Nishikawa, T. and Okabe, S., Jpn. J. Appl. Phys. 21, L95-L96 (1982).

[922] Aoki, M., Nishikawa, T. and Okabe, S., Radiat. Prot. Dosim. 4, 216–218 (1983).

[923] Aoki, M., Nishikawa, T. and Okabe, S., Phys. Status Solidi A 83, 85–91 (1984).

[924] Iacconi, P., Lapraz, D., Keller, P. and Petel, M., Radiat. Prot. Dosim. 4, 219–222 (1983).

[925] Sorkin, B. and Käämbre, H., Phys. Status Solidi 82, K149–K152 (1984).

[926] Bichevin, V., Alseitov, G., Gapanov, M. and Käämbre, H., Phys. Status Solidi A 114, K241-K244 (1989).

[927] Peoglos, V. and Christodoulides, C., J. Phys. D: Appl. Phys. 34, 862–867 (2001).

[928] Akselrod, M.S. and Kortov, V.S., Radiat. Prot. Dosim. 39, 135–138 (1991).

[929] Kortov, V.S., Bessonova, T.S., Akselrod, M.S. and Milman, I.I., Phys. Status Solidi A 87, 629–639 (1985).

[930] Iacconi, P., Petel, F., Lapraz, D. and Bindi, R., Phys. Status Solidi A 139, 489–501 (1993).

[931] Guissi, S., Bindi, R., Iacconi, P., Jeambrun, D. and Lapraz, D., J. Phys. D: Appl. Phys. 31, 137–145 (1998).

[932] Holzapfel, G. and Krystek, M., Phys. Status Solidi 37, 303–312 (1976).

[933] Sakurai, T., Tomita, A. and Fukuda, Y., J. Phys. D: Appl. Phys. 32, 2290–2295 (1999).

[934] Gripp, S., Haesing, F-Wg, Bueker, H., and Schmitt, G. Radiat. Oncol. Invest., 6, 142–149 (1998).

[935] Davis, S. D., Ross, C. K., Mobit, P. N., Van der Zwan, L., Chase, W. J. and Shortt, K. R., Radiat. Prot. Dosim., 106, 33-43 (2003).

[936] Duggan, L., Hood ,C., Warren-Forward, H., Haque, M. and Kron, T., J. Phys. Med. Biol. 49, 3831–3845 (2004).

[937] Khan, F. M., The Physics of Radiation Therapy, 4th Edn. Lippincott, Williams & Wilkins (2010).

[938] Rah, J.-E., Hong, J.-Y., Kim, G.-Y., Kim, Y.-L., Shin, D.-O. and Suh, T.-S., Radiat. Meas. 44, 18–22 (2009).

[939] Nose, T., Koizumi, M., Yoshida, K., Nishiyama, K., Sasaki, J., Ohnishi, T. and Peiffert, D., Int. J. Radiat. Oncol., 61, 945–953 (2005).

[940] Nose, T., Yoshida, K., Nishiyama, K., Sasaki, J., Ohnishi, T., Kozuka, T., Gomi, K., Oguchi, M., Sumida, I., Takahashi, Y., Ito, A. and Yamashita, T., Int. J. Radiat. Oncol., 70, 626–633 (2008).

[941] Aznar, M.C., Andersen, C.E., Bøtter-Jensen, L. , Bäck S.Å.J., Mattson S. and Medin J., Phys. Med. Biol. 49 1655–69 (2004).

[942] Marckmann, C. J., Andersen, C. E., Aznar, M. C. and Bøtter-Jensen, L., Radiat. Prot. Dosim., 120, 28–32 (2006).

[943] Yukihara, E. G., Mittani, J., McKeever, S.W.S. and Simon, S.L., Radiat. Meas. 42, 1256–1260 (2007).

[944] Yasuda, H. and Fujitaka, K., Radiat. Prot. Dosim. 100, 545–548 (2002).

[945] Konnai, A., Ozasa, N. and Ishikawa, Y., Radiat. Meas. 43, 998–1003 (2008).

[946] Olko, P., Czopyk, L., Kłosowski, M. and Waligórski, M.P.R., Radiat. Meas. 43, 864–869 (2008).

[947] Yukihara, E.G. and McKeever, S.W.S., Phys. Med. Biol. 53, R351–R379 (2008).

[948] Aznar, M.C., PhD Thesis, Risø-PhD-12(EN). Available online at http://www.risoe.dtu.dk/rispubl/NUK/nukpdf/ris-phd-12.pdf (2005).

[949] Soares, C., Drupieski, C., Wingert, W., Pritchett, G., Pagonis, V., O'Brien, M., Sliski, A., Bilski, P. and Olko, P., Radiat. Prot. Dosim. 120, 78–82 (2006).

[950] Berni, D., Gori, C., Lazzari, B., Mazzocchi, S., Rossi, F. and Zatelli, G., Radiat. Prot. Dosim. 101, 411–413 (2002).

[951] Johns, H. E. and Cunningham, J. R. The Physics of Radiology. 4th Edn, Charles C. Thomas (1983).

[952] Kalef-Ezra, J. A., Boziari, A., Litsas, J., Tsekeris, P. and Koligliatis, T., Radiat. Prot. Dosim. 101, 403–405 (2002).

[953] Andersen, C.E., Nielsen, S.K., Greilich, S., Helt-Hansen, J., Lindegaard, J.C. and Tanderup, K., Med. Phys. 36, 708–718 (2009).

[954] Andersen, C.E., Marckmann, C.J., Aznar, M. C., Bøtter-Jensen, L., Kjær-Kristoffersen, F. and Medin, J., Radiat. Prot. Dosim. 120, 7–13 (2006).

[955] Trautmann, T., Krbetschek, M.R., Dietrich, A. and Stolz, W., Radiat. Meas. 29, 421–425 (1998).

[956] Trautmann, T., Krbetschek, M.R., Dietrich, A. and Stolz, W., J. Lumin., 85, 45–48 (1999).

[957] Trautmann, T., Krbetschek, M.R., Dietrich, A. and Stolz, W., J. Phys. D: Appl. Phys. 33, 2304–2310 (2000).

[958] Edmund, J.M. and Andersen, C.E., Radiat. Meas. 42, 177–189 (2007).

[959] Miller, S.D. and Endres, G.W.R., Radiat. Prot. Dosim. 33, 59–62 (1990).

[960] Miller, S.D., Radiat. Prot. Dosim. 66, 201–204 (1996).

[961] Regulla, D.F., Health Phys. 11, 419–421 (1972).

[962] Burgkhardt, B., Festag, J.G., Piesch, E. and Ugi, S., Radiat. Prot. Dosim. 66, 187–192 (1996).

[963] Piesch, E., Burgkhardt, B., Fisher, M., Rober, H.G. and Ugi, S., Radiat. Prot. Dosim. 17, 293–297 (1986).

[964] Piesch, E., Burgkhardt, B. and Vilgis, M., Radiat. Prot. Dosim. 33, 215–226 (1990).

[965] Piesch, E., Burgkhardt, B. and Vilgis, M., Radiat. Prot. Dosim. 47, 409–413 (1993).

[966] Becker, K., Solid State Dosimetry, CRC Press, Ch. 4 (1973).

[967] Perry, J.A., RPL Dosimetry, IOP Publishing, (1987).

[968] Croft, S., Radiat. Prot. Dosim., 33, 183–186 (1990).

[969] Lee, J.H., Lin, M.S., Hsu, S.M., Chen, I.J., Chen, W.L. and Wang, C.F. Radiat. Meas. 44, 86–91 (2009).

[970] Miyamoto, Y., Yamamoto, T., Kinoshita, K., Koyama, S., Takei, Y., Nanto, H., Shimotsuma, Y., Sakakura, M., Miura, K. and Hirao, K., Radiat. Meas. 45, 546–549 (2010).

[971] Rah, J.-E, Hwang, U.-J., Jeong, H., Lee, S.-Y., Lee, D.-H., Shin, D.H., Yoon, M., Lee, S.B., Lee, R. and Park, S.Y., Radiat. Meas. 46, 40–45 (2011).

[972] Hsu, S.M., Yang, H.W., Yeh, T.C., Hsu, W.L., Wu, C.H., Lu, C.C., Chen, W.L. and Huang, D.Y.C., Radiat. Meas. 42, 621–624 (2007).

[973] Ugumori, T., Jpn. J. Appl. Phys. 19, 1089–1092 (1980).

[974] Lommler, B., Pitt, E. and Scharmann, A., Radiat. Prot. Dosim. 65, 101–104 (1996).

[975] Horowitz, Y.S., Radiat. Prot. Dosim. 33, 75–81 (1990).

[976] Horowitz, Y.S. and Rosenkrantz, M., Radiat. Prot. Dosim. 31, 71–76 (1990).

[977] Chen, R., McKeever, S.W.S. and Durrani, S.A., Phys. Rev. B 24, 4931–4944 (1981).

[978] McKeever, S.W.S., Radiat. Prot. Dosim. 33, 83–89 (1990).

[979] Horowitz, Y.S., Moscovitch, M. and Dubi, A., Phys. Med. Biol. 27, 1325–1338 (1982).

[980] Moscovitch, M. and Horowitz, Y.S., Radiat. Prot. Dosiom. 17, 487–491 (1986).

[981] Moscovitch, M. and Horowitz, Y.S., J. Phys. D: Appl. Phys. 21, 804–808 (1988).

[982] Rodríguez-Villafuerte, M., Nucl. Instrum. Methods Phys. Res., Sect. B152, 105–114 (1999).

[983] Mahajna, S. and Horowitz, Y.S., J. Phys. D: Appl. Phys. 30, 2603–2619 (1997).

[984] McKeever, S.W.S., J. Appl. Phys. 56, 2883–2889 (1984).

[985] Yuan, X.L. and McKeever, S.W.S., Phys. Status Solidi B 108, 545–551 (1988).

[986] Yossian, D. and Horowitz, Y.S., J. Phys. D: Appl. Phys. 28, 1495–1508 (1995).

[987] Weizman, Y., Horowitz, Y.S., Oster, L., Yossian, D., Bar-Lavi, O. and Horowitz, A., Radiat. Meas. 29, 517–525 (1998).

[988] Horowitz, Y.S., Rosenkrantz, M., Mahajna, S. and Yossian, D., J. Phys. D: Appl. Phys. 29, 205–217 (1996).

[989] Horowitz, Y.S., Belaish, Y. and Oster, L., Radiat. Prot. Dosim. 119, 124–129 (2006).

[990] Weiss, D., Horowitz, Y.S. and Oster, L., J. Phys. D: Appl. Phys. 42, 085113 (2009).

[991] Livingstone, J., Horowitz, Y.S., Oster, L., Datz, H., Lerch, M., Rosenfeld, A. and Horowitz, A., Radiat. Prot. Dosim. 138, 320–333 (2010).

[992] Christodoulides, C., J. Phys. D: Appl. Phys. 19, 1555–1562 (1986).

[993] Haake, C.H., J. Opt. Soc. Am. 47, 649–652 (1957).

[994] Biegen, J.R. and Czanderna, A.W., J. Therm. Anal. 4, 39–45 (1972).

[995] Gartia, R.K., Singh, S.D., Singh, E.D., Deb, N.C. and Mazumdar, P.S., Phys. Status Solidi A 150, 749–753 (1995).

[996] Dingle, R.B., Proc. R Soc. London, Ser. A 244, 456–483 (1958).

[997] Airey, J.R., Philos. Mag. 24, 521–552 (1937).

[998] Chen, R., J. Comput. Phys. 4, 415–418 (1969).

[999] Chen, R., J. Comput. Phys. 8, 156–161 (1971).

[1000] Chen, R., J. Therm. Anal. 6, 585–586 (1974).

[1001] Chen, R., J. Comput. Phys. 6, 314–316 (1970).

[1002] Balarin, M., Phys. Status Solidi A 54, K137–K140 (1979).

[1003] Gorbachev, V.M., J. Therm. Anal. 8, 349–350 (1975).

[1004] Balarin, M., J. Therm. Anal. 12, 169–172 (1977).

[1005] Gorbachev, V.M., J. Therm. Anal. 10, 447–449 (1976).

[1006] Krystek, M., Phys. Status Solidi A 40, K65-K68 (1977).

[1007] Abramowitz, M. and Stegun, I.A., Handbook of Mathematical Functions, Dover Pub. Inc., New York (1988).

Author Index

Subject Index

Thermally and Optically Stimulated Luminescence: A Simulation Approach, First Edition. Reuven Chen and Vasilis Pagonis.   © 2011 John Wiley & Sons, Ltd. Published 2011 by John Wiley & Sons, Ltd.